ENGLISH FIELD SYSTEMS

BY

HOWARD LEVI GRAY, Ph D.

HARVARD UNIVERSITY PRESS:
THE MERLIN PRESS (LONDON)
1959

©1915   COPYRIGHT HARVARD UNIVERSITY PRESS
FIRST PUBLISHED BY HARVARD UNIVERSITY
PRESS AND REPRINTED FOR THE MERLIN PRESS LTD.,
112 WHITFIELD STREET, LONDON W.1.,
BY KRIPS REPRINT COMPANY,
RIJSWIJK, HOLLAND, 1959

PREFACE

For introductions which have given me access to many of the documents cited in this volume I am indebted to the Rev. A. H. Johnson of All Souls, Oxford, and to Mr. Hubert Hall of the Public Record Office. The custodians of various collections of records have been most courteous, notably those in charge of the British Museum, the Bodleian library, the Public Record Office, the archives of the Dean and Chapter of Christ Church, Canterbury, and the estate documents at Holkham Hall. Messrs. J. M. Davenport and J. R. Symonds put at my disposal the enclosure awards of Oxfordshire and Herefordshire respectively. In making revisions and in correcting proof I have relied upon the skill and care of Miss A. F. Rowe of Cambridge. Professors C. H. Haskins and E. F. Gay of Harvard read the unfinished text and offered valuable suggestions; in particular I am under heavy obligation to Professor Gay, whose unfailing encouragement and generous assistance have made possible the publication of these chapters.

H. L. G.

Cambridge, Massachusetts,
August, 1915.

CONTENTS

	PAGE
INTRODUCTION	3

CHAPTER I
THE TWO- AND THREE-FIELD SYSTEM 17

CHAPTER II
THE EARLIER HISTORY OF THE TWO- AND THREE-FIELD SYSTEM . 50

CHAPTER III
EARLY IRREGULAR FIELDS WITHIN THE MIDLAND AREA 83

CHAPTER IV
THE LATER HISTORY OF THE MIDLAND SYSTEM IN OXFORDSHIRE AND HEREFORDSHIRE 109

CHAPTER V
THE CELTIC SYSTEM 157

CHAPTER VI
THE INFLUENCE OF THE CELTIC SYSTEM IN ENGLAND 206

CHAPTER VII
THE KENTISH SYSTEM 272

CHAPTER VIII
THE EAST ANGLIAN SYSTEM 305

CHAPTER IX
THE LOWER THAMES BASIN 355

CHAPTER X
RESULTS AND CONJECTURES 403

CONTENTS

APPENDIX I

A. EXTRACTS FROM A SURVEY OF KINGTON, WILTSHIRE 421
B. EXTRACTS FROM A SURVEY OF HANDBOROUGH, OXFORDSHIRE . 430
C. SUMMARIES OF TUDOR AND JACOBEAN SURVEYS WHICH ILLUSTRATE NORMAL TWO- AND THREE-FIELD TOWNSHIPS 437

APPENDIX II

EVIDENCE, LARGELY EARLY, BEARING UPON THE EXTENT OF THE TWO- AND THREE-FIELD SYSTEM 450

APPENDIX III

SUMMARIES OF TUDOR AND JACOBEAN SURVEYS WHICH ILLUSTRATE IRREGULAR FIELDS WITHIN THE AREA OF THE TWO- AND THREE-FIELD SYSTEM 510

APPENDIX IV

PARLIAMENTARY ENCLOSURES IN OXFORDSHIRE 536

APPENDIX V

EXTRACTS FROM THE SURVEY OF AN ESTATE LYING IN NEWCHURCH, BILSINGTON, AND ROMNEY MARSH, KENT 543

APPENDIX VI

SUMMARIES OF TUDOR AND JACOBEAN SURVEYS WHICH ILLUSTRATE IRREGULAR TOWNSHIP-FIELDS IN THE BASIN OF THE LOWER THAMES 549

INDEX 561

MAPS

		PAGE
I.	Map of England and Wales, showing Important Places referred to and the Boundaries of the Two- and Three-Field System.	*Frontispiece*
II.	Sketch of the Tithe Map of the Township of Chalgrove, Oxfordshire. 1841	20
III.	Sketch of the Enclosure Map of the Township of Croxton, Lincolnshire. 1810	26
IV.	Sketch of the Enclosure Map of the Township of Stow, Lincolnshire. 1804	75
V.	Sketch of a Map of Padbury, Buckinghamshire. 1591	77
VI.	Enclosure Map of Oxfordshire	115
VII.	Sketch of the Enclosure Map of the Township of Kingham, Oxfordshire. 1850	127
VIII.	Sketch of the Enclosure Map of the Risbury Division of Stoke Prior, Herefordshire. 1855	144
IX.	Sketch of the Enclosure Map of the Township of Holmer, Herefordshire. 1855	146
X.	Sketch of the Enclosure Map of the Parish of Marden, Herefordshire. 1819	147
XI.	Sketch of a Townland in Donegal, Ireland, showing the Holdings of three Tenants. 1845	190
XII.	Plan of an Estate of All Souls College, Oxford, lying in the Townships of Newington and Upchurch, Kent. 1593	274
XIII.	Plan of an Estate in the Township of Buxton, Norfolk. 1714	311
XIV.	Plan of an Estate in the Township of Shropham, Norfolk. 1714	312
XV.	Sketch of a Map of the Township of West Lexham, Norfolk. 1575	317
XVI.	Plan of the Open Fields of Weasenham, Norfolk, showing the Location of George Elmdon's Holding. 1600	322
XVII.	Sketch of a Map of the Township of Holkham, Norfolk, showing the Fold Courses. 1590	327

ABBREVIATIONS

C. Inq. p. Mort.	Chancery Inquisitions post Mortem (reign, file, number).
C. P. Recov. Ro.	Common Pleas Recovery Roll.
D. of Lanc., M. B.	Duchy of Lancaster, Miscellaneous Book.
Exch. Aug. Of., M. B.	Exchequer, Augmentation Office, Miscellaneous Book.
K. B. Plea Ro.	King's Bench Plea Roll.
Land Rev., M. B.	Land Revenue, Miscellaneous Book.
Ped. Fin.	Pedes Finium (case, file, number).
Rents. and Survs., Portf.	Rentals and Surveys, Portfolio.

ENGLISH FIELD SYSTEMS

ENGLISH FIELD SYSTEMS

INTRODUCTION

THE term "field system" signifies the manner in which the inhabitants of a township subdivided and tilled their arable, meadow, and pasture land. Although a study of field systems may seem to be primarily of antiquarian interest, the following chapters have been written as a contribution to our knowledge of the settlement of England and to the history of English agriculture. Since these subjects are wide in scope, no attempt has been made to treat either of them fully; yet it may not be impossible to show that a comprehension of the structure and cultivation of township fields is germane to both.

The settlement of England, as every one knows, is a topic relative to which the sources of information are very scanty. To what extent Celtic and Roman influences persisted after the Germanic invasions of the fifth century is inadequately revealed in existing written records.[1] To supplement narrative accounts scholars have had recourse to such indirect sources of information as linguistics, mythology, archaeology, and to later social, governmental, and legal institutions. Since not the least significant among social customs, especially with primitive peoples, is the method adopted in tilling the soil, an understanding of the differences in agricultural practice early manifested in various parts of England may prove of assistance in distinguishing between the groups that retained or occupied and held the several sections of the country.

Perhaps a still more important and more comprehensive subject is the history of English agriculture. Until the nineteenth

[1] The question as to what Germanic groups occupied the several parts of England in the course of the fifth and sixth centuries is ably discussed by H. M. Chadwick, *The Origin of the English Nation*, Cambridge, 1907.

century agriculture remained the chief source of the national wealth of England, and no account of the fortunes of her people that neglects the topic is adequate. No improvements in the arts before the introduction of the factory system affected so large a proportion of the population as did improvements in tillage; and if disastrous changes occurred the men who suffered were the bone and sinew of the nation. What the following chapters have to tell relates to only a single phase of agricultural progress; but, since that phase is the manner in which more and more of the soil was brought under improved cultivation, it has an immediate bearing upon national wealth and individual well-being.

The agriculture and settlement of primitive peoples have been studied with less diligence by English than by German scholars — perhaps a natural outcome of the perception in Germany that an intimate relation existed between the early history of the Germans and the agrarian side of their life. No passages in the writings of classical historians are discussed more frequently than the brief descriptions of these matters found in Caesar and Tacitus. Inherent tendencies toward democracy or toward aristocracy, it is thought, are there to be discerned. Attention, too, has been focused upon the agriculture of the Germans as practiced somewhat later, when they invaded the Roman empire and in their laws gave testimony to their methods of tilling the soil.

Since all such documentary references to early agrarian custom are brief, it has been usual to interpret them in the light of later usages, descriptions of which have an added value in constituting, as they do, records of the age to which they belong. To two of her scholars is Germany particularly indebted for interpretations and descriptions of this kind. During the second quarter of the nineteenth century Georg Hanssen, stimulated perhaps by the pioneer activity of the Danish Oluffsen, set forth in a series of papers the various field systems or types of agriculture existent at one time or another in Germanic territories.[1] In continuation of Hanssen's studies, August Meitzen published in 1895 a more

[1] *Agrarhistorische Abhandlungen*, 2 vols., Leipzig, 1880–84.

comprehensive work.¹ Relying largely upon the plans of township fields as they appeared at the time of their enclosure in the nineteenth century, he interpreted and compared the earlier agrarian arrangements of Roman, Germanic, Celtic, and Slavic peoples. Not merely to field systems, however, did he have recourse, but in the types of village settlement and in the forms of dwelling-house adopted by the peoples in question he found additional evidence for determining the movements of the population of Europe in the early Middle Ages. The task was a vast one and its achievement noteworthy; but the generalizations suffer somewhat from the circumstances that much of the evidence is late in date and that such of it as comes from certain countries, notably France, England, and Italy, is inconsiderable.

For information regarding English field arrangements Meitzen relied mainly upon the lucid account given in Seebohm's *English Village Community*.² In this cleverly written book the author reproduces a plan of the township of Hitchin, Hertfordshire, made at the time of the enclosure of the open common fields about 1816. It is the type of evidence which Meitzen himself was to use extensively and which, despite its recent date, is always of value. Beginning with a description of the features portrayed in the Hitchin plan, features which constitute the so-called three-field system, Seebohm with the assistance of three or four terriers carries the reader back to Anglo-Saxon days, arguing that the open fields of English villages at that time differed in no essential particular from the Hitchin fields of 1816. Behind these descriptions runs the thread of an hypothesis which interested the author more than did the presentation of facts; for it is the thesis of the book that the practically unchanging open-field system of an English township had from Roman days served as the protective shell of a community settled in serfdom upon it.³

---

¹ *Siedelung und Agrarwesen der Westgermanen und Ostgermanen, der Kelten, Römer, Finnen und Slawen*, 3 vols. and atlas, Berlin, 1895. An account of the antecedent literature of the subject is given in vol. i, pp. 19–28.

² Frederic Seebohm, *The English Village Community examined in its Relations to the Manorial and Tribal Systems and to the Common or Open Field System of Husbandry*, London, 1883.

³ Ibid., 409.

In contrast the author sketches the Celtic field system, referring particularly to the aspect of a nineteenth-century Irish township [1] and to the testimony of early Welsh laws. As an account of early agricultural arrangements Seebohm's treatment is deficient in scope, his meagre evidence by no means warranting the inference that the three-field system was prevalent throughout England from the earliest times.

So far as the structure of English village fields is concerned, Seebohm was not the first to make inquiries. Nasse, in a brief monograph, had already examined with some care the Anglo-Saxon evidence to ascertain whether it showed the arable unenclosed and parcelled out among the tenants in intermixed strips.[2] Having satisfied himself that it did, he turned to thirteenth-century documents to inquire whether a two-field or a three-field system was then prevalent. Rogers, as he noted, had surmised that arable lands were at this time usually left one-half fallow each year, and Fleta in the reign of Edward I had implied that the two systems were co-existent. Since Nasse's own investigations revealed to him various instances of three-field husbandry in contrast with only one description of two fields, he concluded that in the thirteenth century the former was " decidedly the prevailing system." [3] This view of Nasse's is what Seebohm, in so far as he wrote of field systems, has made popular. Rogers's conjecture is repeated by Vinogradoff, who, after pointing to nine or ten two-field townships and noting that Walter of Henley as well as Fleta was familiar with both systems, surmises that the two-field rotation may have been " very extensively spread in England in the thirteenth century." [4]

The evidence adduced regarding English field systems thus proves to be somewhat slight — rather too slight to warrant

---

[1] Cf. below, p. 191.

[2] Erwin Nasse, *On the Agricultural Community of the Middle Ages, and Inclosures of the Sixteenth Century in England* (translated by H. A. Ouvry, London, 1872), pp. 19–26. Cf. below, p. 51 *sq.*

[3] Ibid., 52–58. Most of Nasse's citations refer only to a three-course rotation of crops, which does not necessarily imply a three-field system. Cf. below, pp. 44–45.

[4] Paul Vinogradoff, *Villainage in England* (Oxford, 1892), pp. 229–230.

Vinogradoff's summary dismissal of the subject.[1] It reduces, in brief, to a familiarity with the three-field system as practiced in nineteenth-century Hitchin, projected back by the testimony of two thirteenth-century writers and by some twenty references to two- or three-field villages dating mainly from the same century. Yet since Vinogradoff wrote no one has dissented from his pronouncement or taken a further interest in the subject.

One of the problems upon which, as has been intimated, the study of field systems promises to throw light is the development of English agriculture. With this development, so far as it resulted from the innovations of the eighteenth century which had to do with the rotation of crops and the introduction of convertible husbandry, we are not unfamiliar.[2] It chances, however, that these improvements were contemporary with the transformation of England from an agricultural into a manufacturing country, and that for this reason the benefits conferred by the experiments of Thomas Coke and others reached a far smaller proportion of the population than would have been affected had the change occurred earlier. In days when the annual return from tillage and sheep-raising determined the prosperity of the people to a greater degree than when these pursuits were supplemented by the work of the factories, farming assumed more importance. It is with the improvements of the earlier period that the following chapters are more immediately concerned.

In English agriculture interest has always fluctuated between corn-growing and pasture-farming. During the Middle Ages a

---

[1] "The chief features of the field-system which was in operation in England during the middle ages have been sufficiently cleared up by modern scholars, especially by Nasse, Thorold Rogers, and Seebohm. . . . Everybody knows that the arable of an English village was commonly cultivated under a three years' rotation of crops; a two-field system is also found very often; there are some instances of more complex arrangements, but they are very rare, and appear late — not earlier than the fourteenth century " (ibid., 224). The complex arrangement at Littleton, Gloucestershire, that Vinogradoff proceeds to discuss refers to demesne lands, which possibly did not lie in open field.

[2] A good sketch of it is given by W. H. R. Curtler, *A Short History of English Agriculture* (Oxford, 1909), pp. 111–228.

combination of the two was usually effected through the annual communal tillage of a part of the improved arable, and the pasturing of sheep and cattle upon the waste and upon that portion of the arable which during the year in question lay fallow. Of enclosed land held in severalty and available either for tillage or for pasturage there was little. Such as existed was in general to be found among the demesne lands of the lord or in the home closes of the tenants. Whenever, none the less, our records appraise enclosed land they give it a higher valuation than they assign to the open-field arable, an indication that from an early period separable land available for both pasture and tillage was recognized as more remunerative than common arable field.[1]

A corollary of this estimate is that agricultural progress was bound to take one of two directions. It was necessary either that the unenclosed arable of a township should be brought under better tillage while continuing to lie open, or that it should be enclosed and given over to convertible husbandry.[2] From an agricultural point of view the latter procedure was, of course, the wiser and has ultimately been adopted. But there stood in the way of such a transformation serious technical and social difficulties. The enclosure of the old fields implied, as we shall see, a consolidation of the scattered parcels of each holding and a cessation of communal tillage. For a long time the latter step was actually impossible of accomplishment. Mediaeval ploughing demanded a team of eight oxen or horses yoked to a heavy

[1] At Haversham, Bucks, for example, the demesne comprised " c acre terre arabilis iacentes in separali que valent per annum xxx s. iiii d. . . . et centum acre terre que iacent in communi et valent per annum si sunt seminate xvi s. viii d.; et si non sunt seminate nihil valent quia pastura communis est " (C. Inq. p. Mort., Edw. III, F. 45 (20), 9 Edw. III).

[2] The term " convertible husbandry " is used in the following chapters to designate the continuous annual tillage of improved lands under a succession of grain and grass crops. The equivalent German term is " neuere Feldgraswirtschaft " (Hanssen, *Agrarhistorische Abhandlungen*, i. 216 sq.). Although, when once in grass, land thus tilled was usually left so for more than one year, this feature should not be insisted upon in a definition, as is done by W. Roscher (*System der Volkswirthschaft*, 2. Bd., *Nationalökonomik des Ackerbaues und der verwandten Urproductionen*, 12th edition, Stuttgart, 1888, p. 89). Some of Hanssen's illustrations show no series of grass years (cf. pp. 227, 231). Convertible husbandry was sometimes practiced upon open-field lands (cf. below, p. 129, 158).

plough,[1] whereas even a team of four beasts, which was still used in places until the end of the eighteenth century,[2] was beyond the reach of any except the more prosperous tenants. Communal ploughing thus became inevitable, and it was only natural that strips should be ploughed successively for each contributor to the plough team. In this way an antiquated technique of tillage long prevented the consolidation of the scattered strips of the holdings. Added to this difficulty was the social one. Communal husbandry had in its favor the authority of long tradition, a potent force with a timorous and conservative peasantry. In the event of readjustment — the peasant asked himself — would not the strong profit, the poor suffer? Hence there grew up a popular prejudice against the enclosure and improvement of the common fields.

It should not, however, be assumed, as is often done, that agricultural improvement could take place only through enclosure. Certain open-field systems were superior to others, and a substitution of the better for the poorer meant definite progress. While a two-field arrangement, for example, permitted the annual tillage of only one-half of the arable, a three-field one utilized two-thirds of it and a four-field one three-fourths. Moreover, a transition from one to the other of these systems, or to an irregular arrangement of fields, involved no abandonment of intermixed holdings or of coöperative ploughing; and, inasmuch as no tenant had anything to lose by such a change but each was likely to gain by it, friction did not arise. Of the substitution of one system for the other little record is left in complaints before royal courts, in petitions for parliamentary redress, or in the jeremiads of social reformers. Evidence regarding it has to be sought in the records of manorial courts, and especially in terriers and surveys that picture the subdivisions of the arable fields. It was the slow pacific change which most

---

[1] P. Vinogradoff, *English Society in the Eleventh Century* (Oxford, 1908), p. 154.

[2] " Four horses are generally put to a plow, even if the work is a second or third tilth; and on land that has lain a few years the strength is often increased to six horses " (W. Pearce, *General View of the Agriculture of the County of Berkshire*, London, 1794, p. 24).

easily escapes the chronicles but which is no less significant in the annals of progress than are dramatic transformations. Since this phase of the subject has been little studied by modern students, considerable attention will be devoted to it in the following chapters.

The other form of agricultural advance, the enclosure of the township's open arable fields and unimproved common, has attracted much notice even from the end of the fifteenth century. Because it then excited popular discontent and appeared to be conducive to depopulation, it straightway fell under the censure of the Tudor government, which, like the other rising mercantilistic powers, was extremely sensitive on the latter point. Parliamentary enactment was followed by royal inquisition, both concerned primarily with depopulation. Complaint, legislation, investigation, litigation, and revolt continued throughout the sixteenth century and into the seventeenth. Opposition then became somewhat less vocal and less violent, although the process none the less went on. Precisely how much was accomplished during these two centuries in the way of enclosure and conversion of common lands it is difficult to determine. The area seems not to have been great in the sixteenth century, but to have been considerable in a few localities during the seventeenth.[1] What is clear is the persistence throughout midland England, in the middle of the eighteenth century, of great areas of common arable field. During the one hundred and twenty-five years that followed, however, most of this was enclosed by act of parliament, and at the end of the nineteenth century an open-field township in England had become a curiosity.

To this long-continued and much-distrusted process considerable attention has been given by modern students. Scrutton formulated the problem, especially with reference to the enclosure of unimproved commons.[2] Gay has described critically the contemporary literature.[3] He has further examined the findings of the inquisitions of Tudor and Jacobean times, so far as they

[1] Cf. below, pp. 11, n. 1, 101, 107, 149–152, 207, 307–312.
[2] T. E. Scrutton, *Commons and Common Fields, or the History and Policy of the Laws relating to Commons and Enclosures in England*, Cambridge, 1887.
[3] E. F. Gay, *Zur Geschichte der Einhegungen in England*, Berlin, 1902.

are preserved, and on the basis of these has estimated the extent to which enclosure proceeded during the century in question.[1] Relative to the period after 1607 no such comprehensive and scholarly investigation has been undertaken. Miss Leonard's paper on seventeenth-century enclosures is useful for its evidence about Durham;[2] but, since what happened in that county was not representative of the usual course of events, it can form no basis for a generalization.[3] Other testimony concerning seventeenth-century enclosures, as it occurs in the records of the privy council, has been collected by Gonner; and the paper which embodies his results has been expanded into a stout volume by the restatement of much that had already been said on the subject.[4] Touching the enclosures of the eighteenth and nineteenth centuries no better account than Slater's has appeared; but this is hardly satisfactory, for it is based, not upon the most detailed and accurate documents available, but upon more summary ones.[5] Nevertheless, it serves to give a general idea of the extent and location of such arable fields as were enclosed by act of parliament.

In view of the inadequate treatment of the enclosure movement after the days of James I, an attempt will be made in one of the following chapters to outline a more satisfactory method of studying it.[6] The later enclosure history of two English counties will

[1] "Inclosures in England in the Sixteenth Century," *Quarterly Journal of Economics*, 1903, xvii. 576–597; "The Inquisitions of Depopulation in 1517 and the Domesday of Inclosures," Royal Hist. Soc., *Trans.*, new series, 1900, xiv. 231–303; "The Midland Revolt and the Inquisitions of Depopulation of 1607," ibid., 1904, xviii. 195–244. He concludes, "The specific inclosure movement . . . reveals itself as one of comparatively small beginnings, gradually gaining force through the sixteenth century and continuing with probably little check throughout the seventeenth century, until it was absorbed in the wider inclosure activity of the eighteenth century" ("Inclosures in England," p. 590).

[2] E. M. Leonard, "The Inclosure of Common Fields in the Seventeenth Century," Royal Hist. Soc., *Trans.*, new series, 1905, xix. 101–146.

[3] Cf. below, pp. 107, 110, 138.

[4] E. C. K. Gonner, "The Progress of Inclosure during the Seventeenth Century," *English Historical Review*, 1908, xxiii. 477–501; expanded into *Common Land and Inclosure*, London, 1912.

[5] Gilbert Slater, *The English Peasantry and the Enclosure of Common Fields* London, 1907. Cf. below, p. 111, n. 2.

[6] Chapter IV, below.

be examined in some detail, not only for the purpose of ascertaining the extent to which open arable fields persisted within their borders, but also in the hope of discovering what systems of tillage were practiced at the time of enclosing. In so far as it is possible to determine whether these were improvements upon old methods, and whether any relationship existed between them and the tendency toward enclosure, new light will have been thrown upon the history of English farming.

The study of field systems, while it should prove conducive to a knowledge of the phases of agricultural development, is, as has been indicated, related to another aspect of English history. Since the structure and tillage of township fields have roots far in the past, the subject is one that reflects the usages and characteristics of primitive society. For this reason it furnishes acceptable information about the groups of settlers whose fusion in early Anglo-Saxon days resulted in the formation of the English people. Written records of that period being few, investigations and inferences like those which Meitzen made for the continent are pertinent for England. To such the later chapters of this volume are in a measure devoted.

Within the sphere of agrarian studies it is possible to direct attention to types of settlement and to units of land measure as well as to field systems. To the first of these topics no such study has been given in England as Meitzen and Schlüter have bestowed upon Germany.[1] Maitland's remarks and Vinogradoff's examination of Essex and Derbyshire are the only approaches to the subject, and the latter is concerned with the size rather than the structure of village settlement.[2] Units of land measurement, however, have to some extent been considered in two important recent works, whose authors have hazarded certain inferences as to Celtic and Roman influences.[3] Relative to the subject of

---

[1] Otto Schlüter, *Siedlungskunde des Thales der Unstrut von der sachlenburger Pforte bis zur Mündung*, Halle, 1896.

[2] F. W. Maitland, *Domesday Book and Beyond, three Essays in the Early History of England* (Cambridge, 1897), pp. 15-16; Vinogradoff, *English Society in the Eleventh Century*, pp. 269-273.

[3] F. Seebohm, *Customary Acres and their Historical Importance*, London, 1914; G. J. Turner, *A Calendar of the Feet of Fines relating to the County of Huntingdon*

field systems, since no studies have followed those of Nasse and Seebohm, described above, it has for the most part been assumed that either the two-field or the three-field system, or the two side by side, prevailed from the earliest times.[1] Not the least of the aims of the following discussion, therefore, will be an endeavor to show that the field systems of England were by no means uniform, — that no fewer than three distinct types arose, presumably corresponding to as many different influences exerted by the peoples who early occupied the country. No examination whatever of primitive units of measurement will here be attempted, and types of settlement will receive consideration only in so far as they influenced the size of township fields. The structure of villages, a subject which may yet contribute to the writing of early English history, is worthy of an independent monograph.

If we ask what data are available for a description of the types of English field systems, we find that these vary from century to century. The meagre references in the charters of the Anglo-Saxon period barely indicate the existence of open arable fields, without telling the form which they assumed. Only from the end of the twelfth century is the evidence, still brief, at all definite on this point. At that time charters and feet of fines begin, though rarely,[2] to describe in detail the lands which they transfer by mentioning areas of parcels and locations in fields (*campi*) and furlongs (*culturae*). After the middle of the thirteenth century the fines cease to be specific, thenceforth reciting simply the acres of arable (*terra*), meadow, and pasture with which they are concerned; the charters continue to give detailed descriptions until the middle of the fourteenth century, when they too for the most part become formal and jejune.

(Cambridge Antiq. Soc., *Octavo Publications*, no. xxxvii, Cambridge, 1913), Introduction. Cf. below, p. 409.

[1] From this view Meitzen (*Siedelung und Agrarwesen*, ii. 122) vaguely dissents, on the ground that the type of settlement in Kent and elsewhere was Celtic. Gay ("Inclosures in England," pp. 593–594) suggests that differing forms of agricultural practice characterized England from an early period, and Gonner (*Common Land and Inclosure*, p. 125) mentions the possibility.

[2] Perhaps once in a hundred times.

Of far less value than the charters are the manorial extents. Drawn up in considerable numbers in the late thirteenth and early fourteenth centuries, and for the most part embedded in inquisitions post mortem, these documents do not locate the acres of the tenants' holdings in the fields. Occasionally the demesne arable is so described as to show that it lay in a two-field or a three-field township, or was consolidated; but more often it is said to lie " in " several *culturae*, a phrase which leaves us uncertain whether the culturae were open-field furlongs composed of strips or were block-like subdivisions of a consolidated demesne. At times the extents refer to the manner of tilling the demesne; but the implication of such evidence for the history of field systems is uncertain and the interpretation of it difficult.[1]

More serviceable than the extents are the terriers, which appear in increasing numbers from the fifteenth century to the end of the seventeenth. These detailed descriptions of one or more holdings in a township continue the tradition of the most valuable of the fines and charters in tending, like them, to describe freeholds and copyholds rather than demesne. Especially in the seventeenth century are they useful in telling us whether a township was open or enclosed and, if open, what sort of field system it employed.

The obvious defect of all the above-mentioned documents lies in the fragmentary nature of the information which they contain: nowhere do they furnish a complete and specific description of the fields of an entire township. Complete descriptions are to be had, it seems, in only three classes of documents. Two of these are late — the enclosure awards of the eighteenth and nineteenth centuries and the tithe maps posterior to 1836. The awards themselves, though dealing with entire townships, often omit much through their indifference to old enclosures, and frequently they contain no more than casual references to the condition of that open field the disappearance of which they record. They are intent upon becoming authorities for the future rather than sources of information about the past. With the tithe maps and accompanying schedules, which also deal with entire townships,

[1] Cf. below, pp. 43–46, 321.

it is different. These picture exactly the condition of township fields at the time when the rating was made; but, unfortunately for the subject in hand, that time is usually so late that the old field system of the township had already been much transformed. The maps are likely to show considerable arable enclosed and novel field systems in use. Had the tithe maps been made in the middle of the eighteenth century, they would have been a boon to the student; dating as they do from the middle of the nineteenth, they are of only occasional assistance.

A third class of documents, most valuable of all for the purposes of this study, are the manorial surveys (*supervisus*) and field-books[1] of Tudor and early Stuart days.[2] Their completeness and detail, so far as field conditions are concerned, render them a desirable starting-point for any excursus into earlier or later agrarian history. To interpret the more fragmentary material of an earlier time they can be used with particular advantage.

A word should be added regarding township maps other than tithe maps. The earliest of them date from the late sixteenth century and for graphic illustration surpass the surveys. When, however, a township comprised two or more manors, as was usually the case in the southeast, the map often worked out detailed areas for only one manor, merely sketching in the remainder of the township. Such maps are properly akin to terriers rather than to surveys. The rarer ones of the true survey type, giving areas of all strips and plats, were probably made to accompany field-books, as was the excellent one drafted for Sir Edward Coke in 1601.[3]

---

[1] Often calling themselves terriers, or *draggae*.
[2] Documents of this sort were first described by W. J. Corbett ("Elizabethan Village Surveys," Royal Hist. Soc., *Trans.*, new series, 1897, xi. 67–87), most of those cited relating to Norfolk. Recently there has been printed for the Roxburghe Club an excellent series of Wiltshire surveys, entitled *Survey of the Lands of William, First Earl of Pembroke*, ed. C. R. Straton, 2 vols., Oxford, 1909. These and others like them have been successfully utilized for writing the social history of the sixteenth century by R. H. Tawney, *The Agrarian Problem in the Sixteenth Century*, London, 1912.
[3] The Weasenham field-book of 42 Elizabeth, with two maps, preserved in the Holkham MSS.

In the following chapters the plan has been to seek first the characteristics of the field system of a region in those descriptions which, though relatively late, are most nearly complete. Such are the enclosure awards and maps of the eighteenth and nineteenth centuries, and particularly the surveys of Tudor and Stuart times. Earlier evidence is then adduced to discover whether the thirteenth-century situation was a prototype of that of the eighteenth century, or whether there had been change. Before the thirteenth century we shall be on conjectural ground, but some guesses may be hazarded.

This method of trying to ascertain early conditions largely through the use of late evidence is not without danger, and from its ill effects neither Seebohm's nor Meitzen's works are free. Yet there seems to be no other way of approaching clearly the subject in hand, while it is often only by the aid of late survivals that the earlier phenomena can be interpreted at all. The method is therefore adopted with full consciousness of its shortcomings, particularly of the restriction which demands that the projection of any situation into the past be accompanied with provisos. In particular we must not forget that the testimony which survives is only a small fraction of what once existed and what would alone insure certainty. As we approach earlier times our account of the situation must tend to become less of an exposition and more of an argument. We can no longer say, " The evidence tells us thus and so "; we are forced to plead, " Since this was true at a later time and the scanty earlier testimony is in accord with it, may not the known facts be projected into the unknown and unrecorded past ? " Constructive argument and fragmentary testimony thus to a large extent become the basis for a description of early agrarian conditions; but the validity of argument and conclusion may at any moment be tested by the reader who has the known facts before him.

CHAPTER I

THE TWO- AND THREE-FIELD SYSTEM

TWO-FIELD townships left one-half of their arable fallow each year, three-field townships one-third of it. Apart from this the method of tillage employed by both groups was essentially the same and may for the present be called the two- and three-field system. The characteristics of this system have in a general way long been known. No one, however, has ascertained in precisely what way it differed from other field systems, at what time we first get sight of it in England, in what parts of the island it was then to be found, what irregularities it began in course of time to manifest, and what was the history of its last years. This chapter and the three following ones are designed to throw light on these questions.

It is well first of all to determine the fundamental characteristics of the system. Seebohm's description of Hitchin, based upon the tithe map of 1816 and giving perhaps our most concrete picture of a township under a three-field system, is after all not quite complete or accurate. That there were six fields makes little difference, since we know from a court roll that the six were grouped by twos for a three-course rotation of crops. That in one of these fields 48 owners together held 289 parcels of land, each having from one to 38 parcels, is completely deduced from the schedule of the tithe map. Nothing, however, is advanced to show that these 48 owners held corresponding areas in other fields. The map in which the author represents the " normal virgate or yard-land " is, so far as we can see, imaginary.[1] The insertion of a fourteenth-century description of a virgate at Winslow, furthermore, is ingeniously contrived to lead the reader to think that its details applied as well to a Hitchin virgate in 1816; but it will be noticed that the Winslow terrier does not divide its parcels between two or three or six fields. Seebohm

[1] *English Village Community*, p. 27, map 4. The virgate, as will be explained, was the full-sized holding of a villein or customary tenant.

has, in short, grafted the parcels of a virgate of the time of Edward III, the relation of which to " fields " remains uncertain, upon a nineteenth-century tithe map, which has the equivalent of three fields, but fields in which we do not know the distribution of the strips of the several owners.[1] Everything at Hitchin may, of course, have been as one is led to infer. The holdings may have consisted of scattered parcels equally divided among three pairs of fields; the existence of six fields, indeed, makes this probable, or at least makes it probable that such had once been the case. Yet proof of these facts should not be omitted in the description and definition of a typical three-field township. There are instances of townships which had three fields but in which a three-field system did not prevail.[2]

To repair the shortcomings of the Hitchin illustration, and to amplify the description of a type of open field which was undoubtedly once widespread in England, it may be permissible to summarize conditions in certain typical two-field and three-field townships chosen from different counties. In order to make the foundation sure, complete accounts of townships are desirable; and these must, for the most part, be sought for in surveys of the late sixteenth and early seventeenth centuries or in later records.

Since pictorial illustration, as Seebohm knew, is more readily comprehensible than written documents, his happy example may be followed and a tithe map first reproduced. That of a township in eastern Oxfordshire answers the purpose. The village of Chalgrove lies precisely within the area where in 1808 Arthur Young noted the continuance of a three-course husbandry.[3] The tithe apportionment of the township was fixed in 1841, just before its enclosure in 1845; and the map, which is here sketched,[4] indicates all parcels, the areas and tenants being specified in a schedule.

[1] An insert to the Hitchin map does, to be sure, show the scattered strips of William Lucas, Esq., but without areas.   [2] Cf. p. 314, below.

[3] *View of the Agriculture of Oxfordshire* (London, 1809), p. 127. Cf. p. 124, below.

[4] Owing to the reduction in scale, the number of strips in each furlong is not so great as in the original, which measures some six feet by seven. The large irregular blocks of the old enclosures are also not shown; but no other important details are omitted. The map is deposited with the Board of Agriculture in St. James Square.

The area of Chalgrove in 1841 was 2358 acres. Two-thirds of this area was arable, nearly one-fifth meadow and pasture.[1] Much of the latter lay enclosed in three farms, which were situated to the north between the open fields and the common of 140 acres. Probably the farms had at some time been improved from the waste, with perhaps some encroachment upon the common arable fields. When the map was made, however, these fields seem to have been nearly intact. They consisted of about two thousand long narrow "lands" or selions, each containing usually from one-fourth of an acre to one acre.[2] Several parallel lands constituted a furlong or shot, and there were about one hundred furlongs in the township. These differed in shape and size, both features depending largely upon the contour of the land. In consequence the strips varied in length; but a desire to limit their length seems manifest in the frequent appearance side by side of two furlongs the strips of which ran in the same direction. In general the length of a "land" did not exceed that of the English standard acre (forty rods or poles), and there was an undoubted tendency on the part of the acre parcels to conform roughly to the shape of the standard acre. Their breadth thus became one-tenth of their length, that of half-acre parcels one-twentieth, and that of quarter-acres one-fortieth. In other field documents short strips and subdivided strips are often called butts, while triangular or irregular parcels at the end of a furlong are called gores. The map shows the lands of two adjacent furlongs frequently at right angles to one another. In such cases that strip of one furlong upon which the strips of the other abutted served as a turning-ground for the plough when the abutting strips were ploughed, and was called a headland. The lands numbered 755 and 1751 on the accompanying plan are designated in the schedule as headlands, their situation being that just described.

A stream formed part of the northern boundary of the township, and another traversed it near the village. Some of the

---

[1] The schedule appended to the map subtracts the glebe and gives areas in acres as follows: arable land, 1620; meadow and pasture land, 431; wood land, 8; common land (i. e., the common pasture, or waste), 140; homesteads, 48; glebe, 69; roads and wastes, 42.

[2] Cf. the following terrier, p. 22, below.

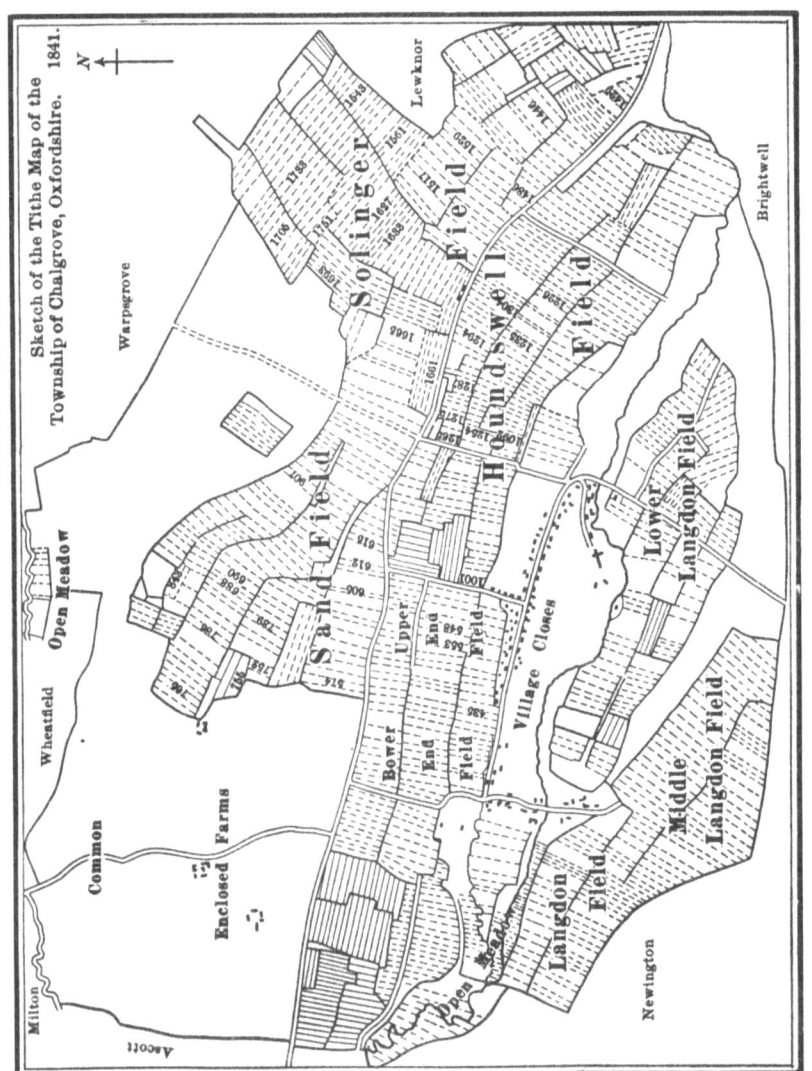

water from the latter was diverted to flow along the village street, rejoining the main brook near the church. Beside both streams were the short strips of meadow which were never ploughed and were elsewhere often called dales. Between the homesteads and the stream were the home closes ("homestalls," "garths," "backsides.")

Thus far the above description might well apply to many an open-field township which was by no means cultivated in accordance with the principles of the two- and three-field system. The characteristic feature of the latter was the further grouping of the furlongs into two, three, four, or six large fields. At Chalgrove there were two groups of fields. The fields of the smaller group to the south of the village are designated in the tithe schedule Langdon, Middle Langdon, and Lower Langdon. With these went certain furlongs toward the northwest, and within them lay much freehold. Indeed, it is not certain that so late as 1841 they were tilled in a strictly three-field manner. To the northeast of the village lay those fields among which the copyholds, the glebe, and certain freeholds were divided. They were without doubt the old three fields of the township, and in 1841 were known as Solinger field, Houndswell field, and Sand field. They adjoined one another and were similar in extent. At the western end of Houndswell field lay two small "fields" named Bower End and Upper End, both pretty clearly appendant to Houndswell field but probably deriving independent names from their proximity to parts of the village called Bower End and Upper End.

How the customary holdings were related to the fields is shown by the following description, transcribed from the schedule.[1] Since this copyhold of John Jones was similar to the glebe and to several other copyholds, it may be taken as typical of early conditions. Although the schedule does not use the term virgate or yard-land, often applied in other documents to customary holdings, the size of this copyhold is about that of the normal virgate, and not improbably represented such a holding: —

[1] Tithe schedule, p. 12.

JOHN JONES. — COPYHOLD FOR LIVES UNDER THE PRESIDENT AND SCHOLARS OF SAINT MARY MAGDALENE COLLEGE, OXFORD

No. on the Map			Acres	Roods	Perches
	BOWER END FIELD				
435	One Land in Acre Hedge Furlong...................	Arable	0	0	24
	UPPER END FIELD				
548	One Land in Harpes End Furlong...................	"	0	1	24
553	One Land in Harpes End Furlong...................	"	0	3	10
	SAND FIELD				
574	One Yard in Lank Furlong.........................	"	0	1	19
605	One acre in Lank Furlong.........................	"	0	3	17
612	One Land in Lank Furlong........................	"	0	2	5
615	One Land in Setts Furlong........................	"	0	1	29
688	One Acre in Little Pry Furlong....................	"	0	3	30
690	One Acre in Little Pry Furlong....................	"	0	3	10
739	One Land in Short Furlong........................	"	0	1	15
752	Two acres in Short Furlong	"	1	1	23
755	One Headacre Land Shooting on Chiswell Common....	"	0	2	39
765	One Land in Great Pry Furlong....................	"	0	1	36
786	One acre in Great Pry Furlong	"	0	3	10
843	Two Lands in Pry Little Furlong...................	"	0	3	1
907	One Acre in Bowsprit Furlong.....................	"	0	2	4
	HOUNDSWELL FIELD				
1001	One acre in Hayes End Furlong	"	0	2	30
1092	One Land in Houndswell Furlong...................	"	0	1	9
1226	One acre in Short Furlong	"	0	2	26
1235	One acre in Short Furlong	"	0	2	29
1254	One Land in Short Furlong.......................	"	0	1	15
1265	One Land in Little Bushes or Rushy Furrows Furlong ..	"	0	1	17
1275	One Land in ditto...............................	"	0	1	22
1287	One Land in ditto...............................	"	0	1	11
1294	One Land in ditto...............................	"	0	1	22
1304	One Land in ditto...............................	"	0	1	12
	SOLINGER FIELD				
1429	One Land in Long Lands.........................	"	0	1	23
1446	One Land in Down Furlong.......................	"	0	1	15
1486	Two Lands shooting on Oxford Way	"	0	2	18
1517	One Land in White Lands.........................	"	0	1	24
1529	Two Lands in ditto..............................	"	0	3	11
1543	One Land in Easington Hedge Furlong..............	"	0	1	11
1561	One Land in Easington Hedge Furlong..............	"	0	1	24
1627	One Land in Woodlands..........................	"	0	1	10
1633	One Land in Woodlands..........................	"	0	1	29
1661	One Land in Rood Furlong.......................	"	0	1	24
1665	One Land in Lower Woodlands...................	"	0	1	34
1693	One Land in Upper Woodlands....................	"	0	1	26
1705	Two Lands in Marsh Furlong......................	"	0	3	9
1733	One Land in Long Snapper Furlong................	"	0	1	29
1751	Headland and Fellow in Long Snapper Furlong.......	"	0	2	23
			21	3	39

It will be noticed that the parcels were distributed with considerable equality among the three fields. Solinger field received 7¼ acres in 15 parcels, Sand field 9 acres in 13 parcels, Houndswell field (with Bower End and Upper End fields) 5¾ acres in 13 parcels. Were the terrier of an earlier date, the irregularity in apportionment would, as will appear elsewhere, probably have been less. The areas assigned to the parcels show approximations to acre, half-acre, and quarter-acre strips; and the locations (numbers on the map correspond with numbers in the schedule) illustrate the scattering of the strips throughout the fields and furlongs. Late though the Chalgrove map and terrier be, they enable us to form a correct and vivid idea of the fundamental characteristics of the three-field system and prepare us to interpret earlier evidence not made graphic by contemporary maps.

As pointed out in the Introduction, the most comprehensive and satisfactory descriptions of English townships are the surveys of the sixteenth and early seventeenth centuries. At their best these note nearly everything that one could wish to know about the manors or townships to which they refer. The metes and bounds, the area of the demesne with its location and the terms upon which it was leased, the number of the freeholders and copyholders, the holdings of each, the rents, fines, and heriots paid, the parcels of land enclosed and in open field, the nature of these, whether arable, meadow, or pasture, the names of the common fields and meadows, — all this, in the most extended of the surveys, a sworn jury of the villagers was called upon to report. The monasteries seem to have originated the custom of making such surveys, for some of the earliest are found in their cartularies of the fourteenth and fifteenth centuries; but the administrators of crown property proved apt pupils, and the most elaborate reports are those relative to crown estates or to manors temporarily in royal hands. During the sixteenth century the latter, of course, included many monastic properties.

So long are the best surveys that it is impracticable to make extended transcripts from them. The information touching field systems is, furthermore, so interwoven with other detail that it is not readily comprehensible unless rearranged and adapted.

For these reasons it seems desirable to print *in extenso* extracts from two surveys, typical respectively of two-field and three-field townships, and to follow these with pertinent field matter abstracted from other surveys. Such is the content of Appendix I.

If there was a difference in the antiquity of two-field and three-field townships, no one will doubt that the former were the earlier. Apart from any question of age, however, the simpler system calls logically for prior treatment. In an excellent series of surveys of the Glastonbury manors in Wiltshire we find pictured the condition of several two-field townships as they were in 9 Henry VIII.[1] The descriptions are particularly minute, the location and area of open-field parcels being always stated. The survey of South Damerham, one of the longest, has been printed by R. C. Hoare in his *History of Modern Wiltshire;*[2] but so inaccessible is this bulky work that it will not be amiss to transcribe a part of the survey of Kington, another of the Wiltshire manors.[3]

After the introduction, the rubric for metes and bounds, and the description of the demesne, this survey makes note of one of the important features of an old English township. It is the common. That of Kington, called Langley Heath, embraced 310 acres, and over it lord and tenants had common of pasture for all cattle throughout the year. In this there was nothing peculiar to a two- or a three-field village. Quite apart from the character of its early fields, nearly every township had such a common and the tenants had rights therein. It would have been more pertinent had we been told about pasturage rights over the common fields; but on that point this survey, like many others, is silent.

The free tenants at Kington were four and most of their holdings were small. Only one held a virgate and paid so much as five shillings rent. One of them was Malmesbury Abbey and another the Prioress of Kington, each answerable for a messuage or two. Similarly John Saunders held in fee a tenement, rendering therefor two geese yearly. The insignificance of the freeholds

---

[1] Harl. MS. 3961.   [3] Cf. Appendix I, below.
[2] (6 vols., London, 1822–44), Appendix II, pp. 40–64.

and the personal distinction of certain of the freeholders are characteristics that will recur.

The customary tenants, or copyholders, on the contrary, were numerous and their holdings were considerable. Six were possessed each of two virgates or "half-hides," seventeen of one virgate, twelve of one-half virgate, and there were two cottagers, each with three or four acres. Besides the two cotlands, six typical holdings have been transcribed,[1] all showing similar characteristics. To each copyholder was assigned a messuage, a yard, a garden, and sometimes an orchard, together with a few closes held in severalty. At Kington the enclosures were larger than in most townships, comprising in general from five to ten acres. After an account of these, we reach in each case the bulk of the holding. This was arable, and for the *virgatarius* (the tenant of a virgate) contained about twenty acres. The *dimidii hidarii* (tenants of two virgates) had some forty acres each, the *dimidii virgatarii* about ten. The arable of each holding, except the last half-virgate, lay in two fields, usually called the North field and the West field, such being the situation of the two relative to the village.[2] Between the two fields the arable of the virgates was pretty equally divided (e. g., 10 acres *vs.* 9½ acres); in some of the larger holdings, however, the lion's share went to the North field (26½ acres *vs.* 20 acres, 24 acres *vs.* 19¼ acres).[3] The parcels ranged in size from one-fourth acre (*perticata*) to two acres, most of them being either half-acres or quarter-acres. A virgate comprised from forty to sixty such parcels. Often more than one parcel of a holding lay in the same furlong. The recurrence of furlong names in the various holdings shows intermixed ownership. There is in the survey nothing about the shape of the parcels, but it is safe to assume that where several acre and half-acre parcels lay in the same furlong they were long and narrow.

[1] In Appendix I. The first two or three of each size have been selected.

[2] In the first holding East field replaces North field; but, as certain of the furlong names are those of North-field furlongs, the East and North fields cannot have been distinct.

[3] The last half-virgater held, along with his half-virgate, some twenty acres of demesne, which lay mainly in the East field.

Since no plan of Kington is available, the appearance of a two-field township may be illustrated by the enclosure map of Croxton, Lincolnshire.[1] As the accompanying cut shows, this rectangular township was in 1810 divided by the highway into an East field and a West field, while to the north lay the sheep-walk.

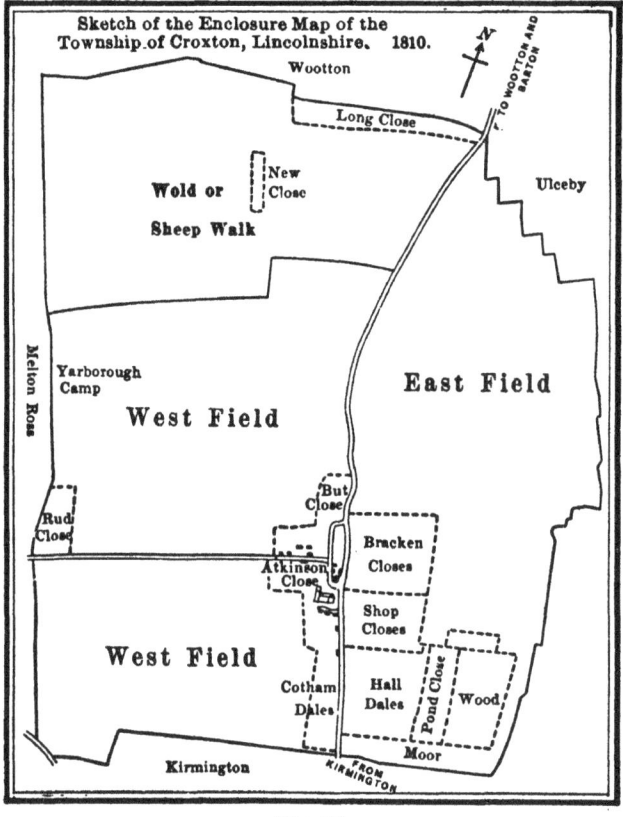

MAP III

Adjacent to the village on the southeast were a few closes, apparently taken from the moor. The two arable fields remained nearly intact and were similar in size. If in imagination we fill them with furlongs and strips, the plan will represent not inadequately the situation described in the Kington survey.

Between two- and three-field townships, as has been said, there was no essential difference in principle. The one divided its

[1] C. P. Recov. Ro., 52 Geo. III, Trin.

arable between two large fields, the other among three. The former tilled one-half of the arable each year, the latter two-thirds, the parts which remained fallow being respectively one-half and one-third. In consequence of having an additional field, the three-field township subdivided each copyhold into three approximately equal parts. This feature is emphasized in another survey, abstracts from which follow the Kington descriptions in Appendix I.

Handborough in central Oxfordshire is a large township which from the thirteenth century has formed a part of the manor of Woodstock.[1] In 1606 it was surveyed as royal property, and the resultant *supervisus* is an excellent illustration of the work of the royal commissioners. The freeholders, who were of the curious sort said to hold " libere per copiam," were much more numerous than the freeholders at Kington. About fifty are named. The two who held most land were persons of quality, viz., George Cole, Gent., with three messuages and $11\frac{1}{4}$ acres, and the heirs of M. Culpepper, Kt., with two messuages and $16\frac{1}{4}$ acres. In both instances most of the acres did not lie in the open fields, and another freehold of ten acres was partly woodland. Sometimes the freeholder was without a messuage, and he might also, as inspection shows, be a copyholder (in the strict sense of the term) who, in addition to a substantial copyhold, held a small parcel of land freely. Such was Roger Brooke, the first on the list. For the most part, however, the " liberi tenentes per copiam " were persons who held merely a messuage and a small close or parcel of land attached. The entire fifty had not a dozen acres in the open fields, and in no instance was there a distribution of acres among fields. At Handborough, as at Kington, the holdings of free tenants are of little value for the study of field systems.

With the customary holdings the case is strikingly different. Almost every one of these supplies information about the open fields. There were forty customary tenements held by thirty-six persons, all tenements except three having messuages. Apart from a half-dozen instances the virgate equivalent of the acreage

[1] A. Ballard, "Woodstock Manor in the Thirteenth Century," *Vierteljahrschrift für Sozial-und Wirtschaftsgeschichte*, 1908, vi. 424.

is given. Each of three tenants held one and one-half virgates, ten others a virgate apiece, the remainder for the most part half a virgate apiece. The normal virgate comprised three or four acres in the common meadow, from five to ten acres of enclosed land, and between seven and ten acres in each of three common arable fields. Often the arable acres of the holdings were almost exactly divided among the three fields (10, 10, 8; 4, 5, 4½; 3, 3¾, 3½). At times, however, there were discrepancies which might give to one field as many as five or six acres more than to another (15, 9, 9; 6½, 11¾, 7; ¼, 3, 1½). The number of parcels into which the arable was divided is not stated, as it usually is not in the surveys of Jacobean days. On the other hand, we are told more about the common meadows than at Kington, and learn that each holding had half-acre or quarter-acre parcels in them. There is further an obvious intention to give information about the pasturage rights of the customary tenants. Nearly always occurs the abbreviation " communia pasture ut supra." But we refer back in vain; for either a folio is gone, or, as is more likely, the folios as they stand at present have been incorrectly rearranged. Toward the end of the survey descriptions of three holdings specify common of pasture " in omnibus Campis, etc.," " in omnibus Communiis, etc.," and "in Einsham heath and Kinges Heath."[1] The first statement, to the effect that there was common of pasture in all fields (*campus* is the usual term for arable field), undoubtedly represents the existent rights.

The somewhat full extracts from the surveys of Kington and Handborough will perhaps serve to make clear the nature of our most detailed evidence about English field systems.[2] For specific and decisive pronouncements Tudor and Jacobean surveys will continually have to be relied upon, and in the light of what they reveal the earlier testimony from many regions of

[1] Cf. Appendix I, below, pp. 434-436.
[2] In one respect the Handborough situation was somewhat unusual. The demesne was farmed, not to two or three or a half-dozen lessees in large parcels, but to some thirty-six persons, nearly all customary tenants. These leaseholds usually comprised less than ten acres each, and frequently lay outside the three common fields in areas called the Great Hide and the Little Hide. The title was " per copiam," and the tenure seems very like copyhold of inheritance.

England will have to be interpreted. At this point it is therefore pertinent to inquire what counties can furnish two- and three-field surveys like those above examined; for an answer to this question will indicate roughly the extent of the system at the end of the sixteenth century.

It is clear that not every holding in a township need be instanced to prove that the arable lay in two or three large open fields. It is equally clear that freeholds, by reason of their smallness, their irregularity, and the social status of their proprietors, were unrepresentative. Descriptions of copyholds, on the other hand, nearly always reflect a two- or three-field system by the approximately equal distribution of their arable between two or three fields; hence ten or a dozen such descriptions from a township will suffice to inform us of the field arrangements existing there. Adaptations of this sort have been made from several surveys and arranged in Appendix I to show the extension of the system illustrated by the surveys of Kington and Handborough.[1]

Tudor and Jacobean surveys of two-field manors most often come from the upland region which begins with the northern Cotswolds and extends to the Channel. Traversing it in this manner, we start with a long Jacobean survey of Upper and Nether Brailes, a township of southeastern Warwickshire. The holdings are estimated in virgates of from eight to twenty acres, all of them divided with precision between North field and South field. There were practically no enclosures save the acre or two attached to each messuage, but there was considerable meadow, some five acres being appurtenant to the virgate. The tenants had stinted common of pasture in as many as nine pastures.

On the eastern slopes of the Cotswolds, just over from Brailes, were many two-field Oxfordshire townships, well illustrated by Shipton-under-Wychwood, a survey of which was made in 6 Edward VI. The customary holdings here usually formed considerable farms of more than one virgate each, the virgate itself containing as many as forty acres. To each farm were attached a

---

[1] The sources from which they are drawn are noted in each case, and the townships to which they refer are located on the map which faces the title-page.

small close and a few acres of meadow. In half of the customary holdings the division of arable acres between the East and West fields was equal; in the other half there was some inequality, usually in favor of the East field.

Two monastic manors of the Gloucestershire Cotswolds, Charlton Abbots and Weston Birt, were surveyed with many others in the time of Edward VI. In both the virgates were large, containing in one township 48 acres of arable and in the other about 40 acres. The division of acres between the North field and the East field of Charlton was even, between the North field and the South field of Weston Birt nearly even. In neither township did the copyholds have other closes than those near the village. With each virgate at Charlton went nine acres in the common meadow, with a virgate at Weston Birt seldom so much as an acre.

The extension of the Cotswold area into Somerset brings us, a little south from Bath, to South Stoke, which in 6 James I was surveyed as one of the queen's manors. Here the enclosures were larger, containing from ten to twenty acres in each holding. Occasionally they had encroached upon the common arable fields, as had those of Lawrence Smythe and Thomas Hudd. Such at least seems to be the inference, since except in these instances the arable was assigned in nearly equal parts to the East and West fields. The meadow, too, had been enclosed. Thus, although the township was obviously one of two fields, there had already begun an attack upon the integrity of the system which we shall see farther advanced in most townships of Somerset.

In the large Dorsetshire township of Gillingham the same change was under way in 6 James I. It had here gone so far that inequality in the division of the arable of a holding between the two fields was frequent. In some holdings meadow and pasture even predominated over the arable; but the general apportionment of the latter to the two fields, South and North,[1] leaves no doubt that a two-field system is described.

Such are typical surveys from six of the counties in which the two-field system was most often apparent. Berkshire, perhaps more extensively characterized by two-field townships than any

---

[1] A third unimportant field occasionally appears.

of them, should be added to the list. The Glastonbury manor of Ashbury was in Berkshire, and of this we have a survey similar in date and character to that of Kington, described above.[1] The upland parts of these seven counties form a compact area in the southwest, characterized by high, bleak down-land not favorable to a developed type of agriculture. Hence in this region the two-field system lingered, little changed, at least until the seventeenth century. We shall see that it was, as might be expected, the prevalent type there at an early date.

There were two outlying areas in which at the end of the sixteenth century it was possible to find two-field townships as unchanged as in the Cotswold counties. One such township was Wellow, in the Isle of Wight. Here, in a Jacobean survey, the customary holdings divided their arable with great consistency and considerable equality between an East field and a West field.[2] Such surveys from the Isle of Wight are, however, so infrequent that a two-field system can hardly be said to have retained much hold upon the island in the days of James I.

It was different with the other outlying area, the so-called wolds of Lincolnshire, where two-field townships were as strongly intrenched as in the Cotswolds or the Wiltshire downs. The Jacobean surveys of Humberston and Alvingham have been chosen for illustration. Both townships had an East field and a West field, and both divided the tenants' arable with marked precision between the two. There was considerable common meadow at Humberston, at Alvingham rather more enclosed pasture. In all respects the townships were of the strictly two-field type.

To show how often the three-field system is apparent in Tudor and Jacobean surveys a longer list of counties than the one just given is required. Among the counties where it rivalled the two-field system were some in which the Cotswold highlands gave place here and there to more fertile areas. Such was Oxfordshire, which has already furnished us the survey of Handborough. Such was Warwickshire throughout most of the valley of the Avon. Such too were the three counties of the south-

---

[1] Harl. MS. 3961, ff. 117–133. The fields were East and West.
[2] Cf. Appendix I, below, p. 440.

west from each of which an example of the two-field system has been drawn, Wiltshire, Somerset, and Dorset. It will be instructive to parallel the two-field surveys already examined with those picturing three-field arrangements in these last three counties.[1]

In southeastern Somerset, where the hills give way to the great plain, lies the large manor of Martock, surveyed in 1–2 Philip and Mary. Four townships were included, Martock, Hurst, Cote, and Bower Henton, and each of the four had its independent group of three fields. Ten of the twenty-nine copyholds at Bower Henton are summarized in Appendix I. Each comprised a messuage, a small close, and an amount of enclosed pasture about equal in area to the arable lying in any one of the three fields. Frequently the survey notes that the enclosure of the pasture was recent. Each copyholder had also from four to six acres of common meadow. The remainder of his holding was arable, divided with little variation among the South, East, and West fields. The recurrence of this characteristic, reproduced as it is in the other townships of the manor, fixes the three-field system upon southeastern Somerset. But the manor was somewhat of an outpost, and we shall not find much similar evidence west of Martock.

Over the county border in Dorset, however, the survey can be matched by a similar one descriptive of Hinton St. Mary in the reign of Elizabeth. Here the enclosures were even more extensive than at Bower Henton, and nearly equalled the area of the open field. Some tenants had enclosures only; but most of them continued to have at least half of their acres in the common arable fields, distributed, though not very evenly, between North field, South field, and West field.

Not dissimilar is the long Jacobean survey of the Wiltshire manor of Ashton Keynes. In it the holdings are estimated in virgates, a circumstance which assures us that they had a long tradition behind them. About one-third of the total copyhold land was enclosed and was largely pasture. Some closes had resulted from encroachments upon the arable fields, the holdings of

[1] Copyholds from all the surveys about to be cited are tabulated in Appendix I.

Joanna Archard and Joanna Syninge having thus decreased considerably their acreage in the East field. Elsewhere, although the distribution of the acres of a holding between East field, North field, and Westham was not so precise as in many townships, discrepancies are not great enough to call in question the existence of a three-field husbandry.

If we make an excursion from the southwestern counties toward the east and north, we shall enter the less disputed domain of the three-field system. Hampshire and Sussex contribute two excellent terrier-surveys of Battle Abbey manors made in the early years of Henry VI. Like the Glastonbury series, they describe each open-field parcel, and the number of these has been indicated in parentheses in the brief summaries given in Appendix I. The small manor of Ansty lay in northeastern Hampshire, and its fields bore the conventional names of South, Middle, and East. The holdings were not estimated by virgates, but nearly every one, except those held at the will of the lord, had its messuage or toft. A few were small, but even these contained an acre or more in each field. It is in two or three of the larger holdings that some unequal distribution appears, an inequality which, so far as we can see, was not compensated for by the possession of enclosed arable. Such occasional deviations from the general practice should not be taken as evidence that a township did not fall within the three-field group. They remind us, rather, that descriptions of several holdings are often needed to give assurance in these matters.

The Sussex survey describes the manor of Alciston as it was subdivided into "borga," a term apparently implying distinct townships. Two of the borga were Blatchington and Alfriston, alike in their field arrangements, of which the descriptions of a half-dozen copyholds and two demesne leaseholds at Alfriston are illustrative. In these we are introduced to a new terminology. Instead of virgates we meet with "wistae," instead of fields with "leynes." Both terms were peculiar to Sussex and occur often in the Battle cartulary. Each wista contained about eighteen acres, and the assignment of its acre and half-acre strips to the three leynes, North, Middle, and South, was on the principle of exact division. Since Alfriston and Blatchington are at the eastern end of the Sussex coastal plain, the three-field system reached at least thus far. Just as the manor of Martock in Somerset, however, was a western outpost, so these townships of the manor of Alciston will prove to be points beyond which it is difficult to discover the existence of the three-field system in southeastern England.

Turning northward, we may add to the description of Handborough briefer accounts of two manors which, like it, lay in the southern midlands. At the end of the sixteenth century there were drawn up for All Souls College, Oxford, maps of its estates in various counties. These are now bound together in volumes known as the Typus Collegii.[1] Among them is a map of Salford, Bedfordshire, accompanied by a schedule which gives names of tenants and areas of the parcels shown on the map. Apart from the glebe and three other small freeholds, the township is assigned to the "tenants of the college grounds." Chief of these was Martha Langford, who had 160 acres of arable and 112 acres of pasture, all enclosed. This was clearly the old demesne. The other tenants represented the old copyholders. In general each had a few acres of enclosed pasture, a few of "pasture and lea ground" not farther described, and a few of "meadow in the fields." But the most of each holding lay in the three open arable fields in many parcels.[2] Brooke, Middle, and Wood were the names of the fields, two of them persisting to the time of the enclosure of the township in 1805. At that date Middle field had been subdivided into Lower and Upper fields, although the total open-field area remained almost unchanged. In 1595 the subdivision of the holdings among the three fields was more consistently unequal than in any other survey yet examined. In the larger holdings fewer acres were assigned to Middle field than to Brook or to Wood field, apparently because the demesne arable lay largely in this field. Five or six of its furlongs were entirely

[1] I am indebted to the warden, Sir William Anson, and to the Rev. A. H. Johnson for the privilege of examining them.

[2] The number of parcels in each holding is noted in the abstract given in the Appendix. A part of the Salford map is reproduced by Tawney, *Agrarian Problem*, p. 163.

demesne, whereas Brook field had only one demesne furlong and Wood field not any. Such concentration of the arable demesne in furlongs, and these furlongs in one field, is unusual; but even this can hardly have affected seriously the three-field character of the township.

A very detailed survey of the township of Welford, Northamptonshire, made in 1602, is preserved in an eighteenth-century copy in the Bodleian. The township comprised two manors, that of William Saunders and that of the " late dissolved monastery of Sulby," then the queen's. The first manor consisted of the demesne and the holdings of several " tenants at will "; the second was in the hands of " ancient freeholders," " new freeholders," and " the Queen's patentees," the last probably representing the copyholders under the monastery. In Appendix I several holdings of three kinds have been summarized in order to show how various tenures fitted into the same field framework. The tenements were rated in virgates. There were no closes except the homestalls, each tenant's holding lying in the open fields, where were also his strips of meadow or " lay ground." Among the three fields, named Hemplow, Middle, and Abbey, the acres of the virgates were, except in a few instances, divided without prejudice. In most respects this survey is a model one, since it gives the names of all furlongs, with the area and location of each open-field strip.

Selected holdings from four northern surveys will complete our three-field itinerary. A Jacobean account of Lutterworth, Leicestershire, illustrates a feature characteristic of many midland and northern field-books, the distribution of several parcels of meadow or " leys " among the arable fields. Here the tenants' strips of meadow, instead of being segregated near a stream, were disposed here and there throughout the arable area. Just as at Welford, certain furlongs which began with arable strips ended with strips of " ley "; and the meadow in each field amounted to as much as one-third of the arable there. In other respects the survey is of the normal three-field type.

Rolleston, a township of eastern Staffordshire, presents the novelty of six fields instead of three. In the Elizabethan survey the first holding groups these by twos, an arrangement that will be found to apply pretty well to most of the other holdings, thus reducing the township to one of practically three fields. In several instances the division of acres was not so exact as that to which we have been accustomed (e. g., 4½, 6, 7; 3½, 6½, 3); yet, if all the holdings be considered, it will be seen that in only about one-fourth of them was there such inequality of division as to make the existence of a three-field system questionable. The remaining three-fourths reassure us on this point, though Rolleston, too, was something of an outpost, for there were not many three-field townships beyond it to the northwest.

Typical of the fields of southern Yorkshire is the Jacobean description of Elloughton. Here the holdings were rated in oxgangs, a single one of which comprised, along with some two acres of meadow, two or three acres in each of the three fields (South-east, Middle, and Milne). The township contained many holdings of about this size and character, although the oxgangs sometimes accumulated in the hands of one tenant to the number of four or more.

In southern Durham, Jacobean surveys record several three-field townships, of which Ingleton was one. In none was there a rating by bovates, and in all the tenants held by letters patent rather than by copy. Each holding had its two or three acres of common meadow and a few additional acres of enclosed meadow. Some of the latter may have been abstracted from the common fields; for when enclosed meadow appears in a holding there is also some inequality in the distribution of arable acres among the fields. Although more remains to be said about this tendency in Durham, the Ingleton acres as they lay in 5 James I had not yet departed far from a three-field arrangement.

From all of the counties which have thus far furnished illustrative surveys of the two- and three-field systems it would be easy to increase the amount of similarly indubitable evidence. There remains, however, one region for which the three-field testimony is relatively slight and for that reason deserving of careful consideration. It comprises the counties of Herefordshire and Shropshire, the greater part of the old Welsh border. As we shall see later, considerable irregularity is visible in the field system of these counties at the end of the sixteenth century.[1] Hence it is pertinent to inquire how clearly a three-field system may be discerned within their limits in Jacobean days. Several surveys need be cited,[2] a course the more necessary since there were few holdings in any township; for it is characteristic of these counties that the townships were only of hamlet size, and that many of them were grouped within one manor.[3]

Perhaps the most unimpeachable testimony to the existence of a three-field system in Herefordshire at the end of the sixteenth century is discernible in a survey of the manor of Stoke Prior. Situated in the northern part of the county, this manor comprised in the days of the survey several hamlets. At Stoke itself all copyholds and freeholds were apportioned to three fields, Blakardyn, Elford's, and Church, although the acres of the last field sometimes have to be supplied from outlying areas pretty clearly connected with it. At Risbury a more exact division of acres than that existing between Muston field, Mere field, and Inn field could not be desired. At Hennor we hear of only one tenant, a freeholder, whose arable acres none the less lay in three fields. Another Herefordshire manor whose members seem to have employed the three-field system was Stockton. In the hamlet of Stockton the number of fields was considerable, but between two of them, Rowley's field and Rade field, each tenant had about two-thirds of his acres pretty evenly divided. All the remaining fields may well be grouped as a third large field, playing this part relative to the other two. One holding, that of William Bach, had precisely twenty acres in each of these three areas. The three tenants in the hamlet of Hamnashe likewise divided their acres among three fields. At Kimbolton, another hamlet of the manor, five fields recur; but, as at Stockton, two of them are each as important as a combination of the other three.

From Shropshire we have only one brief survey which illustrates the three-field system. It describes four copyholds in the fields of Mawley and Prysley, hamlets of the manor of Cleobury. While there was much enclosed pasture here, the arable of

[1] Cf. below, pp. 93 sq.   [2] Cf. Appendix I.   [3] Cf. below, pp. 95, 141.

the holdings, two of which are said to have been virgates, lay equally divided among three fields. Other Shropshire evidence, not less convincing, is of a different nature. It appears incidentally in a specification of boundaries that forms part of an elaborate survey of Morffe forest made in the early seventeenth century.[1] Morffe forest, which contained 3600 acres and was divided into two manors known as Worfield Holme and Claverley Holme, lay near the Severn, between that river and the Staffordshire boundary. Common rights within the forest resided in the townships that bordered it. In assessing these rights the survey states the areas within each township that had valid claims, noting which fields were arable. In several cases these fields were three in number, and their comparatively large size and relatively equal areas make it highly probable that they were in each instance the three common fields of the hamlet in question. The list is as follows: —

Hamlet	Areas in Acres of the Fields "common to" each Hamlet
Bromley	104, 74, 106
Swancote	34, 30, 34
Hoccum Lopp	25, 27, 41, with 15 other acres in two parcels
Barnsley	31 (Windsor field), 5, 28, 33
Roughton	119 (Anesdale field), 134, 103
Wyken	49, 59, 49
Dallicott	58, 32, 38
Hopstone	20, 30, 58 (Ley field)
	34 (Snedwell), 50 (Middle field), 53 (Poole field)
Aston	69, 48, 77

A smaller number of the abutting townships possessed arable fields less regular than those noted above. They belong to a class, numerous in this region, which will be discussed later.[2] If we disregard this class, the foregoing list of fields seems confirmatory of the surveys, and taken in conjunction with them gives

[1] Land Rev., M. B. 203, ff. 305–327.
[2] Cf. pp. 93 *sq.*, below. The hamlets bordering upon Morffe forest which had irregular fields were as follows: —

Hamlet	Areas of the Fields in Acres
Burcott	44 (Mill field), 24 (Woodcroft field), 23 (common to Burcott)
	70 (belonging to Burcott), 35 (belonging to Burcott)
Mose	21 (Bass field), 103 (with 67 acres more)
Sitchhouse	56, 44, 46, 13, 16 ("lately inclosed out of Clarely common field")
Sutton	48 (Home field), 102
Ludstone	43, 6, 48

assurance that the three-field system extended up to the Welsh border.

After this marshalling of typical Tudor and Jacobean surveys from several counties, it should be possible to single out the characteristic features of the two- and three-field system; for only by the aid of such data, as has been said above, can the earlier and more fragmentary evidence be interpreted. The history of English open fields reaches far back of the sixteenth century, and testimony in regard to this earlier time is at hand in the documents described in the Introduction. A method of interpreting them remains to be sought. In drawing up our list of characteristic features we may treat two- and three-field arrangements as a single system that will in due course have to be contrasted with other systems. What, then, is the minimum of information which an early charter, fine, terrier, or extent must supply in order to give assurance that the township to which it refers was cultivated after the manner of Kington or Handborough?

First of all, testimony to the existence of two or three large open fields (*campi*) is essential. If the open-field area was so small that the total amount of it in the tenants' occupation was less than their enclosures, no need existed for the cultivation of the arable in the manner dictated by two- or three-field husbandry. In such cases reliance could be put upon the tillage of the enclosures, and irregularities in the distribution of arable acres among the open fields could thus be corrected. In circumstances like these it is possible that the two- and three-field system may once have been existent but its integrity have been in time impaired. The tenants had perhaps seen fit to change part of their arable to pasture; and the holdings of certain tenants who thus converted a part of their open arable field have been noted at South Stoke, Ashton Keynes, and Ingleton. Such conversion is always a sign of the decay of the original system. The preceding illustrations have shown that the normal enclosed area in two- or three-field townships seldom exceeded one-third of the arable, and usually was much less. Suspicion will therefore attach to any terrier in which the ratio tends to be reversed and closes incline to predominate over the open-field arable in any holding.

Closely bound up with this first characteristic of the two- and three-field system is the further one that the arable acres of a holding were divided with approximate equality between the two or three fields. This is unquestionably the fundamental trait of the system under consideration. It depends, of course, upon the fact that a fixed ratio had to be maintained year after year between tilled land and fallow. Under the two-field system the ratio was one to one, under the three-field system two to one. Any departure from an equal division of the acres of a holding between fields involved shortage for the tenant during the year in which his largest group of acres lay fallow. Increased abundance the ensuing year could scarcely repair the loss to a peasantry which probably lived close to the margin of subsistence. The difficulty would be greater in a two-field than in a three-field township, since a shortage of acres would there be more frequently and more acutely felt. The approximately equal distribution of the acres of a holding between two or three fields must therefore be employed as a crucial test. A single terrier which evinces it constitutes strong testimony to the existence of the system. If, on the other hand, not one but nearly all of the tenant-holdings fail to observe it, the township can scarcely be looked upon as lying in two or three fields. An arrangement of six fields by twos, like that at Rolleston, was only an unimportant modification of the three-field system.

The phrase "tenant-holdings," which has just been used, needs restricting. As the Kington and Handborough surveys show, and as many other surveys would emphasize if they were to be analyzed in full, freeholds are likely to throw little light upon field systems. At least, this is true with regard to townships in which they did not constitute the majority of the holdings. In certain manors, especially in the eastern counties, freeholds assumed such an agrarian importance that they can be relied upon. Elsewhere they were generally small, not largely composed of open-field arable, liable to be without messuage, and frequently in the possession of an absentee proprietor, who was often a corporation or a person of importance. For these reasons they have been seldom cited in the preceding abstracts. Nor can they henceforth be depended upon in either the earlier or the later evidence to disprove the existence of two or three fields. In other words, the fact that a half-dozen freeholds, or even all the freeholds of a township, were not amenable to two- or three-field conditions does not prove that this system was there in disfavor. On the other hand, a single freehold which did divide its arable acres equally between two or three fields is a satisfactory bit of evidence in favor of the existence of the system. Such were the ancient freeholds at Welford, and such was the glebe at Salford. Freeholds, in short, have affirmative, not negative, value. The desirable tenures for our purpose are copyholds, or the leaseholds into which they were sometimes transformed, as they probably were at Welford and Ingleton. Henceforth, therefore, copyholds, whenever available, will be cited in proof or disproof of the existence of the two- or three-field system. Freeholds will be relied upon only in default of other evidence or when their significance is clear.

The superior value of copyholds depends in part upon one of their characteristics which leads in turn to a fourth useful test in the interpretation of field systems. Copyholds were usually rated in virgates or bovates, each of which was responsible for a fixed quotum of rents and services. Probably to avoid inconvenience in the collection of rents and the exaction of services, the virgates and bovates, except again in some eastern counties, remained little changed for centuries. Division appears to have been unusual after the thirteenth century, and consolidation is first apparent in the sixteenth-century surveys. The virgate, therefore, represented a holding of long standing, originally designed to support a peasant family which could muster two oxen for the plough. In Somerset such traditional holdings were sometimes called, instead of virgates, " de antiquo austro." [1] Although the virgates differed in size from township to township, within any particular one they were approximately equal in area, as the foregoing surveys have often shown. For an investigation of the early history of the two- and three-field system no fragmentary evidence is so valuable as the terrier of a virgate. It is

[1] Survey of Kingsbury Episcopi, Land Rev., M. B. 202 ff., 199–253.

the best assurance that there were other similar holdings in the township, and that the acres of all were arranged in the fields much as were the acres about which we are informed. There were, to be sure, unrepresentative virgates.[1] Yet when one considers how many virgate and bovate descriptions were cast in the same pattern, and that pattern perfectly indicative of the field system of the township, the significance of the copyhold virgate terrier is appreciated. While a single terrier may thus go far to establish the existence of the two- and three-field system, more than the terrier of one virgate is needed to disprove its existence. The virgate in question may have been exceptional. Only by the testimony of several irregular virgates from the same region, and preferably from the same township, can it be made clear that the two- and three-field system was non-existent there. Upon this principle several of the following chapters have been written.

In the earlier evidence, however, it seldom happens that we get descriptions of virgates, bovates, or the halves of either. Nor are reasonably large holdings of any sort, whether copyhold, leasehold, or freehold, always described. The acres of early terriers and charters were frequently few in number; and we must ask what confidence is to be put in those grants of land which not only omit an estimate by virgates or bovates, but in addition convey not more than three or four acres? The answer brings us to a fifth characteristic of the two- and three-field system which at this point is more or less decisive. We perceive, in short, that much depends upon the names of the fields. It will have been noted that the names of the fields in Tudor and Jacobean surveys were simple, being usually taken from those points of the compass toward which the fields lay with respect to the village — north, east, south, or west. Often in a two-field manor they were named from opposite points, although at Kington the fields were North and West. The fields again might get their names from the topography of the place, and become Upper

---

[1] For example, the half-virgate of Richard Weller at Handborough, Oxons., that of Robert Sell at Shipton-under-Wychwood, Oxons., that of Joanna Syninge at Ashton Keynes, Wilts, and that of Theron Symes at Welford, Northants. Cf. Appendix I.

and Lower; in a three-field township the third field might become Middle field. Other topographical features sometimes gave the hint. At Salford, for example, Wood field was near the wood and Brook field along the stream. Names like these are what may be called the obvious and usual field names. Accordingly, if in an early charter we discover two acres in three or four parcels lying in the West field of a township and two other acres similarly subdivided in the East field, the probability is that the grant points to a two-field township. In these cases it is always desirable to find a series of such grants (frequently met with in monastic cartularies), and the evidence is more or less convincing as the region is otherwise known to be or not to be one of two fields. Testimony of this sort has been noted in Appendix II, and may be accepted for what it is worth. If the field names appear fanciful, the grant either has been omitted, or has been included only because it is in keeping with what is otherwise known about the region.

Thus far attention has been given only to testimony drawn from descriptions of freeholds or of copyholds (sometimes changing into leaseholds). The third constituent of the manor, the demesne, has not been noticed. It is, in fact, less important than copyholds in helping us determine field systems, since it so often lay without the open fields. Even if it was largely within them it might be irregularly apportioned, as at Salford. If we can be sure, however, that it lay with the tenants' holdings in the open common fields, the even distribution of its arable between two or three fields is as significant a fact as the like distribution of copyholds. Only occasionally do the extents make this point clear. Often they tell us in what field divisions the demesne lay, but frequently these appear to have been numerous. In such cases either the demesne acres were consolidated and the field names refer to large plats, perhaps closes; or, if the acres were not consolidated, we have no clue to the relation existing between the numerous areas named and the field system employed. Such non-committal descriptions have to be disregarded. Sometimes in the extents, however, the demesne arable is said to lie equally divided between only two or three fields, and these bear the usual field names. In such cases we may conclude that the field system is correctly indicated.

Many extents are to be found in a group of documents which for this reason are of significance in the study of field systems. These documents are the inquisitions post mortem, preserved in large numbers among the public records. During a period of about a century (c. 1270–1370) we find inserted in many such enumerations of the property of deceased fief-holders or freeholders extents of their manors. Nearly always the extents are brief, dismissing the demesne acres with an estimate of their annual value; but occasionally a note of explanation is added, and this is the item which relates to field systems. It states that one-half or two-thirds of the demesne may be sown each year, and that when so sown the acres are worth a certain amount. The remaining one-half or one-third, the extent continues, is worth nothing since it lies fallow and — the phrase is sometimes added — " since it lies in common." [1] Thus we are introduced to what might at first sight seem an equivalent of the two- or three-field system, namely, the two- or three-course rotation of crops. Much, however, depends upon keeping the two subjects distinct.

Let it be at once admitted that the existence of a system of two or three fields in any township implies that a two- or three-course method of tillage was there followed. If one-half or one-third of the common arable open-field area lay fallow each year, the parts successively tilled were undoubtedly sown with nearly the same crops year after year. Any series of bailiffs' accounts will make this clear.[2] The reverse of the generalization, however, is not

---

[1] For example, at Corby, Northants, there was a messuage with 180 acres of arable, " unde vi ^{xx} possunt seminari per annum quarum quelibet acra valet . . . iii d. . . . et residuum iacet ad Warectam et tunc nihil valet quia in communi " (C. Inq. p. Mort., Edw. III, F. 44 (6)). Cf. the phraseology in Appendix II.

[2] At Gamlingay in Cambridgeshire, for instance, the demesne lay in three open fields (Merton College map of 1601). A series of bailiffs' accounts from the end of the thirteenth century records the sowing of grains during four years, as follows (Merton Col. Recs., nos. 5355–58): —

Year	Frumentum[et]	Siligo	Drageum[et]	Pisa[et]	Avena
21–22 Edw. I	[illegible]	10 qr. 7 bu.	20 qr. 5 bu.	2 qr.	4 qr.
22–23 "	14 gr. 4½ bu.	10 7	29 3½	2	4 1 bu.
23–24 "	12 4½	11 4	18 6	1 7bu.	3 7½
24–25 "	16 1	9 2	25 2	2 2	6 2

More spring corn than winter corn was required to sow an acre.

equally true. A two- or three-course rotation of crops did not necessarily imply a two- or three-field system. If we have evidence pointing to the former as characteristic of the tillage of demesne lands, or even of the tillage of the entire township, it does not follow that demesne or tenants' holdings had their acres equally divided between two or three large fields. All might have been enclosed and yet a two- or three-course rotation of crops have been found acceptable; for this rotation was adaptable to various field systems. Only in connection with two or three large open fields, intermixed acres, and the annual use of one of the fields as common fallow pasture did it become a constituent part of the two- and three-field system.

With this in mind we may undertake the interpretation of those phrases of the extents which relate to the tillage of the demesne. If the value of two-thirds of the demesne is estimated but the remaining third is said to be worth nothing because fallow, this is insufficient to assure us that the agricultural system was one of two or three fields. Such a statement was applicable to enclosed demesne where the pasturage of the fallow was not deemed to be of value.[1] Again, it is not sufficient to be told, as we often are, that the demesne lands lay in common " while unsown "; for this remark may have referred to the period after harvest, when under various systems these lands would have been thrown open. We must know that the period of common pasturage extended throughout the year.[2] Finally, it must be made clear what fraction of the demesne lay fallow and common. Unless it were one-half or one-third, there is no necessary approach to a two- or three-field system.

[1] To be sure, unsown demesne did sometimes have a definite value as pasture. In several Essex extents, for example, the arable acres were worth 4 d. " quando seminantur, et quando non seminantur valet inde pastura . . . pretium acre ii d." (C. Inq. p. Mort., Edw. III, F. 67 (10), Latchingdon, 17 Edw. III). But it is not quite certain that these unsown acres were fallowed. Their value was rather high for fallow stubble, and some sort of grass may have been grown after the corn years. In general, enclosed fallow was probably worth little and so escaped valuation.

[2] The description of fifty acres of arable at Wrentham, Suffolk, for instance, states that they were worth 2 d. the acre " quum seminantur, et quum non seminantur nihil valent quia iacent in communi" (C. Inq. p. Mort., Edw. III, F. 60 (6), 13 Edw.

Even when all the specifications just insisted upon are met, and we are told that one-half or one-third of the demesne lay fallow and common throughout the entire year and for this reason was of no value to the lord, there remains an element of doubt. Did the fraction in question lie in one of two or three large fields? There is no guarantee that such was the case. Even if, as rarely happens, it be said to lie " in communi campo," the distribution may have been irregular throughout the commonable area. We have seen it so at Salford, Bedfordshire, in the sixteenth century, and yet the preceding specifications could probably have been met in a description of the demesne there. For our present purpose, which is the determination of those two- and three-field characteristics that will enable us to interpret the early evidence, it is sufficient to accept the following working hypothesis: If the arable of the demesne be described in an inquisition-extent as lying one-half or one-third fallow each year, with the fallow acres of no value because commonable, this may be taken as evidence that a two- or three-field system was employed in the township, provided that other testimony shows the system to have been characteristic of the region in question; but if other testimony be against the existence of the two- or three-field system in the county or district in which the township lies, the evidence of the extent will have to be weighed against this other testimony and an independent conclusion reached. Such balancing of evidence must be undertaken in examining the field systems of certain counties of the southeast. In those counties in which there can be no doubt about the general prevalence of the two- and three-field system, the phrases of the extents may be quoted without further discussion. They have been extracted from the inquisitions post mortem for a period of ten years (7–16 Edward III), certain others have been added, and in Appendix II all have been placed last in the collection of early evidence relative to each

III). In contrast with this vague phrase the account of six hundred acres at Lidgate in the same county is entirely specific. Two hundred of them " iacent quolibet tertio anno ad warectam et in communi per totum annum et tunc nihil valent"; the remaining four hundred " iace[n]t in communi a tempore asportationis bladorum usque festum Annunciationis beate Marie [i. e., from September till March]" (ibid., F. 41 (19), 9 Edw. III).

county.[1] By this device their somewhat questionable testimony need not be confused with other authority.

A final characteristic of the two- and three-field system is implicit in these statements made by the extents relative to the demesne. This feature is the existence of common rights of pasturage throughout the year over the field which lay fallow, and, when the other field or fields were not under crops, over them as well. The meadows, too, we know, were thrown open after the hay was removed. Only slight traces of these usages appear in the sixteenth-century surveys. At times after each copyhold entry we are told what were the tenant's rights of pasture,[2] but more often the rights over the arable fields and meadows were assumed to be inherent in the system and were not mentioned. Pasturage rights in the fell, the marsh, or the moor received more attention, especially if such waste land had to be stinted; earlier legal documents, too, especially cases before the courts and agreements between neighboring lords, tell something about these rights of pasture. All this, however, is of no immediate interest in discriminating between field systems. Practically all townships at an early time had their waste, in which tenants had common of pasture. One influence only the waste had upon the tillage of the arable fields, and this arose from the relative size of the two areas. If in any township the waste was extensive in comparison with the open-field arable, utilization of the latter for pasturage might be a matter of little moment, the former sufficing for the cattle and sheep. In consequence, deviation from a strict two- or three-field system in the cultivation of the arable and in the rotation of crops became relatively easy. This aspect of things will claim attention in the counties of the northwest, where for the most part the waste did predominate over the arable. It may also have had much to do with the irregularities which we shall discover in the arable fields of townships situated within forest areas.[3]

Though seldom specifically noticed in the manorial documents, the right of pasturage over the arable fallow was so bound up with

[1] The phraseology of each extent is noted in the transcripts.
[2] Cf. Appendix II.   [3] Cf. pp. 84-88, below.

the nature of the two- and three-field system that it would not be altogether incorrect to call it the determining idea of that system. Why divide the arable into two or three (possibly four or six) large unbroken fields ? Convenience would, of course, be served. It was simpler to have the strips which were to be tilled in a particular year gathered within one-half or two-thirds of the arable area than to have them scattered throughout its entire extent. Yet dissemination of strips was by no means abhorrent to the mediaeval peasant mind. What was really gained by keeping the arable furlongs in a compact area was convenience of another sort. It was the possibility of letting the cattle range without hindrance over a large part of the township. Had any furlongs within a large fallow area been subjected to cultivation while the rest of the area was utilized for fallow pasture, it would have been necessary to fence the cultivated portions. Such an inconvenience was obviated by large and simple boundaries, and the easy utilization of the fallow for pasture was what lay behind a system of two or three comprehensive fields. In East Anglia different pasturage provisions deflected the field boundaries, and with them the field system, from the normal type.

Important as is the relation between common rights of pasture and the two- and three-field system, the records at our disposal seldom enable us to argue from the former to the latter. Since references to common rights of pasture are infrequent even in elaborate sixteenth-century surveys, the less can they be expected in the briefer early documents. It is rather in the direction of disproof that certain items will be of avail. In the East Anglian evidence there are references to pasturage arrangements of a sort not realizable under a two- or three-field system. In consequence of this (and of other circumstances) it will be possible to maintain that the system was not there employed. On the other hand, whenever in the case of two- or three-field townships no information regarding pasturage rights is to be had and no contradictory indications appear, it may fairly be assumed that the sheep and cattle were each year pastured over a large compact arable field.

If this characteristic of the two- and three-field system is seldom perceptible in the early documents, such is not the case with the

THE TWO- AND THREE-FIELD SYSTEM 49

other features which have been noted above. However brief the terrier or grant, it will indicate whether arable open field tended to preponderate over enclosures; it will show how evenly the arable was divided between two or three fields; very likely it will be the description of a copyhold; it may by good chance refer to a virgate, a bovate, or a fraction thereof. If our source of information be an extent rather than a terrier, it may, if it relates to a two- or three-field township, either apportion the demesne acres between two or three fields, or it may state that every second or third year one-half or one-third of the demesne was fallow and had no value because it lay in common. Such are the criteria to be applied in sifting the evidence now to be considered.

CHAPTER II

THE EARLIER HISTORY OF THE TWO- AND THREE-FIELD SYSTEM

RELYING upon the characteristics of the two- and three-field system deduced from the comprehensive evidence of the sixteenth-century surveys, we may now turn to the more fragmentary and, for the most part, earlier testimony touching the system in question. It has been collected and arranged by counties in Appendix II. Much of it is in the nature of terriers of single holdings found in rentals or deeds of conveyance, but only such evidence as satisfies the criteria indicated in the last chapter has been admitted. In particular, reasonably equal distribution of arable acres between two or three fields has been insisted upon. Descriptions of freeholds and leaseholds have been utilized when they give unmistakable information about field systems and when copyholds have not been available. Items relative to small holdings have not been excluded if the acres in question lay equally divided between fields which bore the usual names. Lastly, the statements of the extents concerning fallow and commonable demesne have been appended whenever they appear pertinent. This collection of early evidence ought, it would seem, to enable us to answer certain questions regarding the two- and three-field system. At what time did it first appear in England ? Throughout what territory did it prevail ? Were two-field or three-field townships the earlier ? Were the former sometimes transformed into the latter ? And what were the respective areas appropriated by each group ? Answers to these questions can be secured from Appendix II, although they may not always be so precise as might be desired.

Most unsatisfactory is the testimony regarding the first question — that which asks about origins. The difficulty, as is usual with such queries, arises from paucity of evidence. From the end of the twelfth century, when the feet of fines begin and when grants of land first become specific and descriptive, we have acceptable information; but between the Conquest and the reign of Richard I the charters disdain field detail. So, too, for the most part do those of the Anglo-Saxon period. Since it is very desirable, however, to have some conception about field arrangements at this time, fragmentary evidence may well be attended to.

The testimony of the charters and laws of Anglo-Saxon England relative to open arable fields has been noticed by Nasse and Seebohm.[1] These writers point out that certain suggestive phrases and a few definite statements establish the existence of common arable fields in England long before the Conquest; but neither writer adduces any evidence which shows that the system employed was a two- or three-field one.[2]

Since the charters are more remunerative in information than the laws, we may turn first to them. Such pertinent matter as they contain is usually found in the boundaries of the land which they convey. These boundaries, which follow the Latin body of the charter, are nearly always in Anglo-Saxon. Often they are later than the charter itself, but by how much it is seldom possible to determine. Except for a few brief early ones, they date from the ninth, tenth, and eleventh centuries. Since for the most part they bound large parcels of land — the five, ten, or twenty hides conveyed — they often coincide with the boundaries of a township. Usually, too, they refer to striking features of the landscape — roads, hills, ditches, streams, groves, trees, barrows, and the like; and in so far as this is the case they give no information relevant to our subject.

Certain grants, however, were less extensive than a township, and it might be expected that the boundaries of these would

[1] Nasse, *Agricultural Community*, pp. 18–26; Seebohm, *English Village Community*, pp. 105–117.

[2] Nasse (op. cit., p. 25) was inclined to see a three-field arrangement in King Eadwig's grant of twenty hides to Abingdon monastery (Kemble, *Codex Diplomaticus*, 1216). The specification runs, " Ðis sindon ða landgemaero ðaesse burlandes to Abbendune, ðaet is gadertang on þreo genamod, ðaet is Hengestes ig and Seofocanwyrð and Wihtham." Unfortunately for Nasse's interpretation, it turns out that Hinksey and Witham are two townships just west of Oxford.

immediately reveal the existence of an open-field system. One hide subtracted from a five-hide township should, under a two- or three-field system, comprise many scattered parcels in the arable fields;[1] and the bounding of such a hide should involve a reference to the existence of these scattered acres. Such references, as it happens, are seldom found. Wherefore Nasse and Seebohm have argued that in these cases there grew up the convention of giving the boundaries of the entire township, just as if the latter were conveyed *in toto*.[2] The convention, they explain, would have arisen because the intermixture of acres made difficult any exact definition of boundary. Reversing the argument, they conclude that, if part of a township is described with the boundary phrases employed elsewhere relative to the entire township, this circumstance proves that intermixed acres existed. In all they cite six instances to establish such a usage. Thereupon they infer that the general employment in Anglo-Saxon charters of concise boundaries for relatively small transfers of land is evidence of the wide extension of open-field arable at an early date.

Before this conclusion can be admitted, the six instances from which they argue that a grant of part of a township and another of the complete township employ the same boundaries deserve reexamination. One instance relates to Kingston, Berkshire.[3] Two charters of almost the same date describe respectively thirteen *mansae* and seven *cassati*, the boundaries being alike. We are not, however, left to arrive at the existence of intermixed arable acres by inference; for in both charters we find the preamble, " Ðis sind ða landgemaero [boundaries] to Cyngestune aecer onder aecere." The last phrase, "aecer onder aecere," is so unusual that there might be doubt about its meaning were it not for the explanation vouchsafed in another charter. Three *cassati* at Hendred, Berkshire, transferred in 962, are left without boundaries; but where the *metae* are usually inserted we are told, " Ðises landgemaera syn gemaene sua ðaet lið aefre aecer under

---

[1] Unless, as often happened at a later period, it was consolidated demesne.
[2] Nasse, op. cit., pp. 24, 25; Seebohm, op. cit., p. 111.
[3] *Cod. Dip.*, 1276, 1277 (*c.* 977).

aecer," "The boundaries are common in such way that the arable acres are intermixed."[1] This clarifies the phraseology of the Kingston charters. The preamble to them wishes to tell us that the acres were intermixed. It is equivalent to explaining why the scribe gave the boundaries of an entire township rather than attempt the impossible task of locating scattered acres. We may therefore agree with Nasse and Seebohm in their immediate inference that open arable fields are referred to in the Kingston charters, but we are not obliged to adopt their generalization. It appears rather that, if the boundaries of a township are used to describe a part of the township, this device is explained by a statement about intermixed acres.[2] When such explanation is wanting, inferences as to intermixed acres may be unwarranted.

Another citation of Nasse's and one of Seebohm's are not more happy. The latter is concerned with two charters which relate to Stanton, Somerset, and employ very similar boundaries. One conveys two and one-half hides, the other seven and one-half.[3] But the former distinctly states, " Ðis synt ða landgemaera to Stantune [the entire township]," and after the recital continues, " Ðonne is binnan ðam tyn hydun Aelfsiges [the grantee's] þridde healfe hide." The justification for the use of the boundaries of the township in connection with a part of it is specific: the two and one-half hides lay *within* the ten hides. Nasse's Waltham instance is of the same sort;[4] for the fourteen hides which King Eadmund booked are expressly said to lie " binnan ðam þritigum hidum landgemaero " — *within* the thirty hides whose boundaries are given. At Waltham, as at Stanton, the use of the boundaries of an entire township when a part of the township was to be conveyed appeared so unusual as to need explanation.

Two other groups of charters to which we are referred are not convincing. In 903, as Seebohm points out, King Edward gave to his "princeps" Ordlaf twenty *cassati* at Stanton, Wiltshire; in

[1] *Cod. Dip.*, 1240.
[2] Nasse cites the explanatory phrase of the Kingston charters, but Seebohm refers to it only in a note upon another point (op. cit., p. 112, n. 3).
[3] *Cod. Dip.*, 502 (*an.* 963), 516 (*an.* 965). [4] Ibid., 1134 (*an.* 940).

957 King Edwig conveyed to Bishop Osulf twenty *mansae* at the same place.[1] There is a slight if not a very exact correspondence between the descriptions of the boundaries of the two grants. Assume, as Seebohm did, that the boundaries are the same. Why should each twenty *mansae* (or *cassati*) be looked upon as part of a larger township? Why should they not refer to the same area — perhaps to a township of twenty hides? That the grantees in each charter were different need cause no difficulty. Between 903 and 957 the twenty *mansae* may well have reverted to the crown. The first grant was of the sort which did revert; it had just done so in 903. The boundaries to which Nasse refers at Wolverley, Worcestershire, were probably alike for the same reason.[2] In one charter the king gave two *mansae* to one of his *ministri*, in the other two *mansae* to the cathedral church at Worcester. Both grants, to be sure, occurred within the same year. It is not improbable, however, that there was a speedy reversion and regrant, while the identity of the *mansae* conveyed is insured by the circumstance that the first grantee (Pulfferd) gave his name to the land.

A last instance is cited by Nasse. In the middle of the tenth century eighteen *mansae* and twenty-two *mansae* were conveyed at Welford, Berkshire, with substantially the same boundaries.[3] No phrase explains why this is so, nor do the eighteen seem to have been a part of the twenty-two. Nasse apparently thought them constituents of a forty-hide manor, bounded similarly because arable acres were intermixed. The first part of this assumption seems justifiable. In Domesday Book, Welford is set down as a manor "formerly" rated at fifty hides.[4] What Nasse forgot is that a manor of this size was usually composite, containing within its bounds more than one township. A comparison of the Domesday map with the modern one reveals Welford as such a manor.[5] This being the case, the eighteen and the twenty-

---

[1] *Cod. Dip.*, 335 (*an.* 903), 467 (*an.* 957).
[2] Ibid., 291, 292 (*an.* 866). [3] Ibid., 427 (*an.* 949), 1198 (*an.* 956).
[4] "T. R. E. se defendit pro l hidis et modo pro xxxvii (i. 58b)."
[5] Cf. *Victoria History of Berkshire*, i. 323. Several hamlets near Welford do not appear on the Domesday map, e. g., Easton, Wickham, Warmstall, Clapton, Shefford.

two hides can scarcely be referred to a single township of forty hides. Since, however, the agrarian unit within which arable acres were intermixed was the township rather than the composite manor, these charters tell us nothing about a usage such as Nasse argues for. To be convincing, he should have pointed to a small number of hides (less than ten) bounded with the boundaries of the township within which they are supposed to have lain.

Inasmuch as neither he nor Seebohm cites instances of this sort which are not self-explanatory, it does not seem safe, in cases where we cannot compare the boundaries of fractional and entire townships, to infer that the boundaries of small grants were frequently those of townships. Without this inference it is not possible to argue that the general character of Anglo-Saxon charter boundaries goes to prove the early prevalence of intermixed arable acres in England.

Another aspect of the boundaries, however, better endures examination. This Seebohm pointed out,[1] and this Vinogradoff has emphasized.[2] They remark that in some enumerations are found words and phrases drawn from the open-field vocabulary, phrases which must naturally have occurred wherever the boundary of a township ran for a space along an open arable field. The appearance of these expressions in the description of *metae*, they argue, goes to prove the existence of arable common fields.

Prominent among phrases of this kind is *forierthe* or *heafodaecer*.[3] It was the term applied to the long headland upon which the strips of a furlong abutted, and would scarcely have been used in a region not characterized by intermixed strips. *Garaecer*, or gore acre, the small irregular triangle in the corners of furlongs, also appears.[4] This term was less essentially bound up with an open-field system than was "headland," being applicable to any parcel of land thus shaped; still, it was one of the phrases of the open-field vocabulary, and its use as a landmark may be significant. Relative to *hlinc*, so often found and so strongly

[1] Op. cit., p. 107.
[2] *English Society in the Eleventh Century*, p. 278.
[3] For early instances, see *Cod. Dip.*, 437, 1080.   [4] e. g., ibid., 1080.

insisted upon by Seebohm and Vinogradoff, there seems to be no reason for supposing that in the boundaries it meant anything more than hillside. Such has been and is its usual connotation. Seebohm explains that terraces of hillside arable strips were in the nineteenth century called " lynches ";[1] but the term seems seldom to have had this significance in sixteenth-century surveys or in earlier field documents.[2] Its application to the terraces is probably late, and due to an extension of the original meaning. Among the phrases of the boundaries, therefore, that which most clearly refers to open fields is *heafodaecer*, and the first appearance of this is in the tenth century.

Apart from occasional open-field words which by chance crept into the boundaries, the charters contain a few specific references to the open-field system. Nasse first cited four of them, all in tenth-century grants to Abingdon monastery, and still among our best bits of evidence.[3] Seebohm added one reference,[4] Vinogradoff three,[5] and Maitland four, two of them credible and two doubtful.[6] Ten of these citations, together with nine others, may now be given as embodying the most convincing evidence which the charters of the Anglo-Saxon period proffer regarding open-field conditions: —

[1] Op. cit., p. 5.
[2] The Hertfordshire instance cited below on (p. 377, n. 2) is unusual.
[3] Op. cit., pp. 22, 24. In the following list the four are nos. 1169, 1234, 1240, 1278, in Kemble's *Codex*.
[4] Op. cit., p. 112, n. (*Cod. Dip.*, 1213).
[5] *English Society*, pp. 259, 277, 279 (*Cod. Dip.*, 793, 503; *Cartl. Sax.*, 1130).
[6] *Domesday Book*, pp. 365–366. The credible ones are here given: *Cod. Dip.*, 339, 586. The doubtful ones are from Kent and have not the characteristics of two- or three-field grants: ibid., 241 (*an.* 839), 259 (*an.* 845). Of the latter the first refers to " xxiiii iugeras . . . in duabus locis in Dorovernia civitatis intua [intra] muros civitatis x iugera cum viculis praedictis et in aquilone praedictae civitatis xiiii iugera histis terminibus circumiacentibus. . . . " The boundaries which follow indicate that the fourteen acres formed a single parcel, while the ten acres seem to have been within the walls. The other Kentish charter conveys " xviiii . . . iugera hoc est vi iugera ubi nominatur et Uuihtbaldes hlawe et in australe parte puplice strate altera vi et in australe occidentale que puplice strate ubi appellatur Uueoweraget in confinioque Deoringlondes vii iugero. . . ." The equal division of acres here does indeed suggest a threefold plan, but the awkward location of the three subdivisions with reference to highways rather than fields shows that the arrangement was accidental. Kent, as we shall see, was one of the English counties in which the three-field system did not come to prevail.

Date	Reference[1]	County, Village, and Description
904	339	Worcs., Beferburnan (Barbourn). "Eac hio sellað him be befer burnan þa ludadingwic] ec þaer to sextig aecera earðlondes be suðan beferburnan] oer sextig be norðan. . . ."
953	1169	Berks, Cusanhricge (Currage in Chieveley). Grant of v *cassati*. After the boundaries occurs, "And on ðan gemanan lande gebyrað ðarto fif and sixti aeccera."
958	1213	Berks, Draitune (Drayton). Grant of x *mansae*. "Ðis sind ða landgemaera to Draitune, aecer under aecer."
961	1234	Berks, Aeðeredingetune (Addington in Hungerford). Grant of ix *mansae*. "Ðas nigon hida licggeað on gemang oðran gedallande feldlaes gemane and maeda gemane and yrðland gemaene."
962	1240	Berks, Henneriðe (Hendred). Grant of iii *cassati*. "Ðises landgemaera syn gemaene sua ðaet lið aefre aecer under aecer."
963	503	Wilts, Afene (Avon). Grant of iii *cassati*, "singulis iugeribus mixtim in communi rure huc illucque dispersis."
not earlier than 964	Cartl. Sax., 1145	Wilts, Þinterburnan (Winterbourn). Grant of x *mansae*. "Þis syndon þara fif hida land gemaera Into Þinterburnan be þestan tune syndries landes. . . . Þonne syndon þa fif hida be Eastan tune gemaenes landes on gemaenre mearce spa spa hit þaer to be limped."
966	531	Gloucs., Clifforda (Clifford Chambers). Three-life lease of ii *mansae*: "oðer healf hid gedaellandes and healf hid on ðaere ege."
974	586	Gloucs., Cudinclea (Cuddingley). Grant of i *mansa*. After the boundaries occurs, "and xxx aecra on ðaem twaem feldan dallandes wiðutan."
c. 977	1276	Berks, Cyngestun (Kingston). Grant of xiii *mansae*. "Ðis sind ða landgemaero to Cyngestune aecer onder aecere." Boundaries, "on ða heafodaeceras."
982	1278	Berks[?], Ceorlatun. Grant of v *cassati*. "Rus namque praetaxatum manifestis undique terminis minus dividitur, quia iugera altrinsecus copulata adiacent."
985	648	Hants, Harewillan (Harewell). Grant of xvii *cassati*, "segetibus mixtis."
987	658	Hants, Fearnlaeh (Farleigh). Grant of iii *mansae* at Westwood and iii *perticae* at Farleigh. After the boundaries of Westwood is, "Ðonnae licgeað ða þreo gyrda on oðaere haealfae fromae aet Faearnlaeagae on gaemaenum landae."

[1] The references, with the exception of two to Birch's *Cartularium Saxonicum*, are to the numbers in Kemble's *Codex Diplomaticus*. In volume iii of the *Cartularium* there is a table showing the corresponding numbers in the two works.

Date	Reference	County, Village, and Description
990	674	Worcs., Uppðrop (Upthorp). Three-life lease of ii *hida* to two brothers: "and se ealdra haebbe ða þreo aeceras, and se iungra ðone feorðan, ge innor, ge utter."
c. 972–992	Cartl. Sax., 1130	Northants, Oxanege (Oxney). Possessions of Peterborough abbey. " Ðið utan þan ige sixti sticca landes þet is ameten to xxx aecerum. . . ."
995	692	Gloucs., Dumbeltun (Dumbleton). Grant of "duas mansas et dimidiam . . . praedictum rus, quod in communi terra situm est."
1002	1295	Gloucs., Dumoltun (Dumbleton). Grant of xxiiii *mansae*, " x et vii in occidentali parte fluminis Esingburnan . . . ac duas in orientali eiusdem torrentis climate, sorte communes populari aet Eastune, necnon et v [in] locis silvaticis. . . ."
849 [actually x or xi cent.]	262, App.	Worcs., Coftun (Cofton). Boundary, " up be ðam gemaenan lande."
1050	793	Oxons., Sandforda (Sandford). Grant of iiii *mansae*. " Ðis sind ða landgemaera to Sandforda on ðam gemannan lande."

What is most immediately to be deduced from these nineteen citations is the fact that, while none of them are earlier than the tenth century, there did exist at that time so-called common land. Frequently the passages imply nothing more. The sixty-five acres at Currage were " on ðan gemanan lande,"[1] as were the three roods at Farleigh and the four *mansae* at Sandford. A boundary at Cofton ran " up be ðam gemaenan lande." The Hendred charter, as we have seen, amplifies the term " gemaene " enough to explain that its lands lay " aecer under aecer," a phrase which, along with the mention of " haefod aecer " in the boundaries, must lead us to agree with Nasse and Seebohm in seeing at Kingston intermixed arable acres. " Aecer under aecer " is also used to describe the situation at Drayton.

[1] Nasse, arguing for early convertible husbandry as applied to the waste, sees in this " a certain portion of the common pasturage . . . taken up and applied temporarily to arable purposes " (op. cit., p. 23). Since there is no other reference in Anglo-Saxon documents to convertible husbandry of this kind (Nasse's other citation implies, as many charters do, merely proportionate rights in the waste), it seems better to interpret the Currage phrase as descriptive of five hides of demesne to which sixty-five acres in the common arable fields were appurtenant.

The Addington account of nine hides is valuable in that it further amplifies our conception of " gemaene land." These hides lay " on gemang oðran gedallande," and their " yrðland " was " gemaene." Nothing could more fittingly describe holdings in open field than to say that they lay in the midst of other divided land, with the arable (as well as the pastures and meadows) in common. Gedalland, or divided land, was, then, the technical Anglo-Saxon phrase for intermixed arable acres. Its use may imply that the division of the arable had passed beyond a stage of yearly allotment to one of permanent possession. It may, on the other hand, imply nothing about permanence of possession, but may refer only to the minute subdivision to which the arable had been subjected. Whichever the case, it is a more specific term than " gemaene land," a phrase applicable to common pastures and common meadows as well as to arable.

The implications of gedalland once clear, a brief reference to it in the Clifford charter is self-explanatory. Here one-half of two *mansae* was on an island, and the other half was " gedaelland." But the term seldom occurs in the charters, being of more importance in a well-known passage of the laws. The parcels of the " divided " land were, as Vinogradoff conjectures, probably known as " sticca," sixty of which at Oxney were equivalent to thirty acres.[1]

The term contrasted with gedalland as indicative of ownership in severalty was " syndrig land." In the Winterbourne charter such were the five hides to the west of the village, and pains are taken to contrast them with the five hides of " gemaene land " to the east. The former must have been what would at a later day have been called demesne, relative to which common rights were non-existent.

Latin equivalents of the Anglo-Saxon phrases are easy to interpret. At Dumbleton two and a half *mansae* were " in communi terra," and in another charter two *mansae* were " sorte communes populari," common and in common lot. The phrase " aecer under aecer " got itself translated as " segetibus mixtis " at Harewell. At " Ceorlatun " the circumlocution was longer, " iugera

[1] Cf. the above list, and his *English Society*, p. 279.

altrinsecus copulata adiacent." Clearest was the description at Avon, " acres scattered here and there intermixedly in the common arable field."

At Upthrop we can see the acres getting intermixed. Two brothers so divided two hides there that in all places the elder had three acres, the younger the fourth. The division would scarcely have been described in this way had it looked to the creation of two compact holdings. Instead of this, we may assume that each plot of the two hides was divided and two holdings of scattered parcels created.

Only two of the passages suggest what kind of field system was in use, and these Maitland has already quoted. At Cuddingley in Gloucestershire there were thirty acres of "[ge]dalland " in the two fields, a pretty clear reference to a two-field system. Somewhat more questionable is the other passage. Appended to the grant of a parcel of land within the city of Worcester were sixty acres of arable to the south of " Beferburnan " and sixty to the north. If Beferburnan (Barbourn) was then a hamlet, as it is today, the description would be not unlike many later ones which indicate the presence of two fields by the statement that a certain number of acres lay on one side of a village and the same number on the other side.[1] In the charter of 904, however, the name Barbourn may have designated merely a stream. If so, there is no particular significance in the passage, since land divided by a brook may have been consolidated.

It chances that this Barbourn charter is earlier by fifty years than any other of the list. Indeed, most of our citations date from the second half of the tenth century. If, then, the Barbourn reference be excluded, our first reliable charter testimony touching open fields in England dates from these decades. That we have nothing earlier is perhaps due to the comparative rarity of genuine charters before 950, and to the very brief references to boundaries which the genuine ones contain.

One other feature of the passages quoted is of interest. All refer to townships located within seven counties, and these are counties of the southern midlands. Berkshire, Hampshire, Wilt-

[1] Cf. Appendix II.

shire, Gloucestershire, Worcestershire, Oxfordshire, and Northamptonshire form a compact area, a part of the larger territory within which we shall soon see the two- and three-field system domiciled. The testimony of the charters is, therefore, in accord with that of more detailed but later evidence. Briefly stated, it is this: in seven counties of the southern midlands some twenty charters of the tenth and eleventh centuries testify to the existence of open common arable fields, and one or two of them probably reflect to a two-field system.

Turning to the Anglo-Saxon laws, we find a single passage of first-rate importance relative to open fields, but we find little besides. The passage in question, which has been quoted by Nasse and Seebohm,[1] runs as follows: —

" Gif ceorlas gaerstun haebben gemaenne oððe oþer gedálland to tynanne, 7 haebben sume getyned hiora dael, sume naebben, 7 etten hiora gemaenan aeceras oððe gaers, gán þa þonne þe ðaet geat agan, 7 gebete þam oðrum, þe hiora dáel getynedne haebben, þone aewerdlan þe ðaer gedon sie." [2]

What gives this regulation a unique importance is its date. Ine's laws belong to the end of the seventh century, to the years between 688 and 694.[3] At this time there existed, as the extract shows, common meadow and " other gedalland " which it was the duty of the tenants to hedge. If one of them failed to do his share of the hedging, and cattle destroyed the growing grass or grain, he was responsible to his co-tenants. Such a conception of gedalland corresponds with what we have learned of it in the tenth century. The term was then applied to common intermixed arable acres. The gedalland of Ine's law was not pasture, since pasture would not have been divided. It was " other

---

[1] Nasse, op. cit., p. 19; Seebohm, op. cit., p. 110.
[2] F. Liebermann, *Die Gesetze der Angelsachsen* (3 vols., Halle, 1898–1912), i. 106. " If ceorls have common meadow or other gedalland to hedge and some have hedged their share and some have not, [and if stray cattle] eat their common acres or grass, let those who are answerable for the opening go and give compensation to those who have hedged their share for the injury which may have been done."
[3] Liebermann, "Ueber die Gesetze Ines von Wessex," in *Mélanges d'Histoire offerts à M. Charles Bémont* . . . (Paris, 1913), p. 32. Liebermann recognizes in the above passage " ein Dorf mit Gemeinwiese und Gemengelage der Aecker " (ibid., 26).

than " meadow. It must have been arable. The arable acres must further have been intermixed, else the cattle, once through the hedge, could not have ranged over all of them. Common intermixed arable acres in England are therefore discernible at the end of the seventh century. The law assures us of their existence two centuries before the charters give testimony.

In another respect the law agrees with the charters. Both locate the early arable open field in the western and southern midlands. The counties to which the charters refer were (with the possible exception of Northamptonshire) of West Saxon origin.[1] That the laws of Ine were applicable to the same territory at the end of the ninth century is shown by Alfred's recension of them. Wessex and the southern edge of Mercia were thus the regions within which we see arable open field pretty clearly at the end of the seventh century and quite unmistakably in the tenth.

Apart from their implications regarding the existence of common arable fields, our earliest sources tell us little. No reference to a three-field township is vouchsafed, and only twice (in the tenth century) is there probable reference to a two-field township. But meagre as is the contribution of Anglo-Saxon documents to our knowledge of field systems, that of the first Norman century is not more ample, and we may pass at once to the times of Richard and John.

Only with the definite evidence of the late twelfth and of the thirteenth century do we first come upon townships whose arable fields were clearly two or three. Since both sorts were then reasonably numerous, it is at length possible to ascertain the area throughout which the two- and three-field system prevailed in mediaeval England. Later testimony fills in doubtful stretches of the boundary, until by the sixteenth century the circuit can be pretty well determined. From the available data which have been collected in Appendix II its reconstruction may now be attempted.[2]

In the north the county of Northumberland must for the time be excluded. The three fields which some documents seem to

[1] Chadwick, *Origin of the English Nation*, pp. 3, 5, map facing p. 11.
[2] The result is shown on the map facing the title-page.

disclose there manifest certain questionable features, and will best be discussed in a later chapter which treats of the field system of the Border.[1] In Durham we are on secure ground, although the evidence is relatively late. The survey of Ingleton, which has already been quoted to illustrate the three-field system,[2] is one of a series, several members of which are similar to it. All the townships thus described lie in the southern part of the county in the flat region which stretches from Durham to the Tees. The episcopal city thus becomes the northern outpost of the three-field system.

In Yorkshire, the East Riding and much of the North Riding furnish evidence of the existence of two- or three-field townships. The West Riding is more chary in this respect, for in the mountainous western part the system cannot be discerned.

Keeping to the east, the boundary of two- and three-field tillage follows the coast until, on reaching Boston, it turns inland to exclude the fen country. Parts of the counties of Lincoln, Northampton, Huntingdon, and Cambridge (the Isle of Ely) now fall outside of it, though by far the larger part of each county remains within it. From southeastern Cambridgeshire the line turns sharply to the southwest, follows the hills which separate Hertfordshire from Bedfordshire, passes on along the ridges of the Chilterns through southern Buckinghamshire and Oxfordshire, crosses Berkshire east of Reading, keeps near the eastern boundary of Hampshire, until, on reaching the South Downs, it follows them eastward into Sussex as they stretch on to lose themselves in the Channel at Beachy Head. All the southeastern counties from Norfolk to Surrey, together with a large part of Sussex, are thus excluded.

The western boundary of the two- and three-field area begins in western Dorsetshire, passes north across Somerset including two-thirds of that county, crosses by the forest of Dean into Herefordshire, embraces most of this county and its neighbor Shropshire, passes northeast through Staffordshire and Derbyshire into Yorkshire, where it cuts off the western edge of the county as it continues to Durham. Three areas are excluded

[1] Cf. below, pp. 210 *sq.* [2] Cf. p. 36, above.

on the west and north: Cornwall, Devon, and western Somerset; Wales with Monmouthshire; and the counties of the northwest, Cheshire, Lancashire, western Yorkshire, Westmorland, Cumberland, and possibly Northumberland. Within the boundaries thus drawn lay at least half the soil of England, and the counties comprised are for the most part known as the northern and southern midlands. For brevity, therefore, and because it is not altogether inappropriate, the term midland system will often be employed henceforth in referring to two- and three-field arrangements.

There is one stretch of the boundary just indicated which is not borne out by the citations of Appendix II. This is the link which embraces the counties of Hereford and Shropshire. The early evidence in support of the existence of a three-field system in these counties is relatively so meagre that it seems best to set it forth separately and in detail. It will be remembered that testimony has already been adduced from Jacobean surveys to show the presence of three-field townships in the two counties. Especially at Stockton and its hamlets in northern Herefordshire have three fields been discerned, and the Shropshire hamlets bordering upon Claverley Holme and Warfield Holme appear also to have had rather consistently three arable fields. But little further sixteenth-century evidence is available, and, as we shall see, there were many irregularities in Herefordshire fields at that time.[1] Still later, too, only three or four of all the Herefordshire enclosure awards bespeak three fields.[2] For these reasons early evidence is the more to be desired. The system, if existent, soon began to decline and can have been intact only in its youthful days. What, then, say the early charters and extents?

The Herefordshire evidence is more slight than that from the neighboring county. We have no difficulty in discovering that a three-course rotation of crops was later in favor on demesne lands, but the demesne in question probably did not lie in open field.[3] An extent of Luston, a manor of Leominster priory, how-

---

[1] Cf. pp. 93 *sq.*, below.   [2] Cf. pp. 142–143, below.
[3] The surveyors of the lands of the home manor of the abbey of Dore explain: " And wher also some parte of the arable lands of the sayd demeain . . . is not

ever, states that in 1 Edward III 150 of the demesne acres lay in Tuffenhull field, 140 in Breshull field, and 125 in Wondersback field, a description which seems indicative of a three-field township.[1]

A similar situation may be perceived in a charter of 1273 which transfers the "quartam partem unius virgate terre de Luda que iacet inter agros de Mortuna."[2] This quarter-virgate of Lyde, which lay within the fields of the neighboring hamlet of Morton, had its acres equally divided among three fields: —

"viii acre sunt in cultura que dicitur parve spire ager [the following re-grant adding] quarum v sunt ultra Waribroc, tres vero citra in cultura que dicitur Preostecroft

et vii acre sunt in cultura que dicitur West field

et vii acre sunt in cultura que dicitur Sudfeld quarum iii sunt sub Dodenhulle et iii apud pontem de Ludebroc."

lyck good as the more parte therof is and for that the same arable lands, by all marks as the[y] severally lye, every thyrd year lye fallow, we in consyderatyn therof have valued all the same arable lands togethers in grose al on' hole communibus annis. . . . There be in severall fylds of the sayd demeains of arable lands ccccxx acres valued communibus annis at iiii li" (Rents. and Survs., Ro. 225, 32 Hen. VIII).

[1] John Price, *An Historical and Topographical Account of Leominster and its Vicinity* (Ludlow, 1795), pp. 151 *sq*. In the priory's other townships two- or three-field arrangements are not suggested. At Hope the demesne consisted of 150 acres in Hhenhope and 120 in Brounesfield, improbable names for township fields. At Stockton, in the parish of Kimbolton, the fields were three (or five), but the division of acres among them was unequal. In Whitebroc field were 125 acres, in the field of Conemers and in Alvedon 192, in the field of Redweye and in Stalling 208. Ivington seems at first glance to have had three fields, since the demesne arable comprised 144 acres in West field, 132 in the " field against the Par," and 146 acres in Merrell. It chances, however, that the fields of Ivington are again met with in a fifteenth-century Leominster cartulary, where transfers of six acres and four acres give specific locations (Cott. MS., Domit. A III, ff. 231, 231*b*). Of the six acres, two lay " in campo qui vocatur le merele," two " iuxta parcum de Ivynton," and two " in campo de Brereley qui dicitur Westefeld " in two parcels. Of the four, two were " in campo qui vocatur le Wortheyn " and two " in campo qui vocatur le Stockyng." The fields of the first grant are those in which, according to the earlier document, the demesne was situated. Hence it is disconcerting to learn that West field is a field of Brierley, an adjacent hamlet. When further we find two new fields appearing in the second grant, the three-field character of Ivington, suggested at first, becomes problematical.

[2] W. W. Capes, *Charters and Records of Hereford Cathedral* (Hereford, 1908), p. 23.

Although this description gains in value because the division of a fractional virgate is in question, it must be admitted that two instances would constitute slight proof of the early existence of a three-field system in the county, did they stand alone.

From near-by Shropshire, however, more satisfactory early data are available. There can, of course, be no doubt about the existence of common fields in this county. Later descriptions of monastic properties drawn up in 31 Henry VIII continually locate the arable acres " in communibus campis." [1] Occasionally they give more specific information and mention three fields. At Norton in the parish of Wroxeter there were held at the will of the lord two messuages, two crofts, and " in quolibet campo communi trium camporum ibidem . . . ten dayeserth." [2] Just over the county border in the parish of Gnosall, Staffordshire, Lilleshall monastery had a messuage, a croft, and arable land " in tribus campis communibus ibidem." [3] A copyhold of the monastery of Much Wenlock comprised

" xx acras terre arabilis iacentes in campo ibidem vocato Westwodfeld

xviii acras ibidem in Alden hill feld versus Estp'

xxii acras in campo vocato overfeld. " [4]

In general, however, these monastic properties are described with brevity, and we turn to earlier documents. In 5 Edward III a three-course rotation of crops was employed upon the demesne lands at Ernewood and Hughley, one-third of the acres being sown with wheat.[5] But this proves little. More instructive is the fact that in the fourteenth century the demesne lands of the manors of three of the largest abbeys in the county, Shrewsbury, Lilleshall, and Much Wenlock, were so tilled that one-third of the arable each year lay both fallow and in common.[6] Although such tillage is not conclusive proof that the demesne was distributed among three open common arable fields, it does, we have

[1] Land Rev., M. B. 184, ff. 4, 5, 7, 9, 12, 184f, 190f, 210, 228b, 234, 236, etc.
[2] Ibid., f. 18.   [3] Ibid., f. 15b.
[4] Ibid., f. 134b.
[5] Exch. Anc. Extents, no. 68.
[6] " Tertia pars iacet quolibet anno ad warectam et in communi" (Add. MS. 6165, ff. 37, 43, 51).

seen, make this probable.[1] With regard to the demesne arable of another manor little doubt about the apportionment remains. At Faintree, in 2 Edward I, the jurors say that " in uno campo sunt xxxiiii acre terre arabilis et in alio campo xxx acre et in tertio campo xxvi."[2] Thus in one way or another we get glimpses of considerable tripartite division of the demesne in fourteenth-century Shropshire.[3]

Always more relevant to the study of field systems than the items about the demesne is information about tenants' holdings. Fortunately there are four or five descriptions of early Shropshire virgates or parts thereof. One is contained in an account of the land at Poynton from which Shrewsbury abbey in 13 Henry IV claimed tithes.[4] Most of it was demesne, which lay in furlongs or " bruches " pretty equally divided between Tunstall field, Middle field, and Mulle field. But there is this further specification: —

> " Item de uno mesuagio et medietate unius virgate terre quam Willielmus Bird tenet omnes decimas in le Mullefeild
> Item de medietate unius virgate terre quam idem Willielmus tenet, viz. de Marlebrook furlonge ... tertiam garbam
> Item de tota terra dicte medietatis virgate terre quam idem Willielmus tenet iacenti in le Middlefeild omnes decimas ...

---

[1] Cf. p. 46, above.
[2] C. Inq. p. Mort., Edw. I, F. 4 (14).
[3] It could not always be found in the sixteenth century, however. An extent of the demesne of the home manor of the monastery of Wenlock, made when the property was taken over by the crown, runs as follows (Land Rev., M. B. 184, f. 61): —
" [159 acres in eleven closes.] The nombre of acres lyeing in the comyn fyld
First the West Fyld callyd eadege fyld of arable gronde count' 105 acres valewyd at 4 d. the acre
Item the South fyld count' 95 acres valewyd at 4 d. the acre
Item a leysowe by the myle pole sown with wheat cont' 10 acres [at 6 d. the acre]
Item the further Standhyll lyeing in the comyn feld northward count' 16 acres [at 1 d.]
Item the shorte Walmore dyked and quycksett about cont' 7 acres [at 10 d.]
Item the pole Dame dyked and quycksett about cont' xii acres [at 10 d.]
Item the cawscroft byndyng upon the myll cont' iii acres [at 12 d.]. "
[4] Add. MS. 30311, f. 241.

Item de medietate unius virgate terre quam idem Willielmus tenet in Tonstallfield et de quodam furlongo in Horsecroft Et de una landa terre vocata Longelane in Horsecroft tertiam garbam...."

This description is not entirely lucid. It seems, however, to refer to a single half-virgate, since " dicte medietatis " joins the lands in Middle field with those in Mulle field. If so, the half-virgate lay in the same three fields as the demesne lands from which tithes were due, with apparently something in addition in Marlebrook furlong. Such duplication of three fields in the two descriptions goes far to stamp the township as one in which a three-field system prevailed.

The Shropshire feet of fines occasionally transfer virgates or parts of them. At Darliston the third part of a virgate and four acres were described as " vi acras versus Hethe, vi acras versus Pres, vi acras versus Sanford, et vi acras de essarto sub Northwude."[1] This enumeration wears the aspect of four fields rather than three. Yet it is noteworthy that the three names used to indicate directions are those of townships near by, and this makes it probable that the first three groups of acres may have been situated one in each of three open arable fields. Inasmuch as the fourth group of six acres formed part of an assart, it perhaps represents an early addition to the three original fields. Less irregular was the fourth part of a virgate at Romsley. It comprised " in campo qui vocatur Sandstiele vi acras et in campo qui vocatur Eastfeld viii acras ... et in campo qui vocatur Coldray viii acras et mesuagium quod fuit Roberti Clerenbald."[2] This is the normal virgate terrier of a three-field township, with nothing unusual except perhaps the names of the fields. In neither of these western counties, however, were the field names so direct and simple as in the midlands. They were particularly awkward at The Last. Here a half-virgate disposed its acres so that there were " undecim in campo versus gravam de Lastes, septem in campo versus crucem de Lastes, et novem in campo versus Chetone."[3] Such a relatively equal division of acres

[1] Ped. Fin., 193-2-10, 1 John.
[2] Ibid., 193-2-38, 1 John.
[3] Ibid., 193-3-79, 21 Hen. III.

seems to domicile the three-field system in the northeastern part of the county. If so, this is an outpost toward Cheshire beyond which the system did not much advance. From the southeast of the county we have another important terrier, describing the sixteen acres which were part of a half-virgate at Presthope as "vii acras terre in Arildewelle, v acras terre in Chesterfordfeld, iv acras terre in Hinesmere."[1] The names are scarcely simple, but the division of virgate acres is after the three-field pattern. Finally, the virgate which accompanied a messuage and curtilage in a grant at Shawbury near Shrewsbury comprised

"Sexdecim acras terre campestris in quolibet campo, viz.,
   in campo versus Foret super Crokes forlonge vii acras terre
   et inter terram de Cherletone ... et Cressewalbroke ix acras
     cum particulis ad capita seilonum
Et in campo versus Hadenhale vi acras similiter iacentes in
   Stodefolde
   et quatuor acras terre extendentes inter altam viam et le
     Middelheth
   et iii acras super le Middelheth
   et duas acras terre super Sicheforlonge
Et in campo versus parvam Withiford iiii acras terre ...
   et unam acram terre iuxta Ingriythemedewe
   et iii acras terre super Molkebur'
   cum una forera ad capud dictarum acrarum
   et tres acras terre abuttantes super viam prope gardinum
     domini
   et tres acras terre abuttantes usque ad portam ...
   cum forera ad capud dictarum terrarum. ..."[2]

There could be no more straightforward declaration of a three-field system than this terrier. Only in what has already manifested itself as a Shropshire peculiarity, the predilection for naming fields with reference to adjoining townships, is there any variation from the norm. These illustrations must suffice. They are the best available in proof of an early extension of the midland system toward the Welsh border.

[1] Ped. Fin., 193-3-16, 6 Hen. III.
[2] Add. MS. 33354, f. 81. Earlier than 1254.

With this assurance that the boundary of the two- and three-field system bent westward to include Herefordshire and Shropshire, we may at length return to the entire area comprised within that boundary and attempt to make a discrimination. How much of this extensive territory was, during the Middle Ages, claimed by two-field and how much by three-field husbandry? Hitherto these methods of tillage have been treated as one. A glance at Appendix II, in which an effort has been made to collect early rather than late instances of the occurrence of both, will show that the list of two-field townships is not short. It is, indeed, probably longer than the three-field list, but the number of citations imports little when the finding of them is so hazardous. What signifies is the area over which each method of tillage was extended. As it happens, neither system always dominated large and compact stretches of territory; nearly every county within the boundary above drawn had both two- and three-field townships. Nevertheless, there were preponderances. The southwestern counties were very largely devoted to two-field tillage. Most of eastern Somerset, all the Cotswold area which stretches through Warwickshire, Oxfordshire, and Gloucestershire, all the down lands of Berkshire, Wiltshire, and Dorset, were in the thirteenth century in two fields. Even Hampshire, Buckinghamshire, and Bedfordshire may have been at least half given over to this simpler agriculture, while such was certainly the case with Northamptonshire. Lincolnshire, apart from the fen country, was a two-field county.

A slightly smaller area was characterized by the three-field system at an early time. One finds it prevalent in northeastern Hampshire, in Cambridgeshire, in Huntingdonshire, and especially in the valleys of the Trent and the Yorkshire Ouse. Here it prospered, till its domain came to be the eastern midlands, the north, and the west. In northern Northamptonshire, in Leicestershire, Nottinghamshire, Yorkshire, and Durham, in Staffordshire, Herefordshire, and Shropshire, it was easily supreme. Broadly speaking, the line of Watling Street forms an approximate boundary between the two large areas characterized respectively by the preponderance of two and three fields. Yet we must hasten to make restrictions. The considerable expanse of Lincolnshire in the north remained alien to the three-field system; similarly, in the west, Herefordshire, Shropshire, and Staffordshire showed no two-field affiliations. The subtractions from both areas nearly balance each other and leave the midlands divided into two not unequal parts.

A patent conclusion to be drawn from this localization of two- and three-field methods of tillage is that they were not expressions of racial or tribal predilection. Any attempt to discern in them usages peculiar to Saxons, Angles, or Danes meets at once with grave difficulties. The three-field system preponderated to the northeast of Watling Street. Yet if one should surmise that this is attributable to tribal habits of Angles or Danes, he would at once be reminded that many Lincolnshire townships (with names ending in *by*) had two fields as clear-cut as any situated on the Wessex downs. If, on the other hand, it be suggested that two-field usages were native to the Saxons, the early three-field townships of Hampshire and the three-field character of the Sussex coastal plain are sufficient refutation. In reality, what determined the adoption of the one or the other form of tillage was agricultural convenience, and this in turn depended largely upon the locality and the nature of the soil.

For it must be remembered that between these two modes of husbandry the difference was not one of principle but one of proportion. Under two-field arrangements there was left fallow each year one-half of the arable, under three-field arrangements one-third. The cultivated portion, whether one-half or two-thirds, was sown in the same manner; it was divided between winter and spring grains. Walter of Henley, writing in the thirteenth century, makes this clear: " If your lands are divided in three, one part for winter seed, the other part for spring seed, and the third part fallow, then is a ploughland nine score acres. And if your lands are divided in two, as in many places, the one half sown with winter seed and spring seed, the other half fallow, then shall a ploughland be eight score acres."[1] The distinction between

[1] *Walter of Henley's Husbandry, together with an Anonymous Husbandry*, etc. (ed. E. Lamond, 1890), p. 7.

two-field and three-field modes of tillage reduces, in short, to the utilization of an additional one-sixth of the arable each year. The resort to fallowing, the equitable apportionment of strips to fields, the pasturage arrangements — all the essential features of the system — remained unchanged. A divergence so slight is scarcely one which would evince tribal or racial peculiarities. It would indicate, rather, differing agricultural opportunities as interpreted by men whose fundamental ideas about agriculture were the same. This consideration leads to the inquiry whether the simpler two-field tillage gave place, as civilization advanced, to the somewhat more elaborate three-field one.

As there was little difference in size between the areas within which two-field and three-field husbandry prevailed in the thirteenth century, so the extant evidence does not clearly indicate priority of one over the other in point of time. Of the situation before the feet of fines begin at the end of the twelfth century we know little. Although one or two Anglo-Saxon charters seem to refer to two fields, they constitute no ground for a generalization. Certain inferences, however, are possible in this connection. If we admit that a two-field arrangement was simpler than a three-field one, and discover that at a later time townships sometimes exchanged the former for the latter, we shall not be unready to believe that the three fields which were existent by 1200 may themselves have been the outcome of a similar transformation. Were this the case, the original system of the English midlands should be looked upon as one of two common arable fields. For this reason the occurrence of the transformation at a later time becomes a point of importance.

Two- and three-field arrangements did not, as we have just seen, correspond with tribal usages, but simply with agricultural opportunity. Hence a change from one to the other was a matter of opportunism. As demands upon the soil increased, and as it was observed that the three-field system brought under tillage one-sixth more of the arable each year than did the two-field system, the question must have arisen whether it would pay to change a township's arable fields from two to three. It might, of course, have been argued that in the long run a two-field arrangement was as remunerative as a three-field one. Though more of the soil was left fallow each year, did not the arable repay its cultivators better for the more frequent periods of rest ? Were not the crops grown on land fallowed every other year better than those produced by land fallowed only once in three years ? Such reasoning may at times have got empirical support from the marked prosperity of certain two-field townships. But the general practice told against it. The regions which adhered to two-field husbandry were, on the whole, the bleak, chalky, unfertile uplands; those, on the contrary, which were possessed of better soil and better location came to be characterized by three fields. This can only mean that, wherever natural advantages permitted, men chose the three-field system by preference. The retention of two fields was usually a tacit recognition that nature had favored the township little.

To change from two-field to three-field husbandry was therefore tantamount to making greater demands upon the arable — to taking a step forward in agricultural progress. Some desire for improvement was, of course, bound to come in time; but in a great number of two-field townships it delayed long, becoming operative only in the seventeenth and eighteenth centuries. Surveys, maps, and enclosure awards instruct us as to the character of these late changes, and their teaching is summarized, so far as certain typical regions are concerned, in the two following chapters. In these are described certain townships, particularly in Gloucestershire and Oxfordshire, which during the sixteenth, seventeenth, and eighteenth centuries abandoned the two-field system. What they adopted was not a three-field arrangement, but one of four fields or quarters, the outcome of a subdivision of the old fields.[1] Before the sixteenth century, however, there is no example, within the midland area, of just this method of improvement. If changes took place, the recasting seems to have resulted in three fields. Evidence of such procedure is, therefore, what must be sought, but unfortunately it is the very kind of evidence which, in the nature of the case, must needs be scanty. To chance upon an early and a later reference to the same town-

[1] Cf. pp. 88, 125 *sq.*

ship, one implying the existence of two fields and the other of three, is a rare piece of fortune when single references to field systems are so few. Under these circumstances the following instances seem worthy of consideration.

That township fields were sometimes recast in a manner which involved much surveying and labor is evident from the case of two Northumberland manors. In the middle of the sixteenth century both considered proposals to re-allot parcels in the open fields with a view to the greater convenience of the tenants. One rejected the suggestion because of the difficulties involved; the other undertook the change and we have record of the new arrangement.[1] The instance is relatively late and the system evolved was probably not one of three fields; yet the readiness to undertake a readjustment more difficult than a simple subdivision of two existing fields is noteworthy.

A memorandum of the late fourteenth century from Corsham, Wiltshire, while it does not portray the transformation of two open fields into three, is yet instructive in showing the advent of three-course tillage in a two-field township.[2] It relates to the sowing of 103 acres of demesne arable, of which 47 were in a close and worth 6 *d.* the acre, while 56 were in two open fields and worth 2 *d.* the acre. The open-field acres are described as follows: —

> " Sunt etiam in le Southefeld de dominicis xv acre terre seminate cum frumento hoc anno in diversis particulis
> Item ibidem xi acre terre deputate pro ordo seminature hoc anno unde seminantur ii acre
> Item in le Northefeld xxx acre terre et dimidia que iacent ad warectandum hoc anno in diversis particulis."

This is simple two-field tillage. With the close the case is different: —

> " Est ibidem in dominicis in quodam clauso separabili xiii acre seminate cum frumento hoc anno
> item ibidem in eodem clauso x acre seminate cum drageto

---

[1] *History of Northumberland* (10 vols., Newcastle, etc., 1893–1914), ii. 418, 368. Cf. below, pp. 207–209.

[2] Rents. and Survs., Portf. 16/55.

et ibidem x acre que iacent pro warecto hoc anno
item in eodem clauso in Netherforlong' xiiii acre unde una medietas seminata cum drageto et alia medietas iacet in warecto."

Here the 14 acres in Netherfurlong were tilled as were the common fields, but the greater part of the close had adopted a three-

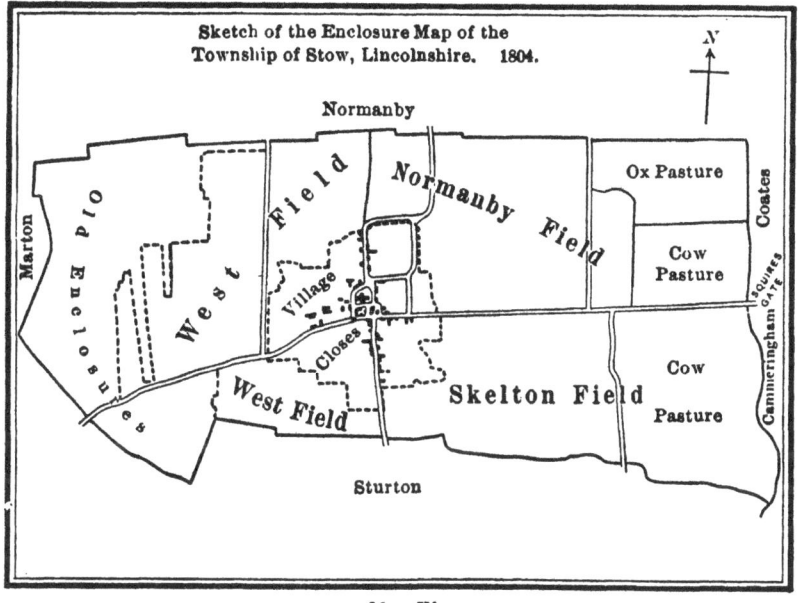

MAP IV

course rotation. One can see that such an example might some day inspire the tenants to make a similar disposition of the open fields.

That at some time a change from two to three fields had taken place in certain townships is suggested by the enclosure maps of the eighteenth century. Now and then three fields are of such a character that two of them seem to have been derived from a single older one. The accompanying plan of Stow, Lincolnshire, is illustrative.[1] If one compares it with the plan of two-field Croxton,[2] one cannot help suspecting that Stow too had once only two fields. Opposite to West field there had been an East

[1] C. P. Recov. Ro., 49 Geo. III, Hil. [2] Cf. above, p. 26.

field, now replaced by Normanby field and Skelton field. We may even conjecture how the shifting of areas had been achieved. West field had been reduced in extent by the enclosure of a part of it and by the setting off of a part for Normanby field; the former East field had in turn been so enlarged by additions from the common that each of the new divisions became approximately equal in area to the shrunken West field.

Much the same transformation can be traced in a plan of Padbury, Buckinghamshire, made in 1591. Here the old fields still retained their original names, East and West; but to the north of East field, between it and the woodland, had appeared a new common arable area called Hedge field. There is no reason to think that the old fields had been reduced in size. Improvement of the waste rather than subtraction from them seems to have been the creative factor in the change, for the names of the new furlongs, which are recorded, often suggest portions of a common.[1]

Although these illustrations do not take us out of the realm of conjecture, several others serve to do so by making it entirely clear that townships once having two fields came to have three. Sometimes the interval between the dates of the documents which picture the two stages of agricultural development is a long one. At Twyford, Leicestershire, a grant to the abbey of Burton Lazars, copied into a fifteenth-century cartulary, relates to two selions in the West field and one rood in the East field. The enclosure award of 1796, however, describes the fields of Twyford as three, Nether, Spinney, and Mill.[2] Similarly the enclosure award for Piddington, Oxfordshire, dated 1758, has reference to three fields, the Wheat field, the Bean field, and the Fallow field;[3] but a charter of 6-7 Henry I conveys to St. Mary of Missenden *inter alia* the tithes from two acres of demesne meadow there, viz., from two acres in Westmead when the West field was sown and from two in Langdale when the East field was sown.[4] At

[1] Cf. on the accompanying map the furlongs called Pitthill, Swatthill, Shermore, Cockmore hill, and Foxholes.
[2] Cf. below, Appendix II, pp. 471, 473.
[3] The award is at the Shire Hall, Oxford.
[4] Appendix II, p. 488.

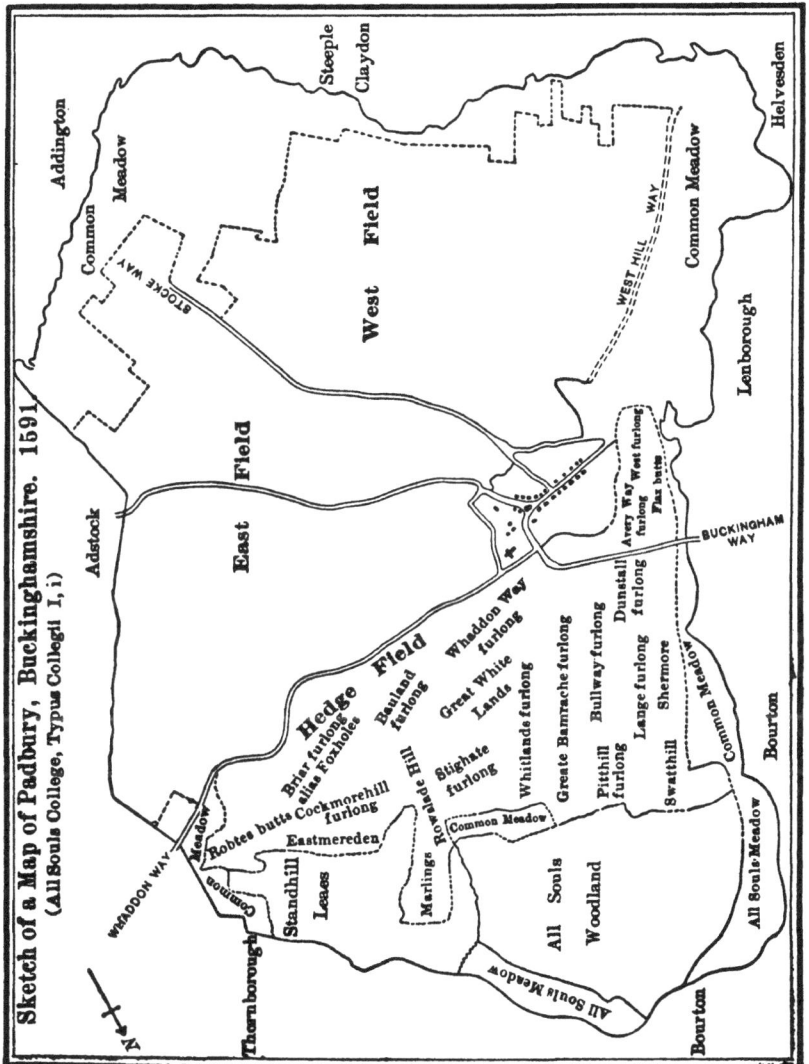

Piddington, as at Twyford, two township fields had at some time between the thirteenth and eighteenth centuries been replaced by three.

Elsewhere it is possible to discover that the transformation took place before the sixteenth century. At Litlington, Cambridgeshire, in 11 Edward III, only one-half of the demesne lands were sown annually, the remainder being of no value since they lay in common. By the time of Henry VIII, however, the demesne arable, so far as it lay in the common fields, comprised 41 acres in Westwoode field, 31 in Grenedon field, and 35 in Hyndon field.[1] In three Northamptonshire townships the period of change is likewise restricted to the interval between the thirteenth and sixteenth centuries. At Holdenby a thirteenth-century charter enumerates 36 acres of demesne arable in small parcels, assigning them in equal measure to the East field and the West field. In 32 Henry VIII another account of the demesne there refers it to West field, Wood field, and Cargatt field.[2] At Drayton a charter of the time of Henry III divides 4½ acres equally between North field and South field, allotting to each five parcels. A survey of 13 Elizabeth, on the other hand, subdivides all holdings with proximate equality among West field, North field, and East field, the respective areas of which were 529, 573, and 414 acres.[3] At Evenley, finally, several thirteenth-century charters convey arable in equal amounts in East field and West field; but a terrier of Henry VIII enumerates in many parcels 47 acres, of which 17 lay in West field, 13 in South field, and 17 in East field.[4]

While these four groups of documents pretty clearly assign the change from two-field to three-field arrangements to an undefined period between the thirteenth and sixteenth centuries, other charters and terriers reveal it accomplished or in process of accomplishment before the fifteenth century. At Long Lawford, Warwickshire, the open fields of the early thirteenth century were two; but a charter copied into a late fourteenth-century cartulary pictures them as three, and divides the numerous parcels of 49

[1] Appendix II, pp. 457, 459.
[2] Ibid., pp. 477, 482.
[3] Ibid., pp. 479, 482.
[4] Ibid., pp. 478, 482.

acres among the three with rough equality.[1] By 1 Henry IV a new third field seems to be making its appearance in an interesting terrier of the lands which the prior of Bicester had in the open fields of his home manor, Market End. According to the enclosure map of 1758 these fields numbered three.[2] In the terrier in question they were also three, but, so far as the prior's lands were concerned, of unequal importance. His acres in the North field numbered 153, in the East field 113, and "in alio campo vocato Langefordfeld" 60.[3] To all appearances an old South field was separating into two parts, with as yet no equitable adjustment of areas. If no positive record survives to assure us that the Bicester fields were once two, no such deficiency attaches to the evidence from Kislingbury, Northamptonshire. Here the fields were East and West, according to what is probably a thirteenth-century charter copied into a fourteenth-century cartulary; but a terrier of 14 Edward III refers to ten acres, of which $1\frac{1}{4}$ lay in the West field, $3\frac{3}{4}$ in the East field, and $4\frac{3}{4}$ in the South field.[4] As at Bicester, the small apportionment of acres to one of the three fields hints at a recent origin. The same situation is perceptible at Houghton Regis, Bedfordshire. A Dunstable cartulary written in a hand of the time of Edward I records the transfer of a half-virgate, eight of whose acres were in an unnamed field and eight in North field. In the same cartulary is entered another grant which refers its acres to North field, West field, and South field.[5] Since the last area receives only one half-acre parcel in contrast with the greater amounts assigned to the other fields ($1\frac{3}{4}$, $2\frac{1}{2}$ acres), here too a new field seems to be making its appearance.

The tendency of two-field townships to change into three-field ones during the late thirteenth or the early fourteenth century is

[1] Appendix II, p. 500.
[2] They were called Home, Middle, and Further. The award is at the Shire Hall, Oxford.
[3] White Kennett, *Parochial Antiquities attempted in the History of Ambrosden, Burcester, and other adjacent parts in the Counties of Oxford and Bucks* (new ed., 2 vols., Oxford, 1818), ii. 185–199. It is not certain that Kennett has transcribed from his original all the furlongs in the East field. His transcript breaks off abruptly and does not record the total here, as it does elsewhere.
[4] Cf. below, Appendix II, pp. 479, 483.   [5] Ibid., pp. 450, 451.

perhaps most unmistakably seen in two other groups of charters. At Stewkley, Buckinghamshire, in 7 Richard I, 80 acres of arable demesne in many furlongs lay *in campo de Suhelt*, and 80 more *in campo del Est*. In a charter copied into an early fourteenth-century cartulary, however, 18½ acres of arable at the same place are described as consisting of 6 in the northern part of the field, 6 in the eastern part, and 6 in the southern part.[1] The precision of the first division is paralleled by that of the second, and is explicable only by assuming a change from two-field to three-field arrangements. At Culworth, Northamptonshire, it seems possible to fix still more definitely the date of a similar change. A long charter of 24 Edward I enumerates 62 acres in many parcels divided between North field and South field. Another grant of 7 Edward III is brief, but none the less apportions to North field one acre, to South field three roods, and to West field one rood.[2] It was apparently during the reign of Edward II that West field first made its appearance.

Finally, we have express statements that three fields were substituted for two. The first relates to South Stoke, Oxfordshire, where in 1366, as an extent notes, two of the three fields were sown annually and the third lay fallow.[3] Somewhat more than a century before this, however, the fields had numbered but two. A plea roll of 25 Henry III records, in a jurors' report relative to a complaint about pasture rights, that " predictus Abbas [John of Eynsham, predecessor of Abbot Nicholas, the defendant] partitus fuit terras suas in tres partes, que antea partite fuerunt in duas partes."[4] The only doubt attaching to this account is the possibility that the lands referred to may have been demesne. Free from any such uncertainty is the record of what happened at Piddletown, Dorset. The township was once in two fields, as we learn from a charter copied into a cartulary of Christchurch priory.[5] In 20 Edward I, however, as the same cartulary narrates, the priory's lands were formally re-divided into three parts.[6] The nature of the field names, the statement that

---

[1] Appendix II, pp. 455, 456.   [2] Ibid., pp. 477, 482.   [3] Ibid., p. 490.
[4] Assize Ro. 696, m. 14a; cited in *Victoria History of Oxfordshire*, ii. 171.
[5] Appendix II, p. 462.
[6] Cott. MS., Tib. D VI, f. 200. " Divisio terre domini Prioris et conventus

the " campus " is divided, the size of the fields, and the fact that not only the bailiffs but also the " other men of the prior and convent " took part in the re-division, make it highly probable that the arable fields of the entire township were recast.

The cumulative effect of all this evidence is to establish the fact that transferences from two- to three-field arrangements in midland townships did take place. Instances have been cited from an area extending from Leicestershire to Dorset and from Warwickshire to Cambridgeshire. The period, too, during which the changes seem most often to have occurred has been determined. It comprises the thirteenth century and the early fourteenth. Of the instances which can be approximately dated, that referring to South Stoke and belonging to the first half of the thirteenth century is the earliest, while the others fall between 1250 and 1350. It is precisely during this most prosperous century of the Middle Ages that one would expect agricultural progress. In midland England, it is quite probable new demands were then made upon the soil leading to numerous changes like those described above.

We thus approach a final question. Since a transformation from two to three fields is discernible in the records that have survived, may not a similar change once have taken place in all townships which, when we know them, lay in three fields? The hypothesis is entirely credible. It is especially so since there were in the thirteenth century no large unbroken three-field areas which would point to an ancient history for that system. Three-field tillage did, of course, come to preponderate in the northern and eastern midlands; but very few counties of that region were

Christi ecclesie de Twynham facta in manerio de Pudelton' per Johannem le Marchaunt et philipum de la Berne tunc ballivos dicti manerii et alios dictorum prioris et conventus fideles anno regni regis Edwardi filii Regis Henrici vicesimo et limitatur campus in tres partes, videlicet.

Primus limes extendit se in regia via de Pudelton usque Cochestubbe et deinde . . . [boundaries] et continentur in parte illa ccxlviii acre de quibus acre iiii^{xx} non sunt digne coli quia steriles et prave sunt.

Et est campus orientalis cum toto campo australi sibi adiuncto medius campus cuius limes incipit apud . . . sselberghe et tendit se . . . [boundaries] et continentur in medio campo in universo clxxvii acre terre et sunt digne coli.

Tertius campus est campus occidentalis. In campo occidentali continentur cc et v acre de quibus acre xxx non sunt digne coli."

without some two-field townships, and much of the three-field evidence is of a date later than the thirteenth century. Hence it is not improbable that the predominantly three-field counties became such during the thirteenth and fourteenth centuries. If so, the system was a derived one, and midland England at the time of the Conquest was a region dominated by two fields.

The questions with which this chapter opened have at length received such answers as accessible data admit of. There is discernible in Anglo-Saxon England an open-field system, which first at the end of the twelfth century reveals itself as one of two or three fields; the territory throughout which this system prevailed was the extensive region known as the northern and southern midlands; the co-existence of two-field and three-field townships within this area as early as 1200 is apparent, but the preponderance of one group or the other in certain parts of it before the sixteenth century is no less obvious; finally, it is certain that to some extent transition from two-field to three-field arrangements occurred during the thirteenth and early fourteenth centuries, and it is not improbable that the three-field system may have been altogether a derived one, arising from an improvement in agricultural method. As the sixteenth century saw both forms of tillage employed, and as further changes had by that time set in, we are naturally led to inquire into the later history of what may henceforth be called the midland system.

CHAPTER III

EARLY IRREGULAR FIELDS WITHIN THE MIDLAND AREA

IT is well known from contemporary descriptions that the large midland area, just described as characterized by the two- and three-field system, showed other forms of open field in the eighteenth century.[1] These were the result of efforts made to reconcile the system with the advancing agriculture of that day. Although we shall have to examine these late innovations, it would be rash to assume that they were the first of their kind until we have inquired whether, as early as 1600, irregularities were apparent in the fields of townships within the midland area.

Such irregularities Tudor and Jacobean surveys show existent before 1610. Since regions favorably situated for agricultural development must have tended to foster them, their appearance in river valleys, frequently fertile and abounding in meadows, would not be surprising. They may be looked for in the neighborhood of the Tees, the Trent, and the Humber; at favored spots along the course of the upper Thames; beside the Severn and its tributaries, the Warwickshire Avon and the Wiltshire Avon; and in the well-watered plains of central Herefordshire or eastern Somerset. In Appendix III have been arranged extracts from surveys illustrative of these early irregularities, many of them from one or other of the regions above mentioned.[2] The field arrangements, however, of the lower Thames, a river basin without the midland area, require separate treatment.[3]

A circumstance other than situation in a favored valley may conceivably have given rise to irregularity of field system. Several tracts of land within the midland area were in early days given over to the royal forests. In course of time settlements encroached upon these, and the new or at least the expanding townships assarted forest land. Was this added arable now

---
[1] Cf. below, pp. 125 *sq.*
[2] The townships referred to are located on the map which faces the title-page.
[3] See Chapter IX, below.

cultivated as were the existent two and three fields? To answer this question an examination of sixteenth-century descriptions is essential; and, since such an examination may perhaps best be undertaken before the more numerous documents from the river valleys receive attention, extracts from surveys of forest townships have been summarized first in Appendix III.[1]

In Oxfordshire, just on the other side of Woodstock Park from Handborough and Bladon, both excellent examples of three-field arrangements, lie three other townships whose field irregularities at the end of the sixteenth century were noteworthy. All five were members of Woodstock manor. Of the three, Stonesfield was least enclosed, and here were to be found three fields apart from " Gannett's Sarte," which contained only freehold. While Church field and Callowe contributed a few acres to most copyholds, the comprehensive arable area was Home field. This, although of little importance to the freeholders, usually comprised three-fourths or more of the acres of the customary tenants. Such an arrangement was, of course, very unlike the normal one and suggests that the first arable to be improved was occupied as a single field. To this, it would seem, two small additions had in time been made for copyholders and one for freeholders. Yet the preëminence of Home field had never been challenged.

Almost as free from enclosure as Stonesfield was Wootton, where the arable lay in North field and West field, both at times called " ends." In only one instance was the virgate holding of a customary tenant divided between them, and this was because two copyholds happened to be in one man's hands. A free tenant, too, had seven acres in West field, one and one-half acres in North field. Each remaining holding was confined to one or the other of the two fields. Wootton was thus, like Stonesfield, so far as the customary tenements were concerned, a township of a single field.

The third member of this group, Long Coombe, had its copyholds considerably enclosed by 4 James I. Though the acres of a large group of " liberi tenentes per copiam " lay more often

[1] All the surveys cited in this chapter are there in part tabulated, the order of their citation being observed.

in the open fields, only one of these fields was of importance. Of West field, in which nearly all tenants, whatever their tenure, had some interest, the total area, including 28 acres of demesne, was 126 acres. In contrast, Over field contained only 28½ acres, shared by a dozen tenants; Land field 34½ acres, held by six tenants; East End, apart from 17 acres of demesne, 42 acres in the hands of four tenants. The other open-field divisions were insignificant and of no interest to the copyholders.

Why there should, at the end of the sixteenth century, have been three townships with markedly irregular fields so near to neighbors with regular fields is not obvious. Situation in a river valley, though true enough, will scarcely explain it, for Handborough and Bladon were even nearer the stream. A more plausible interpretation is suggested by the proximity of the three to Woodstock forest, without the bounds of which the two other townships distinctly lay. If the arable areas of the three were carved from the forest at a relatively late date, the regularity characteristic of older fields may not have been adopted. Pretty clearly Gannett's Sarte at Stonesfield was a recent addition, allotted, as it happens, only to freeholders. Where an assart can take its place independently among the divisions of the arable,[1] it is possible that at an earlier time other divisions came into existence in the same manner.

Further evidence pointing to the same explanation is to be had from a mid-sixteenth-century survey of Ramsden, a township on the southern edge of Wychwood forest, not far from the Woodstock group. Of the nine free tenants here we learn nothing save that they held closes. Besides a cottage, there were five customary holdings, each containing a little enclosed land but for the most part lying in open field, above all in Olde field. Each had about one-half as many acres in Gode field as in Olde field, while there were scattering additional parcels in Shutlake, in Swynepit field, and in two assarts, Herwell Serte and Lucerte. If Olde field is balanced against all the other divisions, three of the holdings can be framed into a two-field system; since, however, the other two cannot be, it is better to class Ramsden with

---

[1] An assart is a recently improved portion of the waste.

the Woodstock group. These four Oxfordshire townships thus seem to confirm the conjecture that location in an old forest area may be a reason for the appearance of irregular field arrangements within the midland territory. Among the fields it is natural that one, an Old field or a Home field, should have been larger and more important than the others.

Another forest region, somewhat nearer the outskirts of the two- and three-field area, was Arden in northwestern Warwickshire. From it we have several Tudor and Jacobean surveys, among the best, since its copyholds are numerous and are estimated in virgates, being that of Hampton-in-Arden. Here, in addition to three inconsequent areas that furnish an occasional acre, four fields frequently recur, or even five, if In field, which is nearly always joined with Mill field, be counted. The township thus bore a superficial resemblance to those of four fields, such as could at this time be found on the lower Avon.[1] Yet the virgate holdings do not well stand the test of quadripartite division. Often they had acres in the four fields, and occasionally a not very unequal number (3, 3, 4, 2½; 4, 4, 4, 3; 6, 6, 8, 5½); but more often one of the fields was slighted (6, 3, 4, 2; 4, 4, 5, 2; 4, 3, 3, 0), and in some cases two fields were altogether omitted (0, 0, 3½, 5½; 0, 0, 4, 7). Since all the irregularly divided holdings were virgates or fractional virgates, and since there were no enclosures recently taken from the fields to account for the irregularities, it is difficult to look upon Hampton as a strictly four-field township.

Near Hampton was the manor of Knoll, comprising various hamlets. At Knoll itself the copyholds consisted very largely of enclosed meadow and pasture. So they did also in the hamlets of Langdon and Widney, the fields of which are not separated in the survey. The freeholds of these hamlets, however, usually contained, along with preponderant enclosures, a few acres of open arable field. For the most part such acres were in Berye field and Seed furlong, but occasionally in Whatcroft and Hen field. The field parcels were often large (4, 5, 11 acres), and no principle of distribution among the curiously-named fields is perceptible.

[1] Cf. below, p. 88.

At Henley-in-Arden the Jacobean holdings were all small. Some of the largest had a few arable acres in Back field, usually less than ten, but with this the tale of unenclosed arable in this township, brief at best, is practically complete.[1] Leaving, therefore, the forest of Arden we may follow the field irregularities that appear to have characterized it into the wooded region which is adjacent on the north.

Much of the county of Stafford was probably in early days an unimproved forested area. In the southeast, indeed, we have found at Rolleston a normal six-field manor situated in the valley of the Trent, but outside the Trent valley fields in the county were likely to be irregular. Though Wootton-under-Weaver was not more than twenty miles north from Rolleston, its fields were five, and the method of tilling them is none too clearly discernible in the survey of 1 Edward VI. One field was too small to stand independently, but the attachment of it to any one of the other four does not result very satisfactorily. Yet a four-field arrangement is more credible than one of two or three fields, if indeed we are to predicate any regularity whatever in the grouping. The enclosures, which were few, explain nothing.

At Rocester, a little farther north, the irregularity is obvious and no conciliatory grouping is possible. Areas, to be sure, are usually given in " lands," but these cannot have differed greatly in size. At best the open fields were small, containing less than seventy-five acres in all, and the tenants at will who shared in them had their " lands," it would seem, much as chance determined.

At Over Arley in the southwestern part of the county, on the borders of Wyre forest, most holdings appear enclosed in the survey of 44 Elizabeth. Only seven tenants shared in the open fields, the area of which was less than fifty acres, and parcels in these fields were located in an even more incidental manner than at Rocester.[2] In general, therefore, outside of the Trent valley

[1] Land Rev., M. B. 228, ff. 40–64. The survey is so simple that it has not been summarized.

[2] Ibid., M. B. 185, ff. 149–156. Two tenants had parcels in Stony field, two in Great field, and five in Godfriesharne.

Staffordshire seems to have been a county tending to show forest irregularities in its open fields rather than the orderliness of the two- and three-field system.

Sufficient illustration has perhaps been given to indicate what deviations from the normal field system were in the sixteenth century to be met with in midland districts recently reclaimed from the forest. It is time to turn to the irregularities which might arise in the fields of the most favorably situated of the old townships, those of the river valleys. Since no part of England within the borders of the two- and three-field area is more endowed with natural advantages than the valleys of the southwest, the basin of the lower Severn and Avon constitutes a suitable region with which to begin our study. To assist us, sixteenth-century surveys of many monastic properties in Gloucestershire are available.[1]

Simplest of the irregularities there visible is the four-field arrangement which several townships had adopted. Since the lower Avon and the slopes of the Cotswolds were in the thirteenth century the home of two-field husbandry,[2] it is not unlikely that each of the old fields had been subdivided.[3] Implying, as four fields undoubtedly did in the sixteenth century, a four-course rotation of crops, this method of tillage brought into annual cultivation three-fourths instead of one-half or two-thirds of the arable of the township.

The surveys of Welford and Marston Sicca, villages lying not far from the Avon, are illustrative, and have in part been summarized in the Appendix. The division of the holdings among four fields was remarkably exact, perhaps an indication that the arrangement was recent. One of the fields of each township was called West field, but the names of the other fields have a ring far from ancient — Sholebreade, Stabroke, Middle Barrow, Natte furlong, Nylls-and-hadland. The copyholds of Admington and Stanton, townships not far away, were divided in the same precise way among four fields, some of which bore more usual

---

[1] Particularly in Exch. K. R., M. B. 39, *temp*. Edw. VI.
[2] Cf. pp. 29-30, and Appendix II.
[3] For later evidence of this procedure, see below, pp. 125-127, where the plan of a four-field township is also sketched.

names.[1] Longney, too, had four important fields, but only the number of " sellions " in each is given.[2] Since the areas of these are uncertain and the number of them in each field was by no means the same for different holdings, the four-field character of the township cannot be established.

Most field arrangements of low-lying Gloucestershire townships were in the sixteenth century not so simple as those just described. No neat four-field grouping is generally apparent. Less than five miles from Marston Sicca was situated Clapton, a member of the manor of Ham. The copyholds of the survey were rated in virgates, but nearly always the fields in which the acres of the virgate lay were surprisingly numerous. Usually as many as six, they might increase to twelve. Fields which appeared in one holding dropped out in another. To Lake field and Lypiatt's field most acres were usually assigned, but either of them was liable to be slighted. There were several common " crofts," Redecroft, Baucroft, Litlecroft, Prestcroft, each shared by several tenants. The township proffers a good illustration of the multiplicity of small fields, how grouped and cultivated we do not know.

Six miles out of Gloucester is Frocester, a township of the plain. About one-half of the area of the customary holdings was enclosed meadow and pasture in 1 Edward VI.[3]; the other half lay in eight small fields. South field and West field received most of the tenants' arable acres, but with no systematic division between them. Neither by joining the smaller fields with them nor by combining the latter apart can one simulate a two- or three-field system. For purposes of cultivation it apparently mattered little in what field or fields a tenant's arable acres lay.

Near Gloucester, too, was Oxlynch, a tithing of Standish, situated on the slopes of the Cotswolds. In the account of its fields nine are named, though four of them seldom. Of the others, Grete

[1] Exch. K. R., M. B. 39, ff. 149, 155. The fields of Admington (Warks.) were Humber, Harberill, Midell furlong, and Nett; of Stanton (Gloucs.), Myddle, South, Honiburne, and North.

[2] Ibid., f. 199. They were named Boinpole, Little, Acra, and Suffilde.

[3] 369 acres out of 707. The demesne comprised 607 acres, of which 136 were in open field.

Combe, Lytelcombe, Stony field, North field, and Dawhill field, at least three usually appear in each of the holdings, which were rated in fractional virgates. Several times, too, the same three recur together, and the distribution of acres among them is not very unequal. It is, therefore, possible that Oxlynch was a three-field township with two groups of three fields. If so, it was somewhat unusual in a neighborhood given over to two-field and irregular-field arrangements. About one-third of each holding was enclosed.

In the southern part of the county on the edge of the Cotswolds, neighbor on the east to townships once clearly in two fields,[1] lies Horton. A survey of 1 Edward VI shows the tenants in possession of 980 acres, of which 302, nearly one-third, were enclosed. There is some uncertainty about the number and names of the open fields. Mershe field and Yarlinge field are clear enough, but there is an " Infeld et alius campus vocatus Ynfeld." Careless spelling may be responsible for the separation of " Endfeld " from the latter. Whatever the identifications, there is no trace of a three-field arrangement in the virgate holdings, and a two-field one is problematical. Three virgates divide their open field between Yarlinge field and Mershe field, disregarding other fields. If In field be joined with Mershe field and the " great felde " with Yarlinge field, other virgates can be subdivided according to a two-field system; but still others cannot, one lying entirely in Mershe field. If Horton had ever been or still in the sixteenth century was a two-field township, it could at least then be convicted of deviations from the norm.

Three or four miles southwest of Horton and distinctly in the flat plain of the Severn is Yate, surveyed at the same time. The proportions of open field and enclosed land were here exactly reversed, two-thirds of the tenants' acres being enclosed, one-third lying in open field. In consequence there was much greater irregularity in distribution among fields than at Horton. Apart from scattering areas, three fields stood out, West field, North field, and Up field, the last necessarily an east or a south field. Although these names suggest an early three-field arrangement,

---

[1] E. g., Hawkesbury and Badminton. Cf. Appendix II, pp. 464, 465.

no holding in the sixteenth century was divided among the three with any semblance of equality, and for this reason we must look upon the township as one alien to the midland system.

Adjacent to Yate on the south is Frampton Cotterell. Here the process, well under way at Yate, was early completed, for in the survey of 1 Edward VI no open-field arable whatever is perceptible.[1] There had been a " Westfeld," to which is assigned a solitary seven-acre parcel together with a two-acre close. Quite possibly a common meadow was still existent, for twice there is mention of such, and some 50 acres of meadow in various holdings are not said to have been enclosed. The remainder of the township's lands, nearly 575 acres, are minutely described as closes. Only about 90 acres were closes of arable, the rest being pasture. Thus completely had the twenty-three substantial copyholders of the township, each possessed of a messuage and upwards of 15 acres of land,[2] gone over to pasture farming. And this had happened, without any evidence of high-handed procedure, in a well-peopled township ten miles distant from Bristol.

To explain what system of tillage was employed on the open irregular fields of the valley townships of Gloucestershire is not easy. William Marshall, who wrote two centuries later but who knew Gloucestershire well, makes suggestive comment. He first notes with scorn the intermixture of the parcels of the several owners. Although this was a feature likely to be seen wherever a common-field system prevailed, Marshall apparently thought the Gloucestershire arrangement more arbitrary than that existing elsewhere. In the fields, he says, the property is not intermixed " with a view to general conveniency or an equitable distribution of the lands to the several messuages of the townships they lie in, as in other places they appear to have been; but here the property of two men, perhaps neighbours in the same hamlet, will be mixed land-for-land alternately; though the soil and the distance from the messuages be nearly the same." Later he gives valuable information about the tillage of the fields. " In the

---

[1] Rents. and Survs., Portf. 2/46, ff. 139 sq.
[2] Except three, who had 4½, 8, and 9 acres respectively.

neighbourhood of Glocester are some extensive common fields . . . cropped, year after year, during a century, or perhaps centuries; without one intervening whole year's fallow. Hence they are called 'Every Year's Land.' On these lands no regular succession of crops is observed; except that 'a brown and a white crop'—pulse and corn—are cultivated in alternacy. The inclosed arable lands are under a similar course of management."[1]

Tillage of this kind, characterized by absence of fallowing and by a varying succession of crops, would go far, if it were practiced two centuries before Marshall's time, to explain irregularities in field systems. Nor is such early practice improbable. Marshall conjectures that the usage was ancient; and the proximity of townships which the tenants themselves had seen fit to enclose, such as Frampton Cotterell, argues that there was abroad a spirit of innovation and a desire so to cultivate fertile land as to get from it the most ample return.

Pertinent evidence regarding irregularities in Gloucestershire tillage as early as the thirteenth century has been pointed out by Vinogradoff.[2] It relates to a custom known as making an "inhoc" (*inhoc facere*). This consisted in enclosing for a year's cultivation a part of the arable fallow which would in the normal course of tillage have lain uncultivated. The anonymous author of a treatise on husbandry written before the end of the thirteenth century knew the custom well.[3] It was the exaction of an added crop from the soil, a demand which could not at that time be made too often. In the instance which Vinogradoff cites, it was thought possible to enclose (*inhocare*) every second year 40 acres out of 174 which were tilled under a three-course rotation.[4] In other words, from the 58 acres which would normally each year have lain fallow, 20 were put under contribution for an extra

---

[1] William Marshall, *The Rural Economy of Glocestershire, including its Dairy* (2 vols., Gloucester, 1789), i. 17, 65–66.

[2] *Villainage in England*, pp. 226–227.

[3] "E si liad inhom il deit ver quele coture il [the provost of the manor] prent en le inhom & de quel ble il seme chescune coture. . . . " (*Walter of Henley's Husbandry, together with an Anonymous Husbandry*, etc., ed. E. Lamond, 1890, p. 66).

[4] *Historia et Cartularium Monasterii S. Petri Gloucestriae* (ed. W. H. Hart, Rolls Series, 3 vols., 1863–67), iii. 35.

harvest. If all the 174 acres were treated alike, each one, instead of being fallowed every third year, was fallowed twice during a period of nine years. Although this instance refers to demesne lands which seem not to have lain in the common arable fields, the other case cited by Vinogradoff makes note of the tenants' interest in the lands which are to be subject to "inhoc," an intimation that these were open arable field.[1]

A usage of this sort would not, of course, immediately affect the integrity of a field system. The old bipartite or tripartite division might still be kept and a survey give no indication of the new custom. But in time the innovation was bound to tell upon field divisions; for these would gradually be shifted so as to reflect the superior tillage, until by the sixteenth century the fields may have become as abnormal as we have just seen them. If these conjectures are correct, the irregularities of the surveys represent an intermediate step between the already improving agriculture of the thirteenth century and the "every-year" lands which Marshall knew.

To the west of Gloucestershire the valleys of the Wye and Lug constitute the largest and most fertile part of Herefordshire. Relative to this county testimony from the sixteenth century and from an earlier period has already been advanced to show that the three-field system was once existent there.[2] It must now be pointed out that alongside three-field townships there appeared in due course others which differed from them. Several Jacobean surveys from Herefordshire manifest characteristics indicative of a departure from normal arrangements. Most striking of these irregularities are the large number of small fields and the break-up of the old tenements.

One of the townships of the manor of Stockton, a manor which has already testified to the existence of the three-field system in the county, betrays in the sixteenth century a tendency toward multiplicity of fields. This is Middleton, about one-fourth of whose area was then enclosed meadow or pasture. The arable,

[1] *Registrum Malmesburiense* (ed. J. S. Brewer, Rolls Series, 2 vols., 1879–80), ii. 186. The "campi" in question were those of Brokenborough, a township on the upper Avon in Wiltshire, but very near Gloucestershire.

[2] Cf. pp. 37, 64–66, above.

which constituted the remaining three-fourths, lay in many common fields, and only by following the names of these does the true complexity of the situation become apparent. In the dozen holdings transcribed in Appendix III about forty field names appear, some of them only once. We might suspect them of being applicable to closes held in severalty, were it not that nearly every one is said to refer to a *communis campus*. Since a tenant's holding was likely to lie in from three to nine of these areas, any attempt to group them according to the three-field system is naturally a hopeless task. The open fields of Middleton had by the end of the sixteenth century got into such a condition that their enclosure cannot have been difficult.[1]

The same multiplicity of fields characterized certain townships of the manor of Ivington, from two of which, Hope-under-Dinmore and Brierley, holdings are likewise summarized. In both, enclosures constituted from one-fifth to one-third of each holding. At Hope three fields, Over, Down, and Priesthey, frequently recur in the survey; but if a holding had acres in all of them it had most in Over field. Though four other fields occasionally appear, they cannot be grouped with the three so as to redress inequalities among the latter. At Brierley there were seven noteworthy fields, among which Gorve field and " le Much Howe " received the largest apportionment of acres. While a few holdings admit of a three-field interpretation, the rest are not amenable to it.

Equally perplexing are the fields of Stoke Edith, which in 40 Elizabeth were described as largely open. The specifications of the survey are not always lucid, parcels being sometimes designated " ridges "; but the holdings transcribed, which can be little questioned, serve to show the open-field areas small, numerous, and indifferent to a three-field grouping.

In certain of the Stoke Edith holdings that are not transcribed there is trace of another tendency characteristic of Herefordshire fields. This is the break-up of old tenements and the dispersion of their parcels among several new tenants. What is meant will become clear by the consideration of a Jacobean

[1] There is no record of their enclosure by act of parliament.

survey of the large manor of Malden,[1] although this is irritatingly complicated. The freehold entries number 35, the copyhold 141, nearly all recording small areas.[2] There were about 70 messuages in the hands of some 57 independent householders.[3] The manor comprised several townships,[4] each apparently with its own fields, although in some cases there may have been no sharp division of common fields among them.[5] The following holding is characteristic of the survey, and serves to emphasize the feature in which we are for the moment interested — the break-up of traditional tenements: —

"Ricardus Grene tenet per copiam datam anno regni regine Elizabethe xxxiv unum mesuagium, pomerium et clausam adiacentem continentem i acram nuper Thome Stead

per copiam ... Elizabethe ii, i acram in Holbach feild in villata de Venne nuper Hugonis Lane

per copiam ... Elizabethe xxix, ii acras in Lake feild

per copiam ... Elizabethe xli, ii acras in Fromanton [another villata] in quodam campo ibidem vocato Holbach feild nuper Thome Wootton

per copiam ... Elizabethe xxv, dimidiam acram de Sockeland iacentem in Nashill feild

per copiam ... Elizabethe xl, ii rodas in Ashill feild nuper Johanis Parsons

per copiam ... Elizabethe xxv, dimidiam acram in Ashill ... in occupatione Johanis Mathie

per copiam ... Elizabethe xxxv, unam acram terre custumarie de Soakeland in Odich feild nuper Willelmi Stephens

[1] Land Rev., M. B. 217, ff. 194–292.

[2] Sometimes the subdivision of the holdings (especially of the larger ones) among the fields is not given. A few of the tenants were gentlemen.

[3] Seven cottages were held by a single person, and six messuages by tenants each already possessed of a messuage.

[4] "Item that there are within the said Lordshippe of Marden viii severall villages or Towneshipps viz. Marden, Fromanton, Sutton, Freene, Wisteston, Vauld, Verne, Fenne, and Marston and that they and everie of them are to doe suyte to the said Courte of the said Mannor . . ." (Land Rev., M. B. 217, f. 290).

[5] Venne and Fromanton, two of the townships which appear in the following holding, seem to have shared in Holbach field.

per copiam . . . Elizabethe iv, unam acram terre custumarie de Sokeland iacentem in Fromanton . . . in campo ibidem vocato Nashill nuper Willelmi Cooke

per copiam . . . Philipi ii et Marie iii, unam acram in Fromanton in quodam loco vocato Odyche nuper Jacobi Greene, patris sui." [1]

During the second half of the sixteenth century Richard Grene is thus seen acquiring ten and one-half acres through no fewer than ten different grants by copy. Starting in the time of Philip and Mary with an acre which had been his father's, he had added parcel after parcel up to the last years of Elizabeth's reign, some acquisitions being customary sokeland, others simple copyhold. These parcels had formerly been in the possession of six tenants, many of whose other acres had also passed out of their hands.[2] Obviously such an agglomerate holding as this of Grene's can instruct us little about the original field system of the townships of the manor, but the bare fact that such tenements were in process of formation proves that the rules of three-field tillage can scarcely have been observed at the end of the sixteenth century. Grene had, to be sure, taken pains to acquire parcels in the three fields which are assigned to Fromanton. Yet there were years when he did not possess them all, and one of these fields (Holbach) was not restricted to Fromanton, since Venne also had interests in it. Grene, further, did not hesitate to acquire two acres in Lake field, which the enclosure map shows to have been at some distance and which was probably not one of the fields of Fromanton.[3] Shifting arrangements of this sort cannot well have been the concomitant of a systematic three-field system.

Another Marden illustration emphasizes what has been said above, and in addition reveals clearly the natural outcome of unstable tenements and a decadent field system. John Carwardyn held by four copies lands which had once been John Heere's, Richard Heere's, and (for the most part) Richard

---

[1] Land Rev., M. B. 217, f. 211.
[2] As is shown in the descriptions of various holdings.
[3] Cf. the sketch of the enclosure map of Marden, p. 147, below.

Danyell's.[1] Among them was half a messuage accompanying half of a virgate of customary sokeland lying in the township of Verne. The half-virgate comprised, besides garden, orchard, and two-acre curtilage,

> " Clausam pasture do novo inclusam extra communem campum vocatum Lawfeild continentem per estimationem i acram dimidiam
> Aliam parcellam pasture de novo inclusam in Senacre feild
> Terram arabilem iacentem in communibus campis de Mawarden cuius quantitatem iuratores ignorant."

This tenant had, it appears, enclosed the part of his virgate which lay in two of the open common fields. The procedure is what might have been expected and permitted when the vitality of the old system had been sapped. Although, as it happens, much of the manor of Marden remained open for two centuries longer,[2] Herefordshire townships in general became enclosed, and the nature of the open fields as displayed in the foregoing illustrations must have been one of the causes contributory to enclosure. Multiplicity of fields and disintegration of the old tenements were transitional phases in the passage from the old system to the new, and the motive prompting the change was probably the same as that effective in Gloucestershire — a desire to cultivate the soil more advantageously than the three-field system permitted.

A region which in situation and soil was well adapted to improve upon a primitive system of agriculture was eastern Somerset. In early days a two-field system had there prevailed, and nearly all townships which appear in Appendix II utilized it for the cultivation of their arable in the thirteenth century. From the southeastern part of the county the Tudor survey of Martock, with its members Hurst and Bower Henton, has been cited in a preceding chapter to illustrate symmetrical three-field arrangements.[3] None the less, Tudor and Jacobean surveys from Somerset which disclose the two or three old fields still intact are exceptional. To show how most records of this date picture

[1] Land Rev., M. B. 217, f. 224.   [2] Cf. p. 32, above.
[3] Cf. p. 140, below.

the original system in various stages of decay, holdings from several surveys have been transcribed in Appendix III and now claim attention.

Adjacent to Martock and situated on the river Parret is Kingsbury, the Jacobean survey of which records that many of the ancient holdings called "de antiquo austro" were largely or entirely enclosed. Such were the three still rated in virgates and "fardells," or quarter-virgates, and such were most of the numerous holdings not here transcribed. Others had still a few or even the majority of their acres in open fields. Of these open fields, the three which were largest and most often recurrent were Byneworth, Kylworth, and Hill field, their total areas being 68, 41, and 38 acres respectively. At the same time the customary holdings at Kingsbury, exclusive of cottages, numbered nearly 70. Obviously the share of any tenant in the arable fields must have been slight, seldom so much as ten acres and usually less than five. Some of the holdings which received most liberal allotments have been transcribed, but even in them there was no distribution of acres among fields that suggests a regular system. The names of certain fields, too, Byneworth, Kylworth, Tunnland, Deanland, were unusual. Kingsbury is thus revealed at the end of the sixteenth century, not only as a parish largely enclosed, but as one that had about it little trace of the system which once characterized the countryside. These features it probably owed to its situation; for it is a river township, and its rich bottom lands must have been early turned to pasturage and improved tillage.

Not unlike Kingsbury was another low-lying manor, that of East Brent, situated nearer the Bristol Channel. Holdings from two of the tithings, which have been transcribed from the Jacobean survey, illustrate the predominance here of enclosed pasture. Of the arable most was enclosed, but some lay in small open fields and appeared in the copyholds sporadically. There were a few acres "super le Downe." Reduced in condition though they were, a West field and an East field still had precedence; in them lay most of the open arable acres, though no longer with two-field precision. The manor was one which had almost forgotten its early days in its adherence to pasture farming.¹

Similarly unmindful of their thirteenth-century condition were two townships nearer the Wiltshire border, not far removed from Bath and Wells respectively. These were Norton St. Philip and West Pennard. In the copyholds of the former enclosures so much predominated that only a half-dozen still had any parcels left in the two fields, which we discern to have been North and South. Since a considerable area in the North field known as " goddes peece " had not long since been converted to enclosed pasture, Norton St. Philip seems already to have devoted itself to the dairy farming of which it boasts today.

West Pennard, a manor once belonging to Glastonbury, was more conservative. Although its holdings were generally about half enclosed and devoted to pasturage, there were in each several acres of open arable field. Often these lay in " Easterne Downe " and " Westerne Downe," so disposed as to tempt one to see in these " Downes " two old fields; but such a conclusion might be hasty, inasmuch as remains of a South field existed and at times some holdings manifested a kind of three-field attitude toward Breach field, Westmore field, and Eastmore field. In view of these contradictions, we can only insist upon the general irregularity of the field arrangements without trying to probe into their past.

Finally, certain townships may be cited to show the two-field system just inclining to decay. On the tongue of high land which borders Sedgemoor in mid-Somerset is situated Curry Mallett, where in 1610 the two old fields, East and West, were still easily recognizable. They had been encumbered, though

---

Two manors of the Earl of Pembroke, surveyed in 9 Elizabeth, lay in this part of the county. South Brent seems to have been entirely enclosed. At Chedzoy near Bridgewater, however, there was considerable unenclosed land, much of it meadow. Seldom did one-half of a holding lie in open arable field, while the fraction might fall to one-seventh and was usually one-third or one-fourth. The fields which most often appear are East and North, the former receiving the greater number of acres. At times there is reference to West field and Slapeland field, but no indications of a regular field system are visible. Cf. C. R. Straton, *Survey of the Lands of William, First Earl of Pembroke* (2 vols., Roxburghe Club, 1909), ii. 471-486, 442-471.

only a little, with such appendages as the Slade, the Breache, and Eyeberie. It was still possible for a tenant to have eleven acres in one, ten in the other, while a small holding, like that of John Polman might even lie largely in the open fields. Yet nearly all copyholders had withdrawn from these fields much of their arable. Departing, however, from the usual practice, they had not converted this into pasture, of which little is described in the survey. Yet it is probable that the " arable " of the enclosures was not without experience of convertible husbandry, and that the copyholders did at times turn their fields into pasture for a year or two. Pasture for sheep or cattle was the less necessary at Curry Mallett since tenants had unstinted common in Sedgemoore.

It will be remembered that Somerset could still in Jacobean days furnish a typical two-field township. Such was South Stoke, situated near Bath, and already described. Not a dozen miles away, Corston had departed from the norm only a little farther. Its two fields were North field and South field, between which several of the holdings, and these large ones, divided their arable evenly enough. Other tenements, however, manifested no equality of division, having many more acres in North field than in South field. Enclosure of a part of the South field may well have been the cause of this, though we are not informed. At any rate, there was in each of these holdings enough enclosed land to redress the balance between the fields, if it may be assumed that some of it had once been a part of them.

One more Somersetshire illustration is pardonable, since it shows the two old fields still existent, though moribund, so late as 1684. The hill township of Bruton in the eastern part of the county was once the seat of a priory. Some ten of the copyholders still in the seventeenth century had acres in North field or South field, the totals being $25\frac{1}{2}$ and 19, with $4\frac{1}{2}$ acres not located. The numerous lessees, holding for life or for 99 years, had in addition 53 acres in North field, 13 in South field, with 47 acres elsewhere.[1] Seldom was there such distribution of acres as

---

[1] There are two or three notes about parcels recently enclosed.

would indicate a two-field system still effective; and the probability that it was so is slight, since five-sixths of the leasehold and two-thirds of the copyhold lay enclosed. Many Somerset townships must at the end of the seventeenth century have been like Bruton, a fact which would account for the comparatively small amount of arable within the county affected by act of parliament.[1]

A Wiltshire township, situated, like most of those above described, in a district favorable for improved or for pasture farming, shows by the Glastonbury survey of 10 Henry VIII that it was already availing itself of its natural advantages. Christian Malford is in the valley of the Wiltshire Avon, where the downs do not yet close in as they do at Bath. Low-lying lands abound. In consequence about one-half of each virgate (and the virgates were large) consisted of closes, but whether these were pasture we do not learn. Altogether the copyholders had 753 acres of enclosed land, in comparison with 68 acres of common meadow and 941 acres in the open arable fields. For a township in the heart of the two-field area these fields were numerous. There was, to be sure, a North field and a West field, though few of the virgate holdings had acres in both of them and some had acres in neither. Other fields were often favored — Little field, Benehul field, Middel field, Wode furlong — and in the most arbitrary manner. A virgater sometimes had more arable acres in one field than he had in all the others, while the dominant field at times varied from virgate to virgate. North field, which in many holdings was not mentioned, contained nearly all the acres of four distinct virgates. Neither uniformity of distribution nor equality of apportionment among fields is anywhere perceptible in the survey. At the beginning of the sixteenth century Christian Malford was as far removed from the appearance of two-

[1] For the entire county Slater (*English Peasantry*, p. 298) cites only forty-one acts relative to open arable field. Of these all except six estimate the areas to be enclosed. Nine thousand acres are said to have been arable, eleven thousand more partly arable, partly pasture. Since the county contains 1,043,409 acres, the open-field arable enclosed by act of parliament was only between one and two per cent of the area of the county. In Oxfordshire, as the following chapter will show, it was about thirty-seven per cent of the county's area.

and three-field townships as disregard of their field conventions could render it.[1]

Before we pass to the north of England, we should not fail to note the decay of a two-field system in the Isle of Wight. The precise division of acres between two fields, characteristic of Wellow,[2] was unusual in surveys from that island. At Niton in 6 James I there were still two fields, but they had suffered from the activity of the tenant encloser, most holdings being half or more than half enclosed. While the acres had been abstracted sometimes from one field, sometimes from the other, East field had shrunken more; in it there remained but 53 acres of copyhold, while West field contained 71 and the enclosures 167½. Of common meadow there was scarcely any.

At "Uggaton" the appearance of open-field names in the survey is so infrequent that it is doubtful whether such fields really survived. One tenant had 2½ acres in South field, two had together 2½ acres in North field, four had 29¼ acres in West field. That was all. Since such data are too slight to build inferences upon, the township should be looked upon as practically enclosed by 6 James I.[3]

Enclosed beyond all doubt were the fields of Thorley, surveyed at the same time.[4] All areas are said to be closes, although the character of these as pasture or arable is not specified. What is interesting here and at "Uggaton" is the goodly array of copyholders whom no evicting landlord seems to have disturbed. At "Uggaton" there were eighteen with from 5 to 68 acres of land apiece, at Thorley fifteen similarly circumstanced. To be sure, these manors were royal ones, upon which evictions could not becomingly have taken place; yet they make it clear that the quiet passage from open fields to enclosures could be effected in the

---

[1] The numerous Wiltshire manors of the Earl of Pembroke, surveyed in 9 Elizabeth, were largely in two or three fields (cf. below, Appendix II, pp. 501–503). Four, however, Bower Chalk, Chilmark, Hilcott, and Stockton, had adopted a four-field arrangement. Two, Berwick St. John and Bedwyn, had irregular fields, due probably to their situation in remote upland valleys. Cf. Straton, *Survey of the Lands of William, First Earl of Pembroke*, vol. i.

[2] Cf. above, p. 31.

[3] Exch. Aug. Of., M. B. 421. Because the open fields were so insignificant, no holdings have been transcribed in Appendix III. [4] Ibid.

Isle of Wight without serious diminution of tenants.[1] Since the copyholders themselves presumably desired such change, the process may be looked upon as a natural one.

Leaving the irregularities of the south and west, we may now inquire whether similar phenomena were to be found at the end of the sixteenth century within the northern part of the midland area. Most important of the river valleys here are those of the Trent and Humber. Just removed from the banks of the latter river in Yorkshire lies Willerby, where nominally the fields were six, though two were small and another very small. If the two be combined, and the smallest be annexed to any one of the others, we shall have in each holding four nearly equal areas. The combination of the two is further permissible, since it is Toffindale, not called a field, which is thus annexed to West field. All the other large areas are designated fields — Lowe field, Kirkegate field, Langland field — while even the diminutive tract is dubbed Ellerylund field. If this grouping be correct, there was here a four-field arrangement like that characteristic of the plain of the lower Avon.

Before proceeding up the valley of the Trent we may turn aside for a moment to another Yorkshire township, that of Breighton on the Derwent. Here the Jacobean survey records five fields of importance — Longland, Borne, South, Car, and Hallmore. Sometimes a tenant had acres in three of them, sometimes in four, sometimes in the five, yet without uniform distribution and in accordance with no system which is apparent by tentative grouping. Some tenants had several acres of enclosed land, but the assumption that these had been taken from the fields does not clear up the subject. Although a four-field arrangement is a little more plausible than any other, so incongruous at times is the distribution of acres that the kind of system employed must remain in doubt.

Passing now from the Humber to the Trent, we straightway reach the fertile Isle of Axholme, where lies the township of

---

[1] One of the first anti-enclosure acts (4 Hen. VII, c. 16) refers to the Isle of Wight as a region suffering from depopulation. Cf. Gay, "Inquisitions of Depopulation in 1517," p. 232.

Owston, of which a Jacobean survey survives. The holdings were small and are not always rated by bovates. The freehold amounted to 247 acres, of which 102 were enclosed and 36 common meadow; of the 239 acres of copyhold, 43 were enclosed and 17 common meadow. Thus less than two-thirds of the tenants' lands lay in open field, and if a holding were a little short in the acres of one field it had enclosed land as a resource. Glancing now at the distribution of the open-field arable of the tenements, we see pretty clearly that the system was, or not long since had been, one of four fields. The larger customary holdings (all of which are shown in the Appendix) were unanimous in dividing their areas among four fields, although the division was not sharp-cut, like that at Marston Sicca or Welford. Despite this laxity and the relatively extensive enclosures, the survey is our best illustration of four-field arrangements in the north.

Farther up the Trent near Nottingham are Lenton and Radford, both of which formed the home manor of Lenton priory. Their fields are described together in a Jacobean survey, the acres of the bovates being frequently distributed among all six of them. Three of the fields were smaller than the others, and some one of them was often not represented in a holding. We might conclude that the arrangement inclined to three main fields with three supplementary ones; yet, if such were the case, groupings to prove it are not easily made. The three fields in each of which the small holding of Andrew Webster had one-half acre are said to be the fields of Lenton. If the other three were the fields of Radford and the two groups were tilled together, the combination should be Beck field and More field, Red field and Church field, Sand field and Alwell field; but neither this nor any other arrangement always works out happily. In each holding there was considerable common meadow, a fact which may account for discrepancies. Only by assuming that parcels of arable in the fields had been converted into meadow,[1] can we group the six fields by twos so as to make the former existence of a three-field system credible.

[1] For contemporary instances of this process, see p. 35, above, and p. 106, below.

Although instances of irregularities like these may be found in the valley of the Trent and its neighborhood, they are less numerous than similar phenomena in the southwest. The northern midlands were more like the southeastern in retaining until the end of the sixteenth century an unvarying three-field character. Farther to the north, however, several interesting irregular fields deserve notice. They were situated to the south and west of Durham, in the territory which has provisionally been designated as the northern outpost of the three-field system.

It will be remembered that this designation was hazarded in connection with the survey of Ingleton. The symmetrical three-field aspect of that township we may see repeated in the descriptions of thirteen of its neighbors.[1] All are taken from a series of Jacobean surveys relative to the extensive manors of Raby, Barnard Castle, and Brancepeth, the members of which are situated for the most part where the moors slope eastward toward the valleys of the Tees and Wear. Those townships lying in the plain of the Tees were the ones in which the three-field system was most intact. Others that lie more on the uplands inclined to enclosure and pasturage. This is particularly true of the members of Brancepeth,[2] to the west of Durham, where the neighborhood of the Wolds may have been responsible for irregularities in field arrangements; for it is not improbable that some arable here was a relatively recent improvement from the waste, akin in this respect to that of forest townships. Yet certain members of the manor cannot be thus classified: Willington, Stockley, Eldon, and East Brandon are near enough to the river Wear to have had a long field history. Conditions at East Brandon, as pictured in the Appendix, illustrate the irregularities of these river townships and show what might have been seen in Jacobean days just outside the gates of Durham.

Closes in the township were few, scarcely more than the acre or two attached to the homesteads. Nor was the intermixed arable in the common fields very great in amount. Several tenants had

[1] Cf. Appendix II, pp. 462–463.
[2] Crook and Billy Row, Thornley, Willington, Stockley, Helme Park, Cornsey, Eldon, East Brandon.

a few acres in Rudenhill, in West field, and in Watergate field, but the arable of others lay entirely elsewhere. William Briggs had eleven acres at Hareham, where he had still more meadow and pasture. Indeed, this brings us to what is perhaps most noteworthy about the survey — the appearance in certain fields of meadow alongside the arable. Lowe field was most transformed by such procedure, for seldom did the tenants retain any arable there. Instead they had large parcels of meadow, sometimes as many as twenty acres; nor does anything indicate that these parcels were enclosed. They seem, rather, to have remained open and to point to a gradual abandonment of arable tillage.

Such an abandonment is more clearly indicated by another survey of this series, that of Eggleston.[1] Eggleston lies well up the valley of the Tees, and still in 5 James I maintained its three fields, East, Middle, and West, among which several holdings were divided with a show of equality. Presumably the fields had once been largely arable. When, however, the survey was made change had begun, though not in the direction of enclosure, of which there was still little. Conversion to meadow had proceeded without it: nearly all the parcels of the various tenants in East field and West field are said to have been meadow; arable still predominated only in Middle field, and even there it had begun to yield. The survey is instructive in showing how naturally conditions arose which must soon have called for enclosure as a matter of convenience.

Eggleston did not stand alone in its early seventeenth-century transformation. Westwick, situated a little way down the river, had begun to make the change at the same time. Apart from large parcels of pasture which each holding had on the moor, from one-third to one-half of the fields (High, Middle, and Low) had become meadow. Whorlton, still farther down the Tees, was making a similar transition, though rather more than one-half of each holding in the three fields remained arable. At Bolam the arable and meadow in the fields (East, West, and North) were nearly equal in amount. At Willington, once more,

[1] Cf. Appendix III. For this and the Durham surveys mentioned below, see Land Rev., M. B. 192, 193.

meadow predominated. Since these townships lie not on the fells, but in the valleys, and since their erstwhile three-field character is clear, we have here an interesting departure from the normal system. It appears that in several places in Durham the old open arable fields were in a state of decay. The tenants preferred meadow and had converted into meadow many of their open-field strips. Pretty clearly the next step was to be consolidation and enclosure. Under these circumstances there could be no occasion for complaint about enclosure as preceding and inducing conversion. The processes were reversed, and the change thereby became more natural.

The enclosure of many Durham townships seems to have occurred not very long after these Jacobean surveys were made. Miss Leonard has described the agreements, enrolled on the register of the court of the bishopric, by which the open fields of upwards of twenty townships were re-allotted between 1633 and 1700. The preambles, she says, often assign as reason for the enclosure the fact that the land "is wasted and worn with continual ploweing, and thereby made bare, barren and very unfruitefull." [1] Doubtless this was a motive with such townships as still lay largely in arable, and we have seen them numerous in the days of James I; but in those townships whose open fields had become largely meadow the desire to complete a process begun must have been operative. If so, we have antecedent changes in the field system as one explanation of the disappearance of open arable fields in Durham.

Our somewhat prolonged progress through the two- and three-field area should ere this have served at least one end. It should have made clear that, even within a territory unmistakably characterized by one type of open field, conditions were not uniform at the end of the sixteenth century. A stretch of forest or of wold might cause marked deviation; still more might a river with its bordering meadows. In the heart of the two- and three-field area departures from the norm were not frequent, but in the outlying counties they occurred often enough to threaten the integrity of the system. As a result, certain districts within the

[1] "Inclosure of Common Fields," p. 117.

boundary which the thirteenth century would have drawn round the two- and three-field area seem in the sixteenth century to detach themselves. Such in particular were the counties of the west, — Herefordshire and Shropshire,[1] parts of Staffordshire, Worcestershire, Warwickshire, Gloucestershire, and Somerset, — the Isle of Wight in the south, and part of the county of Durham in the north. In Tudor days these were regions characterized by innovations in field systems, most of which looked toward the improvement of agriculture. Either the arable of a township had been subdivided in such a way that more of it than before could be utilized for tillage, or large portions of it had been converted into remunerative meadow or pasture. The latter process had at times been accompanied by enclosure, at times not. Even if it had been, there is often no evidence that the tenants had been dispossessed. To ascertain more fully the relation existing between the decadence of the midland field system and the advance of agriculture, especially the enclosure of the open fields, a closer study of what happened in typical counties is essential.

[1] Since we have few satisfactory surveys from Shropshire, none have been summarized. That of the manor of Cleobury shows irregular field arrangements (Land Rev., M. B. 185, ff. 86–97, 21 Eliz.), and it is highly probable that the county differed little in this respect from Herefordshire on the south and Staffordshire on the east.

CHAPTER IV

THE LATER HISTORY OF THE MIDLAND SYSTEM IN OXFORDSHIRE AND HEREFORDSHIRE

IT was pointed out in the Introduction that agricultural progress in England would ultimately demand the disappearance of the open-field system. A form of tillage so inconvenient, so inflexible, so negligent of the productivity of the soil, could not long endure after technical improvement in ploughing had made possible its abandonment and after its social advantages had come to be disregarded.

This significant change, however, it is clear, was not likely to take place suddenly, but improvements in the old system would slowly lead up to it. The probable substitution of three-field for two-field arrangements throughout a large part of the midlands during the thirteenth and fourteenth centuries was only a first step in this advance. Other and later innovations have been disclosed in the preceding chapter. By the sixteenth century, it appears, some townships had already hedged in a part of their arable fields while leaving the remainder open, a piecemeal method of enclosure which seems to have been a kind of experiment undertaken by men who would not yet risk the complete abandonment of open fields. Elsewhere innovation took the form of a multiplicity of fields. To judge from the allotment of tenants' acres among them, these numerous fields could not have been tilled in accordance with two- or three-field arrangements, and in them undoubtedly less arable was left fallow each year than under the normal system. Still other townships remained true to the principles of regularity, but subdivided their two fields into four, of which three were tilled annually. All these changes constitute a step in agricultural progress similar to that which substituted three fields for two. Each in its way sought the ultimate goal, a goal involving consolidation of parcels, enclosure of holdings, abandonment of fallowing, and the employment of convertible husbandry. Together these innovations bring the subject of field systems into immediate touch with the subject of enclosures, a topic, of course, too comprehensive to be treated adequately except in an independent monograph.[1] It can be discussed here only in relation to the final transformation of midland open fields and the accompanying improvements in agriculture.

An understanding of the situation can perhaps best be attained by an examination of typical districts. We should, for example, like to know how the midland system fared in a county where it once prevailed and where open fields longest remained. We should, again, like to know what happened in counties where it once prevailed but where open fields early tended to evince irregularities and decay. In order to approach the subject in this manner, it will be advantageous to examine somewhat in detail the later open-field history of Oxfordshire, a county in which open fields long persisted. An accurate picture of the progress of events there will make clear what went on in most of the counties characterized by the two- and three-field system. A proper corrective will then be introduced by an examination of the earlier decline of the system in Herefordshire, a western county typical in this respect.

An account of the midland system in Oxfordshire between the sixteenth and nineteenth centuries is without doubt best begun by a description of its condition at the time when it disappeared. How long and how extensively, one may ask, did it resist attack, and what innovations had it meanwhile adopted? Such questions are in a measure answered by the parliamentary enclosure awards, few of which, it will be remembered, are earlier than 1755 and few later than 1870. During this century, however, the journals of the commons and the lords are distended with the records of acts authorizing the enclosures which the awards describe.

These acts have been conveniently catalogued by Slater, who has also constructed maps which indicate roughly the areas

[1] Cf. above, pp. 10–11.

affected.[1] Since many acts neglect to give even the approximate areas to be enclosed, his presentation could not attain to any considerable accuracy.[2] This defect can be remedied only by an appeal to the awards, a toilsome undertaking upon which no one has yet ventured. Until it is attempted we shall have to accept Slater's results. For the counties of Oxford and Hereford, however, it has here seemed best to consult all accessible awards, in order that the disappearance of the midland system within limited areas may be described as accurately as possible. This chapter, therefore, stands to Slater's lists and maps for these two counties as the awards do to the acts. It forms a completion of the sketch.

The awards, ponderous as they are, do not always supply such exact information as is desirable, since their form and content varied considerably during the century which saw their preparation. The early ones are relatively brief and uninforming, telling very little about the open fields which they enclosed save, at times, the number of virgates and the total areas. Only toward 1800 did it become usual in Oxfordshire to refer to the ancient field divisions in locating new allotments, and in many instances this was then done cursorily in the text, without notice of the old subdivisions on the plan. Again, a large allotment was often assigned to several field areas without specification of how much belonged to each. Under these circumstances it frequently becomes difficult to tell exactly what was the size and what the arrangement of the old fields. Toward the middle of the nineteenth century the plans accompanying the awards, though large and detailed

[1] *English Peasantry*, Appendix and maps.
[2] Two shortcomings are most noticeable. One, in the appendix, is due primarily to the neglect of the acts to state the areas to be enclosed. The phraseology of the acts, further, is often such that it is impossible to discriminate between arable field and common waste, while the Norfolk acts are deceptive in still another way (cf. p. 305). The second shortcoming appears in the maps, where no attempt is made to distinguish between townships in which there was a large amount of arable field remaining open until parliamentary enclosure and those in which there was little. Townships in which there was any enclosure of arable whatever appear as do those in which there was much. It is questionable whether discrimination in this matter would not have been more acceptable in the maps than are the distinctions by periods which the author has preferred to indicate.

as to the arable strips, often have nothing to say about fields, but require the student to puzzle out the arrangement from the schedule, as has been done in the case of Chalgrove.[1] These late awards, furthermore, like some earlier ones, do not trouble to add up the allotments, but throw that burden upon the investigator. Most annoying of all, however, is the brevity which, in both early and late awards, combines arable and waste without specifying the respective areas of each. For this reason it is often necessary to estimate the extent of the waste, and at times there are no data for such an estimate.[2] The entire enclosure has then to be set down as arable, an expedient which obviously exaggerates the amount of arable that was enclosed. A final difficulty comes in determining the areas of the old enclosures. Seldom are they stated. Sometimes they can be computed from the plan by a comparison of the space there assigned to them with that assigned to the open areas. Again, when it may be assumed that the area of the township has remained substantially unchanged and that no other open common land existed save that described in the award (or awards), the old enclosures may be obtained by subtraction. To ascertain them in this way, we need only deduct the combined area of open-field arable and unenclosed waste from the area of the township. Sometimes, lastly, the allotment for tithes is so described in the award that the part of it made in lieu of tithes due from old enclosures is distinguished from the part made in lieu of tithes due from open fields. Since the former was about one-ninth of the value of the old enclosures,[3] the area of these

---

[1] Cf. above, p. 21.

[2] In the later awards the allotment to the lord of the manor, as such, for his rights in the waste was about one-fifteenth or one-sixteenth of the waste divided. In the tables of the Appendix the area of the uncultivated common has often been computed from this entry.

[3] It is so in the award for Blackthorn, Oxons. At Sandford St. Martin it was one-sixth, at Burford one-fifth. The estimate in question is valid only if no old enclosures had already been exempted from tithes. A divergence between an estimate got in this way and the area obtained by subtracting the total enclosure from the total area of the township may arise because some old enclosures had already been exempted before the award was made. The divergences are noted below in Appendix IV.

enclosures may in such cases be computed. When different computations as to the areas of old enclosures do not agree, the area got by subtracting open-field arable and waste from the entire township has been here adopted. Despite all the uncertainties attendant upon the examination of the awards, a study of them repays the labor, since the information which they yield is far more precise than that to be secured in any other way.

The open fields of Oxfordshire which were enclosed by act of parliament are set down township by township in Appendix IV. The townships are grouped in accordance with the percentage of the area in each which, exclusive of the waste, was thus enclosed. This arrangement amounts to a comparison between the open-field arable and meadow on the one hand and the old enclosures on the other. The assignment of a township to a group has depended upon whether the land to be enclosed, apart from the waste, amounted to more than three-fourths of the township's total area, or to less than three-fourths but to more than one-half of it, or to less than one-half but to more than one-fourth of it, or, lastly, to less than one-fourth of it. The history of parliamentary enclosure between 1758 and 1867 [1], as thus told by the awards,[2] may be summarized as follows, reference being had to the number of townships that fall within the respective groups and to the ratio which the areas of the groups bear to the total area of the county (478,112 acres [3]).

In 89 townships more than three-fourths of the area, exclusive of the waste, was enclosed by parliamentary award during the century in question, and these townships represent 29 per cent of the county's area. In 58 townships, which constitute 22.1 per cent of the county's area, between one-half and three-fourths of the improved area was enclosed. In 28 townships, comprising

---

[1] The Mixbury act dates from 1729, and the Crowell award was made in 1882; but all other parliamentary enclosures of arable fields in Oxfordshire fall between the years mentioned.

[2] In fifteen instances the information is taken from the petitions for enclosure. With one exception the awards in these cases are neither at Oxford nor at the Public Record Office.

[3] This is the area of the land. The area of land and water is 480,687 acres.

12.6 per cent of the county, the arable or meadow affected was between one-fourth and one-half of the respective areas. In 16 townships the fraction sank to less than one-fourth, the townships themselves amounting to 7.55 per cent of the county. For 28.75 per cent of Oxfordshire there is no record of parliamentary enclosure. The townships that fall within the respective groups are indicated on the accompanying plan. (See next page.)

Stated more synthetically, the total amount of open-field arable which the tables show to have been enclosed by act of parliament was 193,781 acres, or 40.53 per cent of the entire county. These figures somewhat overestimate the actual amount, since, as has been noticed, the character of the awards has at times made it impossible to separate arable from waste. Probably the above percentage should be reduced to about 37 per cent. The difference should be added to the percentage which represents the unimproved waste, and which our tables, most defective at this point, show to have been at least 5.83 per cent of the county's area (27,862 acres out of 478,112). Our estimate of the unimproved lands in the county in 1750 thus assigns to them about 9 per cent of its entire surface.

After deducting the open-field arable and the unenclosed commons, we are left with the old enclosures. According to the estimates of the tables these amounted to 256,469 acres, or 53.61 per cent of the county. As was stated above, from townships which represent 28.75 per cent of the county there is no record of parliamentary action; the remaining old enclosed lands (24.86 per cent of the county's area) fall within townships some parts of which were enclosed by award. These large percentages imply, of course, that the enclosure history of the county prior to 1750 is a matter of no small moment. In what way these old enclosures were brought about, what motives lay behind the process, to what extent they represented simply an improvement in agriculture, what relation they bore to field systems, — these are subjects that now demand consideration.

If we turn to those Oxfordshire townships which enclosed their arable without parliamentary act, we shall be able to get some hints, though not always very accurate information, as to how

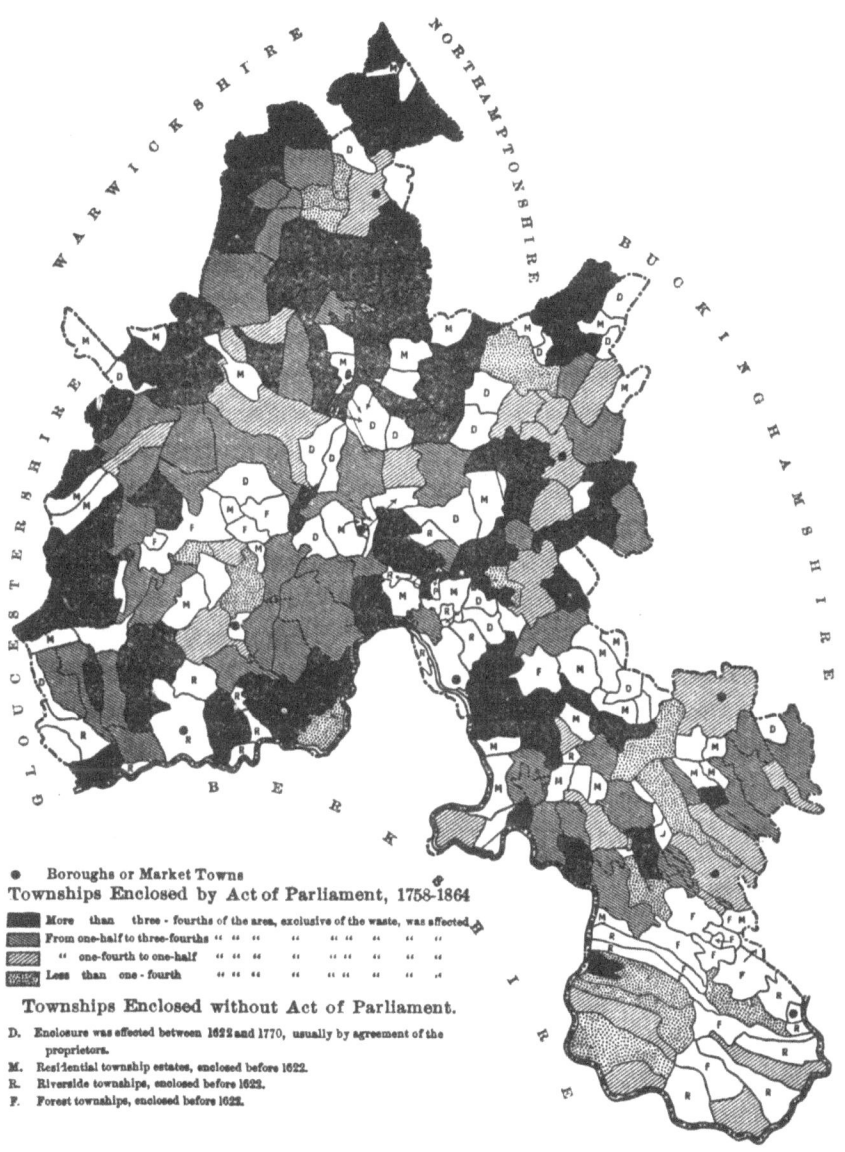

MAP VI

this came about.¹ Formal act and formal enclosure might be altogether obviated if the proprietors of the common fields could agree to keep to their own parcels and renounce the exercise of common rights. This is said to have happened at Ewelm, which has never had an independent enclosure award;² but to make such an understanding possible the open-field parcels must have been to a large extent consolidated. Another method is reported by Arthur Young, who tells of an enclosure devised by a single proprietor. "The parish of Clifton," he wrote in 1809, " thirty-nine years ago was allotted by Mr. Hucks, being a private arrangement of his own. Each farm was enclosed by an outline fence but was not subdivided."³ It was usual, however, to secure for voluntary agreements some formal sanction; and an interesting illustration of this practice, joined with an explanation of how the agreement was brought about, comes from the years when enclosures were authorized by parliament. In 1783 a petition was presented to that body asking its sanction for an enclosure which had been accomplished at Hanwell fifteen years before. The method there employed had been the purchase by the lord of the manor, Sir Charles Cope, of all interests in the open fields except the glebe land and the tithes. Enclosure had then proceeded apace.⁴ Since Hanwell is the site of a castle and a park,

[1] I have, for example, found no record of how the parish of Cuxham came to be enclosed. A map of 1767, which shows much of it still open, has been published by J. L. G. Mowat, *Sixteen Old Maps of Properties in Oxfordshire*, Oxford, 1888.

[2] Leonard, "Inclosure of Common Fields," p. 101, n. 3. The award for Bensington is concerned with certain lands in Ewelm.

[3] *Agriculture of Oxfordshire*, p. 91.

[4] See *Journal of the House of Commons*, petition of 5 February, 1783, "Setting forth, That about the Year 1768 . . . the said Sir Charles Cope being then Lord of the Manor of Hanwell, and seised of the perpetual Advowson . . . to the Rectory and Parish Church of Hanwell aforesaid, and likewise being seised for Life, or in Fee, or some other Estate of Inheritance, of the greatest Part of the Lands of the said Manor, did purchase to him and his Heirs, the Estates and Interests of the several Copyholders, Life, and Leaseholders, and other Proprietors of the Remainder of the said Open and Common Fields, and other Lands, within the said Parish, in order to the inclosing the same; and the Open and Common Fields, Commonable Lands, Cow Pasture, Heath, and Waste Grounds, within the said Manor and Parish of Hanwell, were thereupon inclosed, and have ever since been held in Severalty."

it is natural that most of the township should, as the petition states, have been in the possession of the lord of the manor before the purchases of 1768.

Enclosure by agreement did not necessarily involve the buying out of the tenants. Before the days of parliamentary activity the enclosure of open fields in the large parish of Charlbury was made possible by a deed of agreement drawn up in 1715.[1] It bears fifty-seven signatures with seals, and sets forth that the interested parties are possessed of " several parcels of freehold, leasehold and coppyhold lands lyeing and being in Certain Common fields of Charlbury aforsaid and commonly called or known by the name of the Homefield lands in which said common fields the owners and occupyers of lands therein upon each others Lands there every other year have right of common." Thereupon it is agreed that " each party [is] to enclose his or her own parcel or parcels of land in the said Comon fields at his or her own costs and Charges and to enjoy the same soe Inclosed in Severality."

This deed may have been enrolled in chancery, as Miss Leonard found was the case during the seventeenth century with several similar ones from various parts of England.[2] It is pretty clear, if we may rely upon the notes written on the glebe terriers, that chancery sanction was given to the enclosure of Middleton Stony at almost the same time. Of that parish a terrier records in 1679 that the glebe lay in sixty-four parcels in the open fields. In 1701 a second terrier states that " Part of the Glebe Lands . . . was taken out of the common Field about 15 years agoe and Inclos'd by a generall consent of the inhabitants." Finally a terrier of 1716 explains that the glebe is " all inclosed and a Decree in chancery for a Rate Tythe pap'd Anno 1714 — all parties consenting."[3] Though chancery is said to have authorized only the " Rate Tythe," it is likely that all matters connected with the enclosure were thus sanctioned. There is no later information about open fields at Middleton Stony.

[1] This deed is with the clerk of the peace at Oxford.
[2] "Inclosure of Common Fields," pp. 108–110.
[3] Bodleian, Oxfordshire Archdeaconry Papers, ff. 36, 37.

We have still earlier enclosure agreements from Oxfordshire. In 1667 eight proprietors and three commissioners were parties to a deed by which they divided "all the lands lying in the late open and common fields of Finmere."[1] To the eight were allotted 1273 acres. In 1662 Thomas Horde, Esq. entered into an agreement with the freeholders and copyholders of Aston and Cote, whereby it was declared "lawful at all times hereafter, as well for the Lord of the Manor . . . as for all or any the tenants and owners of lands in Aston and Cote aforesaid, to inclose all or any their respective arable lands there." The interest of the lord of the manor seems to have been the motive force here, since there is special proviso for his immediate action.[2] That the tenants did not fully avail themselves of their privilege is indicated by the fact that most of the common arable field remained open until enclosed by act of parliament in 1855.

Earliest of the extant Oxfordshire enclosure agreements is one relative to Bletchingdon. In 1622 the lord of the manor, the rector, and the tenants drew up a tripartite indenture declaring that "a general division is now intended to be had and made of all and singular ye messuages lands and Tenements. . . . And also of all . . . the Arable Lands, Meadows, Pastures, Heath, Furzes, Commons, Wastes and Wast grounds hereafter mentioned . . . and also of and in all and every the Glebe Lands lying dispersed in the fields of Bletchingdon." Thereupon are enumerated and apportioned some 500 acres of open field and 600 acres of "Heath." Rights of common are renounced, and the enclosure history of the township is brought to an end.[3]

[1] The deed is with the clerk of the peace, Oxford. It is printed by J. C. Blomfield, *History of Finmere*, Buckingham, 1887.

[2] The deed of agreement continues: "Mr. Horde may, as soon as he pleaseth, inclose 54 field acres of arable land lying together in the Holiwell field next to the capital messuage in Cote aforesaid, which said 54 acres is as much arable as usually belongs to two yard-lands in Aston and Cote . . . [and he] may inclose as much of Common called Cote Moor . . . as the proportion of two yards of common shall amount unto. . . . " In other cases he and the tenants are to give in exchange "other lands of as good value as those for which he shall so inclose." J. A. Giles, *History of the Parish and Town of Bampton* (2d edition, Bampton, 1848), Supplement, pp. 8–9.

[3] The indenture is at the Shire Hall, Oxford.

These illustrations may serve to explain how certain townships were getting enclosed during the century which preceded parliamentary activity. No other deeds than those just described are available, but we are not without clue as to what townships made similar changes. The numerous glebe terriers of the seventeenth century,[1] usually dated between 1634 and 1689,[2] show that certain parishes for which there are no parliamentary awards were still open when the terriers were drawn up. Of such there are fourteen instances.[3] Unless awards have been lost, the fourteen townships were enclosed by voluntary agreement, in most cases probably during the century which elapsed between the date of the terrier and the beginning of parliamentary enclosure. Since the series of glebe terriers is incomplete and gives no information about several townships which are later found enclosed, still other enclosures than these fourteen and the six described above may have been effected in the same way.

Several of these seventeenth-century terriers, however, picture the glebe as already enclosed, a circumstance which brings us to a consideration of those Oxfordshire townships in which enclosure occurred before 1634. We are here in a realm of conjecture, but a few surmises are permissible. In the first place, it will be remembered that certain of the townships which in the parliamentary awards had less than one-fourth of their tillable ground in open field lay in the Chiltern region.[4] In no Chiltern township was there much open-field arable,[5] and some of them we shall be prepared to find without indication of any whatever. Such is the case with eight townships on the summits and eastern slopes of the hills.[6] These were upland wooded areas, without

[1] Cf. below, p. 134. [2] That for Waterstock is dated 1609.
[3] Ardley, Broughton Poggs, Cornwell, Cuxham (still open in 1767), Elsfield, Emmington, Glympton, Hardwick (near Bicester), Kiddington, Newton Purcell, Rousham, Steeple Barton, Waterstock, Wood Eaton.
[4] Chakenden, Goring, Ipsden, Rotherfield Greys.
[5] Four other townships that extend from the Chilterns well into the plain — Shirburn, South Stoke, Watlington, and Whitchurch — had between one-fourth and one-half of their improved grounds in open field. No other Chiltern township than the eight mentioned had any open field.
[6] Bix, Kidmore, Nettlebed, Nuffield, Pishill, Rotherfield Peppard, Swyncombe, Stonor.

doubt directly enclosed from the forest. Five other early enclosed townships lay in the plain between the hills and the river,[1] a situation that probably had something to do with the absence of open fields, although two neighboring townships retained such fields until well into the nineteenth century.[2] In general, the Chilterns, apart from certain areas near the Thames, are to be looked upon as a forest region in which enclosure was early and probably coincident with improvement from the forest state.[3]

The condition of such of these townships as lay between the hills and the Thames suggests another reason for early enclosure, namely, proximity to a river. Already situations of this kind have been instanced to explain Tudor and Jacobean irregularities in field arrangements. May they not also have been responsible for a further step — the conversion to enclosed pasture of lands obviously fitted for such use? The three large streams of Oxfordshire are the Thames, the Cherwell, and the Thame. If we run through the list of early enclosed townships, we find that no fewer than nineteen of them were meadow townships lying on or near these streams.[4] Most are of small size, containing from 500 to 1000 acres apiece, a circumstance also conducive to prompt enclosure. There were, of course, many riverside townships which retained open arable fields; but since they were in general larger than the nineteen in question, speedy conversion of all their open fields to pasture would have been more difficult.

We come finally to a group of townships the early enclosure of which is explained by their history. Each has long been notable as the site of a mediaeval monastery, an ancient manor-house,

[1] Eye and Dunsden, Henley-on-Thames and Badgemore, Greys, Harpsden, Mapledurham.
[2] Caversham and Shiplake.
[3] Except in three or four instances the hamlets near Wychwood forest, unlike those of the Chilterns, had open fields and retained a part of them until the time of parliamentary enclosure. Long Coombe, whose field irregularities have already been noticed (p. 84, above), may have been enclosed early, since no enclosure award is forthcoming.
[4] Langford, Radcot, Bampton, Chimney, Shifford, Lew, Yelford, Begbrook, Binsey, Marston, Cutslow, Gosford, Hampton Gay, Newnham Murren, Mongewell, Chippinghurst, Stadhampton, Albury, Tiddington.

or at least as the residence of a county family. In each, furthermore, there is likely to be an extensive park. Notable residences and parks were, to be sure, frequently found where there was enclosure by act of parliament. In these cases their owners had increased the old enclosed area, but had not succeeded in becoming sole proprietors within the townships in question. What a nobleman or a gentleman of consequence in the sixteenth century, however, considered most desirable as a residence was an entire township. The extensive home manors of the monasteries, often provided with well-built dwellings, formed ideal seats for the rising gentry who secured them. As luxurious life in the country became fashionable, each county came to have its large Tudor and Jacobean houses. In Oxfordshire there are thirty-four townships entirely given over to residential estates of this kind. Five are the sites of monastic houses — Bruern, Chilworth, Clattercote, Minster Lovell, Sandford. Seven boast Elizabethan or Jacobean mansions — Chastleton, Cornbury, Fifield, Neithrop, Water Eaton, Weston-on-Green, Yarnton. Elsewhere the houses are of a somewhat later date, but some, like Blenheim and Nuneham Courtenay, are well known.[1] Together the thirty-four constitute 8.4 per cent of the area of the county.

Often two of the reasons above given to explain enclosures earlier than 1634 applied to the same parish. Stonor is in the Chilterns and at the same time is the seat of Lord Camoys, with mansion and park; Cornbury Park, Dichley Park, and Woodstock Park, all notable residential estates, lie within the ancient area of Wychwood forest. At Sandford-on-the-Thames were a preceptory of the Templars and the priory of Littlemore. Just below is Nuneham Park, and above on the bank of the Cherwell is the Jacobean manor-house of Water Eaton. The coincidence of park and stream is natural, since the taste of the sixteenth and seventeenth centuries dictated that, if possible, a mansion be built not far from a stretch of water.

[1] Besides the sites of the five monasteries and the seven Elizabethan or Jacobean mansions, the residential townships were and are Adwell, North Aston, Ascot, Attington, Blenheim, Chislehampton, Crowmarch Gifford, Cuddesdon, Godington, Holton, Holwell, Nuneham Courtenay, Little Rollright, Shelswell, Souldern, Stonor, Thomley, Tusmore, Over Warton, Waterperry, Wheatfield, Wilcote.

The four general reasons advanced to explain enclosures in Oxfordshire — parliamentary activity, voluntary agreement, situation within a forest area or beside a river, and the existence of an ancient residential estate — have accounted for nearly all the townships within the county. For fifteen, however, no explanation is at hand. Most of these are small, a circumstance which in itself favored consolidation and enclosure.[1] In the case of the half-dozen larger ones special causes may have been at work or explanatory data may have disappeared.

So far as it is explicable by two of the foregoing reasons, the achievement of early enclosure was probably a normal development. A favorable situation beside a river was itself an impetus, and voluntary agreement indicates acquiescent tenants. So far, however, as the desire to form a residential estate was responsible for enclosure, high-handed measures on the part of the lord may not have been absent. In townships where this motive came into play, whether directed toward the absorption of the entire area or affecting only a large part of it,[2] investigators should seek for the activity of the sixteenth-century evicting landlord — so far, indeed, as this existed.[3]

If we now return to those townships which in time became the object of parliamentary concern and inquire what agricultural progress they had made before their enclosure, we shall discover that, although in some regions it was negligible, in others it was

[1] Warpsgrove 334 acres and Easington 295 (both in the flat fertile valley of the Thame), Stowood 593 (formerly extra-parochial, near Beckley), Hensington 603 (adjacent to the borough of Woodstock), Widford 549 (adjacent to Burford and now owned by a single proprietor), Ambrosden 600 (the residential part of a parish which once included Blackthorn and Arncot), Prescote 551 (set off from Cropready), and Nether Worton 733. The township of Studley (951 acres) has been transferred from Bucks, and Little Faringdon (1161 acres) lay in Berks, when its open fields were enclosed in 1788. There remain a half-dozen larger townships for the enclosure of which I have no explanation, — Tetsworth (near Thame) 1178 acres, Grimsbury (a rural township set off from the old parish of Banbury) 1218, Middle Aston 894, and, in the southwestern uplands, Crawley 1123 (carved from the old area of the borough of Witney), and Shilton 1596.

[2] The existence of a considerable residential estate is responsible for the preponderance of enclosures in certain townships which, since they later became the objects of parliamentary award, appear in the last two groups of Appendix IV.

[3] On the subject, see Gay's "Midland Revolt" and other papers (cf. above, p. 11, n. 1).

considerable and had manifested itself as a change of field systems. Although relatively few of the maps that accompany the awards give accurate pictures of the old fields, enough of them do so to illustrate the situation. The plans of the Chiltern townships may, for the time, be disregarded. Since the midland system did not prevail in that region, irregular fields, such as these plans show, were to be expected there. We shall return to them later.[1]

In the rest of the county, an area once entirely given over to the two- and three-field system, a diversity of field arrangement had arisen between the sixteenth century and the nineteenth. Occasionally an award or a plan shows the two simple old fields bearing the old names. The Kencott award of 1767 quotes the act to the effect that there were " by estimation in the two open common fields, the East field and the West field, about 731 acres." All the allotments at Hook Norton and Southrop in 1774 were in either the Northside field or the Southside field. At Arncott, in 1816, there were 533 acres in the West field and 555 in the East field. The Taynton award and plan of 1822 have, besides the common, only the two large fields, East and West. These four townships lie in the Cotswold uplands, where the two-field system was once almost universal. That only one-half of their arable was cultivated yearly after the middle of the eighteenth century may seem incredible; yet there is nothing in the awards to show that field conditions had changed since the thirteenth century. We have, indeed, found the Charlbury agreement of 1715 declaring that the owners and occupiers of lands in the open fields " upon each others Lands there every other year have right of common." We have, too, the definite statement of Richard Davis, who made the first report on Oxfordshire to the Board of Agriculture in 1794. " Some open fields," he says, " are in the course of one crop and a fallow, others of two, and a few of three crops and a fallow. In divers uninclosed parishes the same rotation prevails over the whole of the open fields; but in others the more homeward or bettermost land is oftener cropped, or sometimes cropped every year." [2]

[1] Cf. below, Chapter IX.
[2] *General View of the Agriculture of the County of Oxford* (London, 1794), p. 11.

For the most part, however, the two-field system seems to have disappeared in Oxfordshire before the era of parliamentary enclosure. Arthur Young, who in 1809 made for the Board of Agriculture a second and more elaborate report on agrarian conditions within the county, says nothing about it. Yet he does note the continued employment of the three-field rotation, especially on the rich lands west from Thame,[1] an observation that is borne out by the enclosure maps. The tithe map of Chalgrove has been reproduced,[2] and the same district furnishes several three-field enclosure plans. At Thame itself there were in 1826, besides two very small fields, three large ones called Priest End, West, and Black Ditch, while the same plan shows, in the adjacent hamlet of Morton, fields alike in size named Costall, Horsenden, and Chin Hill. The accuracy of this representation is confirmed by an excellent eighteenth-century map of Morton, the strips of which lie in the same three fields, the last being called Little field.[3] A third township affected by the award of 1826 was Sydenham, whose three equal fields were Upper, Forty, and Lower. Near by, with only one intervening parish, are Lewknor and its hamlet Postcombe, each of them in a plan of 1815 showing three fields regularly disposed round the village.[4] Not five miles away is Stoke Talmage, where in 1811 the same neat arrangement was to be seen.[5] Berwick Prior, too, in 1815 retained its three fields.[6] Most striking perhaps of all this group is the township of Crowell, where the enclosure of the arable is the last recorded in Oxfordshire, being delayed until 1882. Yet even at that date Crowell had three open fields, which bore the unassuming old names of Upper, Middle, and Lower.

[1] *Agriculture of Oxfordshire*, p. 127, "On the open field near Thame [the rotation is], (1) Fallow, (2) Wheat, (3) Beans on a very fine reddish loamy sand and the crops great"; p. 123, "On the excellent deap loams between Stokenchurch and Tetsworth, (1) Fallow, (2) Wheat, (3) Beans"; p. 13, "Morton field [next Thame] a stiff loam . . . two crops and a fallow"; p. 133, "On the open fields at Baldons the old course, (1) Fallow, (2) Wheat, (3) Barley, oats," etc.

[2] Cf. above, p. 20.   [3] Add. MS. 34551.

[4] The Lewknor fields were Road, Middle, and Sherburn; those of Postcombe were Clay, Little, and North.

[5] Three equal fields, Westcut, Middle, and Temple Lake.

[6] They were named Marsh, Middle, and Town, and lay compactly to the north-east of the village.

At the beginning of the nineteenth century there were more three-field townships in the region round about Thame than in all the rest of Oxfordshire. Besides those mentioned, Little Milton, Littlemore, Wheatley, Headington, and Islip had each subdivided three fields into six. A little to the north were Beckley, Piddington, and Bicester Market End, also showing at the somewhat earlier date when they were enclosed three fields apiece. In short, if with Oxford as a center a quadrant were to be described from Bicester (fifteen miles to the northeast) toward the east and south, it would include several townships which still, in 1800 or even in 1825, were cultivated in the three-field manner. Since this region was noticeably the last in the county to undergo enclosure, the system seems not to have been wanting in tenacity.

Elsewhere in Oxfordshire few traces of the three-field system appear in the parliamentary enclosure awards, except just to the west of Oxford.[1] What had happened can be read on many pages of Arthur Young's account and verified from the enclosure plans. The change amounted to a substantial improvement in agriculture. Many of the townships had adopted a four-field system,[2] with a four-course rotation of crops, the latter in general being (1) fallow, (2) wheat, (3) beans, (4) barley or oats.[3] Young's illustrations are for the most part from the regions north and west of Oxford, although he cites Garsington, situated in the district in which we have seen the three-field rotation holding its own. The order of the four courses varied only at Deddington, where it was (1) fallow, (2) wheat, (3) barley, (4) peas or beans.

These accounts of four-course tillage are confirmed by the plans and enumerations of the enclosure awards. The region round Oxford is again the one which furnishes most illustrations. At

[1] There seem to have been three important fields at Eynsham in 1802, six at Ducklington in 1839, and six at Curbridge in 1845. These parishes lay close together, ten or fifteen miles west of Oxford.

[2] Davis in the quotation given above says " a few " townships, but the evidence about to be cited seems to show that they were numerous.

[3] Young, *Agriculture of Oxfordshire*, pp. 111-130. So at Bampton, Hampton Poyle, Garsington, Tackley, Wood Eaton, Wendlebury, Bicester Kings End, Kidlington, Kelmscott.

Hampton Poyle, where Young noted a four-course rotation, the award enumerates five fields — Lower, Bletchingdon, Grettingdon, Collet, Friezeman's Well — but the last one was probably small. At Standlake, though several areas are named in the award of 1853, four fields stand out, North, South, Church, and Rickland. At Culham four fields are specified on the plan — North Middle, South Middle, Ham, and Costard — and they are shown to be relatively equal in size. Frequently the divisions were no longer called fields, but had come to be known as " quarters." At Hailey, in the parish of Witney, one of the four open-field areas was in 1824 called Home field, but the others were Crowley quarter, Middle quarter, and Witney quarter. At Kingham, on the western edge of the county, the plan, drawn in 1850 and sketched in the cut on the next page,[1] indicates six quarters; but the glebe terrier of 1685 shows that only four of these quarters (Ryeworth, Withcumbe, Brookside, Broadmoor) were at that time important, a fifth not being mentioned and a sixth consisting of " every yeares Land."[2]

Division by quarters is particularly characteristic of northern Oxfordshire. This region, which lies round Banbury, is possessed of a fertile soil known as " redland " and adapted to improved cultivation. If but one of Young's illustrations of four-course rotation, that of Deddington, comes from here, the reason is that nearly all townships hereabouts had already been enclosed when he wrote, having been among the first in the county to apply to parliament. Their awards, however, make it clear, by references to field divisions, that four quarters were existent at the time. At Sandford near Tew the old fields, North and South, had seen appear beside them two large " quarters " fully as important, called Down and Beacon. At Wardington in 1762 South field had become South field quarter, and ranked along with Ash, Spelham, and Meerhedge quarters. Near by, the township of Neithrop, a rural division of Banbury, had before 1760 divided its open fields into four quarters — Thoakwell, Lower, Forkham, and Greenhill.[3] Seldom was the nomenclature of the old fields

[1] The award is at the Shire Hall, Oxford. [2] Cf. above, p. 92.
[3] Frequently the relative areas of the quarters cannot be ascertained, since in

retained, and the names applied to the quarters indicate that the formation of them was recent.

In several instances the awards picture nothing so comprehensible as a four-fold division; instead of this, furlongs acquire

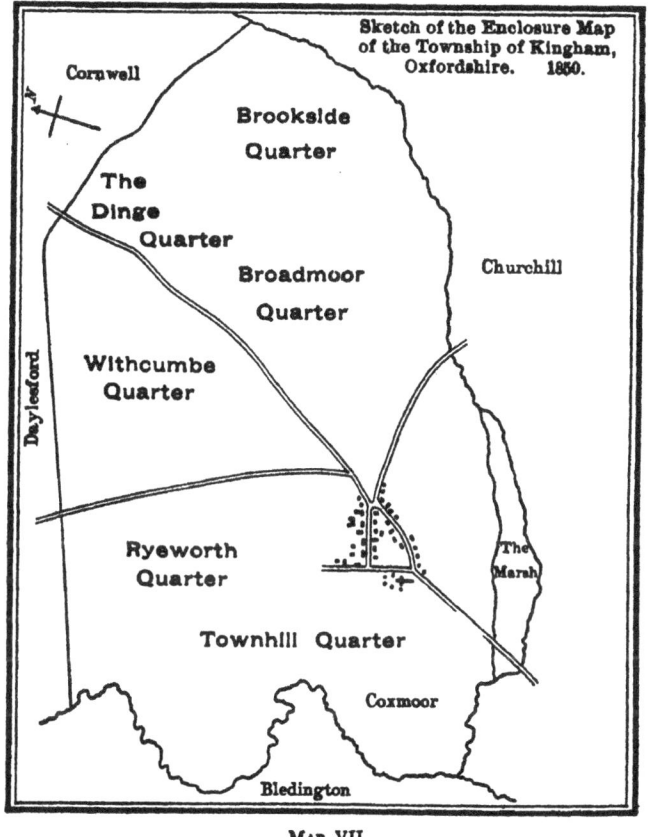

MAP VII

prominence, and are the principal areas of which cognizance is taken.[1] At Sibford Ferris in 1790 the allotments lay in the

early awards the plans take no account of antecedent conditions and a single large allotment sometimes extended into more than one quarter. At Cropredy, for instance, the total allotment was 1582 acres. To Sir William Boothby, Bart., were assigned, in lieu of 33¾ yard-lands, 961 acres in Howland quarter, Hackthorn quarter, Oxley fields and quarter. Elsewhere a quarter called Heywey frequently appears.

[1] So, for example, at Drayton near Banbury, Fritwell, Spelsbury, Stonesfield, Little Tew.

furlongs named Shroudhill, Stonewall, Seven Acres, White Butt, Middle, Boyer, Townsend, Blackland, Longman's Pool, Gore, Bush, Pitch, Church, etc., as well as in several other areas not called furlongs (e. g., Wagborough, and Long Stone Hill). When large divisions of the arable have become thus obscured, it is natural to find in some townships the number of quarters increasing, since these too may have ceased to retain agrarian significance and may have become largely topographical. At Burford in 1795 they numbered seven, at Duns Tew in 1794 eight, and at Neat Enstone in 1843 eleven.[1] The result of disintegration of this kind was often a bewildering array of field names, in which fields, quarters, furlongs, and nondescript patches were indiscriminately mingled.[2] At Kidlington in 1815 the allotments lay in nine fields, four furlongs, and a half-dozen miscellaneous areas. At West Chadlington in 1814 the more important open-field areas were Lower field, Lockland quarter, Crosses Quarry quarter, Gardens quarter, Banks quarter, Blackmore Brakes quarter, Cockcroft Stone quarter, Green Benches quarter, Broadslade quarter, Ashcroft furlong, Cooper's Ash furlong, Standalls Pit furlongs, Quarry furlong, Berry Hill, the Down, Broadslade Mill Hill, Thornwood, Great Lands, and Lone Land Hill. These areas were presumably grouped in some manner for a regular rotation of crops, but the inability to locate allotments more simply shows that large field divisions had become obsolete.

Under such circumstances the grouping of the many quarters and furlongs could, for the sake of improved tillage, be easily changed by decree of the manorial court. How this was done may be illustrated from three court rolls of Great Tew, a township on the edge of the redland district.[3] The rolls date from the autumns of 1756, 1759, and 1761, nearly a decade before the

[1] Hull Bush, Abigals, Sturt, Batlodge, White Hill, Windmore Hedge, Whores; Berry Field, Tuly Tree, Tomwell, Whittington, Ridges, Sands, Red Hill, Lands; Hore Stone, Great Stone, Long Lands, Lady Acre, Heythrop, Leazow Hedge, Folly, Crook of the Hedges, Long Weeding, Sheepwalk, Slate Pits. All are called quarters.

[2] So at Westcot and Middle Barton, Bloxham, Church Enstone, Iffley, Milcombe, Swinbrook, Wendlebury, Wigginton.

[3] P. Vinogradoff, "An Illustration of the Continuity of the Openfield System," Appendix, *Quarterly Journal of Economics*, 1907, xxii, 74–82.

enclosure of the open fields of Great Tew in 1767. The first two are not clear about field divisions or the rotation of crops. The third, dated October, 1761, is specific, enumerating the open fields in eight divisions, as follows: (1) Huckerswell, (2) Between the Hedges, (3) Upper Barnwell, (4) The Lower side of Woodstock way beyond the Brook, (5) Gally Therns and the old Hill, (6) Park Hill and Great Oxenden, (7) Upper Oxenden, plank pitts, ten Lands . . . Wheat Land, Broad and picked Castors, Hollowmarsh Hill to Alepath, (8) Alepath to the Great Pool and the West field from Alepath and Woodway Ford. The first of these divisions was to be subject to an eight-course rotation beginning the following spring and observing the following succession: (1) turnips, (2) barley with grass seeds, (3) hay, (4) sheepwalk, (5) oats, (6) fallow, (7) wheat, (8) peas. The second division was to begin the same rotation a year later, the third two years later, and so on throughout the series. Eight presumably equal divisions of the open fields were, in short, arranged for an eight-course rotation of crops.

That this arrangement was not new in 1761, but that certain of the areas mentioned were at the earlier dates sown precisely as they would have been had the specified rotation been in force, is suggested by four items in the first two rolls. Thus, in the spring of 1757 Upper Barnwell was destined for spring grain and grass seeds, while Between the Hedges was the clover quarter. Eight years later, as we have seen, the same crops were assigned to these areas. According to the second roll the Upper Oxenden group was in the spring of 1761 to be sown with barley and grass seeds, while Park Hill and Great Oxenden were in 1760 to be "lay'd down with rye grass and clover" (i. e., mowed for hay). The specifications of the third roll were to the effect that the same situations should prevail in the respective divisions eight years later. An eight-course rotation and the subdivision of the open fields into eight parts thus seem to antedate 1756, the date of the first roll, but by how much we cannot say.

These arrangements amounted to the introduction of a second four-course rotation besides the one described by Arthur Young. The normal four-course succession, it will be perceived, appears in (5) oats, (6) fallow, (7) wheat, (8) peas. After this came the new rotation, (1) turnips, (2) barley with seeds, (3) grass mowed, (4) grass pastured by sheep. The innovation lay in the direction of turnip and grass cultivation, half of the wheat crop having been replaced by hay, and the sheep being pastured on a sward rather than on fallow ground. Upon the fields the effect of this eight-course arrangement was the formation of small areas, no longer even formally named "quarters." Once, however, Between the Hedges is referred to as the "Clover quarter," and elsewhere we hear of the "Turnip division." If the other townships of northern Oxfordshire which in the awards have such confusing field divisions owed them to the same cause, an eight-course rotation of crops or something similar must have been well known at the very time when parliamentary enclosure was beginning.

All this, of course, implies marked improvement in open-field agriculture. Although the eight-course rotation may perhaps be looked upon as a special refinement, not widespread, there can be no doubt about the extent and importance of four-field and four-course arrangements. They probably constituted the prevalent method of open-field cultivation in Oxfordshire in 1750. The enclosure history of the townships in which they prevailed seems, moreover, to warrant a generalization. So long as the three-field system maintained itself intact, landlord and tenants were inclined to rest content and allow the fields to remain open. For this reason the district southeast of Oxford round about Thame, which clung to its three fields until into the nineteenth century, was the last part of the county to undergo enclosure. Where, however, four-field and eight-field husbandry had come to prevail, as they had in the north and west of the county, enclosure was favored. The tenants, already in advance of the inhabitants of three-field townships, were prepared to outstrip them still more. As soon as parliamentary facilities were offered, acceptance was general; within two or three decades nearly all of the north and west became enclosed. Great Tew itself yielded in 1767, thereby revealing, it would seem, a connection between the eight-course rotation upon its much-divided fields and this early enclosure. Elsewhere in the north the complex fields and the four-course rotation with which we have become familiar suggest a similar explanation of the relatively early dates which characterize the enclosure awards of this region.[1]

Having discovered that by the later eighteenth century many Oxfordshire townships of the northwest had discarded the two-field husbandry once practiced there, we are led to inquire what period is responsible for this improvement. Such a query involves a consideration of seventeenth-century field arrangements. Of these we have, fortunately, a contemporary account which, if not a model of style, is yet instructive. In Robert Plot's *Natural History of Oxfordshire*, published in 1677, one chapter treats of the tillage employed on the various soils of the county. Before quoting this, however, it will be helpful to note the characteristics and boundaries of the soils themselves.

Arthur Young's description is best.[2] According to him, the fertile "redland" of the northern townships near Banbury is one of the best soils of the midlands. It extends over the wedge-shaped area that protrudes between Warwickshire and Northamptonshire, and constitutes about one-sixth of the entire county. South of it there stretches from the Cotswolds eastward to Buckinghamshire a broad belt of less desirable soil, for the most part a limestone and known as "stonebrach." Its southern boundary runs from Witney to Bicester, and it comprises a third of the county. South of this again is a belt of miscellaneous loams including the valleys of the Thames, the lower Cherwell, and the Thame. This constitutes another third of the county, reaching to the Chilterns on the southeast. The latter, one-sixth of the county, have a chalky soil, not ill adapted to certain crops.

Following these divisions, which he too recognizes, Plot begins his description with an account of the tillage of clay soils, most numerous in the north. It will be seen that he has primarily in mind a four-course rotation of crops, precisely that described by Young one hundred and thirty years later: —

"And first of Clay, Which if kind for Wheat, as most of it is, hath its first tillage about the beginning of May; or as soon as

---

[1] Cf. Appendix IV.   [2] *Agriculture of Oxfordshire*, p. 3.

Barly Season is over, and is called the Fallow, which they somtimes make by a *casting tilth*, i. e. beginning at the out sides of the Lands, and laying the Earths from the ridge at the top. After this, some short time before the second tilth, which they call *stirring*, which is usually performed about the latter end of June, or beginning of July, they give this Land its manure; which if Horse-dung or Sheeps-dung, or any other from the Home-stall, or from the Mixen in the Field, is brought and spread on the Land just before this second ploughing: But if it be *folded* (which is an excellent manure for this Land, and seldom fails sending a Crop accordingly if the Land be in tillage) they do it either in Winter before the fallow, or in Summer after it is fallowed. . . .

" After it is thus prepared, they sow it with Wheat, which is its proper grain . . . and the next year after (it being accounted advantagious in all tillage to change the grain) with Beans; and then ploughing in the bean-brush at All-Saints, the next year with Barly . . . ; and then the fourth year it lies fallow, when they give it Summer tilth again, and sow it with Winter Corn as before. But at most places where their Land is cast into three Fields, it lies fallow in course every third year, and is sown but two; the first with Wheat, if the Land be good, but if mean with Miscellan, and the other with Barly and Pulse promiscuously. And at some places where it lies out of their *hitching*, i. e. their Land for Pulse, they sow it but every second year, and there usually two Crops Wheat, and the third Barly, always being careful to lay it up by ridging against winter; Clay Lands requiring to be kept high, and to lie warm and dry, still allowing for Wheat and Barly three plowings, and somtimes four, but for other grains seldom more than one. . . .

" As for the Chalk-lands of the Chiltern-hills . . . when designed for Wheat, which is but seldom, they give it the same tillage with Clay, only laying it in four or six furrow'd Lands, and soiling it with the best mould . . . and so for common Barly and winter Vetches, with which it is much more frequently sown, these being found the more suitable grains. But if it be of that poorest sort they call white-land, nothing is so proper as raygrass mixt with Non-such, or Melilot Trefoil. . . .

"If Red-land, whereof there are some quantities in the North and West of Oxford-shire, it must have its tillage as soon in the year as possibly may be, before the clay. . . . This never requires a double stirring. . . . Nor is the Sheep-fold amiss either Winter or Summer. . . . This Land, like clay, bears wheat, miscellan, barly, and peas, in their order very well, and lies fallow every other year, where it falls out of their hitching. . . .

"In some parts of the County they have another sort of Land they call Stone-brash, consisting of a light lean Earth and a small Rubble-stone, or else of that and sour ground mixt together. . . . These Lands will also bear Wheat and Miscellan indifferently well in a kind year, but not so well as clay, sourground, or red-land; but they bear a fine round barly and thin skin'd, especially if they be kept in heart: They lie every other year fallow (as other Lands) except where they fall among the Peas quarter, and there after Peas they are sown with Barly, and lie but once in four years. . . .

"There is a sort of tillage they somtimes use on these Lands in the spring time, which they call streak-fallowing; the manner is, to plough one furrow and leave one, so that the Land is but half of it ploughed, each ploughed furrow lying on that which is not so: when it is stirred it is then clean ploughed, and laid so smooth, that it will come at sowing time to be as plain as before. . . .

"Lastly, their sandy and gravelly light ground, has also much the same tillage for wheat and barly, as clay, etc., only they require many times but two ploughings. . . . Its most agreeable grains are, white, red, and mixt Lammas wheats, and miscellan, i. e. wheat and rye together, and then after a years fallow, common or rathe-ripe barly: so that it generally lies still every other year, it being unfit for hitching, i. e. Beans and Peas, though they somtimes sow it with winter Vetches."[1]

This account makes it clear that in 1677 a four-course, a three-course, and a two-course rotation of crops were in use in different townships of Oxfordshire. The relation between the four-course

[1] Robert Plot, *Natural History of Oxfordshire* (Oxford, 1677), pp. 239–244.

and two-course rotations also becomes apparent: the old two-field townships had subdivided their fields and had begun to sow one-half the former fallow field with pulse, i. e., with peas, beans, or vetches. This procedure came to be known as "hitching" the field. A township which remained in two fields practiced little or no "hitching," and even in four-field townships certain poorer lands sometimes "fell without the hitching," i. e., were not sown in the pulse year. The particular rotation (wheat, beans, barley, fallow), always recounted by Young, is thus explained. It was the natural outcome of sowing one-half of the fallow field of a two-field township.

Thus prepared by Plot's account, we may turn to such descriptions of seventeenth-century fields in Oxfordshire as are available. Happily, there exists for this county, as for many others, a series of glebe terriers, a single parish often furnishing two or three such documents.[1] The dates range from 1601 to 1685, with occasionally a terrier for the sixteenth century and many from the early eighteenth. Most frequently they are dated about 1634 or 1685. Since some parishes do not appear and many terriers are not easy to interpret, no complete classification for the county can be attempted. Yet even an incomplete series shows in a general way the field usages most in favor in Stuart days.[2]

In fourteen parishes, the glebe is said to have consisted of crofts.[3] Six of these lay in the Chiltern region, and several of the others were riverside or residential townships. Many terriers, however, picture the original two or three open fields.

Seventeen townships retained the two-field system, the field names being for the most part such primitive ones as East and

[1] The Oxfordshire terriers have been gathered into one volume, now in the Bodleian (Oxfordshire Archdeaconry Papers, i). A second volume contains the Berkshire series. Terriers for other counties are usually to be found in the archives of the diocese within which the county lay.

[2] In Appendix II, under "Oxfordshire," the description of the glebe as it lay in two-, three-, or four-field townships is summarized.

[3] Bix, Caversham, Harpsden, Ibston, Rotherfield Greys, Rotherfield Peppard (all in the Chilterns); Begbrook, Bletchingdon, Broughton (near Banbury, the glebe being enclosed c. 1700), Goddington, Lillingston Lovell, Minster Lovell, Pirton, Over Worton.

West, North and South. All were in the north and west of the county. In them the abandonment of two-field agriculture seems to have occurred between the period of the terriers and that of the awards, since only Kencott is known to have retained its two fields until 1767. Three were enclosed without act of parliament, we know not how.[1] For several the parliamentary awards give no field detail, most of them being of early date.[2] In three instances, however, we discover from the awards that the old two fields of the terriers had disintegrated before enclosure. At Asthall the plan of 1814 bears the names of many small fields, six at least; at Duns Tew in 1794 there were six quarters; at Tackley the references in 1773 are to furlongs and quarters only. Had the awards which contain no field detail been as specific as these three, they would probably have disclosed a similar situation, and have made it quite clear that the definite abandonment of the primitive agriculture by even the least enterprising of two-field townships occurred between the middle of the seventeenth century and the middle of the eighteenth.

Several glebe terriers, of course, picture the continuance of the three-field system, the point of interest here being the location of the townships. All are near Oxford, mainly to the east, but partly to the west near the Thames, the region which we have already seen characterized by three fields in the nineteenth century. It was not there that agricultural advance was to be expected.

For evidence of such progress we turn to eight of the terriers dated about 1680.[3] In them the division of the arable into four quarters, later to become so frequent, is already apparent, and they illustrate the four-field arrangements which Plot, writing in 1677, had in mind. All are from the northern part of the county.[4] Certain other terriers for townships of the northwest have not this neat quadripartite division of the glebe, but in them also

[1] Ardley, Broughton Poggs, Glympton. The enclosure of Middleton Stony has been explained from the glebe terriers themselves (see p. 117).
[2] Alkerton (1777), Alvescot (1797), Steeple Aston (1767), Brize Norton (1776), Cottisford (1854), West Shutford (1766), Westwell (1778).
[3] Cf. Appendix II, pp. 493-494.
[4] The case of Kingham has already been cited and illustrated (above, p. 126).

acres are apportioned in such manner as to indicate that a two- or three-field arrangement was no longer satisfactory.[1] Where there are four quarters or fields, the names of these are curious and local, obviously of recent origin. At Somerton in 1634 precise designations had not yet been adopted, it being necessary to call the fields second, third, fourth, and to locate them with reference to highways;[2] but at least it is clear that four-field arrangements were known to Oxfordshire early in the seventeenth century.

This fact, taken with the testimony of the preceding chapter, seems to warrant the generalization that a four-field system, making its appearance in the English midlands during the sixteenth century and the early seventeenth, was employed more and more in the course of the latter century and in the early eighteenth. This transformation marks the second important stage in the development of open-field husbandry in the midlands. The first occurred when, in the thirteenth and fourteenth centuries, three fields were substituted for two in many regions, among others in eastern Oxfordshire. Elsewhere, as in northwestern Oxfordshire, no change then took place, the primitive two-field system remaining intact. These two-field regions, however, were those which, in this second period of development, ceased to be dormant.[3] No longer did they allow one-half of the arable to lie fallow each year, but they reduced the fraction to one-fourth. Not content with this, they sometimes went farther, introducing an elaborate rotation of crops and a complicated field system in natural approach to the still more scientific principles em-

[1] Steeple Barton 1685, Charlbury 1635 (cf. the enclosure of 1715, above, p. 117), Churchill 1722, Cornwell 1614, Heyford ad Pontem 1679, Kidlington 1634, Swerford 1614, Wood Eaton 1685.

[2] So early as 1622 there is evidence in the indenture that records the partition of the open fields of Bletchingdon (Cf. above, p. 118) of four small quarters beside a much larger West field.

[3] Other two-field regions underwent the same transformation as northwestern Oxfordshire. The surveys of Welford, Gloucestershire, and Owston, Lincolnshire, have already been described to illustrate four-field townships. Similarly in four fields at the time of their enclosure were, for example, East Hanney, Berkshire (C. P. Recov. Ro., 49 Geo. III, Hil.), Massingham, Lincolnshire (ibid., 45 Geo. III, Mich.), Green's Norton, Northamptonshire (ibid., 47 Geo. III, Trin.).

bodied in the practice of convertible husbandry upon enclosed lands.

To the adoption of these principles they speedily came as soon as parliamentary enclosure offered facilities. For among other things which the Oxfordshire evidence has illustrated is the circumstance that townships which had four or more fields were the most prompt to get enclosure awards passed. By the end of the eighteenth century there was little open field left in the northwestern part of the county. Not so, however, with the region southeast of Oxford. In this stronghold of three fields enclosure was long delayed. Whether throughout midland England in general the three-field system acted similarly as a protective shell for the preservation of open-field arable cannot be determined without further investigation.[1]

Our study of Oxfordshire, which may end here, should have served to illustrate various aspects of the decay of the two- and three-field system. As to those townships of the county whose enclosure antedated parliamentary activity, a perception that most of them were situated in forest areas or along streams, or were desirable as residential estates, will perhaps serve to explain considerable early enclosing activity. Further, the achievement of this step by voluntary agreement has been instanced in order to indicate the legal methods first employed. In the case of townships enclosed by act of parliament, the awards have enabled us to discover what fraction of the county still remained open arable field up to the time when this finally disappeared. The awards, too, assisted by glebe terriers, have disclosed what transformations the two- and three-field system had undergone since the days of its earlier simplicity. Only in one part of the county, it appears, and that the section given over

[1] To judge from the dates of the acts for enclosure, certain of the old two-field counties — Lincolnshire, Gloucestershire, Warwickshire — were prompt to avail themselves of the new facilities, while ancient three-field counties, like Bedfordshire, Cambridgeshire, Huntingdonshire, long remained indifferent. But there are enough apparent exceptions to make one hesitate to generalize. Hampshire, Leicestershire, Nottinghamshire, Yorkshire, all three-field counties, inclined early to parliamentary enclosure, while Berkshire, once two-field, showed no haste. Cf. Slater, *English Peasantry*, Appendix.

to three-field husbandry, had there been no change since the fourteenth century; and that part, as it happens, was the one least inclined to undertake enclosure. Elsewhere in the old two-field territory a four-field system, or one still more advanced, had arisen as a testimony to the best efforts of open-field husbandry to achieve efficient agricultural method.

For those midland and southern counties in which the two- and three-field system once prevailed there is abundant evidence of its long-continued existence,[1] and the enclosure history of these counties did not in general differ greatly from that of Oxfordshire. In certain western counties, however, where there were pretty clearly two or three fields at an early time, similar long life was not granted. In the preceding chapter it was pointed out at length that marked irregularities in field arrangements had already appeared there in the sixteenth century, and that enclosure was frequently in progress. It remains to inquire how much open field remained to be enclosed by act of parliament.

The region in question comprised the forest area which extended over the northern parts of Warwickshire, Worcestershire, Staffordshire, and Derbyshire. It reached westward and southward to include the fertile valleys of the Wye and Severn, passing thence into the low-lying stretches of Somerset. Throughout large parts of the eight counties within this region there was a tendency from the sixteenth century onward to increase the area under pasture. The relatively small extent of the arable left to be affected by parliamentary activity can be roughly gauged from Slater's list of acts and areas.[2] In the valley of the Severn and in the plain of Somerset several townships procured awards, but the amount of arable enclosed by each award was seldom great. Elsewhere the acts were less numerous. As Slater records them, there were twenty-nine for Herefordshire,[3] seven

[1] Slater, *English Peasantry*, Appendix.
[2] Ibid. They are assigned to Gloucestershire, Herefordshire, Shropshire, and the counties mentioned above.
[3] *English Peasantry*, Appendix. His Herefordshire list includes at least three acts that should have been omitted. The Wigmore petition of 1772 distinctly states that it is concerned with 600 acres of "common wood," not with 600 acres of common arable, as Slater has it. The award for Bredwardine (with Dorston),

for Shropshire, and for the northwestern parts of Warwickshire, Worcestershire, Staffordshire, and Derbyshire, eleven, six,[1] five, and five respectively. Since Herefordshire furnishes as many awards as any part of this region, unless it be Somerset, and since it lies well to the west, its open-field history may be relied upon to illustrate conditions and changes within the territory above defined.

The contrast with Oxfordshire, whose field transformations we have been following, is marked. Though Herefordshire is the larger county of the two (537,363 vs. 483,614 acres), and not inferior in the extent of its fertile fields, its parliamentary enclosures of arable were not more than 31 in contrast with Oxfordshire's 158. The awards recording them which are preserved and accessible correspond with sixteen of the acts listed by Slater,[2] and add five that he does not mention. There are, however, ten petitions and acts which mention common fields but for which the subsequent awards are missing.[3] In no instance do these petitions or acts give areas, and how much confidence should be put in the mention of common fields in a routine formula, especially when the specification of the common wastes precedes, is uncertain.[4] The Bredwardine petition, for example, mentions

makes it clear that no arable was in question. At Byford the award allotted only a common, although it provided for the exchange of certain strips of arable. On the other hand, the list omits five townships for which we have awards concerned with the allotment of arable, viz., Wellington, Humber and Stoke Prior, Holmer, Pembridge, and Madley.

[1] Kidderminster, Wolverley, Overbury, Ombersley, Alvechurch, Yardley.

[2] Most of them are at the Shire Hall, Hereford, and I am indebted to J. R. Symonds, Esq., clerk of the peace, for permission to examine them. Of the thirty-four there preserved, fourteen relate to commons only. The Marden award, most important of all, is kept at the village of that name, but there is a copy at the Public Record Office.

[3] Perhaps some of them are, like the Marden award, to be found in the parishes to which they refer. The townships or parishes with which they are concerned are Bodenham, Shobden, Bishopston and Mansell Lacy, Steepleton, Allesmore, Eardisland, Clehonger, Stretton Grandison, Norton Canon, and Puttenham.

[4] These petitions usually recite " certain Commons, Wastes, Common Fields and Commonable and Open Lands " (*Journal of the House of Commons*, 31 January, 1811, Eardisland). When arable is prominent the phrase runs, " several Open Fields, Meadows, and Pastures " (petition for the enclosure of Tarrington, ibid., 9 February, 1796).

Date of Award	Township	Open Arable Field Enclosed	No. of Parcels	Meadow Enclosed[1]	Waste Enclosed
1779	Winforton	[379][1]	...	...	...
1797	Much Marcle	c. 622½[2]	...	...	304
1707	Wellington	611	...	38	72
1799	Tarrington	378	...	75	...
1803 {	Leintwardine Burrington, Aston, Elton, Marlow[3]	252	...	...	1902½
1804 {	Yarkill Stoke Edith West Hide Weston Beggard	1216½[4]	...	193	...
1806	Castle Frome	183	...	146[5]	7
1812	Kingston	197	...	...	...
1816	Mordiford	258½	...	48½	191
1816	Ledbury	{ 29½ 132[6]	...	...	107
1816	Estnor	54	...	...	91
1817	Aymestry	c. 302½[7]	...	...	...
1819 {	Marden, Sutton St. Michael Sutton St. Nicholas, Withington, Amberley, Preston Wynn	2371	...	1017	250
1826	Much Cowarne	536	...	...	10
1829	Lingen	78½	...	...	543
1854	Bosbury	81½	108	21½	...
1855 {	Humber Stoke Prior	149 57	109 51	... ...	... ...
1855	Holmer	297	259	...	...
1856	Ullingswick	290½	488	...	...
1862	Pembridge[8]	73½	82	...	...
1862	Pipe and Lyde	13½	16	...	...
1863	Madley	381½	298	...	...
		8565½		1539	3477½

[1] This is enclosed land over which certain persons had rights of common.
[2] Probably some of this area was waste.
[3] Of the open field 53 acres, of the waste 1201 acres, lay in these four townships.
[4] Of the open field 398 acres were in Yarkill and 138½ acres in Stoke Edith, but there is no information as to how the remainder was divided among the townships.
[5] A small part of the meadow is in Bishops Frome, Much Cowarne, and Evesbatch.
[6] " Recently inclosed from the open field," and still in strips on the plan.
[7] Some of this may have been common waste. It was divided among the four townships of Upper Ley, Nether Ley, Covenhope, and Shirley.
[8] Manley field and Manley Lower field, in the parish of Pembridge.

"common fields," but the award shows that only certain "hills" and "heaths" were in question. Although some of the ten missing awards were probably not dissimilar, it will be safe to assume that at least a small amount of arable was allotted by them. The results of an examination of all the available Herefordshire awards that relate to arable are tabulated in the schedule on the preceding page.

The open-field arable and meadow in Herefordshire which the existing awards show to have been enclosed by act of parliament amounted to 10,104½ acres. For the ten missing awards perhaps some 3000 acres more should be added, but at most not more than 2½ per cent of the total area of the county was affected. All this is in marked contrast with the late eighteenth-century situation in Oxfordshire. There the open fields of a dozen townships would have equalled in area all the open-field land in Herefordshire.[1] There 37 per cent instead of 2½ per cent of the area of the entire county was still unenclosed arable.

The foregoing list also makes it clear that the open arable fields in any township were not extensive. Those of the Marden award, which seem largest, belonged to several hamlets, and the second large area, that assigned to Yarkhill, was apportioned among at least four townships. In no other place were more than 650 acres re-allotted, while 200 to 300 acres was a usual amount, which in turn often had to be divided among the constituent townships of a parish. The Aymestrey award of 1817 apportions its open field, already small, among four such townships, and compares these areas with the far more extensive old enclosures. In both respects it is typical of Herefordshire conditions: —

Township	Open and Common Fields and Waste Lands Acres	Old Enclosures Acres
Shirley	13¼	252
Upper Ley	93	375¾
Nether Ley	181½	718¼
Covenhope	14½	676
	302¼	2022

[1] J. Clark, who in 1794 published a *General View of the Agriculture of the County of Hereford*, says parenthetically (Appendix, p. 1), "since a great part of the county

Where nineteenth-century open fields thus constituted so small a fraction of the total improved land of a township we should like to know about their appearance. Such information might assist in determining whether they had originally been limited in extent, or whether they had in course of time decreased in area and, if so, in what manner. The aspect of the fields may, in short, explain why they were so diminutive when enclosed. In this way it may become clear why Herefordshire had only one-fifteenth as much open field for the parliamentary encloser as had Oxfordshire.

The enclosure plans of Oxfordshire townships usually reveal the open fields as large compact blocks of arable lying near the village and often surrounding it. The plan of Chalgrove, which has been reproduced, is in no way exceptional. Often, as there, two or three enclosed farms lay in remote parts of the township. Sometimes enclosures had severed the open fields into two or three parts, but the parts at least remained large compact blocks. This state of things is precisely what is seldom to be found in the Herefordshire plans. The three-field townships which were once existent in the county, and which must have had fields that were more or less compact, had clearly survived in not more than four or five places.[1]

One of these survivals was at Sutton, where 700 acres were allotted by the Marden award of 1819. The three fields, bearing the becomingly simple names of Upper, Middle, and Lower, would have graced any Oxfordshire township. They lay just to the east of the village, were nearly equal in size, and, except for a strip of old enclosure between two of them and a patch of the same in the third, formed a compact arable area. This plan is one of the few bits of evidence looking toward the long-continued existence of a three-field system in the county.

still remains in this state," i. e., common field. The vagueness and the incidental character of the remark render it unworthy of much attention, especially since the report in general is unsatisfactory.

[1] Yet the rotation in 1794 was a three-course one, according to the reporter to the Board of Agriculture: "In all the common fields and in that district called Wheatland the rotation is (1) fallow, (2) wheat, (3) beans" (Clark, *General View*, etc., p. 18).

Another apparent survival of three fields, only a little less convincing than the one at Sutton, was to be seen at Ullingswick as late as 1856, when 290 acres distributed in 488 parcels were enclosed. The land lay at a little distance from the village, surrounding it on three sides in three rather large compact areas called Wood field, Broomhill field, and Bebbury field. Between each two was a tongue of enclosures, and all three fields show jagged edges where closes had eaten in. Still, on the surface at least this is a recognizable survival of a three-field township.

A parliamentary enclosure which, for Herefordshire, was both large and early occurred at Wellington in 1797. We learn the names and areas of the common fields in which the 611 acres of arable lay, but no plan tells us of their shape or location. Conjecture may none the less be based on the schedule, which runs as follows: —

	Acres		Acres
Orrington field	77	North field	87½
West field	168	Hope field	140¾
Hither Adzor field	57½	Mill furrows	5
Farther Adzor field	26½	Moor Croft	7¼
Thatchley Lands field	41		

The important fields here were West, North, and Hope (simple names) while the other fields, Orrington, Adzor, and Thatchley Lands, could easily have adapted themselves to a tripartite arrangement. Nor does the Tarrington schedule of 1799 forbid a three-field grouping. Its 378 acres of arable lay, except for five small patches, in seven areas. To be sure, only three of the latter, and these not the largest, are called fields; yet, if " Radlow " be accounted a field and two of the so-called fields be combined, a grouping into three equal areas becomes possible.[1]

Only one other enclosure schedule hints at three ancient fields, and that rather vaguely. In the Kingston award of 1812, 197 acres are allotted, of which 116 lay in Brooke field, 64 in Chrise field, and 17 in Kipperley field. The last two areas were adjacent, but were somewhat separated from Brooke field. The three were situated relative to the village much as three fields would have

---

[1] Radlow 113 acres, Lower Field 93, Church Hill 36, Long Croft 17, East Field 55, Mickle Field 25, Willsill 22, five small parcels 17.

been, but Kipperley field was so shrunken that we cannot base any argument upon the Kingston situation. Since the parliamentary plans and schedules of only these five townships suggest the survival of three compact, open, arable fields in Herefordshire, we may now give attention to a marked change which most of such records portray.

It will be remembered that one of the townships in which three fields were seemingly intact in the sixteenth century was Risbury,

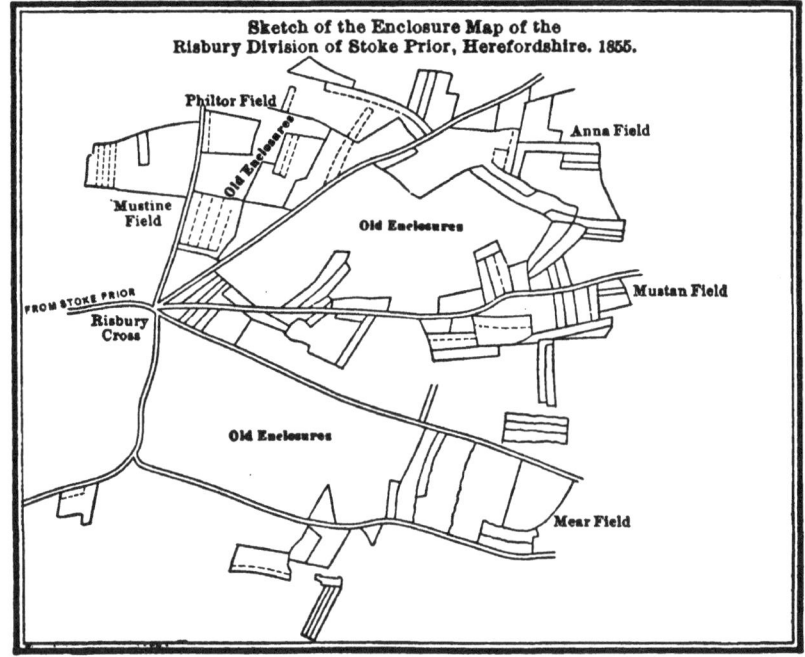

MAP VIII

a member of Stoke Prior. Fortunately, the enclosure award of Stoke Prior is accompanied by two carefully-drawn plans that locate the arable area which is to be enclosed, indicating the open-field strips and giving the names of the fields. These plans are far more representative of the condition of Herefordshire common fields at the end of the eighteenth century than are the accounts of the more compact three-field areas already noticed. One of them relates to fields lying in the parish of Humber, the other to fields in the Risbury division of Stoke Prior. The Humber fields were connected with the two hamlets of Priddleton and Puddlestone. Priddleton field, so called, was an isolated patch containing some ten acres in seven parcels; the Puddlestone parcels were perhaps four times as numerous and lay largely in Sparrow Hill field, though a Fair Mile field is mentioned.[1]

More comprehensive is the plan of the Risbury division. As the preceding sketch shows, the strips there were scattered throughout three rather extensive fields called Mear, Mustine, and Anna, while at one edge they ran into Philtor field.[2] They numbered about one hundred, and their area was about 150 acres. The aspect of the important fields (the three called in the Jacobean survey Meer, Mustine, and Inn field[3]) as they reappear here illustrates what had been happening in the interim of two and a half centuries. From the plan, which looks more like the terrier of a single estate than the representation of township fields, the one thing obvious is that enclosure had been eating into the old commonable areas on all sides. Much enclosing had taken place in the middle of the fields, until of the trio there remained only skeletons to which the old names could be appended. Any three-course tillage must long since have disappeared, and the isolated strips must have become a source of annoyance to their proprietors, who numbered a dozen or more.

So important was this process of piecemeal enclosure in bringing about the decline of Herefordshire open fields that another illustration may be permissible, especially as it also exemplifies an aspect of the field system of the county which first became clear in our examination of Jacobean surveys — the multiplicity of the fields.[4] A plan of the common fields of Holmer, sketched in the accompanying cut, pictures them as lying in two groups, one to the northeast, the other to the southwest, of the village.[5] To the northeastern group were attached nine names, — Hill field, Patch Hill field, Munstone field, the Butts, Stoney furlong,

---

[1] The total area of this group of strips was 58 acres.
[2] The plan is at the Shire Hall, Hereford.
[3] Cf. above, p. 37.
[4] See above, pp. 93–94.
[5] The plan is at the Shire Hall, Hereford.

Churchway field, Hopyard Piece, Ten Acres, and Pinacres, — while the southwestern group comprised West field, Lower West field, Rotherway field, Moor field, Crow Hill, Bobble Stock, and Sickman's field. Quite apart from the odd pieces, such as Bobble Stock, these subdivisions were numerous for an area of 297 acres. Four areas in one group and five in the other are distinctly called fields, the average size of a field being thus not more than 25 or

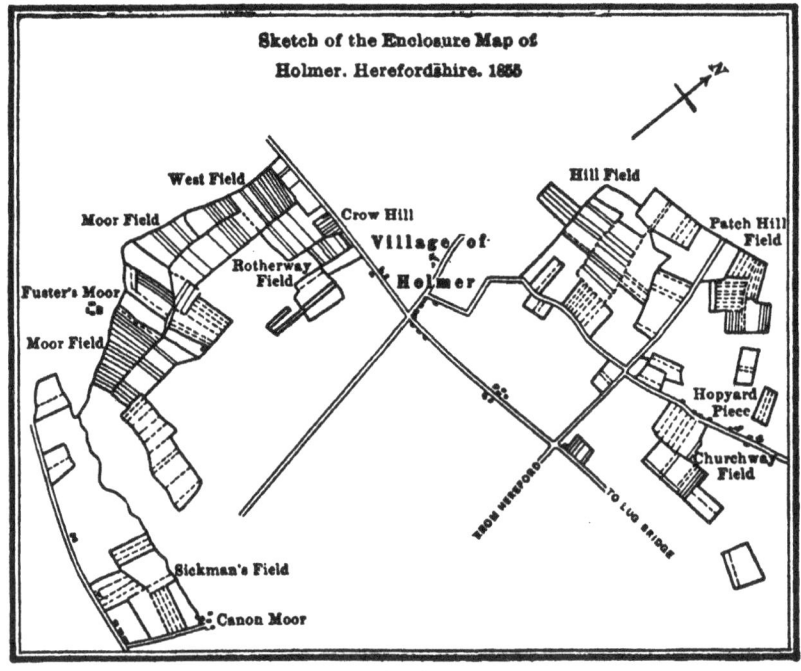

MAP IX

30 acres. Enclosure had, to be sure, wasted them, as at Risbury; but they can never have been very large. Either early fields had been much subdivided, or the system was irregular, as it so often showed itself in Jacobean surveys.

In many of the Herefordshire awards a multiplicity of fields, like that shown in the Holmer plan, is a striking feature. At Marden, a parish of many hamlets, some 1000 acres were enclosed in 1819. These lay in forty-six fields and patches, several of them being small plocks or crofts. Most of the fields contained from three to forty acres each, though in eight instances the

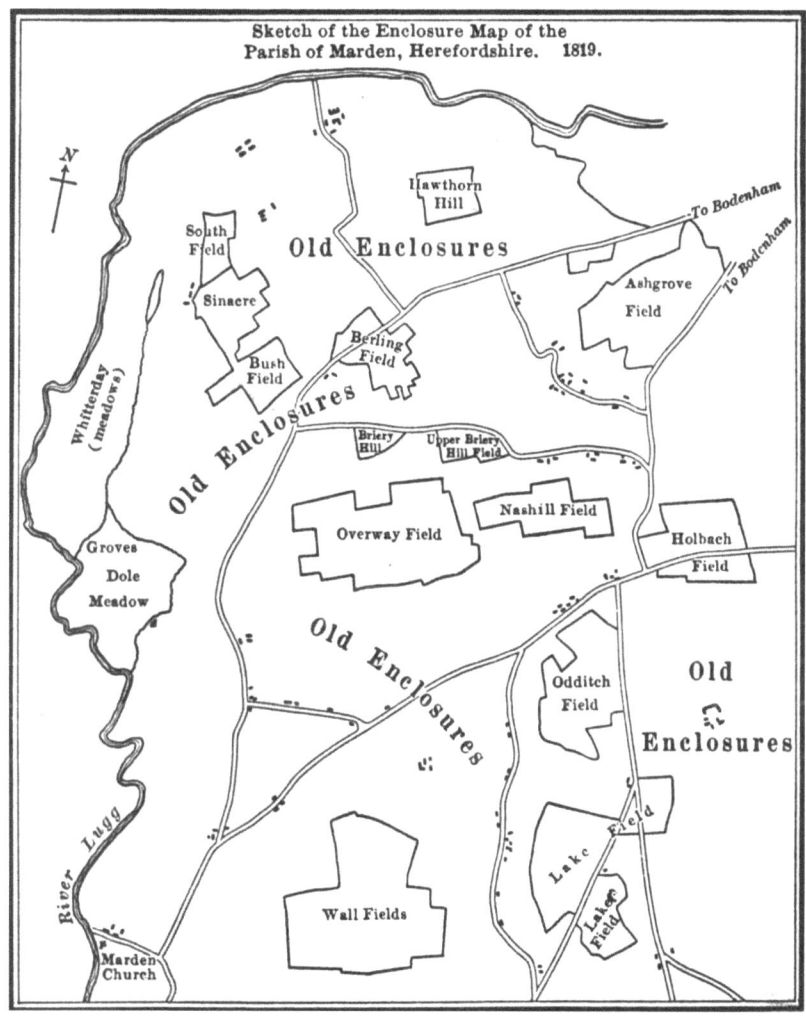

acreage rose above 40 and once reached 125.¹ These fields, but not the strips of which they were composed, are located on a plan which is here sketched, and were, it becomes evident, distributed like islands throughout the entire area of the parish. At an earlier day they probably had some logical grouping relative to the half-dozen hamlets of which the parish was composed, but this is no longer discernible. Indeed, it will be remembered that the Jacobean survey showed tenants acquiring acres from the fields of the various hamlets.² By 1819 enclosure had broken the edges of the fields, separating them more and more from one another and sometimes, as in the case of South field, nearly obliterating them. When one considers that this is the parish which in all Herefordshire retained the largest area of open field at the time of its enclosure, one readily surmises that multiplicity of fields and piecemeal enclosing were signs of decay from which the staunchest townships in the county had seldom been exempt.

To multiply illustrations of numerous fields is easy. The 564 acres enclosed at Much Cowarne lay in some thirteen fields, apart

	Acres		Acres
¹ Little Horn Field	5	Hill Field Crofts	4
South Field Croft	5	Hill Field	35
South Field	10½	Upper Brierly Hill	13
Hawthorn Hill Field	13½	Roads Orchard	3½
The Twenty Acres	13	Kineton Field	10½
Portland Head	11	Little Field	7
Butchers Plock	4	Ninewells Field	3
Upper Vauld Field	25	Venns Green Pasture	2
Apothecary's Croft	5½	Holbach Field	50
Ashgrove Field	88	Nashhill Field	49
Pale Croft	9	Overways Field	102
Meggs Corner	4	Hare Furlong	4
Venn Field	7	Little Field near Paradise	2½
Venn Field Croft	7	Odditch Field	63½
Lower Vauld Field	12½	Holbach Plock	2
Greathorne Pasture	3	Little Wall Field	16
Chestern Field	3	Great Wall Field	47½
Sinacre Field	38	Lower Wall Field	39½
Bush Field	32	Gott Wall Field	33
Lingens Hook	1	Mill Croft Field	5
Burling Field	52	School Wall Field	11
Lower Brierly Hill Field	10	Newfoundland Field	7
Carry Lane Croft	5½	Lake Field	125

² Cf. above, pp. 95–96.

from many smaller areas.¹ At Madley the field names applied to 381 acres numbered thirty, while at Much Marcle as many as forty were required to locate 622 acres. The Yarkhill award, which, to be sure, makes allotments in other townships as well as in Yarkhill, avails itself of thirty-eight such names. Instances like these show how typical are the descriptions of Holmer and Marden. They make it clear that throughout the county the open fields of the era of parliamentary enclosure were for the most part small, numerous, more or less isolated, and considerably eaten into by piecemeal enclosures.

The relatively small amount of arable enclosed in Herefordshire by act of parliament necessitates one of two explanations regarding the earlier history of open common fields there: either they were never extensive, or the majority of them disappeared without special act. In choosing between these alternatives one should remember that parliamentary activity in the county began late. The first extant award relative to open arable field dates from 1797, and none of the ten missing awards were earlier. By that year parliamentary enclosure in Oxfordshire had run half its course. In view of the fragmentary condition of the Herefordshire fields when they first appear in the plans and schedules, it is scarcely credible that no enclosing had been going on throughout the preceding fifty years. Since appeal to parliament seems not to have become the vogue until 1797, the natural explanation is that would-be enclosers were getting on very well without it. The simple method of enclosure by agreement, one may surmise, was known and practiced.

For such a conjecture we have further justification. In 1779 parliamentary sanction was sought for the abolition of common rights over 379 acres of enclosed lands at Winforton.² The ownership of these closes resided in the lord of the manor, but thirteen other persons were seized of the rights in question. The meadows were " commonable at midsummer yearly," certain pastures at Lammas day, and several arable fields " when rid or

---

¹ Great field, Wheatland field, Elms field, Quarry field, Walnut Tree field, Claypit field, Twenty Acres, Birley field, Batch field, Henacre, Stream field, Psalters field, Perry field.

² The award is at the Shire Hall, Hereford.

cleared of their respective crops of grain and hay." Since these rights were estimated by the award as equivalent to one-fourth of the yearly value of the 379 acres, the claimants were allotted about one-fourth of this area. Obviously we are here dealing with common arable and meadow lands which had been enclosed at some time before 1779 without the extinction of common rights. Agreement there doubtless had been between the lord and the tenants, but no separation of interests.

One other parliamentary act relative to a Herefordshire township, said to be the earliest of its kind,[1] provides for the enclosure of lands at Marden in 1608.[2] The reason assigned to justify its passage is that there may be " better provision of meadow and pasture, for necessary maintenance of husbandry and tillage "— the same reason which many sixteenth-century surveys from Herefordshire, Gloucestershire, and Somerset might have advanced to explain the considerable departures from normal open-field tillage which they manifest. This act recognizes and legalizes what was apparently a usual procedure in this region.

These two parliamentary sanctions given to enclosure by private agreement in Herefordshire, standing as they do a century apart, are significant, since they suggest that the process of which many traces can be seen in the later plans had been of long duration. The process was one of piecemeal enclosure by private agreement, and it remains to inquire whether much open-field arable disappeared before it. On this point some light is thrown by the condition, in the time of Henry VIII, of certain hamlets formerly in the possession of Wigmore monastery. For each of them there is a brief survey telling the number of tenants and the areas of their holdings. Nearly always the holdings were largely in open field, the situation, briefly stated, being as follows:[3] —

[1] Leonard, "Inclosure of Common Fields," p. 108. Miss Leonard states that the act enabled the commoners to enclose a third of their lands.

[2] "An Act for the better Provision of Meadow and Pasture, for necessary Maintenance of Husbandry and Tillage, in the Manors, Lordships, and Parishes of Marden, alias Mawarden, Bodenham, Wellington, Sutton St. Michael, Sutton St. Nicholas, Murton upon Lugg, and the Parish of Pipe, and every of them, in the county of Hereford" (*Journal of the House of Lords*, 12 May, 5 Jas. I).

[3] Land Rev., M. B. 183, ff. 2–24. The account of the demesnes, which were seldom large, is not transcribed.

LATER HISTORY OF THE MIDLAND SYSTEM 151

Township	Messuages	Enclosed Acres	Arable Acres in the Common Fields	Common Meadow
Marlow.....	4	7½	60, 60, 60, 27	
Whitton.....	6	8¾	60, 46, 21, 40, 60, 38	30½ acres
Ratlinghope .	8	13½	30, 10, 22, 14, 24, 22, 10, 36	45 " dayesmath "
Letton......	{ 6 { 1	14½ 14½	30, 53, 59, 21, 25, 30	{ 8 acres { 2½ "
Adforton....	{ 9 { 1	16 17½	18, 24, 7½, 18, 40, 12, 6, 35, 16 6	{ 28 acres { 5¾ "
Yatton......	3	6½	26, 40, 30	4 " dayesmath "
Lye	4	38½	60, 30 30, 2½	

In all these townships the open-field arable was far more extensive than the enclosures, but, considerable as it was in Tudor days, relatively little of it remained to be enclosed by act of parliament. Only three of the townships appear in the awards.[1] Of these, Marlow and Whitton are in the parish of Leintwardine, the award for which was drawn up in 1803. In it reference is made to Marlow, where a large common of 392 acres is allotted, but only five acres of common field (in Little Marlow field). To the township of Leintwardine itself is assigned common field amounting to 197 acres. Since the hamlet of Whitton is not more than a mile distant from the village of Leintwardine, a part of the area enclosed probably came from the fields of Whitton; yet such part can have been no very large fraction of the 265 acres which were open arable field there in Tudor days. The third township of the list which appears in the awards is Lye, where in 1817 an area of 274 acres was enclosed. How much of this was arable is not made clear in the award, nor can it be determined what ratio the four monastic holdings, specified above, bore to the entire township; but, even if it be assumed that they remained unenclosed, the total open-field arable and meadow affected by act of parliament in the seven townships of the foregoing list did not exceed 300 acres. Since in Tudor days they had contained upwards of 1200 acres, the area enclosed without parliamentary intervention was about 75 per cent of the total.

This fraction may not, of course, be applicable to the county as a whole. On the other hand, there is no reason for assuming

[1] Ratlinghope is in Shropshire, but there is no record of its enclosure in the acts or in the enclosure awards at Shrewsbury.

that the hilly, forested, northwestern corner in which were situated the townships above noted had relatively more open field in Tudor times than had the amiable plains round Hereford. Since the Leintwardine and Lye awards are in their areas entirely representative of parliamentary enclosures in Herefordshire, the fraction which seems to reflect conditions in the northwest may not after all be inapplicable to the entire county. At least it becomes probable that a very considerable amount of arable open field, once existent, disappeared without leaving record of itself in parliamentary act or award; and one can scarcely avoid the inference that private agreement and piecemeal enclosure were operative in this process.

At what period between the days of Henry VIII and those of George III the decay of the old fields was most rapid is not easily ascertainable. It cannot have been before the beginning of the seventeenth century, since the Jacobean surveys show no marked encroachments upon the arable. Surveys later than these are not to be had, though glebe terriers might throw light upon the subject, as they have upon similar matters in Oxfordshire.[1] Until information from them or from some other source is forthcoming, the decades during which the old field system fell most rapidly into decrepitude must remain in doubt.

Why piecemeal enclosure was so much more prevalent in Herefordshire than in Oxfordshire can only be conjectured. In general during the sixteenth century the western counties appear to have been much more inclined to pasture farming than were the midlands. To judge from the respective values assigned to arable, meadow, and pasture in the contemporary surveys, this preference implies progress. An acre of pasture was usually worth at least half as much again as an acre of arable, and an acre of meadow was easily worth twice as much.[2] Conversion of the arable, therefore, meant an increase in values and income. The fact that

---

[1] I have not been able to examine the glebe terriers for Herefordshire, and do not know to what extent they are available.

[2] According to a survey of 1 Edward VI, the open-field arable at Horton, Gloucestershire, was worth from 6 *d*. to 12 *d*. the acre, the enclosed pasture from 1 *s*. 6 *d*. to 3 *s*., and the meadow from 3 *s*. to 5 *s*. (Rents. and Survs., Portf. 2/46, ff. 92–104).

this advantage, patent in all the surveys, was realized only in the west suggests that conversion may for some reason have been easier there than in the midlands.

Thus we are brought back to the conditions described in the preceding chapter. There were disclosed, especially in the forest areas and the river valleys of western England, deviations from the two- and three-field system. Most noticeable of these were such irregularities in field arrangements as made it uncertain whether either a two-course or a three-course rotation of crops was still practiced. If neither was in force, there can have been little reason for maintaining the integrity of the arable fields — unless, indeed, a four-course system was adopted, as happened on the lower Avon. In general in the river valleys, including those of Herefordshire, and near the moors of Somerset, irregular fields, themselves often indicative of progress, must easily have yielded to enclosure. At Frampton Cotterell in Gloucestershire they had done so completely before the days of James I.

One other feature of Herefordshire fields must have been favorable to innovations. This was their relatively small size. As has been noticed, a Herefordshire parish usually consisted of several hamlets,[1] each with its group of fields in which seldom so many as ten tenants had holdings of any size. Frequently the tenants numbered less than a half-dozen. Obviously the situation in a township of this nature was very different from that existing in a township of Oxfordshire, where there were nearly always more than ten tenants and sometimes as many as thirty. From so large a group consent for enclosure could be got only with difficulty, whereas by the half-dozen Herefordshire tenants it might readily be conceded. If this conjecture be justifiable, the form of settlement which prevailed in the western counties had its influence upon the open-field history of the region.

[1] The parish of Marden is a good illustration. Reference to the modern map shows six constituent hamlets or townships, — Marden, Wisteston, Vern, Venn, Vauld, and Fromanton. The name of the last hamlet is supplied from the Jacobean survey, a document which tells us that the *manor* of Marden comprised also the township of Sutton with its hamlet Freen (cf. above, p 95, n. 4).

At this point it will be necessary to conclude our study of the two- and three-field system. This method of tillage has been followed from Anglo-Saxon days to the latter part of the nineteenth century. The area throughout which it prevailed has been defined as the northern and southern midlands — the territory from Durham to the Channel and from the Welsh marches to the fens. In its primitive Anglo-Saxon form the system seems to have been one of two fields. As soon, to be sure, as we get full evidence from the beginning of the thirteenth century, three-field townships are apparent. The discovery, however, that two-field arrangements sometimes gave place to three-field ones has encouraged the belief that such transformation was perhaps responsible for the existence of the three-field system. The period to be credited with this first step in agricultural advance is the thirteenth century and the early fourteenth. From that time on, all the more fertile townships of the midlands, especially of the northern and western portions, were in three fields.

So they remained, it seems, for about two centuries. When the curtain next rises upon midland fields as they appear in the surveys of the sixteenth and early seventeenth centuries, another transformation has begun. Although most townships still remain in two or three fields, more complex arrangements appear here and there, especially in forest areas and in river valleys. Sometimes strips of meadow have substituted themselves for strips of arable in the otherwise normal fields. Sometimes the division of tenants' acres among fields is incomprehensible, even though the fields are few. Sometimes the fields have become numerous and admit of no grouping which adjusts them to the traditional system. Sometimes much piecemeal enclosure has taken place and the open-field arable is visibly in a state of decay. Very often, finally, a new regular system, one of four fields, has replaced the two-field arrangement, and so has brought into annual tillage an additional quarter of the township's arable.

These changes, it is obvious, were evidences at once of the decay of the old system and of an advance in agricultural technique. To study them at closer range we have given attention to the enclosure history of Oxfordshire and Herefordshire. The latter county has shown itself particularly notable for the extent to which piecemeal enclosure went quietly on within its borders, a procedure which seems to have been facilitated by its numerous and small fields. In Oxfordshire it was different. More than one-half of the county, to be sure, became enclosed before 1750, but the causes of this seem to have been the fertility and residential desirability of certain townships. In other townships a greater or less amount of open arable field survived, the total constituting more than one-third of the county's area. This surviving open-field arable had in part undergone certain changes, particularly the substitution of four fields for two; and the extent of such transformation in Oxfordshire, indeed, suggests that it constituted the most important step in systematic agricultural advance made by the midland system since the fourteenth century. In conjunction with certain refinements upon itself, it was the last endeavor of open-field husbandry to till the soil in the most remunerative manner possible. In this attempt it failed, being unable to equal the advantages offered by enclosure and convertible husbandry.

Thereupon set in an epoch of parliamentary enclosure which, continuing from the middle of the eighteenth century for rather more than a hundred years, left England a country substantially devoid of open arable fields. The progress of this late enclosure in Oxfordshire and Herefordshire has been followed in order to make clear what material is available for an extended study of the subject, and to emphasize the distinction between those counties in which the two- and three-field system was firmly entrenched and those in which it yielded easily to formal or informal enclosure. The first group comprised the counties of the eastern, central, and southern midlands; the second included counties or parts of counties lying to the north and west in a belt of territory which stretched from Durham to Somerset. In the latter group piecemeal enclosure went on more rapidly than it did in the former, a circumstance that constitutes the most striking differentiation within the entire two- and three-field area. Next to it in suggestiveness is perhaps the readiness manifested by four-field and eight-field townships to avail themselves of the opportunity to enclose their open fields by act of parliament. Behind all the differences, however, which in course of time manifested themselves within the two- and three-field area was the unity of origin and character that marked off the midlands from the other parts of England which are now to be considered.

CHAPTER V

The Celtic System

BEFORE examining the field arrangements of the north and the west of England, we shall do well to glance across the border to see what method of cultivation was employed by peoples of Celtic descent. Phenomena otherwise perplexing may thereby become intelligible.

Of the three Celtic divisions of the British Isles, Scotland furnishes perhaps the most specific information as to how the soil was tilled in the eighteenth century. Among the Scottish reporters to the Board of Agriculture in 1794 were two or three men whose habits of thought led them to go beyond the formal answer to the queries propounded and write scholarly accounts of the situation, past and present. If to their descriptions be added the briefer notes of the other reporters, the composite picture leaves little that is vague about the later history of the Celtic system in Scotland. In particular it makes clear the nature of runrig, the relation of which to the three-field system of England has never been well set forth.[1]

A striking feature of Scottish agriculture before 1794, and one upon which the reports are practically unanimous, is that most of the arable, as well as the meadow and pasture, lay unenclosed. Near gentlemen's seats only were enclosures to be seen. While the reporters wrote, the process of enclosing was making headway, especially in the southeast; and often matters had got to a point where a ring fence had been built about the farm, although no

---

[1] Slater has a chapter on the subject, and quotes at length Alexander Carmichael's description of the Hebrides (*English Peasantry*, ch. xv). He has not utilized the best information contained in the reports to the Board of Agriculture, nor is his contrast of runrig with English common fields adequate (cf. below, pp. 171-172). Seebohm had apparently not read the reports.

subdivisions had been made. A half-century earlier nearly all of Scotland, they say, lay open.[1]

As a lucid description of the tillage of an open-field Scottish farm, James Anderson's account, written with reference to Aberdeenshire, cannot be excelled:[2] —

"Throughout the whole district," he writes, "the general practice that has prevailed for time immemorial is to divide the arable lands of each farm into two parts at least, Infield and Outfield. The in-field, as the name implies, is that portion of ground which is nearest to the farmstead; and usually consists of about one-fifth part of the whole arable ground in the farm. This is kept in perpetual tillage; and the invariable system of management was, and still is, with few exceptions, to have it divided into three equal parts to be cropped thus: First,

---

[1] Cf. the following reports, each entitled *General View of the Agriculture of the County* [in question]: —

Aberdeen, p. 40, "But if by commons be understood uninclosed fields [i. e., not heath or waste] . . . then the greatest part of the county might be accounted such"; p. 59, "The old corn lands near Aberdeen . . . [are] for the most part open and uninclosed."

Southern Perth, p. 60: "Three-fifths at least of the whole arable land is open . . . and on some farms no fence is made except a ring fence around the whole."

Argyll and West Inverness, p. 26: "There is but little of it [the country] inclosed, and that which is only by feal dykes; . . . the tenants, from want of sufficient inclosures, cannot protect turnip and sown grass and thereby have been discouraged . . . to raise these articles."

Annandale (co. Dumfries), app. iv, p. xxiii: "There was scarce an inclosed field thirty years ago in Annandale, unless on the mains or manour place of a gentleman, and they were not at all frequent. There was no such thing at a much later period as a divided or inclosed farm, with any sort of fence, occupied by a farmer."

Dumbarton, p. 19: "Till about thirty or forty years ago, none of the country was inclosed, except a few fields adjoining to gentlemen's seats . . . [but] inclosing has been daily going on. One-third of the county, however, is yet open, or but roundly inclosed; that is, the farms are inclosed, but not subdivided."

Berwick, p. 45: "Almost the whole or two-thirds, at least, of the lands of the lower district, are now inclosed, and a considerable part of the arable lands of the higher district."

Orkney Isles, p. 252: "The land is almost wholly in open fields."

Midlothian, p. 34: "Even so late as thirty years ago, there was hardly a farm inclosed in the whole county."

[2] *General View of the Agriculture of the County of Aberdeen* (Edinburgh, 1794), pp. 54 *sq*.

Bear [barley], with all the dung made by the beasts housed on the farm laid upon it. Second and third, oats: then bear again; and so on in the same unvarying rotation. [For bear, the earth was turned over upon the stubble in the winter, a process called "ribbing." At the end of April, after harrowing, the dung was spread, the soil lightly ploughed, and the crop sown.] For oats the ground is ploughed as soon after the grain is cut down as possible; often some parts of the ridges are ploughed the day the corn is cut down. . . . It is impossible to form an idea of the foulness of the crop. . . . It is by no means uncommon to see one-half the ridge (usually that side which lies to the east or north) cut up for green food that year it is in bear, no grain being to be seen among it. . . .

"That part of the farm called out-field is divided into two unequal portions. The smallest, usually about one-third part, is called *folds*, provincially *falds;* the other larger portion is denominated *faughs*. The fold ground usually consists of ten divisions, one of which each year is brought into tillage from grass. With this intent it is surrounded with a wall of sod the last year it is to remain in grass, which forms a temporary inclosure that is employed as a penn for confining the cattle during the night time and for two or three hours each day at noon. It thus gets a tolerably full dunging, after which it is plowed up for oats during the winter. In the same manner it is plowed successively for *oats* for four or five years, or as long as it will carry any crop worth reaping. It is then abandoned for five or six years, during which time it gets by degrees a sward of poor grass, when it is again subjected to the same rotation.

"The *faughs* never receive manure of any sort; and they are cropped in exactly the same manner as the folds, with this difference, that instead of being folded upon, they are broke up from grass by what they call a rib-plowing about midsummer; one part of the sward being turned by the plow upon the surface of an equal portion that is not raised, so as to be covered by the furrow. This operation on grass land is called faughing, from whence the division of the farm takes its name. It is allowed to lie in this state until autumn, when it is plowed all over . . .

and is sown with oats in the spring. It produces a poor crop and three or four succeeding crops still poorer and poorer; till at last they are forced to abandon it by the plough after it will scarcely return the seed. It is deplorable to think that . . . such a barbarous system . . . should have been, from local circumstances, continued for several centuries."

From every part of Scotland come similar accounts of the division between infield and outfield. The variations in detail are slight, having reference largely to the rotation of crops and to the proportions existing between the various sorts of land. No other report makes a distinction between folds and faughs, the entire outfield being usually described as Anderson describes the folds. In East Lothian the outfield was divided into five, six, or seven *brakes* (instead of ten folds and ten faughs), the number depending upon the quality of the soil.[1] In Ayrshire " no dung was ever spread upon any part of it. The starved cattle kept on the farm were suffered to poach the fields from the end of Harvest till the ensuing seedtime."[2] Contrasted with the outfield was the infield, which in Dumbarton comprised about one-fourth of the farm.[3] Sometimes the rotation of crops upon the infield extended over four years instead of three. In Ayrshire a year of ley intervened between the crop of barley and the two crops of oats.[4] In the Carse of Gowrie and in East Lothian one-fourth of the infield " was dunged for pease [and] . . . the second crop was wheat, the third barley, the fourth oats." In southern Perthshire, along with the usual rotation, a crop of peas or beans might be introduced between the oats and the barley, or barley and oats might alternate in two-course rotation.[5] The reporter for Annandale explains what part of a farm the system brought under annual cultivation. The quantity of infield land, he says, " was proportioned to the number of cattle wintered and housed on the farm. An acre of land might be dunged for each five or six cattle. . . . A farm that could fold five acres of Outfield land [from which three crops of oats were then taken], and

---

[1] *East Lothian*, p. 48.
[2] *Ayr*, p. 9.
[3] *Dumbarton*, p. 44.
[4] *Ayr*, p. 9.
[5] *Southern Perth*, p.22. The introduction of the peas or beans was deemed an improvement.

could manure as many of Infield [from which one crop of barley and two crops of oats were then taken], had in all [each year] twenty-five acres of oats and five acres of bear." [1]

In the Highlands the poorer soil introduced slight modifications. William Marshall's description is in substance as follows.[2] The valleys were separated from the hills by a stone fence called the " head dyke," or by an imaginary line or partition answering to it and running along the brae or slope. Within the head dyke lay the more productive or greener surface, the black heathy brows of the hills being left out as " muir." The muir was an addition to the farm peculiar to the Highlands, since the portion within the head dyke comprised what was elsewhere called infield and outfield. The description of these divisions runs much as usual. Some patches, however, which were " too wet, too woody, or too stoney to be plowed, are," Marshall notes, " termed meadow and are kept perpetually under the scythe and sickle for a scanty supply of hay, being every year shorn to the quick and seldom, if ever, manured." Other patches constituted permanent pasture. " The faces of the braes, the roots of the hills, the woody or rough stoney wastes of the bottom; with a small plot near the house, termed ' door land ' (for baiting horses upon at meal times, teddering a cow, etc.) are kept as pasture for cattle in summer and sheep in winter, the sheep and generally the horses being kept during summer above the head dyke upon the muir lands." In estimating the average amount of each kind of land on a farm on the sides of Loch Tay, Marshall brings to light the principal difference between the Highland farms and the more level ones, whether of the north or of the south. The farms of Loch Tay, he states, " contain on a par about twenty acres of infield, fifteen acres of outfield [both tilled as elsewhere], ten acres of meadow, thirty-five acres of green pasture, with about ten acres of woody waste — in all, about ninety acres within the head dyke, and about two hundred and fifty acres of muir or hill lands." The infield and outfield which were more or less available for tillage thus constituted only a small fraction of the total

---

[1] *Annandale* (co. Dumfries), app. iv, p. xxii.
[2] *Central Highlands of Scotland*, pp. 29 sq.

area of the farm, instead of all of it, as elsewhere. Apart from the extensive but not very valuable stretches of permanent muir, pasture, and meadow, a Highland farm was like any other in Scotland.

Up to this point in the description Scottish agriculture shows slight resemblance to the two- and three-field system of the English midlands. The arable fields were, to be sure, open, and the best of them, the infield, was subject to a three-course rotation; but the three courses involved continuous cropping and knew nothing of the fallow year. With the outfield, the larger part of a Scottish farm, there was nothing in a midland township to correspond, and its alternation of five years of tillage with five years of recovery was far removed from midland methods. We come now, however, to a characteristic of Scottish agriculture which seems to ally it with the common fields of England. This feature is *runrig*, or *rundale*, the subdivision of a holding into strips or ridges intermixed with those of other holdings.

The existence of ridges has already come to light in Anderson's account, where he refers to the unproductiveness of the northern halves of the ridges of infield during the year in barley. Ridges may, of course, comport with almost any field system in which there is no cross-ploughing. They are a device for drainage, and were commended by the reporters when they were straight, not too high, and so arranged as to drain the furrows properly. In Scotland, as it happened, they had got out of hand, and, according to the reporter for East Lothian, the following shape of the ridge was universal: " Anciently almost every ridge in this country was from 18 to 22 feet broad and sometimes more; they had curves at each end, somewhat in the form of the letter S; and these ridges were always twice, and upon strong lands generally three times, gathered from the level of the ground." [1] This report is confirmed and explained by another from Midlothian: " It was formerly the universal practice to form the land into high and broad ridges, commonly from 36 to 48 feet wide and elevated at least three feet higher in the middle than in the furrows; but this mode, which perhaps was consistent enough with

---

[1] *East Lothian*, p. 51.

the heavy, cumberous six-horse ploughs then employed, is now disused since the introduction of the two-horse plough, which has of late been general in this county." [1]

The long ridges were called *riggs* or *dales*, the short ones *butts*. The riggs contained from one-fourth to one-half of an acre each, the butts less. As in the midland system, headlands were to be found, and the acres were gathered into " shots." All these features were apparent in 1599, as the following enumeration of six acres, part of a husbandland at Eymouth, Berwickshire, shows: —

" One acre containing three rigs lying in that shot called the Schuilbraidis, sometimes occupied by Patrick Huldie, maltman

other three acres, sometime occupied by John Johnstone, merchant, of which one is in Over Bairfute, called the Heidland acre, half an acre containing three butts adjacent in the Over Welsteil, half an acre containing two rigs and a rigend in the Blackcroft, and the other acre containing two daills in the Hilawbank

another acre containing two daills and a rig lying on the west side of the said Hilawbank, sometime occupied by Robert Gotthra . . .

and the other acre, containing three rigs of land, lying in Nather Bairfute." [2]

The transition from ridges to runrig is made for us in Sir John Sinclair's disdainful account of Caithness. " In order to prevent any of the soil being carried to the adjoining ridge," he writes, " each individual makes his own ridge as high as possible, and renders the furrow quite bare, so that it produces no crop, while the accumulated soil in the middle of the ridge is never stirred deeper than the plough." Here at length is intermixed ownership or occupation; and Sir John leaves the matter in no doubt. " The greater part of the arable land in this County," he continues, " is occupied by small farmers, who possess it in *run-rig* or in *rig and rennal*, as it is here termed, similar to the common fields of England, a system peculiarly hostile to improvement.

[1] *Midlothian*, p. 55.
[2] Hist. MSS. Commission, *MSS. of Col. D. M. Home* (1902), p. 214.

Were there twenty tenants and as many fields, each tenant would think himself unjustly treated, unless he had a proportionate share in each."¹ Of the Orkneys, too, he writes, " Much land that formerly lay in the state known in Scotland under the name of run-rig land has been divided, but much still remains in the same situation . . . a source of constant dispute."² At the other end of Scotland, in Berwickshire, runrig was at least a memory. The reporter notes that " the common fields, runrig, and rundale lands in the county were all divided previous to any attempt to improve them by inclosing."³

Certain passing remarks of other reporters indicate more exactly the nature of the intermixed property, and at the same time point to its prevalence throughout Scotland. Most illuminating of all is the report from southern Perthshire by James Robertson, D.D. " The husbandry of the particular district under consideration," says he, " was in a most wretched condition, even so late as fifty years ago. The land was always occupied in run-rigg by the different tenants on the same farm and sometimes by coterminous heritors. The houses were in clusters for the mutual protection of the inhabitants, and the farms were universally divided into out-field and in-field except in the neighborhood of the larger towns."⁴ The intermixed strips of the several tenants, we now perceive, were those of a single farm, and the method of tillage called runrig had the farm as its unit. Robertson's further comment makes the matter clearer. Discussing production and population, he uses this illustration: " No man will venture to say, that a farm of fifty acres in the hands of four tenants, who have each a horse in the plough, and their ground mixed in run-rig, will produce the quantity of subsistence, which the same farm can do in the hands of one man, who has both money and industry to cultivate the ground. With respect to

[1] *Northern Counties and Islands*, p. 207.
[2] Ibid., 227.
[3] *Berwick*, p. 50.
[4] *Southern Perth* (1794), p. 22. In the second edition (*General View of the Agriculture in the County of Perth*, 1799), the author adds that there were clusters of farms " even to the number in some cases of six or eight ploughs of land in one hamlet."

population, where is the difference, whether the other three farmers live on the farm or in an adjoining village ? "[1] Elsewhere, writing of runrig as an obstacle to improvement, he continues: " But in our times nothing can be more absurd, than to see two or three, or perhaps four men, yoking their horses together in one plough and having their ridges alternately in the same field, with a bank of unploughed land between them by way of boundary. These diminutive possessions were carried to such a length, that in some parts of Scotland, beyond this county, the term a *horse's foot* of land is not wholly laid aside.[2] The land is like a piece of striped cloth with banks full of weeds and ridges of corn in constant succession from one end of a field to the other. Under such management, all these people must have concurred in one opinion with regard to the time and manner of ploughing every field, the kind of grain to be sown, and the season and weather fit for sowing, and whether they and their horses were to be employed or idle. Even so late as thirty or forty years ago, this practice prevailed, not only over the greater part of the county of Perth, but with very few exceptions over all Scotland. Since that period it has been gradually going into desuetude . . . and must soon disappear, except where the landlord is as much of a Goth as his tenants."[3]

In verification of the important fact that runrig applied to the arable strips of the tenants of a single farm, who were seldom more than six in number, we have the explicit statement of two other reporters. Fullarton writes of Ayr: " The arable farms were generally small, because the tenants had not stock for larger occupations. A plough-gate of land, or as much as could employ four horses, allowing half of it to be ploughed, was a common sized farm. It was often runridge or mixed property; and two or three farmers usually lived in the same place, and had their different distributions of the farm in various proportions, from 10 to 40, 60, or 100 acres."[4] Again, from Annandale, in the west, comes the comment: " It may have been from the same ideas of

[1] *Southern Perth*, (1794), p. 65.
[2] According to the author's note, this was " the sixteenth part of a plough-gate."
[3] Ibid. (1799), 392. [4] *Ayr*, p. 9.

common danger, and to call attention to the general safety, that so much of the corn lands lay in run-rigg or in run-dale property; and that almost every farm was run-dale in the corn-lands, and common in the pastures among four, six, eight or sometimes more tenants."[1] Lastly, the reporter from Dumbarton notes, "In some places the old system . . . is yet retained, [and] a mixed farm of little more than a hundred acres is subdivided, *stuck-runways*, among five or six tenants."[2]

Sometimes, however, the tenants of a farm might come to number distinctly more than six or eight. Not, to be sure, the normal contributors to the plough, as the rhetorical phrase of Sir John Sinclair might suggest;[3] but the increase was due rather to the addition of crofters, or cottagers, so well described by Marshall in his account of the agriculture of the Highlands. "This extraordinary class of cultivators appear to have been quartered upon the tenantry after the farms were split down into their smallest size; the crofters being a species of sub-tenants on the farms to which they are respectively attached. Besides one or two ' cows holdings ' and the pasturage of three or four sheep, they have a few acres of infield land (but no outfield or muir), which the tenant is obliged to cultivate; and they in return perform to him certain services, as the work of harvest and the casting of peats, the tenant fetching home the crofter's share. And still below these rank the Cotters, answering nearly to the cottagers of the southern provinces; except that, in the Highlands, they are attached, like the crofters, to the tenants or joint-tenants, on whose farm they reside; receiving assistance and returning for it services."[4] Robertson tells of similar holdings of cottagers in southern Perthshire: "Without taking notice of small possessions, which are called *pendicles*, because they are small portions of the land allotted by the farmer to cottagers, labourers and servants, which in some places is still the practice; the extent of what may be called farms, where one or more ploughs are yoked, is from 30 to 400 acres."[5] Elsewhere he says, "Many

---

[1] *Annandale* (co. Dumfries), app. iv, p. xxii.
[2] *Dumbarton*, p. 15.
[3] See above, p. 164.
[4] *Central Highlands*, p. 32.
[5] *Southern Perth*, p. 57.

instances might be pointed out where all the tenants of several ploughs and a number of cottagers are huddled together in one hamlet."[1]

The phrases "tenants of *several ploughs*" and "where one or *more ploughs* are yoked" introduce a new complication. Thus far we have been told of the farm of one plough, whose tenants, besides crofters, were usually three or four, but might be six or eight. Their settlement, which was clearly the typical Scottish farm, was correspondingly small. If, however, the ploughs of a settlement sometimes increased, so too must the population have increased, the tenants to a plough remaining constant. For this larger aggregate of lands and tenants a special term was sometimes reserved. It was called, par excellence, a township. Although Marshall speaks without differentiation of "the nominal farms or petty townships,"[2] Robertson makes the distinction. In outlining the obstacles to improvement he begins with "townships," and under this rubric proceeds: "A number of ploughgates ['farms' in the first edition] in one village or several tenants about one plough, having their land mixed with one another is a great bar to the improvement of any country. [Although they have disappeared where cultivation has made progress] in some districts they still remain and the blame is to be attributed to the landlord. Wherever a stranger sees four or six or eight ploughs of land, possessed perhaps by double that number of tenants and perhaps a cottage or two annexed to each plough, all huddled together in one village, he instantly judges that the proprietor is destitute of understanding. . . . However necessary these hamlets were for the mutual aid of the inhabitants in rude ages and unsettled times . . . in the happy days in which we live such clusters of houses are no longer necessary."[3] Immediately after this the author notes as the second obstacle to improvement the existence of runrig. "This," he says, "is a species of the former evil *upon a smaller scale*," and he continues with the description, already quoted, of the two or three or four men who yoke their horses in one plough team.[4]

[1] *Southern Perth*, p. 117.
[2] *Central Highlands*, p. 32.
[3] *Perth* (1799), p. 392.
[4] See above, p. 165.

There were, then, settlements larger than the farm of one plough, settlements consisting of six or eight ploughs and of twenty or thirty tenants and cottagers. Strictly speaking, these were the townships, although the term was doubtless applied to the farm. Indeed, there can have been no sharp line of demarcation between farm and township. It may have been simple enough to call a settlement of one plough-gate a farm and one of six plough-gates a township; yet which term was to be applied to a group of tenants who maintained three or four ploughs? Sharp distinctions must have faded away, till the terms farm and township tended to become confused.

One thing, however, seems clear enough: Scottish units of settlement inclined to be small. Usually they comprised not more than a half-dozen tenants tilling together less than 100 acres of land. Such in all strictness was the farm. If the number of ploughs multiplied and the tenants, apart from crofters, increased to a dozen, the arable might expand to 300 or 400 acres. In general, however, we shall not be wrong in calling the group of tenants' houses a hamlet and the unit of settlement a hamlet-farm.

All this is in contrast with the method of settlement usual in the English midlands. There the township often contained a thousand acres or more and the tenants numbered from twenty to one hundred.[1] The ratio of one to four may not very inaccurately represent the relation between Scottish and English units of settlement in point of size. In other words, the fields of the smaller Scottish hamlet-farms were perhaps only about one-fourth as large as the fields of the smaller English townships, and the same was true of the fields of larger townships in both countries. It happened, of course, that the largest Scottish township-fields were as considerable as the smallest ones of the English midlands. Furthermore, the ratio did not hold good for all parts of England where the midland system prevailed. In Herefordshire, for example, the townships frequently had no greater area than those of Scotland, and yet a three-field system was employed there. None the less, the contrast is for the most part valid and

[1] Cf. the areas of the townships of Oxfordshire given in Appendix IV.

is of importance. Hamlets and small fields were peculiar to Scotland, villages and large fields to the English midlands.

A single feature remains to be added to the picture of a Scottish hamlet-farm, one which appears in certain changes made by James Robertson in the second edition of his report on Perthshire. After repeating that fifty years ago all farms were occupied in runrig, and after pointing to the inconvenience of the intermixed ridges, he continues: " And to add to the evil, one farmer possessed this year what his neighbor did possess the former. Not only farms but in some instances estates were divided in this manner, especially where a property fell into the hands of co-heirs. The first deviation from run-rig was by dividing the farm into Kavels or Kenches, by which every field of the same quality was split down into as many lots as there were tenants in the farm . . . [and] the possessors cast lots (or Kavels in the Scottish dialect) for their particular share. (Kench signifies a larger portion of land than a ridge.) This was a real improvement so far as it went; every farmer had his own lot in each field, . . . reaping the benefit of his industry, which by the run-rig husbandry he could not enjoy, owing to the exchange of ridges every year. Kavels still exist in the Stormont, and in some other parts of the county in a certain degree, and almost universally in village lands. In the latter they are unavoidable; in the former they are regularly exploded, as the old leases fall."[1] In his description of Inverness-shire Robertson amplifies this statement about the annual exchanging of ridges. " In some parts of the Highlands," he writes, " I have seen the land first-ploughed without leaving any boundaries except the furrow betwixt the ridges; then the field was divided by putting small branches of trees into the ground to mark off every man's portion before the field was sown. No man knew his own land till the seed was to be cast into the ground and it became impossible for him to have the same portion of land any two successive years."[2]

We are at length in a position to summarize the principal characteristics of the Scottish agricultural system as it appeared

[1] *Perth* (1799), p. 61.   [2] *Inverness*, p. 335.

in the eighteenth century, and as it had probably existed for some time. The unit of the system was the farm, an area apparently comprising from thirty to four hundred acres, but usually less than one hundred, and requiring for its cultivation a plough of four horses, or at times more than one plough. The tenants were in general from two to four, although the number might increase to six or eight, apart from cottagers attached to the farm. Tenants and cottagers lived together in a cluster of houses, and their horses were joined to form the plough or ploughs. The acres of the farm were divided into infield and outfield, the former tilled year after year with the assistance of manure, the latter ploughed, part by part, for some five years and then allowed to revert to grass for at least as long a period. The arable was divided into strips, long, narrow, and sometimes serpentine. The strips of a tenant were not contiguous, but were separated one from another by the strips of other tenants, an arrangement known as runrig. Sometimes the allotment of strips did not take place until the ground was ready for the seed, and in such cases a tenant was not likely to receive the same strips in successive years.

Nearly everything except the intermixture of the strips of the several tenants was different from the English two- and three-field system with which we have become familiar. The size of the farm as compared with that of the English township, the number of tenants, the infield and the outfield, the method of tillage, the annual re-allotment of strips — all differed. Slater, in getting at the distinctive feature of runrig in contrast with the English open common field, concluded that it resided in the last of these characteristics — in the annual re-allotment of strips. The persistence of such a custom, furthermore, seems to him to have facilitated enclosure, since the tenants, when they finally dissolved their plough-partnership, must have tended to allot their lands with regard to convenience, and must have assigned to each of their number, not several scattered strips, but one parcel or at least few parcels. No resort to act of parliament or to the creation of a commission would thus be necessary to effect enclosure.[1]

[1] *English Peasantry*, pp. 174–175.

There may be some truth in this conjecture as to the consequences of the long persistence of the annual redistribution of strips. Robertson's account of the first steps taken in getting rid of runrig shows that such fluidity made easier the beginnings of a more convenient arrangement.[1] Yet in many places the custom of annual re-allotment cannot have persisted so long as coöperative ploughing and the old intermixture. The other reporters do not speak of it, and Robertson elsewhere is careful to limit his statement by saying that " these ridges were *in some cases* frequently exchanged."[2] What generally gave the first impetus toward consolidation was not the practice of annually re-allotting strips, but the falling-in of the leases and the action of the landlord. Disregarding, however, the effect of annual redistribution upon the beginnings of consolidation, we can scarcely look upon the usage as the most distinctive feature of Scottish runrig. Had the practice been in vogue under English two- and three-field husbandry as we have come to know it, the latter would still have been very different from the agriculture of Scotland. More characteristic of the latter were the size of the farm or township, its occupation by co-tenants or co-heirs, the manner in which it was tilled, and the distribution of the tenants' acres throughout the arable fields.

Before considering these features, however, as manifestations of a Celtic type of field system, we shall do well to examine such information touching them as comes from Wales and Ireland. Some of it is earlier and some of it more specific than the Scottish evidence.

When reports from Wales were made to the Board of Agriculture in 1794, no open-field arable lying in common was to be found in certain counties.[3] Much waste land in the principality[4]

[1] Cf. above, p. 169.   [2] *Inverness*, p. 334.
[3] *Brecknock*, p. 37: " There are no common fields in this district." *Carmarthen*, p. 21: " I do not know of any considerable extent of open common field land in the county." *Denbigh*, p. 11: " There are no common arable lands in this county."
[4] *Brecknock*, p. 39: " One half of the district, containing on the whole 512,000 acres, is waste lands." *Cardigan*, p. 29: " The greater part of the low lands is pretty well inclosed; but hilly and exposed situations are mostly open." *Carmarthen*, p. 20: " About two-thirds of the county is inclosed." *Glamorgan*, p. 42: " The waste land in this county is considerable; computed to amount to upwards

remained unimproved, sometimes not because of its poor quality but because of the inertness of the occupiers. The arable and pasture were usually described as enclosed.[1]

Against this background of enclosures and unimproved wastes there were to be discerned, however, certain patches of common arable field. The reporter from Flintshire wrote: " There are no common fields, or fields in run-rig in this county, as I am informed, except between Flint and St. Asaph and it is intended to divide and inclose them. The difference in rent between open and inclosed fields is estimated at one-third. . . . From the appearance of the fences in this county, inclosing has been very general many years ago."[2] Thus in northeastern Wales the remnants, at least, of common fields lingered, their value was estimated relative to that of enclosed land, and the writer thought it probable that existent closes were made within living memory. On the western coast another instance was noted by the reporters from Cardiganshire: " The only tract like a common field is an extent of very productive barley-land, reaching on the coast from Aberairon to Llanrhysted. This quarter is much intermixed and chiefly in small holdings."[3] The tract in question is some ten miles in length. Farther along the coast at the southwestern extremity of Wales, is St. David's. Here again the reporter for Pembrokeshire noted and explained the existence of common fields: " In the neighborhood of St. David's

of 120,000 acres; upon which common without stint is exercised by the occupiers in the vicinity of such waste land." *Carnarvon*, p. 15: " A great part of Carnarvonshire is still unenclosed." *Denbigh*, p. 11: " There are . . . several commons of very considerable extent." *Flint*, p. 2: " Although some small portions of the waste lands have lately been divided and inclosed, yet there are many thousand acres still left in their original state, which are capable of being converted into arable and pasture lands. And although all the waste lands or commons in North Wales are denominated mountains, yet many of them are as level as a bowling green; and in this county they are, in general, not more hilly than the arable lands nor is the soil inferior in quality, were it well cultivated."

[1] *Merioneth*, p. 8: " The lands in this county are mostly enclosed, the sheep walks excepted." *Montgomeryshire*, p. 9: " The cultivated parts of this county are mostly inclosed, and the fences are in general old, consisting of an intermixture of hawthorn, hazel, crab, etc., as in Flintshire."

[2] George Kay, *Flintshire*, p. 4.

[3] T. Lloyd and the Rev. Mr. Turner, *Cardigan*, p. 29.

considerable tracts of open field land are still remaining, which is chiefly owing to the possessions of the church being intermixed with private property; and the want of a general law to enable the bishop and clergy to divide, exchange and enclose their lands."[1] The situation and the explanation of it are reiterated, finally, by the reporter from Glamorgan. "The land in tillage, or appropriated to grazing," he wrote, "is generally inclosed; open or common fields are rarely met with in South Wales. It is a mode of occupation practiced there in some few instances where ecclesiastical and private property are blended."[2]

Such is the sum of the Welsh evidence contained in the reports of 1794 relative to common arable fields. Three occurrences of such fields are noted, one in the extreme northeast, the others in the south and west on or near the coast. For the phenomenon in south Wales we are told that ecclesiastical properties were answerable; but there is nothing to indicate that such was the case in Flintshire, while on the coastal stretch of Cardiganshire the intermixed properties were chiefly small holdings, apparently not ecclesiastical. If, as seems probable, these ecclesiastical properties were glebe lands, their scattered parcels suggest that at some earlier time all holdings may have been similarly constituted and that the glebe parcels were the last to be exchanged. About the nature of the open fields we learn little. The Cardigan stretch was "very productive barley-land," while the district between Flint and St. Asaph was more hilly but not ill adapted to agriculture. In contrast with this small amount of common field, the central and northwestern parts of Wales are said to have been entirely enclosed, so far as improved lands were concerned.[3] To discover whether the eighteenth-century patches were due to exceptional causes operative only on the borders of the principality, or whether they were survivals of what had once been a

[1] C. Hassal, *Pembroke*, p. 20.
[2] J. Fox, *Glamorgan*, p. 41.
[3] Of Carnarvonshire, in the northwest, the reporter writes (p. 15), "A great part is still uninclosed"; but he does not state whether the unenclosed lands were arable or waste. Probably he refers to waste lands, since he continues: "The old fences appear to have been finished in a very imperfect manner. They consist chiefly of dry stone walls and earthen banks."

universal phenomenon, we turn to the surveys of Tudor and Jacobean times.

As it happens, these surveys refer for the most part to the very regions which have just been noted as retaining open fields in the eighteenth century. They are concerned with large lordships in Pembrokeshire in the southwest and Denbighshire in the northeast. From other counties of Wales evidence is scanty, save for one acceptable survey from Anglesey. The testimony from the first two regions, which, to judge from the liberal sprinkling of English place-names, were less purely Welsh, may be examined first.

The intermixture of the parcels of the holdings in Pembrokeshire, described in the eighteenth-century report, is confirmed by a note prefixed to the survey of the royal lordship of Haverfordwest, made in 21 James I. " Also whereas the Landes of theise Tenements doe lie devided amonge the Tennants in small parcells lyeng intermixedlie wherebie the Tennants cannot make full profitt of theire tenements and thereby they are the lesse valuable in the lettinge; It were verie convenient in our opinions for his highnes proffitt and for the benefitt of the Tennants that by viewe of a Jurie in everie Mannor or by some direction from your Lordship the land were viewed and by exchange made entire as neere as maie be, or sorted in such partes as the tennantes maie enclose and therebie make theire beste proffitt. And wee holde it conveynient that for all exchaunges to be made of anie peeces of land betwixte the Tennantes for conveyniencie, that the same be made in writinge and presented at the next Courte to the Stewarde to be Recorded, and that Notwithstandinge the exchaunge the auncient landshares and meares betwixt the peeces be preserved."[1] In determining the value of a ploughland the surveyors state further that they have had " regarde to the goodnes of ech mans holdings and whither it laye togethers or dispersed."[2] No doubt can exist, then, about the intermixture of parcels here; and, since there is talk about ancient land-

---

[1] Land Rev., M. B. 206, f. 39. The lordship included the manors of " Camros, St. Issmells, Rock, Pull, and Staynton."

[2] Land Rev., M. B. 238, f. 37.

shares and meres and enclosing, it is evidence that the parcels must have been in open field or at least intermixed in large enclosures.

The procedure which is recommended above by the surveyors was in 1593 well under way at Carew or New Shipping, some ten miles distant. In the list of demesne lands, for the most part closes, we hear of the following: —

" One acre lying in the closure which lyeth on the north side of the myll pond; it lyeth among other lands; it was taken from the tenement that nowe John hillen holdeth and added to the demains of New Shippinge; this land is errable or pasture ground. . . . Item iiii acres lyinge in the foresaid close, whereof iii acres lyeth togeather in one peece and one acre at the end thereof, all arrable or pasture ground . . . ; it [the iiii acres] was sometyme belonging to the tenement that now John mertyn holdeth. . . .

" Parcells of grounde taken from tenements in newton and added to the demesne of New Shipping: fower acres arable or pasture; five of like ground; three acres of like errable . . .; two acres of like errable . . .; two acres of like errable . . .; two acres of like errable. . . . Memorandum, all these . . . parcells of grounde are newly enclosed in one closure which close lieth on the north side of the said mesuage of newe shippinge. . . .

" Lands taken From newton annexed to New shippyng: Item three acres situate in the fielde or crofte on the north side of Carewe bridge sometimes belonginge to a tenement in Newton in the occupation of Henry Saunders consisting of errable or pasture grounde . . . ; Item two acres in the saide feelde or croft taken from the tenement wherein John woodes now dwelleth beinge errable or pasture grounde . . . ; two acres in the saide croft sometimes belonging to a tenement wherein Richard Bowen now dwelleth of like errable or pasture grounde."[1]

Near by, at Sagestown, certain lands of the queen were thus described by the surveyors: —

" John Benion occupieth the tenemente and xvii acres parcell of the saide xxv acres; and as to viii acres, the residue, iiii of them

[1] Land Rev., M. B. 260, ff. 217, 219b.

lie togeathers in an open fielde on the easte side of the said townshippe of Sageston havinge the highway that leadeth from Carewe to temby on the south side; two other acres lie togeathers in the same fielde neare a place called the haies, these vi acres are pulled from the forsaide tenemente and are anexed to a tenemente in the occupation of one griffith Froine; one acre and a half lyinge togeathers in the said fielde nowe holden by John Gibbe and John Thomas; One acre the residue lyeth in a fielde on the weste side of Sagiston neare the church way taken from the said tenement and anexed to the demaine lands of the castle. . . . Memorandum. insteade of the vi acres annexed to Froines tenemente . . . there is vi other acres taken from the saide froines tenement and added to the demaines of the castle they lie in the fielde on the west side of Sagiston neare the church way beinge errable or pasture. . . . "[1]

Of these parcels in New Shipping and Sagestown it will be noticed that the second group, once open, had been enclosed upon consolidation, that the last group apparently still lay in open field, while the first and third groups had lain intermixed in fields already enclosed. These two groups show that intermixture of tenants' parcels in Wales does not necessarily imply that the parcels in question were in open field. Strips of more than one tenant sometimes lay within the same close. It will be noticed further that the intermixed parcels above described as newly enclosed were arable or pasture. The situation is one which could as well have arisen from the subdivision of a close of arable or pasture among several heirs as from the enclosure of an open field.

If in any particular Pembrokeshire survey which has come down to us we try to discover the number and extent of the open fields or of the closes containing intermixed parcels, we shall find only a few of them. In the survey of St. Florence, made in 1609, much land is described as pasture or enclosed arable, while only the following field names recur more than twice, with parcels of the size indicated held in each place by different tenants:[2] —

[1] Land Rev., M. B. 260, f. 222.
[2] Land Rev., M. B. 206, ff. 227–243. The areas are in acres.

Bloody acre, 1, 1, 2¼, 1
in Blackhill fields or at Blackhill, 4½, 8, ½, 1¼, 7, 9½
at Burrows, 1, 1, ¾
at Ladyland, and at Langstone "in Ladyland field," 4, 1, ½, 2, ½, 1¼, 1¼, 1¾
at Middlehill, 5 (in open field), 4¾, 1, 5½
in Cherrieland, ½, ½
at Honnyland, 4, 4
in the East field of Flemyngton, 9, 6.

Flemington is the township to the west, and apparently had its East field. Since the other localities have not perpetuated themselves on the ordnance map, they were probably fields rather than hamlets. The total area at St. Florence throughout which the parcels of the tenants were intermixed appears therefore to have been about seventy-five acres.

A few miles to the north lay the lordship of Narberth, of which we have a survey made in 7 James I.[1] The lordship comprised, besides Narberth, the townships of Templeton and Robeston. At Templeton there was no open common arable, all holdings consisting of "arable land enclosed" and "mountain ground." The Narberth holdings were less uniform. For the most part they were, so far as described, either closes or "arable and pasture at Middle hill." Three tenants at least had "arable not enclosed," in amounts of from six to fourteen acres, but no further description of these unenclosed acres is vouchsafed.

Robeston was the township which, of the three, seemed most inclined to intermix the parcels of the tenants. Of this we are assured by no definite statement, but the assignment of small parcels to the same field division can scarcely be interpreted in any other way. Particularly noteworthy is the case of four tenants, each of whom had exactly the same series of small parcels in nine localities. Four times is repeated the following list of fractional acres: ¾ in hill park close, ¾ in woodways close, ⅝ (or ⅜) in Hookesmeade, ¾ in Blind will, ⅜ at Utter hoke, ⅛ above the haies, ¼ at Narbert waie, ¼ at Langstone, ¼ at Lynacre. What had taken place was a division, among the four tenants, of plots of land

[1] Land Rev., M. B. 206, ff. 118–186.

containing respectively 3, 3, 2, 3, 1½, ½, 1, 1, 1, acres, and a mingling of small parcels had been the result. Such intermixture does not imply the existence of open field, since before subdivision the areas may have been closes, and in two instances are said to have been. Indeed, "parkes" and closes at Robeston were numerous, a sign that the township was largely enclosed. Some further intermixture of the same sort there may have been, especially in the following localities, where acre- or fractional acre-parcels were in the occupation of different tenants (except in case of those connected by +):—

Castlecroft, 1, 1½ (arable), 1½ (2 parcels), ½, ¾, 1, 1½ (2 parcels of arable), ½

at Two Acres and Little Two Acres, 1, 1, 1, 1, 1, 2 (3 parcels), 1¾, ½, 1

Stubby land, ½, ½, ½

in or at Woostland, ½ (arable), ¾, ¾, ½, ¼, ¾ (arable)

Shortlands, 2, 2, ¾

upon the Hill (arable), 1 + ¼, ⅛, 2½ + 1¼, 1, 1½ (2 parcels), 1 + ¼, ⅛, 2½ + 1¼

in the Vran (arable), 3 (3 parcels), 1 (2 parcels), ½, 1½, 3 (3 parcels), 1 (2 parcels).

At best, the total area of the tenants' parcels which were intermixed at Robeston was probably not more than eighty acres. This amount differs little from that just estimated for St. Florence. Since these two Pembrokeshire townships, of all those described in the Jacobean surveys, inclined most to intermixed holdings, we may conclude that at the end of the sixteenth century the county had its arable largely enclosed. Some intermixed land was to be found; but at times it lay within closes, and in certain instances it pretty clearly arose from the subdivision of parcels among a group of tenants. It seems never to have predominated in a township, and probably seldom exceeded one hundred acres.

From the Pembrokeshire surveys we may turn to those of Denbighshire in the northeast, especially to some that come from a region in which the place-names are even less Welsh than those of Pembrokeshire. This is a part of the valley of the Dee, ten miles above Chester and adjacent to the English county. Wrexham is the largest town of the district, and its open field, as pictured in John Norden's survey of 1620,[1] has been briefly described by A. N. Palmer,[2] who follows the history of the butts and quillets to the present day. Norden's survey, like several others antedating it, refers to the lordship of Bromfield and Yale, a lordship so extensive as to be subdivided into seventeen manors containing 62 townships or hamlets. Excellent and detailed as is this description, it is not more so than one of some seventy years earlier preserved at the Public Record Office.[3]

For the most part both surveys are concerned with townships and hamlets entirely enclosed. Such, for example, in Norden's survey are Brymbo, Esclusham, Bersham, Moreton Anglicorum, all of which are described in full, with specification of closes.[4] There are, however, three or four townships which in both surveys show certain traces of open field. These traces are very slight at Holt, being confined to three fields, each divided between two freeholders, and to a fourth in which six freeholders have parcels of arable or pasture.[5] They are most numerous at Wrexham, at Pickhill and Siswick, and at Issacoed, a division in which the principal hamlets were Sutton and Dutton. The earlier survey is henceforth quoted.

At Wrexham we find, what is very rare elsewhere, the term " common field." John Hower had, besides a messuage, garden, and pasture close of an acre, " ii acras terre arabilis iacentes in communi campo dicte ville." David Middleton, along with four tenements and eighteen acres of pasture in seven closes, had another tenement, a close of pasture, and " xii seliones terre iacentes in communibus campis villarum vaure Wryxham et Waghame continentes [with the close] viii acras terre arabilis et pasture."[6]

[1] The survey is printed from Harleian MS. 3696, in *Archaeologia Cambrensis, Supplement of Original Documents* (1877), vol. i, pp. cxi sq.

[2] *The Town, Fields, and Folk of Wrexham in the Time of James the First*, Wrexham, etc., [1884].

[3] Land Rev., M. B. 249, the entire 210 folios. The survey as a whole is not dated, but the most recent leases and copyholds are c. 39 Henry VIII.

[4] *Archaeologia Cambrensis, Supplement*, etc., vol. i, pp. ccii sq.

[5] Land Rev., M. B. 249, ff. 8–22.

[6] Ibid., f. 68.

Usually the "communi" is omitted and "in campo" occurs alone.[1] Nine other tenants resembled the two mentioned in having a few acres or selions in the fields of Wrexham Vaur.[2] Since closes are sometimes included in the areas given, the total amount of open-field arable at Wrexham cannot be exactly determined. It comprised about one hundred selions, varying in size from $\frac{3}{14}$ to $\frac{6}{7}$ of an acre, and it can hardly have exceeded sixty acres. The amount is not large for the middle of the sixteenth century, nor can it, of course, have increased by Norden's time.

In the survey of Pickhill and Siswick the term "butts" for the most part replaces "seliones," and each butt contained from a quarter-acre to a half-acre.[3] Of the two holdings which incline most to open arable field, one has about one-fourth of its area so described, the other about one-half.[4] Elsewhere in the survey

[1] For example, "William ap Maddoc et Robert ap David ap Gruff ap Robert" held three messuages, three closes containing eleven acres, and "xiv seliones terre iacentes in campo ville predicte [Wrexham Vaur] continentes iii acras terre" (Land Rev., M. B. 249, f. 72b).

[2] Their holdings comprised the selions of the preceding note, together with the following eight entries: —

"ii acras terre in diversis selionibus . . . et unam sellionem
ii Eruas terre continentes dimidiam acram et unam rodam
vii seliones terre . . . continentes per estimationem vi acras terre arabilis
viii seliones terre . . . continentes iv acras terre
quinque clausas et xii seliones . . . continentes . . . vii acras terre . . . et vii seliones [no area]
cum octo clausis et diversis sellionibus . . . continentibus per estimationem xx acras terre
v selliones [no area]
xii selliones continentes per estimationem vi acras terre" (ibid., ff. 65–74).

[3] Ibid., ff. 124–130.

[4] The two are as follows: —

"Jenkyn ap Jenn ap David nativus ut dicit tenet ibidem unam ceparalem clausam pasture vocatam Ibryn Istrowe alias Stonyclose continentem per estimationem iii acras pasture
et unam clausam prati continentem ii acras et dimidiam ibidem
ac unam aliam parcellam terre arabilis vocatam Estymarowe continentem per estimationem iiii acras pasture
Et v butts iacentes in Kay Jenkyng continentes i acram terre
et iii butts in dole Seswyke continentes i acram
et iiii butts iacentes in dole Seswyke
et vi butts iacentes ibidem continentes ii acras et dimidiam terre arabilis

closes very largely predominate.[1] As at Wrexham, the total open-field arable did not amount to more than sixty acres, and was probably less.

> Et ii pecias terre ceparalis continentes dimidiam acram terre iacentem iuxta brynstonoc
> [et] ii ceparales clausas continentes per estimationem iii acras terre ibidem et i acram et dimidiam in quadam clausa vocata Kay parva."

> " Maddoc ap Roberti ap llywelyn tenet in pychell unum tenementum et viii clausas terre in ceparali continentes xii acras pasture et arabilis
> et i et dimidiam acram prati vocatam gwerlozh ekeyveney
> et unam peciam prati continentem dimidiam acram prati iacentem in prato vocato gwerne estymavall'
> et xv lez butts iacentes in dolebikill [' dolbykelfeld ' is crossed out]
> et xi alias continentes per estimationem iii acras terre
> et ii pecias terre arrabilis iacentes in campo vocato ystymarowe continentes i acram terre
> et in le maysegwyn i peciam continentem i acram et dimidiam
> Et in campo vocato Oldymawre i peciam pasture continentem i acram
> Et in campo vocato Frythe iii butts continentes tertiam partem 1 acre
> Et in campo vocato maysmawre iiii lez butts ibidem continentes i acram et dimidiam
> Et in campo vocato Skythery unam peciam continentem dimidiam acram terre arabilis
> Et in campo vocato Ekeyveney unam peciam continentem dimidiam acram terre arabilis
> Et in clauso vocato Ekeyvya unam parvam peciam terre " Land Rev., M. B. 249, ff. 128,128b).

[1] The following are the only other indications of open-field arable. Except where bracketed together, the parcels are in different holdings: —
" vii butts in dollgough
{ xiv butts in campo de Keynistneth continentes iii acras
{ iv butts continentes i acram
{ xxiv le butts iacentes in dole gowgh et urencregog continentes iii acras terre
{ i acram in massewell
{ sex parcellas continentes xv butts terre in campis de Pychyll
{ unam sellionem terre vocatam heyle
iii acras terre arabilis iacentes in le butts
xii butts continentes vi acras terre
xviii butts continentes iii acras terre
vi butts iacentes in bryng cregoch continentes dimidiam acram terre et i rodam
ix butts in le bullowgh ald corñ continentes per estimationem ii acras terre
iii butts in dole gowgh
xix parcellas et . . . butts terre arabilis continentes per estimationem iii acras terre
vii lez butts continentes ii acras terre " (ibid., ff. 124-130).

Longest of all these surveys which reveal open field is that of the division of the lordship which is called Issacoed and which contained nine hamlets.[1] The descriptions are often non-committal, but the total, once more, did not exceed fifty acres.[2]

[1] Land Rev., M. B. 249, ff. 147-164.
[2] The following list seems to embrace all the open-field parcels, those of the same holding being bracketed together: —

{ iiii butts in Gillowistayth continentes dimidiam acram
iii acras in le grodyer ac iiii lez butts iacentes in alio Grodyer
i acram in dolblythy
v butts in quadam clausa vocata dollevellen
ii pecie terre iacentes in Caystabell continentes i acram
ii butts in cargrose continentes i rodam terre
xi butts in le grodyer continentes i acram et dimidiam
ii butts in alio grodyer
ii butts in berthyer continentes dimidiam rodam
viii butts in Errowe continentes i acram terre }

{ vii butts in le mersshegwyn
xi butts in dyttonbrayne continentes per estimationem ii acras terre }

ix acras terre arabilis iacentes ibidem [in Ditton] in diversis parcellis
dimidiam lez butts iacentem [in] panthulog continentem i acram
ii lez butts continentes dimidiam acram terre in Sutton

{ in diversis parvis peciis terre iacentes in dole Sutton circa iii acras terre
i acram in le grodyer }

quinque diversas parcellas terre iacentes in communibus campis [of Sutton]
    continentes per estimationem iii acras terre

{ iii butts vocatas tyre y Kauboth continentes dimidiam acram terre
ii parcellas terre in dole Sutton continentes i acram terre
i acram et i rodam per estimationem in Kayrkewle }

{ vii butts [in] iii parcellis in le goidra continentes ii acras
i acram et dimidiam prati iacentem ibidem
i rodam prati iacentem in doll vha }

{ v butts in doll utha et i rodam et i butt in doll utha
i parcellam in dollissa continentem dimidiam acram terre }

{ vi butts et unam parvam clausam . . . continentem i acram et dimidiam
    terre
ii butts in Kaystabell continentes i rodam terre
i butt iacentem in Kaystabellissa continentem i rodam
vii butts in Kayglase continentes ii acras terre
iiii butts in drowestole ussa
iiii butts in grodyer dytton
vii butts iacentes infra Gillough isstathelogg . . . }

ii clausas pasture et certis terris [sic] in communi campo de Sutton continentes x acras terre iacentes in Sutton
i acram terre [in] duobus parcellis in communibus campis ibidem [Horseley].

To the survivals of common fields in the lordship of Bromfield should be added slight traces found in the survey of the adjacent manor of Ruthin.[1] Although this is much concerned with messuages and with small holdings which are nearly always enclosed, common fields are mentioned two or three times.[2] In a holding at Llammirock there was appurtenant to a house and garden " terra arabilis in communibus campis vocatis tir y cech," the tenant paying a total rent of 6 d. In Ruthin itself there was held by lease a messuage, three closes containing eleven acres, and

" terra arabilis in communi campo vocato Pantmigan continens per estimationem xii acras
    terra arabilis in predicto campo continens per estimationem ii acras."

In a survey so long as this one such common fields are almost lost.

Thus far only surveys from the English parts of Wales have been examined. Nearly all hamlets in which intermixed parcels have been found bear English names, and even in these the amount of common arable was surprisingly small. One might surmise that in purely Welsh surroundings no common fields whatever were known. So far as our evidence goes, however, the situation seems not to have differed from that already described. A certain amount of intermixed ownership is visible in places where it can scarcely be attributed to English influence.

Best of the Jacobean surveys of Welsh regions are those from Anglesey. Frequently these descriptions speak only of noncommittal " parcels," but occasionally we discover that the parcels which constituted a holding lay scattered. Such was the case with certain of the lands of John Lewys, armiger, at Cliviock, which are described as follows: —

" ii parcelle terre arabilis sparsim iacentes in quodam campo vocato Dryll y Castell . . . continentes ii acras v rodas
ii parcelle terre arabilis sparsim iacentes in quodam campo vocato Glodissa . . . continentes iiii acras . . .
una parcella terre arabilis iacens in quodam campo vocato dol Gledog continens iii acras et dimidiam

[1] Land Rev., M. B. 239, ff. 125–181.   [2] Ibid., ff. 167, 175.

terra arabilis iacens in campo vocato Bryn y gwyddal continens iiii acras." [1]

In the rather long account of the lands held by lessees in the manors Hendre, Rosfaire, and Mardreff, in the Anglesey commot of Menai,[2] one may find holdings which comprised, for example: —

"Domum mansionalem . . . cum parvo crofto . . . et
Sex parvas clausas continentes per estimationem ix acras
Sex alias parcellas terre arabilis iacentes in gallt Beder continentes per estimationem i acram
Tres alias parcellas terre arabilis iacentes in carreg y gwydd continentes per estimationem i acram i rodam
Unam aliam parcellam terre arabilis ibidem iacentem iuxta viam ad Carnarvon continentem i rodam." [3]

The holdings in general inclined, much as did this one, to have parcels in three or four localities, called *kesu mawr*, *kesu bychan gallt bedr*, and *carreg y gwydd*. It is not clear whether the first two were hamlets or fields; but *carreg y gwydd* is once called a "campus," [4] and five times acres are said to lie "sparsim" in *gallt bedr*.[5] These two areas were thus presumably characterized by intermixed ownership, eight tenants having parcels in *carreg y gwydd* and seventeen in *gallt bedr*.[6] Some intermixture of parcels in Anglesey thus seems demonstrable, although we learn nothing about the shape of the parcels, their relation one to another, or the method by which they were tilled. The character of the open field is far from clear, and the descriptions of the free-

[1] Land Rev., M. B. 205, f. 135.
[2] Ibid., ff. 25–30.
[3] Ibid., f. 30.
[4] Ibid., f. 28.
[5] e. g., "Sex parcelle terre arabilis iacentes sparsim in gallt beder continentes ii acras."
[6] The areas in acres were as follows: —
  carreg y gwydd, 3 ("acras terre arabilis iacentes sparsim"), 1, 2, 8 (7 parcels of arable, ¾, ½, 1¾ (4 parcels of arable), 2¾
  gallt bedr, 3 (1 parcel), 3 (6 parcels of arable), 2 (3 parcels of arable "sparsim"), 2 (5 parcels of arable), 1 (6 parcels of arable "sparsim"), 1 (2 parcels of arable), ½ + 1 (7 parcels of arable) + 1 (meadow), 1 (2 parcels "sparsim"), 1 (6 parcels of arable), 2 (6 parcels of arable "sparsim"), ¼, ½, 4 (8 parcels of arable), 1 (6 parcels of arable), 2 (4 parcels of arable), 2 (6 parcels of arable "sparsim"), 3 (5 parcels of arable "sparsim").

holds of these manors add little to our knowledge. Since in them closes are sometimes designated as such, we may perhaps be justified in inferring that parcels not so designated were unenclosed and non-adjacent.[1] Considered in its entirety, accordingly, the evidence from Anglesey points to the existence there in Jacobean days of holdings which consisted to some extent at least of scattered parcels of arable lying in open fields.

In other parts of Wales than those already considered traces of open field are slight. The twenty-six tenants of the manor of Eglowis Kymin in Carmarthen had in 7 James I only " parks " or closes.[2] In a survey of Gower and Kilvey, Glamorganshire, made in 1665, there were many " closes " and " parcels," but little to indicate open field. Of doubtful significance are the " three other parcells, called fields, leying intermixt with the lands of the said George Lucas," containing three acres.[3] In the six large volumes of *Cartae et alia Munimenta quae ad Dominium de Glamorgancia pertinent*, edited by G. L. Clark, we find many descriptions of closes, but only two or three revealing open fields. In the fief of Landbither four acres are specified, of which

[1] A certain freehold, for example, consisted of a house, a garden, and " quatuor parcellas terre arabilis vocatis Cay pen y kevn insimul continentes . . . v acras l perticatas

unam parcellam terre et pasture, unam peciam vocatam cay bach, et alteram . . . iacentes super quandam clausam vocatam Cay y weyrglodd continentes . . . ii rodas xiv perticatas

unam parcellam terre arabilis vocatam y rerw dew continentem . . . i acram

unam aliam parcellam terre arabilis iacentem in quadam Clausa vocata Cay r llo continentem . . . i acram

et aliam parcellam terre arabilis vocatam y dalarhir continentem . . . xxx perticatas

et aliam parcellam terre arabilis iacentem in loco vocato cay y felin continentem . . . i rodam xxxv perticatas

et clausam terre arabilis et pasture saxosam continentem . . . v acras

et parcellam terre arabilis et prati vocatam bryn llin continentem in toto . . . ii acras

unam aliam parcellam terre arabilis iacentem in quadam clausa vocata Penrhyn fadog continentem . . . ii rodas x perticatas " (Land Rev., M. B. 205, f. 16).

[2] Land Rev., M. B. 258, ff. 1-17.

[3] *Archaeologia Cambrensis, Supplement of Original Documents*, i. 270.

" duae acrae iacent in cultura qui vocatur Kayraryan . . .
et una acra iacet in cultura qui vocatur Kayrpistèl
et una acra iacet in cultura qui vocatur Hendref."

This may be open field or it may not be. Somewhat more suggestive of it are the 4¾ acres of arable which were conveyed along with one-third of a messuage in Landoghe; [1] but abundant testimony of this sort is by no means forthcoming.

Our examination of Jacobean surveys from Wales has brought to light only a relatively slight extent of open arable field in which the parcels of the tenants were intermixed. In each of the two townships of Pembrokeshire for which areas can be estimated it did not exceed one hundred acres, and it was not greater in the Denbighshire townships. In purely Welsh regions little more than the existence of common arable fields in Jacobean days can be determined.

A reason for the insignificance of such fields, together with testimony to their earlier prevalence, is to be found in Owen's description of Pembrokeshire, written in 1603. Explaining why winter corn is so little grown in that county, the author remarks: " One other cause was the use of gavelkinde used amonge most of these welshmen to parte all the Fathers patrymonie amonge all his sonnes, so that in proces of tyme the whole countrie was brought into smale peeces of ground and intermingled upp and downe one with another, so as in every five or sixe acres you shall have ten or twelve owners; this made the Countrie to remayne Champion, and without enclosures or hedging, and wynter Corne if it weare sowen amonge them should be grased all the winter and eaten by sheepe and other cattell, which could not be kept from the same: . . . this in my opinion was one cheefe cause

---

[1] " Dimidia acra iacet in loco qui vocatur Votlond inter terram . . . et terram . . . et caput ejus occidens extenditur usque ad feodum de Denaspowys

due acre et dimidia iacent apud Langeton inter terram . . . et terram . . .

dimidia acra iacet inter terram . . . et terram . . .

dimidia acra iacet scilicet in Votlond inter terram . . . et terram . . .

dimidia acra iacet in loco appellato Morewithe Stlad . . .

una roda iacet in parte boreali prati quondam Alexandri " *Cartae*, etc. (Cardiff, 1910), iii. 722.

they restrayned sowing of wynter corne but as nowe sythence the use of *gavelkinde* is abolished for these threescore yeares past [by statute of 34 and 35 Henry VIII c. 26, secs. 36, 64] in many partes the grounde is brought together by purchase & exchanges and headging & enclosures much encreased, and now they fall to the tillinge of this wynter corne in greater aboundance then before." [1]

From this it is clear that parcels of holdings in the Welsh parts of Pembrokeshire had once been intermixed and unenclosed but that the abolition of transmission by gavelkind had encouraged consolidation and enclosure. The reference to gavelkind suggests that it was a determining principle in the Welsh field system, and at once calls to mind the part played by co-tenants in Scottish agrarian arrangements. Before following out these suggestions, however, we shall profit by attending for a little to Irish conditions; and we shall naturally inquire first whether Irish units of settlement were, like those of Scotland and Wales,[2] of the hamlet type surrounded by small arable fields.

In Ireland the units of settlement are and long have been the townlands, but in seventeenth-century surveys they assume various names and are variously grouped into larger units. Since many of these units were more or less artificial, subserving purposes of rating or assessment, like English hides or virgates, it is always necessary to keep apart the actual from the artificial units. The size and shape of an actual Irish townland of the nineteenth century is illustrated by any section of the six-inch ordnance survey map; and the areas of the eight towns which Seebohm has reproduced from county Monaghan, and which range in size from 35 to 165 acres with an average of about 90 acres, are entirely typical.[3]

Seebohm has gone farther, and identified these eight townlands by means of their names with eight *tates* of a survey of 1607. The tate was primarily a unit of rating, whereas the Latin term for townland was *villata*. Sometimes, as in the instance cited

[1] George Owen, *The Description of Penbrokshire*, (ed. H. Owen, Cymmrodorion Record Series, 3 pts., London, 1892–1906), i. 61.

[2] The places referred to in the survey of the lordship of Bromfield and Yale were usually hamlets, the arable fields of which were inextensive.

[3] *English Village Community*, p. 224.

by Seebohm, villata and tate corresponded in extent;[1] elsewhere they did not, more than one villata being sometimes included in a tate.[2] Again, in county Tyrone, where the townland was equivalent to the " balliboe " and contained about sixty acres, it was itself a unit of rating.[3] In county Fermanagh, however, it once more differed from the artificial units. Usually each of the seven baronies into which this county had been artificially divided contained seven and one-half *ballybetages*. Each ballybetage in turn contained four quarters, each quarter four tates, each tate 30 acres, " contrey measure." In consequence the barony comprised 30 artificial quarters, or 120 tates.[4] Elsewhere we learn that the first of the baronies, Knockenyng, was six miles in length by three in breadth, " wherein are 24 townes."[5] The townlands in county Fermanagh therefore corresponded with none of the artificial units, although they were not far removed in size from the quarters. In this barony of Knockenyng their average area was 150 acres. In Donegal the villata was equivalent to the quarter, and, since $7\frac{3}{4}$ quarters are said to have contained about 1000 acres,[6] the townland here too comprised on the average about 125 acres.

[1] *Inquisitionum in Officio Rotulorum Cancellariae Hiberniae Repertorium* (Rec. Com., 2 vols., 1826–29), ii, co. Monaghan, no. 2 (1609): " Tres vil' sive precincte terre vocate ballibetaghes . . . que . . . continent quasdam minores parcellas, villatas sive particulas terre vocate tates, viz. Ballileggichory continet 1 tate vocatam Ballileggichory, 1 tate vocatam Mullaghbracke " [etc.; sixteen tates are named].

[2] Ibid., no. 4 (1619): " Jacobus O'Donelly nuper abbas nuper monasteri sanctorum Petri et Pauli de Ardmagh ac conventus . . . seisati fuerunt . . . de separalibus villis, villatis sive hamlettis et terris vocatis Mullaghegny, Reagh, Aghnelyny, Edenaguin et Broaghduff, cum suis pertinentibus continentibus i tate . . . ac de villis, villatis sive hamlettis ac terris vocatis Knocknecarny et Urny, cum suis pertinentibus continentibus i tate."

[3] Ibid., Tyrone, no. 5 (1628): " King James did grant unto James Claphame . . . all the lands in the severall townes, etc. following, i. e., Cloghogall being 1 towne or balliboe of land, Creighduffe being 1 towne-land," etc. These townlands consisted of three *sessiaghs* each. The uniformity of the subdivision and of the sixty-acre area (ibid., 1661, no. 19) are what suggest that the *towne* is here a unit of rating rather than one of settlement.

[4] Ibid., pp. xxxiii–xl.
[5] Ibid., p. xviii.
[6] Ibid., Donegal, no. 9 (1620).

In a survey of 32 Henry VIII we are told the areas of the townlands, and learn in addition how many peasant households each contained. Two descriptions run as follows: —

" Villata de Balnestragh. William Dyxson tenet scitum manerii vocati Balnestragh super quod edificantur duo castra . . . terra dominica continet lx acras terre arabilis et i acram prati iacentes in villata de Balnestragh. . . .

" Et [dicunt] quod sunt infra eandem villatam vii messuagia cxv acre iii rode terre arabilis x acre communis pasture ac ii acre more in occupatione Donaldi O'Daylye, Donaldi Holloghan et aliorum. . . . Et sunt iii cottagia. . . .

" Et [dicunt] quod sunt in Villata de Ballerayne vi messuagia cxxxiii acre terre arabilis xx acre communis pasture et xx acre more quas Mauricius O'Nayry, clericus, Ricardus O'Morrye et alii tenentes ibidem occupant. . . . Et sunt x cotagia. . . . "[1]

From these instances we may conclude that Irish units of settlement were in size much like Scottish townships. Their areas, averaging from one hundred to two hundred acres, were perhaps a little greater and their tenants may have been a little more numerous. From the point of view of the English midlands, however, both forms of settlement coalesced into what may be called the Celtic type. Instead of the large village we find the hamlet; instead of extensive arable fields, the restricted areas of the farm, the townland, or the petty township. Wherever in England hamlets and small townships appear as the prevailing type of settlement, Celtic influence is to be suspected. Within the three-field area we have already seen such, notably in the border counties, Herefordshire and Shropshire. If, however, Celtic influence determined the form of settlement and the size of the townships there, it did not prevent the superposition of a three-field system upon the arable. Since such a system was not Celtic, a further effort should be made to determine what was its Celtic correspondent.

We have seen that a salient feature of the ancient agriculture of Scotland and Wales was the intermixture of the parcels of the tenants. Known in Scotland as runrig or rundale, this feature

[1] Rents. and Survs., Ro. 934.

was there to be seen at the end of the eighteenth century in farms or townships which had not been improved. Although in Wales it was far less usual in the eighteenth century, or even at the end of the sixteenth, traces of it have been discerned at both periods. An item from the early seventeenth-century survey of Robeston in southwestern Wales has disclosed how rundale might arise, nine parcels of land there having been divided among four tenants, with resultant intermixture; and a contemporary

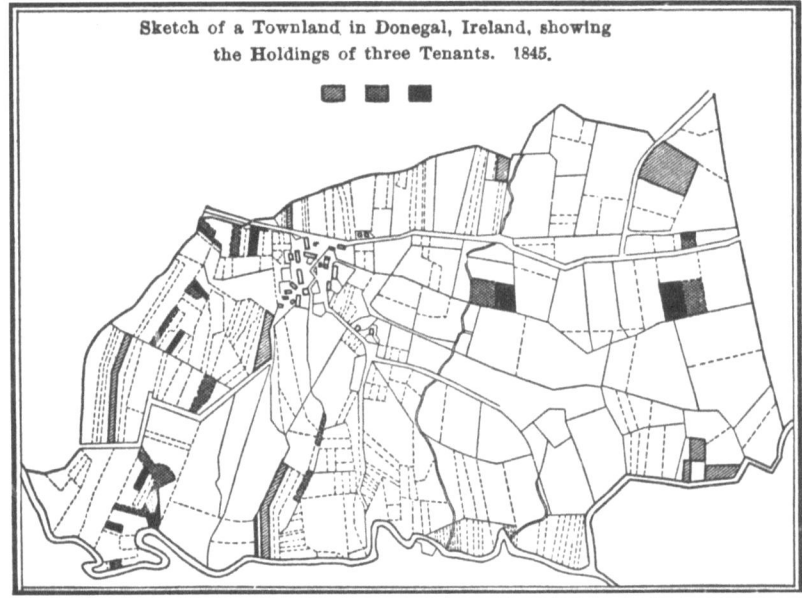

Map XI

account of Pembrokeshire relates that such subdivision and intermixture were still more prevalent at an earlier time, attributing the phenomenon to the custom of transmitting land by gavelkind. From Scotland we learn that Scottish runrig was characteristic of farms held by co-tenants and of lands held by co-heirs. The reporters imply that it was an ancient custom, and excuse it as a concomitant in earlier days of the grouping of peasants in villages for purposes of defence. Since historical explanations were with them only remarks by the way, a further examination of the occasions which gave rise to runrig and of the antiquity of the phenomenon is desirable.

THE CELTIC SYSTEM

Perhaps the most pertinent testimony on these points comes neither from Scotland nor from Wales, but from their more purely Celtic neighbor, Ireland. This evidence, too, is of a more recent date than that hitherto cited. It is embodied in the report of the so-called Devon Commission, made to parliament in the middle of the nineteenth century.[1] From this report came (apparently at second hand) the plan which Seebohm used to illustrate the intermixed strips of the tenants of an Irish townland.[2] The plan itself, which is herewith once more reproduced, is instructive, but the accompanying explanation, which Seebohm omits, is still more so. It runs as follows: —

"Fig. i shows the condition to which subdivision of holdings has brought a neglected townland in Donegal, containing 205 statute acres. The whole was occupied in one farm two generations ago; it then became divided into two farms, and those two have been since subdivided into twenty-nine holdings, scattered into 422 different lots. The average arable quantity of each holding is four acres, held in fourteen different parts of the townland; the average quantity of pasture per farm is three acres, held in lots in common. The largest portion of arable held by any one man is under eight acres; the smallest quantity of arable in any one farm is about two roods. The pasture being held in common cannot be improved. . . . They had been in the habit of subdividing their lands, not into two, when a division was contemplated, but into as many times two as there were qualities of land in the gross quantity to be divided. They would not hear of an equivalent of two bad acres being set against one good one, in order to maintain union and compactness. Every quality must be cut in two, whatever its size, or whatever its position. Each must have his half perches, although they be ever so distant from his half acres. And this tendency is attributable to the conviction of these poor ignorant people, that each morsel of their neglected land is at present in the most productive state to which it can be brought."[3]

---

[1] *Evidence . . . [on] the State of the Law and Practice in respect to the Occupation of Land in Ireland*, 4 vols. (*Parl. Papers*, 1845, vols. xix–xxii).

[2] *English Village Community*, p. 228.

[3] *Parl. Papers*, 1845, xix, app., p. 59.

The most surprising thing in this account is perhaps not the excessive subdivision which resulted in 205 acres being cut up into 422 lots held intermixedly by twenty-nine tenants; it is rather that a compact farm had been thus transformed within two generations — a fact which Seebohm neglected to note. The cause of the subdivision and the manner in which it had been made are indicated in the quotation. Co-tenancy had been responsible. This custom demanded that the heirs of a tenant receive equal parcels of each quality of his land, no matter how widely distributed the plots of the same quality may have been. The tangle of strips and plats shown on the map was the result.

Such an account corresponds with what has already been noted relative to a Scottish farm or townland. There, too, co-heirs were often the tenants who held their lands in runrig. In both countries other tenants, not heirs of the original holder, must at times, through purchase or otherwise, have substituted themselves for some of the co-heirs. But the principle is plain and the rapidity with which results could be achieved is startling. With such a tradition at work, both countries must necessarily at one time or another have had many a townland as much subdivided as were the open fields of the English midlands. Testimony to the prevalence of runrig in Scotland before the middle of the eighteenth century has been given. Something more should be added regarding Ireland.

When the Devon Commission made its report in 1845 runrig had pretty nearly disappeared in certain parts of the island. The following quotations are respectively from Antrim, Down, and Londonderry, three counties of Ulster: —

"Are there many farms near you held in rundale, or in common?" "Very few. . . . There are none on the Ballycastle estate. . . . I do not know more than a dozen cases in my range. I consider it a very objectionable system."

"Are there any persons holding in common or in joint tenancy?" "Very few. I do not know any at present. I had a property some time ago under me which was in rundale." "In what state were the tenants?" "Very bad indeed; but I divided it all."

"There are no farms held in rundale for some years past; I remember when it was the practice. [Yet] farms are a good deal subdivided among the members of a family which is a bad system." [1]

In Roscommon there was more of rundale, and again we are told of the custom which produced it: —

"Are there many farms held in common or in joint tenancy?" "Yes, a good many." "What is the condition of the people occupying them?" "Principally very poor persons. There are none others in my neighborhood. . . . [The system] is decreasing . . . it is very much the habit of the lower orders to divide their holdings, and give to their sons and sons-in-law a portion of their holdings, which leaves the holding little enough to support them and pay their rent. . . ." [2]

The best account of the getting rid of rundale was given by Marcus Keane, Esq., who was land agent for about 60,000 acres in or near county Clare. This large area was owned by twelve proprietors (principally by three), but was occupied by a great number of tenants. Few holdings were larger than fifteen acres. Since Keane had occasion to divide "many thousand acres," there must have been relatively more of rundale in county Clare than in the northeast of the island. His description of the situation and of his own activity is as follows: —

"The farms were hitherto (and are up to this day, where the changes have not been made) held by tenants in several different divisions scattered over the district, some . . . being as far as a mile distant from other divisions. In some cases one man held so many as ten, twelve, or fourteen different divisions, and it has been my business to go through the estates and divide them out again, giving each tenant his holding in one lot of a convenient size and extending to the high road. . . . [At first there was opposition] but of late the people themselves wish to have the changes made. . . . There was one case of a large farm of 1000 acres held among 200 tenants nearly, and they gave me much opposition. It was two years before I completely satisfied them all and satisfied myself. . . . And among the tenants upon

---

[1] Parl. Papers, 1845, xix, nos. 130, 99, 131.   [2] Ibid., xx, no. 430.

many thousand acres, whose farms I have so divided, I do not know more than two or three who complain." [1]

Proceeding, Keane corrected the testimony of the Rev. Timothy Keily, who had stated that on one farm nineteen or twenty houses had been levelled. Again we perceive the extreme subdivision of a small township and the process by which it had come about. " The fact is, that only eleven families were turned out, and fewer than eleven houses were levelled. . . . only one [tenant] had so much as five acres; the remaining ten had [together] less than twenty acres. . . . The person who had five acres was never known as a tenant, but was the younger son of a tenant who had divided his land without permission . . . most of them were persons who had divided their holdings, or had been brought in by such persons without permission . . . the whole farm contains 185 acres of arable land besides bog, and there are left on it twenty-six tenants, making an average of less than eight acres to each; only one tenant of these has more than twelve acres of arable, and that man has not thirteen acres." [2]

These descriptions of Irish farms in the nineteenth century confirm the Scottish reports of a half-century earlier and assist in explaining them. In Ireland, as in Scotland, the farm or townland was occupied by several tenants. The arable was in rundale, the parcels of a tenant being considerably scattered and intermixed with those of other tenants. What is new in the Irish account is the description of the rapidity with which the subdivision could be achieved. In the Donegal townland two generations had been sufficient to transform an undivided area of 205 acres into 422 separate lots held intermixedly by twenty-nine tenants. In the last quotation a townland of 185 acres was deprived of eleven tenants because they had, not long before, become tenants through unwarranted division. The witness from Roscommon commented on the frequency with which the "lower orders" divided their holdings among their sons and sons-in-law.

The Irish evidence thus supplements the Welsh and Scottish by accounting for the appearance of rundale. Rundale was pri-

[1] Parl. Papers, 1845, xxi, no. 1063 (14–16).   [2] Ibid., xxi, no. 1063 (4).

marily due to the custom of transmitting land to co-heirs and giving to each a share in parcels of every quality. In a brief time this practice might transform a compact farm or townland into a congeries of ill-compacted holdings, and, once transformed, a farm had little chance of regaining its earlier semblance except by the falling-in of the leases or by the action of the landlord.

Since in Scotland at the end of the eighteenth century runrig was considered ancient, it becomes pertinent to inquire whether transmission of holdings to co-heirs was a custom found in Celtic countries in earlier times. The Jacobean description of Pembrokeshire notes that its effects were at that time beginning to disappear. Hence we turn with expectation to a Welsh survey of the Tudor period which gives suggestive information. The two following descriptions of holdings, which are typical, illustrate how a transmission to several co-heirs, presumably resultant in runrig, had recently taken place at Eskirmaen.[1] To judge from the rents, which elsewhere in the survey are 2 $d.$ the acre, the first holding must have contained about 35 acres, the second about 150: —

" John ap griffith henry howell ap henry david ap meredydd griffith lloyd Morgan ap meredydd ap griffith lloyd Isabell merch griffith ap meredydd ap griffith lloyd Maude merch griffith ap meredydd ap griffith lloyd Gwenllian merch griffith ap meredydd ap griffith lloyd	tenent certas terras et tenementa ibidem que nuper fuerunt henrici ap griffith lloyd.
" Johannes dny ap gwilym Gwalter Redd ap meredydd ap gwilym Gwalter Johannes ap Jenñ ap gwilym Gwalter howell ap Jenñ ap gwilym Gwalter Griffith ap Morgan ap gwalter Gwalter ap Morgan ap gwalter Johannes ap Owen ap morgan David ap Owen ap morgan Gwalter ap Henry morgan	tenent certas terras et tenementa que nuper fuerunt Gwalter ap Jenñ llm."

This Tudor survey with its holdings in the occupation of several heirs finds a prototype in another and earlier Welsh survey — a rate-book of 8 Edward III, known as the Denbigh extent.

---
[1] Rents. and Survs., D. of Lanc., Portf. 12/4 (15–19 Hen. VII), ff. 27b, 28b.

Although this has regard primarily to the assessment of rents, the number of persons bearing the same family name who have become responsible for the return from a particular *lectum*, *Gavelle*, or *Wele* (the terms are used interchangeably) testifies to the widespread transmission of land to co-heirs. The structure of the *villata*[1] of Wigfair in the commot of Ysdulas will make this clear.[2] The *villata* in question consisted of eight *lecta*, the first of which was divided into three smaller *lecta* or *gavellae*, while the first of these in turn comprised three gavellae or *weles*, each having several tenants. If we attend to only the first of the subdivisions in each instance, the account runs as follows: —

" Villata de Wyckewere cum Hamellis de Boydroghyn et Kylmayl consistebat temporibus Principum ante Conquestum in octo lectis unde vi lecta fuerunt in omnibus locis predictis. . . . Et de hiis vi lectis

[I] unum lectum fuit penitus in tenura liberorum quod vocatur Wele Lauwargh' ap Kendelyk.

[II] Secundum lectum consistit videlicet due partes in tenura liberorum et tertia pars in tenura Nativorum quod lectum vocatur Wele Morythe.

[III] Tercium lectum consistit videlicet due partes in tenura liberorum et tertia pars in tenura Nativorum quod quidem lectum vocatur Wele Peidyth' Mogh'.

[IV-VI] Cetera tria lecta de predictis vi lectis fuerunt integre in tenura Nativorum, unde primum lectum vocatur Wele Breynt' secundum lectum vocatur Wele Meyon et tercium vocatur Wele Bothloyn.

[VII-VIII] Et duo ultima lecta . . . fuerunt tantumodo in villa de Boydroghyn et consistunt penitus in tenura Nativorum, unde primum lectum vocatur Wele Anergh Guyrdyon et secundum lectum vocatur Wele Thlowthon. . . .

[1] The *villata* of fourteenth-century Wales was a far larger unit than the Irish villata or townland of the seventeenth century, referred to above (pp. 187-189).

[2] *Survey of the Honour of Denbigh, 1334* (ed. P. Vinogradoff and F. Morgan, London, 1914), pp. 210-212.

THE CELTIC SYSTEM 197

[1] De primo lecto [Wele Lauwargh' ap Kendelyk] ...
fuerunt tria lecta seu tres gavelle videlicet
 [a] Wele Risshard ap Lauwargh'
 [b] Wele Moridyk ap Law' et
 [c] Wele Kandalo ap Lauwargh'. ....

 [a] De Wele Risshard ap Lauwargh' fiunt tres gavelle videlicet
  [1] gavella Madok ap Risshard
  [2] gavella Kendalo ap Risshard et
  [3] gavella Ken' ap Risshard.
   [i] Gronou ap Madok Vaghan, Eynon Routh' frater eius, Heilyn ap Eynon ap Risshard, Heilyn ap Gron' ap Eynon, Bleth' et Ithel fratres eius et Heilyn ap Eynon Gogh' tenent gavellam Madok ap Risshard integre, redd. de Tung' inter se per annum ... [8 d. + 12 d. + 6 d. + 7½ d. + 6 d.]. Et faciunt cetera servicia cum aliis liberis istius commoti in communi, de quibus patebit in fine istius commoti inter communes consuetudines &c."
   [ii] Seven men, of whom three are brothers of two others, hold Gavella Kendalo ap Risshard " integre."
   [iii] Thirteen men, of whom seven are brothers of five others and one is a guardian of one other, hold three-fourths of Gavella Ken' ap Risshard, and one-fourth is escheat to the lord.

The first of the lecta was in the hands of the descendants of a certain Lauwarghe, from whom it derived its name. To his three sons, Risshard, Moridyk, and Kandalo, it had passed as three lecta or gavellae. The three sons of Risshard, named Madock, Kendalo, and Ken', had in turn received their father's share as three gavellae, and their cousins had inherited similarly. Thus far there had been subdivision of the original lectum. Thenceforth these units allotted to the grandsons of Lauwarghe did not undergo formal subdivision. Yet each was by no means transmitted to a single heir; one of them might come to be held by as many as thirteen co-tenants. For the most part, each group of co-tenants, to judge from the names of its members, was descended from one of the grandsons of Lauwarghe, but for the future its bond of union was its joint responsibility for the rents and services due from the gavella which it now held " integre." In this manner did co-tenancy arise.

Thus by somewhat devious ways the custom of transmitting a holding to co-heirs has been followed from Scotland and Ireland in the eighteenth century to Wales in the fourteenth. It is, as Seebohm has shown,[1] a usage apparent in early Celtic law, and from primitive times can scarcely have failed to influence the field system of a hamlet. The subdivision that went on in the Donegal township during two nineteenth-century generations had without doubt often occurred at an earlier time in Ireland, in Scotland, and in Wales. Probably the early usage was to make the allotments for a year only; such a custom, as we have seen, was still observed in Scotland as late as the eighteenth century. In Wales, however, permanent allotments may have taken place before the sixteenth century, since Owen, describing Pembrokeshire, declared that the extreme subdivision of the lands of the Welsh in that county was due to the custom of transmission by gavelkind, a custom itself made illegal by a statute of Henry VIII.[2]

Whatever may have been the time when the subdivision among co-tenants came to be made for periods longer than a year, there is little doubt about the manner in which it took place. Each heir, the Irish account declares, demanded a portion of all qualities of land within the townland. As a result, small scattered parcels became the constituents of each allotment or holding. Certain of these parcels were of course arable, and so far as this was the case it was more or less necessary that the tenants should share in the ploughing; seldom can one of the co-heirs have had

[1] *English Village Community*, pp. 193–194.   [2] Cf. above, p. 187.

enough horses or oxen for a plough team. Coöperative ploughing must, in short, have been a custom complementary to the subdivision of holdings among heirs. Further, in so far as the parcels were arable and ploughed with a common plough they would tend to be, not block-shaped, but long and narrow, for such was the shape of the unit ploughed by the heavy plough. Pasture subdivided among heirs might fall into parcels of any shape; arable would in its nature separate into strips like those described by the Scottish reporters to the Board of Agriculture.

The appearance of runrig can thus be explained as due to the custom of subdividing arable land of different qualities among co-heirs. This custom and its effects constitute the second of the distinctions which differentiate the field system of Celtic countries from that of midland England. The first difference we have found in the markedly smaller size of Celtic townships. It has now become clear that the intermixed holdings of central England had one history, those of Scotland, Ireland, and Wales another. In the English midlands, virgates consisting of scattered strips had been fully formed when they were first described in the thirteenth century; after that they underwent little change through subdivision, the integrity of the virgate almost never admitting of fission into more than four parts. In Celtic countries, on the other hand, subdivision of a townland or a township sometimes first arose as late as the eighteenth century, and no limits were set to the lengths to which it might go. The distinction is fundamental for the comprehension of runrig, and explains the greater flexibility of its open-field arrangements.

In a general way, however, the furlongs of open arable field cultivated in accordance with Celtic runrig presented an aspect not very different from that of an English midland township. We must therefore hasten to note two other distinctions between midland and Celtic arrangements, those, namely, which resided in the methods of tillage employed and in the grouping of the parcels of the tenants' holdings.

Relative to Welsh tillage the Denbigh survey of 1334 twice mentions a three-course rotation of crops; but in both instances the reference is to demesne lands and the usage was apparently of recent introduction since these lands had been " converted " to it.[1] Elsewhere in the survey there seems to be a tacit understanding that old Welsh methods of tillage prevailed. What these had come to be in Pembrokeshire in 1603 Owen tells us: " The part of the sheere inhabited by the welshmen as before is saied, followinge their forefathers husbandrie regard more the tillage of oates then of the former graines. . . . [After folding cattle upon parcels of land from March to November] this lande they plowe in *November* and *December*, & in *March* they sowe oates in yt and have comonlie a goodlie Cropp, then they followe these landes with oates seaven eight or ten yeares together till the lande growe so weake & baren that it will not yeald the seede: and then let they that lande lie for eight or ten yeares in pasture for their Cattell." [2] Such tillage is like that of the Scottish outfield, and since there is no mention of continuously tilled infields we may conclude that it represents primitive Celtic usage.

This tillage of Scottish or Welsh outfields was, of course, far removed from English midland methods. To take crops of oats for a succession of years from land which had been prepared by a preliminary dressing of manure, and then to turn the exhausted fields over to fallow pasture for another succession of years, was unknown in the valley of the Trent. More like midland practices was the tillage of the Scottish infield. On this there was often a three-course rotation of crops; but the tillage differed in that the three crops were all spring grains, the cultivation was continuous, and the absence of fallowing was compensated for by annual manuring. Such advanced practices must have been innovations in Scotland, probably not much antedating the seventeenth century.[3] In English counties which may in early times have had a Celtic field system this particular development probably

[1] " Et sunt in dominico de terra arabili conversa in tress eisonas . . ." (*Survey of the Honour of Denbigh*, pp. 4, 230.)

[2] *Description of Penbrokshire*, i. 61–62.

[3] An account of " two husbandlands " at Lymouth in Berwickshire, dated 1651, gives detail for the infield and the outfield separately; two other descriptions of fractional husbandlands at the same place, earlier by a half-century, make no distinction between infield and outfield lands. Cf. Hist. MSS. Commission, *MSS. of Col. D. M. Home* (1902), pp. 220, 212, 214.

seldom took place.¹ From the limitations of outfield tillage another escape, it seems, was devised and some approach to midland methods was made.² But this happened so early and the traces of outfield cultivation in England are so slight that the contrast between Celtic and midland tillage, sharp as it was in reality, is not very helpful in estimating Celtic influences in England.

On this account it is desirable to determine for the Celtic system the attribute which we have so often found pertinent in midland England — the grouping of the parcels of a tenant's holding within the arable area. In the fragmentary Welsh fields of the sixteenth century such grouping would tell us little. Where only a few tenants had each a parcel or at best a few parcels in the common arable field, the location of the parcel or parcels imports little, since the tenant's reliance was upon his closes. If the grouping of parcels is to be important, the parcels must constitute the major part of the holding.

So they did in Scotland and without doubt in Ireland. In the latter country, whenever townlands were subdivided each tenant desired a share of each quality of land. The location of the parcels of a holding was thus dependent upon the number and location of the different qualities of land to be divided. There can scarcely have been thought of dividing the townland into two or three equal compact fields. Indeed, it would have been impossible to do this unless nature had given to the township only two or three qualities of land in compact areas, and there would have been no occasion for doing it unless a fixed two- or three-course rotation of crops was to be established. The map of the Donegal townland, which has been reproduced above, shows no such division. The strips there assigned to three tenants were not scattered throughout the arable; in fact, in about one-half of it no one of the three tenants had any strip whatever. The field system evolved by Irish co-heirs and co-tenants in the eighteenth and nineteenth centuries was clearly not that of the English midlands.

In Scotland the succession of crops itself prevented the subdivision of the outfield into three equal parts. Only about

¹ But cf. below, p. 232. ² Cf. below, pp. 221, 225–226, 271.

one-tenth of it was brought from grass into tillage each year, the remaining tenths being similarly treated in succeeding years. It is possible, however, that the infield may have met with tripartite division, since a three-course rotation of crops was usual there. Yet no advantage could have been gained by the marking out of three compact areas. All crops were spring grains and no furlongs lay fallow; rights of summer pasturage, the main pretext for the tripartite division of English midland fields, were non-existent. Nothing would have been sacrificed if the furlongs which in any year were devoted to barley had not been contiguous. Nor do the documents divide Scottish infields by hard and inflexible boundaries into three equal compact areas; the parcels of a holding are not, for example, assigned to East field, North field, and South field. Absence of division by fields thus becomes a concomitant of runrig and one of its distinguishing marks. It will prove important when the question of Celtic influence in England arises. Terriers from counties where such influence is suspected should, if the suspicion be correct, show no grouping of their parcels by fields.

Before the subject of Celtic field arrangements is dismissed, it should be pointed out that the subdivision of arable in the manner of runrig was not, at any one time or place, an essential characteristic of the system. If the explanation of the origin of runrig above given be correct, such subdivision was rather an accident. Farms, townships, townlands, which are found divided in the eighteenth century may well have been undivided a few generations earlier. Landlords may at times, on the expiration of leases, have taken certain townships in hand and reconsolidated holdings; in the recompacted areas subdivision may once more have been permitted and the cycle again have run its course. Regarding Celtic countries, then, no sweeping statement can be made as to the precise aspect of the townships at any particular time. Some of them may have been entirely in the hands of one or two tenants, with no runrig manifest; others may have been much subdivided.

The latter sort would in turn have assumed different aspects in so far as arable or pasture predominated. If the township were devoted to pasturage, closes, more or less irregular in shape, would appear. In the Donegal townland the pasture was " held in lots in common." The map shows that the plots of pasture in the occupation of a tenant were about as much severed one from another as were the strips. Yet there was less reason for their being so and remaining so, since pasture is not so diverse in its qualities as arable and there was no question of common ploughing. One can, therefore, imagine co-heirs subdividing a pasturage township on broader lines than they would have thought applicable to one largely arable. For these reasons it is not improbable that such a township sometimes broke apart into closes which may have been to some extent consolidated.

Probably this is what happened at times in Wales. There in the sixteenth century township after township consisted of closes,[1] those of a holding being frequently contiguous. The principality seems at that time to have been much more of a stranger to runrig than was Ireland or Scotland, a circumstance best explained by the Tudor prohibition of transmission by gavelkind and by the hypothesis of an early predominance of pasture. In Scotland, as we know, runrig prevailed in the first half of the eighteenth century, and the situation in Ireland was without doubt similar. The reason must have been that arable was, or had been, relatively more extensive in these countries than in Wales. If this supposition be correct, the different aspects assumed by the fields of Celtic countries are only natural developments of a flexible field system.

We are left, accordingly, with four distinctive characteristics of the Celtic field system. In the first place, the open arable fields were small, a necessary corollary of the small size of the townships; in the second place, they frequently consisted of the intermixed strips of several tenants, but this intermixture was variable, originating with and depending upon the extent to which subdivision among co-heirs or co-tenants had proceeded; in the third place, the rotation of crops, so far as we know it, was not winter corn, spring corn, fallow, but something quite different, viz., a succession of spring crops followed by several fallow years,

[1] *Archaeologia Cambrensis, Supplement of Original Documents*, vol. i *passim*.

or an unbroken succession of three crops of spring corn upon land manured once every three years; finally, the tenants' parcels were not divided between two or three large arable fields, and there is no evidence that fields of this sort ever existed.

The influence of such a system in England is not altogether easy to trace in the documents at hand. One of its four characteristics will be of little assistance. The continued subdivision of holdings, farms, or townlands among co-heirs or co-tenants, perhaps the most striking feature of runrig, is only occasionally perceptible in western England. In general it seems to have given way before the more rigid requirements of the English manorial system, which preferred that the rent of a holding should be paid by relatively few tenants. Nor have we many instances to show that the English counties which bordered Scotland or Wales favored a rotation of crops different from that which prevailed in the English midlands. In this matter, too, non-Celtic influences were early dominant.

The smaller size of Celtic townships is a feature which is reflected in several English counties. Useful as it is, however, in tracing Celtic influence, it yields in utility to the last of the four characteristics, the arrangement of tenants' parcels in the arable fields. Where Celtic influence was felt, the parcels, we shall find, were closes, or irregular plats, or arable strips in runrig. Closes or plats may be expected to predominate in regions situated near Wales and seemingly devoted early to pasturage; arable and the attendant runrig may be expected on the Scottish border. In no instance will there be a division of the arable into two or three large fields with a distribution of the parcels of holdings between them. Evidence on this point, so far as the terriers are concerned, will be largely negative. Only rarely will a terrier so clearly locate the strips of a bovate or a virgate as to render it probable that these strips were closely grouped within one part of the arable area of a township and hence not amenable to distribution throughout fields.[1] Elsewhere we shall have to be content with such negative testimony as results from the omission, in all available descriptions, of those field divisions

[1] Cf. below, pp. 208–210, 235–237, 245.

which midland terriers more or less frequently contain. Hence it will never be possible to say regarding an English county, "Here is clearly the field system of Scotland or Wales or Ireland." We shall rather have to conclude: "Its fields lack the positive attributes of English midland fields, just as the fields of Celtic lands do. In their negative characteristics they are Celtic." In so far as this conclusion is convincing, Celtic influence in England will have been established. This prefaced, we may begin our examination of the field systems of such counties of northern and western England as did not fall within the boundaries of the two- and three-field system. Since Scotland introduced us to the Celtic system, the counties of the Scottish border may occupy us first.

CHAPTER VI

THE INFLUENCE OF THE CELTIC SYSTEM IN ENGLAND

Northumberland

THE history of Northumberland open fields was nearly completed before the period of parliamentary enclosure. The reporters to the Board of Agriculture in 1794 declared that the parts of this county " capable of cultivation " were " in general well enclosed by live hedges," the only exceptions being " a small part of the vales of Breamish, Till, and Glen," where enclosure was then in progress. They noted further that lands which were or might be cultivated by the plough constituted two-thirds of the county, an area equal to nearly twice that of Oxfordshire.[1] Of acts of parliament earlier than 1760 Slater found two relative to Northumberland, enclosing respectively 1300 and 1250 acres of arable; of acts later than that date he discovered but six.[2] Two of the latter do not distinguish between arable and common, in three others the amount of arable to be enclosed is estimated at 380 acres altogether, while at Corbridge only did the open arable field amount to as much as 945 acres.[3] Parliamentary enclosure of common fields in Northumberland after 1760 is practically negligible.

The earlier acts, those of 1740 and 1757, point to the completion of an enclosure movement which had been in progress for a century and a half. Some information regarding this process may be obtained from the monumental *History of Northumberland*, since the contributors, in their accounts of the various parishes, refer at times to the enclosing of the open fields. North

[1] J. Bailey and G. Culley, *General View of the Agriculture of the County of Northumberland* (London, 1794), p. 50: " Lands which are or may be cultivated, 817,200 acres; mountainous districts improper for tillage, 450,000 acres."

[2] *English Peasantry*, p. 294. The two earlier acts relate to Gunnerton (1740) and West Matfen (1757).

[3] *A History of Northumberland* (in progress by the Northumberland County History Committee, vols. i-x, Newcastle, etc., 1893-1914), x. 143.

Middleton and Broomley, it appears, remained open until the beginning of the nineteenth century, undergoing enclosure in 1805 and 1817 respectively.[1] To judge from what happened in several other townships, however, the movement was most pronounced during the seventeenth century and the early eighteenth. Not only were the open arable fields of Seaton Delaval already enclosed in 1610, but articles of agreement looking to the enclosure of Cowpen were drawn in 1619, and the process was under way at Dilston in 1632.[2] The tenants at Earsdon signed their articles in 1649, the same year in which the re-allotment at Preston was completed.[3] At Backworth the open fields disappeared in 1664.[4] The Ovington and Rennington enclosures, however, were delayed until the next century, the former being the work of commissioners appointed in 1708, the latter being asked for in 1707 but not carried out till 1720 and 1762.[5] At Newton-by-the-Sea and Embleton the open fields disappeared a little later still, in 1725 and 1730 respectively.[6] From such items, insufficient as they are, it seems not improbable that the greater part of the common arable fields of the county had been enclosed by the end of the first quarter of the eighteenth century. If so, Northumberland in its enclosure history resembles Durham, but differs markedly from the midlands.[7]

Even in the sixteenth century the transformation of Northumberland fields had begun. At Lesbury, on December 6, 1597, the tenants resolved at the manor court that they would, "between this and the 1st of March next, procure a survey of the South field in Lesbury, and that every tenant [should] have his land laid in several, and the same to dyke in convenient time after the said survey."[8] Clarkson, who made a survey of the township of Tuggal in 1567, intimates that it was largely if not wholly

[1] *Archaeologia Aeliana*, new series, 1894, xvi. 138; *History of Northumberland*, vi. 143.
[2] *History of Northumberland*, ix. 201, 325; x. 276.
[3] Ibid., viii. 244, n. 3; ix. 4.
[4] Ibid., ix. 40.
[5] *Archaeologia Aeliana*, new series, xvi. 129; *History of Northumberland*, ii. 159.
[6] *History of Northumberland*, ii. 45, 98.
[7] Cf. above, p. 107, and Chapter IV.
[8] *History of Northumberland*, ii. 424.

enclosed. After explaining that it had been "divided at the greate suite of the Bradfords, who, havinge the moste parte of the towne in ther hands, wolde not agree with the other tennants in ther ancyent orders but with thretnings overpolled and trobled the said tennants in th' occupacion of ther grounde," he proceeds to consider measures "that his lordship may have also the said towne planted in the anncyent orders with the same number of tenant cottigers, smithe and cotterells, to have ther groundes severallie enclosed by themselves, wherfor they dyd lye in common, as well to the great strengthe of the towne as comodetie to them all."[1] Thus cautiously were proposals for enclosure advanced in the days of Elizabeth.

Perhaps the best conception of the earlier condition of Northumberland open fields and of the changes in progress at the end of the sixteenth century can be got from documents relative to Long Houghton. In this township, which lay on the coast and was the property of the duke of Northumberland, a survey was undertaken in 1567 to rectify mistakes made in the rearrangement of a few years earlier. The introduction to the survey explains what had been the state of affairs before the rearrangement. "The arable lande . . . [of Houghton Magna] lyeth for the moste parte nighe [the] sea syde, and is donged with the sea wracke . . . and, because of the greatnes of the said towne, the towne is now dividit in two partes, for that they were xxvii tenants besyde cotteagers, havynge alwayes and in every place every one tenant one rige by [him]sellfe, and so consequentlye, from ryge to ryge, that every tenant had one rige, then the first did begyn to have his a ryge for his lot agayne, and so by rygge and ryge it was in every place devidit amonge them to the great chardge and laboure of everye one of the said tenants: althoughe the same partition did geve to every tenant like quantite of all sortes of lande, yet it was so paynefeule to them and ther cattell that for the moste parte the said tennants did never manure ther

[1] Farther on he notes that "at the late partic[i]on . . . the churche landes nowe in the tennure of Rolland Foster were layed altogether," and that certain crofts contained "xii rigges before the particion of the towne" (*History of Northumberland*, i. 351, 353).

grounde threwgly; wherby they did fall in great povertie; and also ther severall grounde, called their oxen pasture . . . was in breaffe tyme over eatyng and maide baire of fedyng."[1]

Such is the picture of an open-field township in which the distribution of parcels throughout a large area had become intolerable. To relieve the situation an unusual remedy had been devised. A short time before the survey of 1567 the arable area had been divided into a northern and a southern part, and the parcels of each husbandland (as a holding was called) had been confined to one or the other of these two divisions. To accomplish such an alignment parcels had been exchanged but not to any extent consolidated, as a later survey of 1614 makes clear. One of the fully described furlongs of this last survey, called "Bastie lands," is transcribed by the historian of the parish. In it each of thirteen husbandlands on the south side then had parcels, which usually contained about half an acre apiece.[2]

Similar divisions of townships into two parts seem to have been not infrequent in Northumberland.[3] A survey of Acklington made in 1702 shows that one had there been accomplished, dividing the $17\frac{1}{2}$ farms into $8\frac{1}{2}$ on the north side and 9 on the south side.[4] At Lesbury, which adjoins Long Houghton, a division was proposed when the latter township was divided, and the matter was put into the hands of the surveyor who has already been quoted. In this case, however, he pronounced against the division, chiefly on the ground that an equally good water supply could not be had for both parts. His account begins as follows: " It wer not good that this towne wer devyded into thre [farther

[1] *History of Northumberland*, ii. 368.

[2] Ibid., 378. The survey of 1567 was undertaken to adjust minor details. In the earlier division the tenants of the north side had got the poorer lands, and the boundary between the common pasture of the farms on the one side and the arable lands of the farms on the other was unsatisfactory.

[3] At Rock before 1599, according to a map of that date, there had been a rearrangement of farms as follows (ibid., 128): —

" Belonginge to 5 Farmes on the North Barne in arable, meadow, and pasture, 214 acres
"   " 7   "   " " south side   "   "   "   "   "   " 301 "
"   " 5   "   " " moore   200 "
"   " 7   "   " " moore   280 " "

[4] Ibid., v. 372.

on he says "two"] severall townes, althoughe yt ys a greate towne, many tenants and cotteagers, every tenant having his lande lyeinge rigge by rigge and not in flatts nor yet in parcells of grounde by yt selfe, so that therby the labor of the tenants and their cattell ys muche more, to the greate dystruction of the said tenants." In verification of this description, a survey of 1614 tells of a furlong (South Brig haugh) in the West field which contained 4 acres, 3 rods, 26 perches, in eighteen strips held by fourteen tenants.[1]

If we inquire what field system held together the widely-scattered parcels of Long Houghton before the rearrangement in the middle of the sixteenth century, the answer must be sought in a map of 1619, which is closely associated with the survey of 1614. Although the township had by this time been divided into a northern and a southern half, the boundary between the halves crosses what was clearly an older division into fields. These fields were three and they were unequal in size. Old and new arrangements by fields and by halves distributed the unenclosed arable as follows: —

South field, 99 acres on the North side, 276 acres on the South side

West field, 181 acres on the North side, none on the South side

East field, 242 acres on the North side, 302 acres on the South side.[2]

Although there is here the suggestion of an early three-field arrangement, the inequality in area between the fields is a questionable circumstance. Especially great is the discrepancy between the 181 acres of the West field and the 544 acres of the East field. Furthermore, if the midland division was known and after the rearrangement did not lose favor (there is no indication that the strips were consolidated or the method of tillage changed at the time), it seems strange that within fifty years three new fields had not taken form on the north side and three on the south. Of such, however, there is no trace in the map of 1619, which

[1] *History of Northumberland*, ii. 418, 425. [2] Ibid., map facing p. 368.

clings rather to an antiquated division. We are thus led to conclude that no three-field system prevailed at Long Houghton in 1619, and that the three "fields" of them ap were never really such, but only convenient topographical names for different parts of the township's arable.

To get further information regarding the possible existence of a three-field system in Northumberland, we turn to other surveys made in the days of Elizabeth or James I. Many of them, accompanied by maps, exist in the archives of the earl of Northumberland, but the authors of the county history have seldom transcribed the information which might be at once decisive. They have not, except in one instance, given the distribution of the acres of the tenants' holdings throughout the fields, an omission which greatly increases the difficulties of the investigator at this point.[1]

A notable feature about several of the maps and schedules which describe the townships belonging to the duke is their insistence upon a division of the arable into three or four fields. Round the village of Acklington, a map, probably made in 1616, shows three fields, North, South, and East, but it gives no areas.[2] The plan of Clarewood and Halton Shields, dating from 1677, pictures two groups of three fields but is equally reticent about their areas.[3] On the Tuggal map of about 1620, what remained of the fields amounted to 71 acres in South field, 64 acres in Whittridg field, and 118 acres in Hedglaw field.[4] At Rock, too, according to the map of 1599, there were "remaines" of three fields — Earsley field containing 84 acres, Rockley field 70, Arksley field 131.[5] The survey of Bilton, completed in 1614, assigns to three fields, also shown on a map of 1624, areas which give to South field 176 acres, to East field 138, and to North field

[1] Unfortunately, I have been unable to examine the documents at Alnwick Castle.

[2] *History of Northumberland*, v. 376.

[3] Ibid., x. 389. Similarly, there is record of three fields, North, Middle, and Low, at Ovington, but no information about their respective areas or the apportionment of the tenants' holdings (*Archaeologia Aeliana*, new series, 1894, xvi. 129).

[4] *History of Northumberland*, i. 342. To Whittridg field should probably be added 26 acres in Townsend flat and 17 acres in Glebeland.

[5] Ibid., ii. 128.

216.¹ At Rennington, where the fields seem to have suffered no diminution from their original size, there were, in 1618, 89 acres in South field, 248 in West field, 146 in North field, 6 in Orchard, and 29 in Barelaw field.² It will be noticed that at Bilton and Rennington, those townships in which the fields were most intact, the areas of the three fields were distinctly unequal.

Other townships divided their arable into four parts. At Shilbottle the fourth part, which was smaller than the others, apparently had no close connection with them. It was known as " The Fower Farmes called the Head of Shilbotle " and contained 200 acres; but its four tenants had together only 56 acres of pasture lying in the other three fields. The latter were known as North, Middle, and South, their areas being 347, 268, and 350 acres respectively.³ Were it not for the " Fower Farmes," this division would wear somewhat the aspect of a three-field township.

Elsewhere the four fields bore conventional names, but their acres were unequal. The Lesbury fields, which, as we have seen, were in 1567 proposed for division, numbered four in 1614. Of these the West field, not shown on a map of ten years later, contained 110 acres, while the other three, Northeast field, East field, and South field, were much larger, comprising respectively 395, 246, and 287 acres.⁴ No combination here would evolve into anything like a three-field arrangement except the union of West field with East field, and even this, apart from the situation of the two, does not obviate considerable discrepancy. Slightly more symmetrical were the fields of South Charlton in 1620. Three of them included meadow, and the subdivision gave to North field 142 acres of arable and 11 of meadow, to East field 122¾ acres of arable, to Middle field 58½ acres of arable and 38 of meadow, to West field 147 acres of arable and 8½ of meadow.⁵ By combining the arable of East field and Middle field we should get a total only greater by about thirty acres than the area of each of the other fields, a not impossible three-field arrangement. At Lucker the four fields were less amenable to a three-field

---

¹ *History of Northumberland*, 451–452, 456.   ² Ibid., 156–157.
³ Ibid., v. 416, 427, 429 n.   ⁴ Ibid., ii. 416 sq.   ⁵ Ibid., 307.

grouping, nor could they well have maintained themselves as four fields. Their names, too, were unusual. To Quarrell field were assigned 72 acres, to West field 97, to Bank quarter 158, and to Gawkland quarter 57.[1] Here, as in the maps and terriers of several other townships, the same phenomena appear. Despite what superficially seems to be a simple three- or four-field arrangement, the inequality in the apportionment of the arable among the fields raises the question whether the subdivision has more than topographical significance.

There is, of course, a simple criterion in such cases, one to which resort has often been had. It is the distribution of the acres of a holding among the township fields. Only the inadequacy of the transcripts in the otherwise elaborate county history makes necessary inferences from other data. Yet a few terriers of the desirable kind are discoverable. The best refers to Rennington, a township in which, as has been noted, the West field contained 248 acres, the North field 146, and the South field 89. Since the terrier is a part of the survey which states these areas, one would expect some correspondence between them and the apportionment of the acres of the holding in question. Yet scarcely any appears, the terrier assigning to the three fields 21, 2, and 10 acres respectively.[2] West field and South field thus receive more than their due, while North field is markedly slighted. The terrier of a holding at Elford, made in

[1] *History of Northumberland*, i. 234.

[2] Ibid., ii. 157. The specifications of "Trestram Philpson's farme" run as follows: —

	Acres	Roods	Perches
House and garth	1	2	38½
South field arrable	10	0	27⅛
Orchard "	0	1	8¾
West field "	21	0	5½
North field "	8	3	11¾
Barlawe feild "	2	0	0⅛
In the West feild meadowe	2	1	11⅛
In Twenty acres	1	3	30¼
In Cowde close	2	3	32
In Gowlands Croke poole	1	1	16⅝
In the Meadow Dayles	1	3	20¼
In the Orchard Layninge	0	2	16⅜
Eight gaytes in the Oxe pastures	19	2	28
	—	–	—
Some of acres	74	3	8½

1621, has similar characteristics. There the arable lay largely in three " quarters," North, East, and West, doubtless the principal divisions of the township fields; yet to North quarter were assigned 8 acres, slightly more than were given to the other quarters together, these receiving respectively $3\frac{7}{8}$ and $3\frac{1}{2}$ acres.[1] From Corbridge come particulars of the distribution of the demesne acres among four fields. In West field were 26 acres, in East field 112, in North field 84, and in Little field 25.[2] Since in the detailed list of " riggs " there is no separation of Little field from North field, it is possible that the two were tilled as a unit. If so, this composite field becomes as important as East field, but the insignificance of West field is only the more emphasized. Finally, the terrier descriptive of a holding at Great Felton in 1585 is concerned with only an East field and a West field. None the less, it fails to divide its acres evenly between the two, assigning to one 15 acres and to the other 5.[3] In general, we are thus led to conclude that the acres of a Northumberland holding, whether apportioned to two, three, or four fields, were not disposed as they would have been in a normal township of the midland area.

In some Northumberland terriers of Tudor and Jacobean days there is discernible a tendency to group fields along with other

[1] *History of Northumberland*, i. 287. This holding of John Chaundler is thus described: —

	Acres	Roods	Perches
" The house and scite	0	0	30
Six butts of arable land lying among other lands in a croft there	2	1	10
Fowertene several parcells of arable land which lie on the North Querter containing	8	0	35
Thirteen parcells of arable land lying on the East Quarter	3	3	30
Other parcels in West Quarter	3	2	0
A small parcel lying in East Meade	0	0	35
Another small parcel	0	0	20
3 beaste gates in the Ox Pastures	..	..	..
Total	18	0	2 "

[2] Ibid., x. 124–130.

[3] Ibid., vii. 252. Besides the tenement and a croft containing a half-acre, the holding comprised: —

" 2 closes in the east field of Felton . . . together of 12 acres
11 selions in the same field ' super moores pett ' of 2 acres
At Chamley gappe 1 acre
In the west field parcels called ' Botons peace,' ' le lawe' et ' le hedlandes,' together of 5 acres
1 close of pasture . . . called ' le birkeclose ' of 8 acres."

divisions of the arable area. Whether this be the fault of the maker of the survey or whether it points to the minor importance of fields is difficult to say. A Brotherwick terrier from the survey of 1585 suggests the latter explanation.[1] Most of the selions (excluding those recently held by Thomas Pinne) lay in North field or in South field or adjacent to the "Lang-rigges." If all those in the list which fall between North field and South field be looked upon as lying in North field, the total much overbalances the number left for South field. If, on the other hand, the two fields stand independently toward the other areas, no three-field grouping is apparent, even if "Lang-rigges" be exalted into a field.

A tendency to neglect division by fields in enumerating the parcels of a holding appears in one of the Northumberland surveys which has been printed in full.[2] The survey is concerned with the open fields of two townships, Tynemouth and Preston, but it is incomplete. Only such fields as are about to be re-allotted receive attention. It is possible that all the unenclosed arable at Tynemouth was redistributed, but of this there is no certainty. If it was, only two fields, North and South, existed there and they were somewhat unequal in area.[3] The Preston fields are confessedly not described in full. Only "so much as was now presented to be divided" appears in the total, which comprised

[1] *History of Northumberland*, v. 258: —
   16 selions of arable land in the North field
   14 south of the "Lang-rigges"
   4 in "Whyte-lees"
   3 "super le Lang-rigges"
   2 "by the Hall-well"
   10 in the South field
   5 "iuxta le snake hole"
   12 "in the Crokes, formerly held by Thomas Pinne".

[2] "The Terraire or Accompt of Measure of certain Lands lying within the Territories of the Mannor of Tinemouth and Preston, 1649," *Archaeologia Aeliana*, new series, 1887, xii. 173 *sq.*

[3] "Of the Particon of Tinemouth —

	Acres	Roods	Perches
The Quantity of the South Feild of Tynemouth	188	1	9
The Brocks contains	30	2	20
In the North Feild on the upper side of Monkseaton way	51	1	32
In the North feild more East from that and more Northerly	206	1	30 "

183½ acres in North field, 137¼ in South field, and 161½ in Miller Leazes. This area was re-allotted to five copyhold " farmes," each containing 53 acres, but the former relation of these farms to the fields is not indicated. Such relationship is stated only for certain old freeholds, which are, however, not very satisfactorily described.[1] In them something is usually allotted to West field and to Miller Leazes, but there is considerable obscurity about North field; it will be noticed that, of the many rood parcels that were " next the Rake," only one is located in that field. For the most part the strips are assigned to such areas as Dikan Dubb or the Long Dike, and it is impossible to group them by fields. If terriers like these be typical of the Northumberland surveys which were made in such considerable numbers preparatory to the re-allotment of holdings in the early seventeenth century, the surveys either contain little useful information about fields or show the acres of the holdings irregularly distributed.

It may be urged, however, that we are here dealing with relatively late field arrangements, that in Northumberland the old system, whatever it was, had by this time begun to decay. The very ease with which a re-allotment of parcels was brought about, as at Long Houghton before 1567, testifies, one may say, to the laxity of the old ties. Laxity there pretty clearly was, and it

[1] For three of them the details run as follows, the areas being in roods: —

Location in Preston Field	Robert Ottway's Freehold	Robert Spearman's Freehold	George Millburn's Freehold
In the West Feild	1, 1 and a butt, 4, 4, and 2 banks	1	3 butts
In Shedletch	1, 3	2, 2 riggs	6
Att Moor Dike	2, 1 and a bank	....	1, 1 and a bank
At Long Dike	1, 1, 1, 1, 1, 1, 1 short headland	1, 1, 1, 1, 1	....
Att Dikan Dubb	6, 3, 1	3	4
Next the Rake	3, 1, 1, 1, 2, 4, 1	1, 3, 1, 1, 2	6 (in the North field), 2, 1, 1, 1 lee rigg
Att Morton Way	1 headland	1, 1, 1	....
In the watery Reens	3, 1 headland	2 butts	....
In the Burnetts	3	1	....
In the miller Leazes	....	1	2
In the Garland meadow	....	1 and a meadow Spott	....
In the Hundhill	....	4	....
In the New Close	....	....	3

only remains to inquire whether earlier evidence hints at the strict observance of more inflexible rules and divisions.

Numerous terriers of Northumberland holdings earlier than the sixteenth century are to be found in the feet of fines and in the monastic cartularies, especially in the record of a survey of the lands of Hexham abbey. All these documents agree in showing the prevalence of intermixed strips,[1] which were often no more than one-fourth of an acre in size. They agree further in seldom referring to fields, and are almost unanimous in never dividing the strips of a holding between two, three, or four fields. The parcels in question are assigned to divisions which in a midland area would have been called furlongs, shots, or quarentines, but in Northumberland were usually designated rigs, dales, flats, laws, and occasionally even fields. A terrier of 1479, which describes a husbandland containing 27 acres of arable attached to a tenement at Chollerton, illustrates most of these peculiarities.[2] It is

---

[1] At "Mulefen" the 12 acres which accompanied a messuage lay in 11 parcels; at "Copum" two tofts were transferred along with 15 acres in 8 parcels; at "Berewik et Bitewurth" 10 acres were divided into 6 parcels (Ped. Fin., 180–3–22, 11 Hen. III; 180–4–61, 19 Hen. III; 180–5–107, 30 Hen. III). Two deeds transferring 7 and 12 acres at "Thrasterston" enumerate 7 and 16 parcels respectively (*Chartulary of Brinkburn Priory*, Surtees Soc., 1893, pp. 43, 45). At Thockrington, about 1280, 20 acres of arable lay in 33 small parcels (*History of Northumberland*, iv. 401).

[2] James Raine, *The Priory of Hexham* (Surtees Soc., 2 vols., 1864–65), ii. 30. The description runs as follows: —

> "Et idem [Hugo Colstane] tenet xxvii acras terrae arabilis pertinentes tenemento praedicto, quarum ii acrae et dimidia acra ex parte australi rivulae de Eriane . . .
> Et super le Kilnflate ex parte occidentali dimidia acra
> " " Horslawspule dimidia acra Schothalghbankys
> " " Nithre dimidia acra
> " " Overschotlaubankes dimidia acra
> " " le Blaklaw ex parte orientali ejusdem dimidia acra
> Et ex parte occidentali ejusdem i acra
> " " " orientali le Lons . . . ane i roda
> " " " orientali Bronslauemedoue i roda
> Et super Bronslawflate ex parte occidentali dimidia acra
> Et ex parte boriali Bronslawmedoue buttando super eodem iii rodae
> Et super le Canonflatte . . . i acra et dimidia
> " " " lez buttes iuxta le dyk . . . dimidia acra
> " " " lez hevedlandes de Brouneslawflatte . . . i roda
> Et ad capud del Maynflatt . . . dimidia acra
> Et ex parte australi le Crosse . . . i roda
> Et super Holmersbank . . . dimidia acra
> Et in Harlawhop buttando super le Messeway dimidia acra
> Et ex parte occidentali iuxta le Harlaw dimidia acra

closely followed by a similar terrier describing a holding of 28 acres, the parcels of which are located with the same explicitness in the same field divisions.[1] Apparently the husbandlands of a Northumberland township in the fifteenth century shared in all furlongs after the manner described at Long Houghton a century later,[2] and this without reference to any three-field arrangement.

In contrast with the rather impressive bulk of negative evidence in the early terriers pointing to the non-existence of the midland system in Northumberland, a few items seemingly suggestive of it deserve notice. In a terrier of a half-carucate at Whalton the 52½ acres "in campo ejusdem villae" are described in such order that, if the first two items be added together, four thirteen-acre groups result. It is noteworthy, however, that the name of only one field appears, that the half-carucate consisted of five relatively large blocks of land rather than of scattered strips, and above all that four of these blocks lay to the west of the village, two of them being carefully located in the West field.[3]

> Et super le Mesiway super eandem i roda
> Et ex parte orientali le lonynghed i roda
> Et super Aldchestre ex parte australi iii rodae
> Et ex parte occidentali super le Stobithorn i roda
> Et super Morelaw ex parte orientali ejusdem . . . dimidia acra
> Et ex parte boriali de Dueldrigge dimidia acra
> " " " orientali le Smythehopside dimidia acra
> Et in medio Craustrige dimidia acra
> Et ex parte orientali de Westraustrige i acra
> Et super Estraustrige ex parte occidentali i roda
> Et ex parte orientali Fartirmerethorne . . . i acra
> " " " australi terrae praedictae i roda
> Et inter Faltermere et lez Merlpottes i acra
> Et ex parte boriali lez Merlpottes i roda
> Et in medio Waynrig iii rodae
> Et ex parte orientali le Brereryg i roda
> " " " orientali lez Hudesrodes i acra
> Et inter lez Kornhilles i roda
> Et in medio le Milnrig dimidia acra
> Et super Fulrig iii rodae
> Et in Swynburne-feld ex parte boriali le crosse i acra."

[1] Raine, *Priory of Hexham*, ii. 32.
[2] Cf. above, p. 208.
[3] Raine, *Priory of Hexham*, ii. 39. The description runs as follows: —
> "Tenent etiam in campo ejusdem villae dimidiam carucatam terrae, viz., lii acras terrae et dimidiam, quarum
> Super Lindeslawe ex parte occidentali ejusdem villae iacent iiii acrae
> Et super le Flores ex parte occidentali prope dictas acras ix acrae
> Et in le Westfeld inter Walwyk et Leverchild xiii acrae vocatae le Burnflatt
> Ex parte australi molendini ibidem xiii acrae
> Et super le Farnelaw ex parte orientali villae ejusdem xiii acrae."

These circumstances scarcely accord with midland arrangements. No more do the names and the allotment of acres in another thirteenth-century terrier, dated 30 Henry III.[1] Of the twelve acres which, according to this, accompanied a toft and were subtracted from four bovates at Billingworth, $4\frac{1}{4}$ are referred to East field; but the assignment of $2\frac{1}{2}$ to a field called Hypelawe and of $5\frac{1}{4}$ to a field called Horchestres-and-Bereacres destroys the symmetry of any three-field arrangement, quite apart from the fact that the names are unusual.

Three early charters hint at two-field usages but without giving definite assurance. At Whittonstall six acres were in the thirteenth century described as "duas in tofto et crofto . . ., et in campo apud orientem iuxta spinam dimidiam acram, et iuxta viam . . . dimidiam acram, et in campo versus occidentem iii acras."[2] A division which thus gives to the East field one acre and to the West field three acres is corrected in a Cramlington grant of the twelfth century, according to which thirty acres were so situated that there were " xv in una parte villae et xv in alia ";[3] still, these vague localities are not fields. An early grant which does locate its dales in two fields transfers $14\frac{3}{4}$ acres at Leighton, describing them as follows: —

"In campo occidentali totas illas duas mikel dales et totas illas duas fair dales, quas Syuuardus et Robertus filii Stephani tenuerunt, cum toto prato in transverso marisci
    et totam Thirndale Roberti cum prato
    et totam Halledale Syuuardi cum prato
et in campo orientali totas illas duas Horthawedales
    quas praedicti homines tenuerunt
    et ii dales totas in Prestesflat quas Thomas de Clenil tenuit. . . ."[4]

---

[1] " Quatuor acras et unam rodam que iacent in campo qui vocatur Estfeld . . . duas acras et dimidiam que iacent in campo qui vocatur Hypelawe . . . quinque acras et unam rodam que iacent in campo qui vocatur Horchestres et Bereacres " (Ped. Fin., 180–5–113).

[2] *History of Northumberland*, vi. 182, n. 3.

[3] Ibid., ii. 226 n.

[4] [J. T. Fowler], *Chartularium Abbathiae de Novo Monasterio* (Surtees Soc., 1878), p. 85.

Two fields are here, but the chronic irregularity of division is also present, especially so far as the lands of Syuuardus and Robert are concerned. Thus the early two-field evidence for the county is hardly more satisfactory than the three-field evidence has proved.

From all the Northumberland testimony relative to fields only one item points clearly to three-field husbandry. This occurs in an account, written in or about the year 1596, of the expulsion by Robert Delavale, Esq., of the tenants of Hartley and Seaton Delaval, two townships near Newcastle-upon-Tyne.[1] In both, it is stated, each dispossessed tenant had been able to till " 60 acres of arable land, 20 in every feild." Such even division of holdings among three fields is something hitherto not met with in the Northumberland evidence, and seems at first sight to constitute straightforward and convincing testimony that a three-field system existed in the county. Before this conclusion is admitted, however, the seemingly decisive passage should be more closely examined to see whether it admits of any other interpretation.

In the first place, the assignment to these townships of husbandlands precisely similar in size and divided in precisely the same manner suggests that the writer, whose subject was in no way related to field systems, was mentioning the tenants' holdings only incidentally and in a very general manner. Even in the most typical of midland townships the acres of the copyholds were not divided with this precision among the fields. If we ask for more specific evidence about the subdivision of a copyhold at Hartley or at Seaton Delaval, we find that the editor of the county history has been obliged to make inferences in the one terrier of which he gives an account. At Hartley, William Taylor had, it appears, 105 acres which lay in three groups of shots or furlongs. One group was assigned to the South field and one to the North field, but either the third group was not assigned to any field or the attribution is missing through injury to the manuscript.[2] Although the editor conjectures that a West field was in question, he gives no reason for his belief, nor

---

[1] *History of Northumberland*, ix. 124, 201.   [2] Ibid., 122.

does he make note, as he so easily might have done, of the areas of the parcels which fell within each group. Even at Hartley, therefore, we are left with something of the uncertainty which has thus far attended Northumberland maps, terriers, and surveys.

A better reason than this ambiguity, however, for thinking that the Hartley and Seaton Delaval statements do not unquestionably imply the existence of a three-field system is the possibility that the author, speaking as it were parenthetically, may have been referring to a three-course rotation of crops. This method of tillage, as is explained below, might appear where the open-field furlongs were not grouped into three compact fields.[1] From occasional items there is reason to think that in Northumberland a three-course rotation was employed, at least upon demesne lands. Nine large consolidated parcels at Hextold were in 1232 so tilled that 51½ acres were sown with wheat and rye, 78 with oats, and 50 were " de terra wareccanda." [2] Although the division here into three parts was not precise, it was approximate. Regarding other demesne lands no uncertainty exists, and it is furthermore obvious that they might lie in common. At Hepscott, for instance, an inquisition describes 88 acres of demesne " de quibus tertia pars iacet in warecto et pastura eiusdem warecti nihil valet per annum quia iacet in communi." [3] Though we have no corresponding information regarding the rotation of crops which was usual upon tenants' land, it may well have been at times a three-course one. If so, the Hartley and Seaton Delaval statements perhaps refer to such a situation, and the term "field" is used carelessly in place of the more exact " seisona."

If this seem an over-refinement of explanation, and if it be urged that a three-course rotation upon tenants' lands was not far removed from a three-field system,[4] the extent of the negative evidence from Northumberland must once more be insisted upon. Similar avoidance of three-field indications is not characteristic

---

[1] Cf. below, pp. 321–325.   [2] Raine, *Priory of Hexham*, ii. 96.
[3] C. Inq. p. Mort., Edw. III, F. 2 (17).
[4] The difference was, however, pronounced. Cf. below, pp. 321–325.

of the testimony from a midland county. Even if it be admitted that a three-field system at times appeared in Northumberland, it seems equally clear that an alien influence early made itself felt and differentiated the county from its southern neighbors. A ready conjecture would designate such an influence as Celtic, and evidence supporting the surmise is not wanting.

One aspect of this evidence is the character of the terms used in describing Northumberland open-field strips and field divisions. At Long Houghton the parcels lay " rigge by rigge,"[1] and some terriers enumerate " riggs " without giving areas.[2] The description, further, of a husbandland at Chollerton, which has been already quoted, shows how frequently the names of the furlongs ended in " rig."[3] This nomenclature was, of course, the substructure of Scottish runrig.

More decisive, however, than terminology is the appearance in Northumberland of the Scottish method of tillage. A description of this in 1599 refers to what was perhaps at that time the persistence of an antiquated usage, but it is particularly instructive as indicating the character of primitive husbandry in the county. It occurs in an account of the queen's demesne lands at Cowpen, but relates as well to the lands of freeholders and lessees: —

" At the layenge forth of any decayed or wasted corne feilde, and takinge in any new feildes of the common wastes in liewe thereof, everie tenaunte was and is to have so much lande in everie new fielde as everie of them layde forth in everie wasted or decayed corne feilde, or accordinge to the rents of everie tenaunte's tenement in such place and places as did befall everie of them by their lott; and so hath everie of the quene's tenauntes within the towne of Cowpon aforesaide, as well leassors, tennaunts at will, as freeholders, contynewed the occupacion of all their arable lands by partinge by lott as aforesaide; and that after the layenge oute of everie wasted corne feilde within the

---

[1] Cf. above, p. 208.

[2] For example, the terrier of the demesne lands at Corbridge (*History of Northumberland*, x. 124).

[3] Cf. above, p. 217.

feldes and territories of Cowpon aforesaide, everie so wasted and layde oute corne felde nowe is and ever was reputed and used as the quene's common wastes there are, until the same lately layde oute corne feildes or any of them be by generall consente of neighbours taken in, parted, and converted to arable lande or medowe again; . . . [many tenants] affirme they alwayes so had used and enjoyed the same parted landes tyme out of mynde of man." [1]

This description might well apply to the Scottish outfield, described at length in the preceding chapter. In Northumberland, as in eighteenth-century Scotland, large parcels of land were temporarily reclaimed from the waste, reduced to tillage for a series of years, and then allowed to revert to waste again until they had in a measure recovered their fertility. In a newly-improved area each tenant had a share similar to that which he had had in the "decayed corne feilde" simultaneously "layde forth" or abandoned. Just how this procedure went on is illustrated by the following provision of a Corbridge court roll of 1594: "*Item* it is agreed at this courte for the devideinge of the land in Dawpathe, that betwen this and the next fawghe it shalbe equallie parted by the consience of xii men." [2] Apparently the re-allotment of a furlong about to be brought once more under cultivation was entrusted to a committee of villagers. Such a method of tillage accounts for the dispersion of a tenant's strips and explains the persistence of such dispersion.

Of immediate interest, however, is the probable relation of this practice to a three-field arrangement. Unless the arable area of a Northumberland township is to be thought of as entirely surrounded by a tract of waste, the permanent division of the arable into three equal compact parts is difficult to imagine in connection with the type of cultivation just described. Assume, for instance, a three-field arrangement of the arable, with the waste lying in one part of the township — an arrangement usual in the midlands. Assume further that a furlong was to be "decayed," or allowed to drop out of cultivation. If this furlong

[1] *History of Northumberland*, ix. 324.   [2] Ibid., x. 270.

were adjacent to the waste, it might be replaced by another contiguous to the field in such way that the integrity of the latter would in a measure be maintained. But if the furlong lay in a field remote from the waste, how could it be replaced without destroying the compactness of the field in question? Furthermore, the abandonment of furlongs *within* the arable area would under any circumstances make impossible the persistence of *compact* arable fields. Any field would always contain "decayed" areas, and the term "field" could at best be applied only to one-third of the entire area of the township, composed in turn of certain furlongs under cultivation and of others abandoned for a series of years. Since such a field would be very different in appearance and in mode of tillage from a midland field — would, in short, be a "seisona" — there is no reason why the term "field" should have been used in Northumberland with its midland significance. Its non-appearance in the documents, or its use in them merely to indicate topographically a part of the township or to designate one of the furlongs, at once becomes explicable. The infrequent use of the term in early charters, furthermore, is a guarantee that the field arrangements of the midlands did not extend to Northumberland.

That a system of Celtic type long persisted in the county is apparent from certain evidence offered before chancery in a suit relative to lands in North Middleton as they were prior to their enclosure in 1805. The fourteen ancient farms, which comprised about 1100 acres of arable, meadow, and pasture, were thus described: "These farms are not divided or set out, the whole township lying in common and undivided. . . . The general rule of cultivating and managing the lands within the township has been for the proprietors or the tenants to meet together and determine how much and what particular parts of the lands shall be in tillage, how much and what parts in meadow, and how much and what parts in pasture, and they then divide and set out the tillage and meadow lands amongst themselves in proportion to the number of farms or parts of farms which they are respectively entitled to within the township, and the pasture lands are stinted in the proportion of 20 stints to each farm."[1] Although by the nineteenth century the boundaries between arable, meadow, and pasture may have become more flexible than they were at an earlier time, there can be little doubt that the method of allotment here described was a survival of the principle that all land newly taken under cultivation should be equitably apportioned among the husbandlands.

If it be impossible to look upon Northumberland tillage as identical with that of the midlands, there is, on the other hand, no difficulty in seeing how it could transform itself into the latter with ease. Were the cultivation of the arable in any township to become more intensive, the period of years during which a furlong could be allowed to revert to waste would have to be decreased. The ratio might become two years of productivity to one of fallow;[2] with such a rotation once adopted, only the laying together of the fallow furlongs would be wanting to make the system one of three compact fields. If it may be assumed that this step was at times taken before or during the sixteenth century, some of the questionable indications of a three-field system which have been cited in this chapter are perhaps entirely authentic. At least there is opportunity, if any one thinks the evidence sufficient, for attributing to Harley and Seaton Delaval the practice of three-field agriculture.[3]

Viewed in all its relations, Northumberland thus becomes a transitional county, having affiliations on the one hand with Scotland, on the other with the midlands. Despite the nominal division of the arable of its townships into fields, a division sometimes apparent in maps and terriers, the absence of an equal apportionment of the acres of the holdings among these fields has led us to doubt the midland character of the latter. Apart

[1] *Archaeologia Aeliana*, new series, 1894, xvi. 138.

[2] Seebohm, in his latest book, remarks that the co-aration of the waste described in the Welsh laws of the tenth century " is an embryo form of the more advanced open field system of the settled agricultural village community. It is only necessary," he continues, " to extend the corn crop over a wider area and to subject the strips to a permanent rotation of crops, and the result would be holdings with scattered and intermixed strips and the *vaine pâture* over the stubble " (*Customary Acres and their Historical Importance*, London, 1914, p. 6).

[3] Cf. above, pp. 220–221.

from this negative testimony to the absence of a two-, three-, or four-field system within the county, the nomenclature employed relative to fields and the method of fallowing strongly suggest a Scottish connection. Entirely Scottish was the temporary improvement of tracts of waste land, followed in turn by the abandonment of them to their original state.

Scarcely have Celtic characteristics been discerned, however, before Northumberland fields are seen to have been cultivated in a manner which was not precisely that employed in Scotland at the end of the eighteenth century. It is not clear, first of all, that there was a permanent infield, and still less is it clear that there was continuous tillage of any part of the arable which would make possible such an infield. All cultivable land seems to have been treated in the same manner — tilled, probably, under the rotation of two crops and a fallow. At times a new furlong was improved from the waste, subjected to the usual cultivation for a series of years, and then allowed to revert to waste as another furlong was substituted for it. In Scotland, give-and-take of this sort was limited to the outfield; in Northumberland, it seems to have been applicable to all lands which at any time were brought under the plough.

Another way in which a township of Northumberland differed markedly from one of Scotland was in its size. The surveyor of Long Houghton remarked upon " the greatnes of the said towne ";[1] subsidy lists frequently point to the existence of a not inconsiderable number of tenants;[2] sixteenth- and seventeenth-century plans usually show a single large settlement within a large township area;[3] and, finally, the modern map reveals Northumberland as a county of villages rather than one of hamlets. In Scotland, on the other hand, the townships, as we have seen, were usually small and the settlements in general had not a half-dozen houses. Northumberland, so far as concerns the area of its townships, was allied with the English midlands rather than with its northern neighbor.

---

[1] Cf. above, p. 208.
[2] See, for example, *History of Northumberland*, ii. 236, 365, 414, 472.
[3] For example, ibid., 368, 413, 452.

Such are some of the reasons for looking upon the county as a region which in regard to its settlement and field system was transitional between the Celtic and midland areas. Originally perhaps, except for the size of its townships, it inclined to be Celtic, but as cultivation of the soil became more intensive the three-field system practiced toward the south may have been in a measure adopted. Scarcely, however, had this taken place when the process of enclosure began, and with more rapidity than in the midlands the history of open fields in Northumberland came to an end.

Cumberland

IN the period of parliamentary enclosure few open arable fields, it seems, remained in Cumberland. Slater cites only five acts which mention them, and of these but two specify the acreage.[1] The reporters to the Board of Agriculture in 1794 subdivided the county into 350,000 acres of lakes and mountains, 150,000 of improvable common, and 470,000 of old enclosures, making no rubric for open arable fields.[2] These last had, however, been existent a half-century earlier. Eden, writing in 1794–96, declared that in each of six parishes tracts of cultivated common field ranging in area from 100 to 3000 acres had been enclosed within the preceding fifty years.[3] In the case of four parishes he added brief descriptions. At Croglin, he wrote, " a great part of the arable land still remains in narrow crooked dales, or ranes "; at Cumrew " the grass ridges in the fields are from 20 to 40 feet wide, and some of them 1000 feet in length " ; the greater part of Castle Carrock " remains in dales, or doles . . . which are slips of cultivated land belonging to different proprietors, separated from each other by ridges of grass land " ; the cultivated land of Warwick " formerly, although divided, lay in long slips, or narrow dales, separated from each other by ranes, or narrow ridges of land, which are left unplowed." [4]

[1] Twenty acres at Torpenton and 240 at Greystock (*English Peasantry*, p. 256).
[2] J. Bailey and G. Culley, *General View of the Agriculture of the County of Cumberland* (London, 1794), p. 9.
[3] Sir F. M. Eden, *The State of the Poor* (3 vols., London, 1797), ii. 45–93. The parishes were Ainstable, 400 acres; Castle Carrock, 100; Croglin, 100; Gilcrux, 400; Warwick, c. 1100; Wetheral, 3000. [4] Ibid., 65, 67, 68, 92.

Glebe terriers drawn up a century earlier (about 1704) illustrate at length these descriptions.[1] Sometimes the parcels of the glebe were not numerous and comprised but few acres. At Addingham there were 7½ acres in five parcels, at Hayton 6¼ acres in four parcels, at Castle Carrock 7¼ acres in seven parcels. Elsewhere parcels were more numerous and the total areas greater. At Hutton-in-the-Forest twelve parcels contained 18½ acres; at Melmerby (besides 12 acres enclosed) twenty-two parcels contained 14 acres; at Skelton thirty-one parcels contained 32 acres. Typical of these terriers, and instructive as to the size of the strips, the subdivision of them into riggs, and the names of the open-field areas in which they lay intermixed, is the description of the glebe at Orton.[2] Apart from parcels of moss and rights of pasture over the moors, the parson had sixty-three riggs and one butt of arable, with various small pieces of meadow at the ends of these and certain *raines* or strips of turf

[1] See Cumb. and Westm. Antiq. and Archaeol. Soc., *Trans.*, new series, 1910, x. 124 *sq.*

[2] Ibid., 1893, xii. 137 (also new series, 1910, x. 124). The specifications run as follows: —
> " In the West field in the Croft 11 Riggs with a Head Rigg, 3 acres [This parcel and each following one is bounded.]
> In Low Croft or East Roods 4 Riggs with a Raine between them and a piece of Meadow at the North End, 1 acre
> In the West Roods 4 Riggs, one acre . . . with a rigg of John Robinson's between them
> At the Croft Head two large Riggs . . . 1 acre
> At the Parson's Head two long Riggs, one acre . . .
> In Crossland two Riggs, 1 acre . . . with a piece of Meadow at the South end . . .
> In the Shaws three Riggs, one acre with a piece of Meadow at the low end . . .
> In the Organ Butts two small Riggs, half an acre . . .
> In Inglands two Riggs, one acre with a small piece of Meadow at the low end . . .
> In Sheep Coats two Riggs, one acre with a broad Raine between them and a piece of Meadow at the low end . . .
> In Crabtreedale two Riggs, one acre with a piece of Meadow at the low end of them . . .
> In Grayston Butts two Riggs, half an acre . . .
> More in Grayston Butts two Riggs, half an acre . . .
> In the Shaws more two Riggs, half an acre . .
> Glebe in Orton Rigg Field. In ye West end four Riggs, half an acre . . .
> At the Parson's Lees eight Riggs . . ., two acres with a Daywork of Meadow at the North end
> Glebe in Woodhouses Field
> In Bredick two Riggs, half an acre . . .
> Underbricks, a butt . . .
> Upon the Bank or Priest bush three Riggs with a piece of Meadow at the North end . . .
> In the East Field four Riggs, three roods with a piece of Meadow at the North end . . .
> In Great Orton Moss a large parcel of Moss
> In the Flatt Moss another great parcel of Moss
> Common of Pasture for all the Parson's cattle with four Dayswork of Turf upon all the Moors of Orton within the Parish."

between them. The riggs lay in twenty divisions of the field and contained about 19 acres, from two to four riggs constituting an acre. Four fields are mentioned, but they take their places along with such curiously-named areas as " the Shaws " and " Underbricks." No grouping of strips by fields is perceptible.

Descriptions like these at once establish the former existence of open-field intermixed strips in Cumberland. The period at which they were consolidated and enclosed cannot be here investigated. Slater accepts Wordsworth's conjecture that a movement in this direction was not " general until long after the pacification of the Borders by the union of the two crowns." [1] What is without doubt is that in 1665 the estimated areas of certain townships, apart from common pasture, could be described as " Inclosed Ground — meadow, pasture, and arable." In this list are assigned to " the Lawnde or close of Heskett " 2500 such enclosed acres, to the hamlets of Serbergham and Scotby 750 and 700, to Gamblesby 1870.[2] Early in the reign of James I the twenty-five tenants at Plumpton Park had enclosed their holdings, save that five had an interest in Le Haythornefields." [3] By the middle of the seventeenth century enclosed townships were therefore easy to find.

Leaving aside the date of enclosure, we may refer at once to Tudor and Jacobean surveys in order to determine, if possible, what was the nature of Cumberland open fields. Sometimes, it appears, all holdings were in meadow, as in the mountain township of Matterdale;[4] again, as at Cokermouth, we learn that there were arable acres " in communibus campis," but we learn no more.[5] Elsewhere, however, certain features that seem to have been characteristic of the field system of the county are discernible, and of these the first is the grouping of rather small fields round correspondingly small hamlets.

In determining the areas of townships we are likely to be misled if, retaining the midland point of view, we give attention

[1] *English Peasantry*, p. 258.
[2] Land Rev., M. B. 258, ff. 64–65. Hesket was one of Eden's open townships, but there is a High Hesket and a Low Hesket.
[3] Land Rev., M. B. 213, ff. 1–10.
[4] Land Rev., M. B. 212, f. 270.   [5] Exch. K. R., M. B. 37, ff. 4–8.

merely to the area assigned to a Cumberland manor. In the midlands, manor and township tended to coincide, the latter being relatively large and comprising a single settlement also relatively large. A different situation has come to light in Herefordshire. There a manor comprised several townships, each containing a small settlement, more properly a hamlet than a village. Cumberland units were like those of Herefordshire: the manor was composite, the townships were small, the settlements were hamlets.

No survey shows these features better than one of Holme Cultram, made in 2 James I.[1] At that time this old monastic manor was divided into four quarters, called Abbey, St. Cuthbert's, Loweholme, and Eastwaver. The tenants of each of the four are mentioned in alphabetical order, and their holdings are located, with statement of areas. The names used in locating holdings turn out upon examination to be those, not of fields, but of several contiguous hamlets which lie to the west of the village of Holme Cultram. A summary for the Abbey quarter is as follows: —

Name of Hamlet	Number of Tenants	Total Area in Acres
Swinestie	10	116
Sowter field	15	184
Aldeth	9	96½
High Loese	13	129½
Abbie Cowper	13	203
Sanden House	13	168½
Browne Riggs	6	220

The quarter, which itself was only the fourth part of the manor, thus broke in turn into seven townships, the largest comprising only 220 acres. Since the holdings are described as "arable, meadow, and pasture," a part of each township must be set aside as non-arable. We thus have an agrarian situation in which the units of settlement comprised not more than fifteen tenants and the arable area contained usually less than 150 acres.

Not dissimilar were the hamlets and fields of the manor of Hayton. A map and schedule of 1710 describe the "infields" as comprising 1478 acres, the common or waste 3178 acres.

[1] Land Rev., M. B. 212, ff. 307-389.

Within the infields, according to the map, were six hamlets — Hayton, Fenton, Edmond Castle, How, Faugh, Headsnook. Hayton and Fenton gave their names to quarters which contained respectively 440 and 528 acres, the one being occupied by forty-five " toftsmen," the other by forty-three. A quarter probably embraced the lands of more than one hamlet; for, even if it is not clear that Edmond Castle was included in Hayton quarter, there can at least be no doubt that How fell within Fenton quarter. The improved land of either How or Fenton must therefore have comprised about 200 or 300 acres, an area somewhat larger than that of the Holme Cultram hamlet-fields.[1]

The size of other Cumberland townships may be discovered in a survey of 1608 which relates to the " Castle Soake and Demaines of Carlisle."[2] Enough of the place-names can be identified on the modern map to make it clear that locations are by hamlets. The " Standwicks freehold," to which are assigned fourteen free tenants and 153 acres, was no other than the township of Stanwix, a hamlet just across the river from the Castle. The fields of Currock, Blackwell, Upperby, and " St. Nicholas Hill " are grouped together. In them sixteen free tenants had 192 acres and nine customary tenants 99 acres, about one-fourth of the total area being meadow and pasture. Other hamlets were Almery Holme with twenty-one tenants in possession of 51 acres, and Wery Holme with thirty-one tenants possessed of 130 acres. The fields of no hamlet in the survey contained so many as 300 acres, and usually a far smaller number. This illustration, together with the two preceding ones, may suffice to determine our conception of Cumberland settlement. We must think of the county as peopled by groups of from five to thirty tenants dwelling in hamlets round which the arable fields were seldom 300 acres in extent, and often not above 50 or 100 acres.

From this first characteristic of Cumberland fields we pass to a second — the distinction occasionally noted between infield

[1] Cumb. and Westm. Antiq. and Archaeol. Soc., *Trans.*, new series, 1907, vii. 42 *sq.*
[2] Land Rev., M. B. 212, ff. 129–158.

and outfield. In the Hayton map of 1710 already referred to the arable is designated " infields " in contrast with the encircling waste. More striking is the name given to one of the hamlets of the manor. On the edge of the infield next the common was a tiny settlement called Faugh, and elsewhere on the map of Cumberland the same place-name is to be found.[1] It is, of course, the term which in Scotland was applied to that part of the outfield brought under occasional cultivation. The situation of Faugh on the Hayton map at a point where infield and outfield meet suggests a settlement due to the permanent improvement of the waste. In other Cumberland documents we learn further that a holding might consist of specific amounts of both infield and outfield. In a survey of Fingland made in 36 Elizabeth each of the eight tenants had " 16 acre terre arabilis in Infield et 10 acre terre arabilis in Outfield." [2] That the outfield was arable and was allotted in specified amounts implies an improvement of the waste before the end of the sixteenth century. This confirms our conjecture as to how the hamlet of Faugh may have arisen, and suggests that the situation which was characteristic of eighteenth-century Scotland was a transitional one in sixteenth-century Cumberland.

Further light is thrown upon the appropriation of the outfield by two surveys of Soulby, a hamlet of the manor of Dacre.[3] In 9 Elizabeth Soulby was occupied by ten tenants, each of whom had a messuage with from five to seven acres of arable and meadow adjacent thereto. Besides this, there was assigned to each one acre of meadow " apud Bradhoomyre," two acres of arable "apud le Tofts," two of pasture "in Sourelands," and two of pasture " apud Fluscoo." In another survey of some forty years later the pasture in Sourelands and Fluscoo had become arable or arable-and-meadow, while a fifth area, called Woodend and Crakowe, had appeared.[4] In this last area each tenant had 2½ acres of pasture or of pasture-and-arable. The two surveys suggest that there were appurtenant to the tenements at Soulby four

---

[1] For example, a hamlet of Ainstable is called Faugh Heads.
[2] Land Rev., M. B. 212, ff. 81 sq.
[3] Land Rev., M. B. 213, ff. 26–29b.   [4] Ibid., ff. 47–48.

or five large parcels of the outfield or waste, each of which had been divided with precision among them and thenceforth appeared in the surveys, sometimes as pasture, sometimes as arable. Such a description, of course, applies rather well to certain furlongs of a Northumberland township, but even more accurately to the Scottish "folds" or "faughs," those divisions of the outfield which were brought under crops for a number of years and then allowed to revert to pasture for a corresponding period of time. It should be added that both Soulby and Fingland were small townships, each containing less than two hundred acres and each having not more than ten tenants.

A field situation not unlike that perceptible in these townships is described in a somewhat confused Elizabethan survey of Lazonby.[1] Besides noting the acre or two adjacent to each tenement, it recounts a large number of field names — more than fifty. Since half of them are mentioned in connection with only one tenement apiece and are applied to but small areas, they must have referred to parcels of land in the possession of single tenants. Fifteen other field names recur two or three times, and in these areas, which seldom contained so many as five acres, two or three tenants shared. In the following field divisions a greater number of tenants had parcels: —

Field Name	Number of Tenants	Total Area in Acres
Outelayerclose	27	87
Le Holme	13	15 + 10 (enclosed)
Le Holmebushes	16	2
Redmore (arable)	19	$17\frac{3}{4}$
Halling (meadow)	13	$13\frac{3}{4}$
Le Linge	9	$15\frac{1}{2}$
Keld head (meadow).........	8	$8\frac{7}{8}$
Kelderdales (meadow)	4	$5\frac{1}{8}$
Galloberg	5	$9\frac{1}{2}$

In these larger areas the shares of the tenants inclined to be more or less equal. Holdings in Le Holmebushes were usually $\frac{1}{8}$ of an acre, in Outelayerclose $2\frac{1}{2}$ or 5 acres, in Halling and in Redmore $\frac{3}{4}$ of an acre, in Gallowberg $1\frac{1}{2}$ acres, and in Le Linge 2 acres. The equality of partition and the character of the names

[1] Land Rev., M. B. 212, ff. 1–7.

suggest that here too there had been improvement of the waste in which many tenants had shared. " Ling " is a term applied to a common;[1] Le Holmebushes bespeaks a brewery; Outelayerclose is reminiscent of outfield. It is not improbable that at Lazonby there was practiced the temporary appropriation of cultivable land, perhaps followed by its reversion to waste — a procedure suggested at Soulby and Fingland and well known to Northumberland and Scotland.

Although this characteristic seems discernible in Cumberland tillage, the location of the acres of a holding one to another has not yet become apparent. At Soulby the half-dozen acres of each tenant's infield appear to have been consolidated, since they are described as having been " adjacent " to the tenements.[2] At Lazonby the undivided areas may have been similarly situated, but we cannot tell. A survey of Ainstable made in 19 Elizabeth assists a little in elucidating this important point.[3] To each of the three constituent hamlets of "Southeranraw," Ruckroft, and Castledyke it assigns some half-dozen tenants, with holdings of about ten acres apiece in the respective hamlet-fields; but regarding the position of these acres we learn nothing. The remaining tenants seem to belong to the hamlet of Ainstable proper. Although sometimes the holdings here are not located, at other times they are said to have lain largely in South field or Kirk field. When this is the case, each was, except in one instance, entirely within one or the other of these fields.[4] Sometimes, too, the acres of a tenant of one of the other hamlets lay wholly or partly in South field. Now, South field and Kirk field were pretty clearly not hamlet-fields attached to different hamlets, but were the two fields of a single township. Nor can the acres which fell within them be looked upon as enclosed,

[1] Cf. below, p. 326.
[2] Once the account adds that they lay to the south (*ex austro*), once that they were enclosed, twice that they were called Lyngarth.
[3] Land Rev., M. B. 212, ff. 7-12.
[4] In South field were three tenements of 5, 4½, and 4 acres respectively; in Kirk field there were six containing in all 24 acres; one tenement had 1½ acres in South field and 2 in Kirk field; another had 5 acres in Kirk field and 2 in Low field; one tenement of 10 acres lay in Low field. The acres of six tenements are not located.

since the survey takes pains to distinguish its enclosures,[1] and at times states that a tenant's acres were scattered.[2] The conclusion, then, must be that, if a Cumberland hamlet had two open fields, the acres of the holdings were not divided between them but lay distributed throughout one of them. This implies further that the dispersion of parcels was not very great, since it was possible to gather all those of a holding within an area described as one field.

A similar situation is pictured in part of a description of the manor of Bromfield entitled "The Survey of lands in Alenbye now in the tenure of Jenet Shaw widoe and Michell fawcon."[3] The first rigg which each tenant held is said to have lain in the East field, and the four following riggs were presumably in the same place. Thereafter one butt, two riggs, and two "Ingdailes" are definitely said to have been in this field, and the location of only three butts is left uncertain. Without much doubt the parcels of the two tenants lay almost entirely within the so-called East field of Alenby.

A like tendency toward segregation rather than wide distribution of the parcels of a holding appears in certain glebe terriers. At Hutton in the Forest the twelve strips of which the glebe

[1] For example, "John Gibson tenet unum tenementum et unam clausam eidem ibidem adiacentem . . . continentem ii acras terre, prati, et pasture, et unam peciam terre in South field."

[2] Appurtenant to one holding was a messuage, an acre close, and eight acres of arable, meadow, and pasture lying " diversim in campo ibidem vocato Southfield."

[3] Add. Char. 17163, 1 Eliz. The specifications run as follows: —

    "Ayther of them one Rigg in the estefeyld called Ingdales
    Ayther of them a Rigge called totteryge
    Ayther of them another Rigge called lange smele Rige
    Ayther of them one Rigge upon borwe
    Ayther of them A wawcaye Rige
    Ayther of them one but in the same feyld called udge on butt
    And Ayther of theym one Rige in the said feld called grige
    Also Ayther of theym haith one Rige of medo lying in the este field in one plays called the mire Doyle conteyning by estimation two parts of one acar
    Item two Ingdailes lying in the newe Inge in the same contening by estimation one half Acar belonging Evenlye betwyn the said tenants
    Item Ayther of the sayd tenants haith one but called the crosse but, et Ayther of theym haith one wheat but lying on the weste syde of Alenbye mill
    Item Ayther of them haith one Dryebut of the weste syde
    Item Ayther of them haith one cowegate in the griff Ing als leckryge
    Also there is comen of pasture and turf graysce for there Rate of the comen of Alenbye."

was composed (a total of 18¼ acres) were described in 1704 as
"all . . . butting on the pasture,"[1] a situation which precludes
their distribution throughout the entire arable area. In another
terrier inserted at the end of the register of Wetheral is a list
of the parcels of land which in 1455 belonged to the prior at
Warwick, a village near Carlisle.[2] A glance at the description
will show that many parcels either lay in or abutted upon "Les
Halfakyrs," and that Les Halfakyrs in turn abutted upon the

[1] Cumb. and Westm. Antiq. and Archaeol. Soc., *Trans.*, new series, 1910, x. 126.
[2] J. E. Prescott, *Register of the Priory of Wetherhal* (London, 1897), p. 374. The specifications run as follows: —

"Terrae de Morehouse jacentes in diversis locis infra Dominium de Warthewyk pertinentes Priori de Wedyrhale . . .
Imprimis predicti juratores praesentant et dicunt quod sunt infra dictum Dominium
i acra vocata le Toftlandakyr cuius unus finis abuttat super Bromlands et alius finis versus Lynstock
Item dimidia acra terrae cuius unus finis abuttat super les Bromlands et alius finis versus Lynstock
Item iii rodae de les Bromland buttantes super terram quae vocatur le Bromylcroft
Item i roda et dimidia terrae buttantes super altam viam et super les Bromlands
Item le Tendlatheakyr buttans super altam viam et super communam de Warthewyk
Item i roda terrae jacens super Roclyfbank et buttans super le Skewgh
Item i acra terrae jacens super Roclyfbank et buttans super le Skewgh
Item iii acra terrae jacentes super Roclyfbank et super dictum Skewgh
Item i acra parcella de les Halfakyrs abuttans super Henry-holme et super les Halfakyrs
Item dimidia acra terrae parcella de les Halfakyrs abuttans super Henry-holme et super les Halfakyrs
Item i acra terrae parcella de les Halfakyrs abuttans super Warthewyk-wath et super les Halfakyrs
Item le Shouptreflat continens ii acras parcella de les Halfakyrs abuttantes super Rotclifyate et super les Halfakyrs
Item ii acrae parcella de les Halfakyrs abuttantes super altam viam et super aquam de Eden
Item dimidia acra parcella de les Halfakyrs abuttans super altam viam et super aquam de Eden
Item dimidia acra parcella de les Halfakyrs abuttans super altam viam et super Mydleholmewath
Item i acra terrae vocata le Goteakyr jacens in longitudine per aquam de Eden
Item i roda terrae vocata Strawfordrode abuttans super aquam de Eden versus castellum de Lynstok et super les Bothomrodes
Item ii acrae terrae vocatae Grastanflatt jacentes super les Shortbutts versus aquam de Eden
Item le Stockflatt continens v acras terrae abuttantes super le Soketflatt et super altam viam
Item le Pittflatt continens ii acras terrae abuttantes super altam viam et super le Goteakyr
Item dimidia acra terrae abuttans super altam viam et super le Syke vocatum Whetland syke
Item ii acrae jacentes super le Butbrome et super les Halfakyrs et super altam viam
Item ii acrae terrae abuttantes super terram de Aglunby et super terram vocatam Fulla-lands
Item i acra et dimidia terrae vocatae Fulla lands abuttantes super altam viam et super les Halfakyrs et super Fulladub
Item i acra terrae vocata Stanbryglands."

alta via and the *aqua de Eden*. Every parcel in the list except the last one and the three in Roclyfbank is thus described fully enough to be brought either into immediate contact with the *alta via* or the *aqua de Eden*, or into contact with some parcel which touches upon one of them. The chain becomes continuous, except for four parcels about which we are insufficiently informed and which at best contain only one-sixth of the total area. Unless the entire open arable field of Warwick abutted upon the *alta via* and the *aqua de Eden*, we may safely conclude that the prior's acres lay segregated in one part of it.

Early and late terriers thus concur in segregating to some extent the parcels of a holding. Perhaps not too much should be made of this feature, since we are not well informed of the precise extent of the fields to which the foregoing terriers relate. Yet one aspect of the subject seems clear, — the grouping of strips which prevailed at Ainstable, Alenby, Hutton, and Warwick was not consistent with a two- or three-field system. Whether the parcels of arable were markedly segregated or not, their distribution throughout two or three large fields is not at all perceptible.

One should formulate no conclusion, however, without giving attention to earlier testimony. Little of this is to be found in the feet of fines, but a few instructive thirteenth-century terriers are embedded in the cartularies of Holme Cultram, Wetheral, and St. Bees.[1] Noteworthy is the unanimity with which these terriers locate their parcels by furlongs, without any attempt at grouping them by fields. At Wetheral, for example, the 4 acres that accompanied a house and croft consisted of nine such parcels, and another grant of 1¾ acres refers the parcels to eight localities.[2] Sometimes the specifications are full enough to show that the localities were not after all remote from one another. This was the case with 10 acres and 3 perches which St. Bees acquired at Rotington.[3] All parcels except the first lay adja-

---

[1] The cartulary of Lanercost priory I have not been able to examine.

[2] Prescott, *Register of Wetherhal*, pp. 136, 141.

[3] Harl. MS. 434, f. 169 (a late thirteenth-century cartulary). The specifications run as follows: —

"Due acre et dimidia iacent in meysigwra inter moram et campum quod vocatur Kenelflat Item una acra que vocatur garebrad iacet iuxta terram que vocatur Kirkeland . . .

cent to " Kirkeland " or to " Wynnefoth," which were in turn connected by the fourth parcel. If we knew that the terrier referred to a tenant's holding, we should have clear evidence bearing on the segregation of parcels. That we do not points to our need of terriers that describe bovates, the unit in which lands were rated in Cumberland.

Happily such terriers are available from five townships. At Melmorby, in the eastern mountains, the bovates were undivided plats. One is described as " illam bovatam quae iacet propinquior terrae Adae filii Henrici versus orientem ";[1] two others are " illas quae iacent inter terram Beatae Mariae Karleoli et Littilgilsic."[2] Near the western coast at Blencogo much the same plat-like character must have pertained to " duas bovatas terre . . . iacentes propinquiores porte ex occidentali parte ville."[3] Since, however, the two had been given as a dower (*in liberum maritagium*) and were not accompanied by messuages, they may have been demesne lands. At the mountain hamlet of Caber two bovates are somewhat similarly described, about 1240, as comprising one parcel of land probably rather large, three professedly small, and a parcel of marsh.[4] From Warwick, whence we have already had the terrier of the Wetheral lands in 1455, comes the description of a bovate which was given to the priory soon after 1175. It consisted of " quinque acras

> Item dimidia acra iacet in fridaylandes et extendit se . . . a bercaria usque ad . . . Kirkeland . . .
> Item una acra et septemdecim perticate iacent iuxta Bercariam . . . inter terram que vocatur Wynnefoth et . . . Kirkeland
> Item una crofta . . . que continet in se unam acram et sex perticatas iuxta . . . Kirkeland
> Item tres acre et tres Rode et dimidia iacent inter Bernardhou et Brezhou et extendunt se in longitudine de Wynnefoth usque ad seberth."

[1] Prescott, *Register of Wetherhal*, p. 289.
[2] Ibid., 291.
[3] Harl. MS. 3911, f. 57*b* (an early fourteenth-century copy).
[4] " Totam terram a superiori parte Mussae ad Neubussehill sicut le silkette descendit a predicta Mussa usque ad viam ad Surflatende et sicut dicta via tendit usque . . . [etc., bounded at length], et in Bacstaneygle et in Bochum duas acras et dimidiam, et quandam partinunculam terrae quae vocatur le Gare . . . et ab angulo fossati de Communa duas acras terrae in latitudine versus Mussam . . . et totam meditatam Marisci Scalremanoch versus meridiem " (Prescott, *Register of Wetherhal*, p. 283).

in Westcroft, et duas acras in Graistanflat, et unam acram iuxta holm cum prato ad predictam terram pertinente."[1] These four early bovate terriers show no marked distribution of parcels. One describes the acres as lying in five places (including the marsh), another assigns them to three, while two terriers imply that the bovates were consolidated.

Different is the fifth terrier, superior to the others in that the bovate which it describes was appurtenant to a messuage. The land lay at Tallantire, and was granted to St. Bees late in the thirteenth century. It consisted of twelve parcels, but there is little indication how these were situated relative to one another.[2] Only two lay in the same area, Biggehove, the others being in different furlongs. In the absence of descriptive locations we cannot tell how widely these furlongs may have been separated; but at least they formed a group distinct, with one exception, from the group in which nine other acres in Tallantire lay. The latter are described in a grant which appears to have been contemporary with the other one, since to some extent the names of the witnesses are the same; like the first, its acres were attached to a toft and probably constituted a holding. If such were the case, we have two tenements in the same township, each comprising several parcels, but parcels which in only one instance lay in the same furlong (Bighou).[3] While this does not

[1] Prescott, *Register of Wetherhal*, p. 121.

[2] The description (Harl. MS. 434, f. 161) runs as follows: —

"Unam bovatam terre ad mensuram Rode viginti pedes continentis . . . cum tofto et crofto integro et toto prato ad illam bovatam terre . . . pertinentibus . . . videlicet,
in crofto duas acras et dimidiam rodam et quatuordecim perticatas
apud biggehoue versus occidentem unam acram et unam rodam et quatuor perticatas
ibidem versus orientem duas rodas et dimidiam
apud thuahouel unam acram
sub Wartheholis unam acram et unam rodam et vigintinovem perticatas
apud routhelands unam rodam et tresdecim perticatas
infra vias de Warthehol' et Karliol duas acras et unam rodam et dimidiam et quinque perticatas
apud heyberhe unam acram et viginti tres perticatas
apud leuedibuthes dimidiam acram et quatuor perticatas
ad Sandrig tres rodas et dimidiam rodam et octo perticatas
ad hildirflath unam acram et unam rodam et dimidiam rodam et quatuordecim perticatas
in cultura a molendino versus aquilonem in quinta et sexta selione
versus orientem duas rodas et dimidiam et sexdecim perticatas."

[3] Ibid., f. 161b. The description runs as follows: —

"In crofto eiusdem domus tres rodas et octodecim perticatas
In hagwrinron cum prato ibi iacente quinque rodas et unam perticatam

prove that the parcels of each group were segregated, it suggests that such may have been the case.

Other early terriers of the cartularies relate to groups of acres or fractional acres which are not designated as bovates or tenants' holdings. They incline, like the Tallantire terriers, to locate parcels in several furlongs which are not brought into relation with one another and are never grouped by fields. Hence they furnish little information, except to make clear that more or less scattered strips were the constituents of early Cumberland fields and to emphasize the absence of a two- or three-field grouping.

At this point our evidence comes to an end. The nature of Cumberland open field has been ascertained only in its broader aspects; yet these are perhaps sufficient to determine certain affiliations. It has been pointed out that the field arrangements of Northumberland in the sixteenth century and at an earlier time manifested Scottish characteristics, though various descriptions concur with the map in disclosing other characteristics not Scottish. In particular did the size of the townships differ from what was usual across the border. Nor is it clear that the arable of a Northumberland township was ever divided sharply into infield and outfield, each tilled in the Scottish manner. On the contrary, a larger stretch of cultivable land was probably kept under more continuous tillage than in Scotland. The field system of the county seems, in short, to have had midland as well as Scottish aspects. Cumberland, on the other hand, appears to have inclined more to Celtic usages.

In the first place, there nowhere occur in Cumberland surveys and terriers suggestions of a two- or three- or four-field grouping such as are often found, though not well substantiated, in Northumberland documents. If, by chance, mention is made in a Cumberland terrier of an East field, there is small likelihood of finding further reference to a West field or a Middle

    Ad Braidron unam acram et tres rodas et decem perticatas
    In thorfinesakyr unam acram et tres rodas et octodecim perticatas
    Ad bighou tres rodas una perticata minus
    Ad blaakepot unam rodam et triginta duas perticatas
    Super Banks unam acram et triginta perticatas
    Ad viam que ducit ad capellam Sancte Trinitatis tres rodas octodecim perticatas."

field. It is not merely that the absence of an equal division of the acres of a holding among such fields leads to a distrust of the agrarian significance of the latter, as in Northumberland, but it seems clear that symmetrical fields seldom or never existed. Nor is the infrequent appearance of fields due to paucity of documents; for Cumberland surveys and terriers are not less numerous than those usually available from a midland county. Instead of adopting the midland arrangement, the acres of a holding seem even to have manifested a tendency to concentrate within one part of the arable area of a township. If we have insufficient evidence to prove that this was usual, its occasional occurrence is none the less contributory to a disbelief in the extension of the midland system to the county.

Apart from the intractability in Cumberland and probably in Northumberland, of the acres of a holding relative to a systematic field arrangement, we have from both counties positive proof of Scottish affiliations. Briefly stated, it is that, in both, portions of the waste were after the Scottish manner temporarily tilled and then allowed to revert to pasture. For Northumberland the evidence of this practice consists of certain descriptive statements, for Cumberland of inferences drawn from sixteenth-century surveys. But whether the Scottish division between compact outfield and infield was maintained in Northumberland there is reason to doubt. In Cumberland, on the contrary, it was perhaps more persistent, if one may judge from the phrases of the Fingland terrier.[1] Such a persistence, were we assured of it, would constitute a second point of difference between Cumberland and Scottish agrarian arrangements on the one hand and those of the midlands and Northumberland on the other.

We are better informed, however, regarding a third dissimilarity—that, namely, which inheres in the size of the townships. As has been pointed out, Northumberland townships were large, those of Cumberland small, as were also those of Scotland. Often the total area of these small townships was not more than one-fourth of what was usual in the midlands or in Northumber-

---

[1] Cf. above, p. 232.

land. Whether, then, the size of township fields or the method of their tillage be considered, Cumberland appears more Celtic than any other county of England thus far examined. To the south, however, lies a stretch of territory in which the Celtic population long withstood the Anglo-Saxons, and in which, therefore, phenomena not unlike some of those already described in this chapter may be apparent.

Lancashire

SINCE Lancashire was once joined with Cumberland in the old Celtic kingdom of Strathclyde, we shall expect to find in the two counties similar agrarian conditions. There should be discernible in Lancashire, as in Cumberland, few surviving open common fields in the eighteenth century, but at an earlier time a certain number of small ones in which the parcels of the tenants had no systematic midland arrangement.

Slater found in Lancashire no common fields enclosed by act of parliament, although there are numerous acts affecting commonable waste.[1] The report submitted by John Holt to the Board of Agriculture in 1794 estimated that nearly one-half of the area of the county was waste — 508,000 acres out of 1,129,600. " There are," he says, " but few open or common fields at this time remaining; the inconvenience attending which, while they were in that state, has caused great exertions to accomplish a division, in order that every individual might cultivate his own lands according to his own method; and that the lots of a few acres, in many places divided into small portions, and again separated at different distances, might be brought together into one point. . . . The inclosures or fields are in general very small, so much so as to cause great loss of ground from their number and the space occupied by hedges, banks and ditches. "[2] All this bespeaks piecemeal enclosure of common fields, perhaps long continued.

---

[1] *English Peasantry*, p. 255.
[2] *General View of the Agriculture of the County of Lancashire* (London, 1794), pp. 49, 52.

A concrete illustration of the early prevalence of enclosures is at hand in a detailed survey of the large manor of Rochdale, made in 1626.[1] This estate, situated in the southeastern part of the county, included some twenty-four hamlets and had an area of 41,828 acres. Somewhat more than one-fourth of the manor remained in open common waste at the time of the survey, but the remainder lay almost entirely in closes. At times there were parcels of pasture which, being newly divided, were not yet enclosed.[2] Intermixed arable strips were nowhere to be found. Thus, to a large tract of land on the edge of the moors — a tract which may never, to be sure, have had much open-field arable — the eighteenth-century description was applicable a hundred and fifty years earlier.

There is, however, no difficulty in finding traces of open common arable in the seventeenth century. The rental of the houses and lands of Edward Moore at Liverpool, drawn up in 1667–68 and unconsciously offering a striking comment on the later development of that port, frequently attaches to the houses " several lands [e. g., ten] in the field." Elsewhere it is the " town field," but we get no further detail.[3]

An instructive document illustrative of early seventeenth-century conditions in Lancashire is an account, drawn up in 1616, of the "Appropriate Parsonages or Rectories of Blackbourne and Whaley . . . possessions and Heriditaments belonging to the Archbishopricke of Canterburie."[4] Since there belonged " unto the said Rectory the Moietie of the Lordship of Blackbourne," the townships included in the enumeration extended over an area of at least 200 square miles in the northeastern part of the county.[5] With one exception the land described in

[1] Henry Fishwick, *Survey of the Manor of Rochdale* (Chetham Soc., 1913), pp. xiii, xv.

[2] Ibid., 240, Whitworth hamlet: "A parcel of pasture . . . lying open amongst the rest of the Copyholders in the Trough containing statute [measure], 10 acres, 3 roods." Areas held by other copyholders " in the Trough " are given.

[3] Thomas Heywood, *The Moore Rental* (Chetham Soc., 1847), pp. 19, 23, *et passim*.

[4] Exch. K. R., M. B. 40, ff. 24-46.

[5] The townships were Samelsbury, Overdale, Walton, Downham, Church, Haslingden, Burnley, Colne, Clitheroe, besides Blackburn and Whalley.

each township lay in closes, often many in number. The exception was the township of Altham in the parish of Whalley, where the glebe lands comprised a long list of open-field selions.[1] A note

[1] Exch. K. R., M. B. 40, ff. 40 sq. The description runs as follows: —

"Now followeth ye parsonage Gleabe of Whaley lying within ye towneship of Alvetham alias Altham, viz:

Item . . . in a feild of the Eyes toward Simonston certaine lands called Calved Eyes with a parcell betwene the divisions of the waters

Item in the same Eyes towards Alvetham alias Altham, certaine lands called little Eyes neere the Milldam . . . and so descending into Calder

Item at a place called the Bronckhouses a Mesuage which sometimes Adam of Aspden held

Item there nere the greene gate fower Selions butting upon the way

Item in the same feild in a certaine place called the farthings fower selions

Item in the same feild a selion like to a headland nere the house which sometime John of Boncke held

Item nere the said mesuage six selions butting upon the said messuage

Item in the same feild two selions

Item in the Mitthom twoo selions nere the syke with meadou in both the ends

Item in the west parte of Nether East feild a Selion with a geron towards the west

Item in the same feild six selions iacentes divisim

Item in the Over east field xi selions iacentes divisim

Item in the feild of Hoghton in the Blackcroft xviii selions iacentes divisim amongst percells of the oxegangs

Item in the same feild six Butts iacentes divisim

Item in the same feild of Hoghton at the Rishy flatt thirteene selions iacentes divisim

Item in the same field ten selions iacentes simul which is called the Barriers

Item in a certaine feild called the Hanflatt contayning in length xxxvii selions and on the other side ix selions with a way lying to the said feild toward the wood

Item in the same field of Hoghton the Walllands conteyning xi selions and in another place in the same feild xxi selions

Item in the feild above the Hall a mesuage which is called Hannehousteed and four selions extending themselves from the said mesuage to the Barne of the Mannor of Altham

Item a certaine place called Hannecroft in the same feild conteyning six Butts

Item on the west parte of the Mannour a certaine Messuage which Roger de Ornesden sometime held with all Selions abutting upon the said Mesuage and two selions whose ends are extended neere the said Messuage

Item in the feild of Milnecroft two Selions iacentes divisim

Item in the feild of Lordshall six Selions iacentes divisim

Item in the feild of Tonnested twenty-one Selions iacentes divisim

Item there at the Hartstalgreve and Firme ten Selions iacentes divisim

Item in the Nethertonnsted xi selions iacentes divisim

Item in the West Eyes xvii selions with three Butts and a geron

Item beyond the water towards Reved five selions iacentes divisim

Item a certaine Mesuage on the east part of the Mesuage with the priests house

Item in the Brichholme a certain Croft

Item a certaine place neere the Manner gate for a Tyth barne

Item in the greate meadow under the Lords hill in the East part of the said meadow in breadth lxiii feete lying together whose longitude is extended from the Lords hill to the Hay of the Kerre

Item in the middle of the same feild from a part of the old Cawsey xxvii feet in latitude and from the other part of the said Cawsey xviii feet in latitude extending itself in longitude as before

Item further in the same feild toward the west two Selions

Item in the same meadow toward the Meneeage the moytie of all that parcell called the Meneeage dividing it equally with the Lord alternis vicibus

at the end of the list records that certain of the field names were ancient and that enclosure had recently been in progress.¹

As to the field system which underlay this elaborate description only tentative conclusions can be drawn. Since four messuages appear, it is possible that three or four holdings were thrown together. If the selions described between the first and second messuages belonged to one holding, they lay very largely in " the field of Hoghton." Here were 59 selions and 6 butts, whereas to none of the other five fields in this group were assigned more than 11 selions, Hanflatt being probably a close. This implies considerable segregation. After passing the second messuage new fields appear, none of them containing very many selions, save Tonnested and Nethertonnsted with 21 and 11, and West Eyes with 17. If, on the other hand, the messuage succession has no significance and parcels in the same field belonging to different messuages are thrown together, we can only say about the field system that it seems to have been entirely irregular.

While Altham yields only this terrier, Warton situated to the west on the coast boasts a complete survey of 7 James I.² Enclosed land is here carefully distinguished from common field, the latter being said to lie " in communibus campis," or " in Warton field," or " in le Townefield." Although an occasional parcel of common meadow is singled out for specific location, this almost never happens to the arable, except to eight parcels " in le Bonetowne " and four " super le towne." The field system cannot therefore be ascertained, and it only remains to

> Item in the furthest part of the said meadow which goeth towards the Milne croft in the East part two selions lying neere the Lords hill with a certaine round parcell of the same meadow nere the hedge of the same Selions
> Item in the west parte of the end of the said meadow a gereon and beneath that two Selions
> Item at the Hartalstall greve ten Selions
> Item all the meadow of Altham in Symonston Eighes with all the errable there."

¹ " Item the said Jurors do further find present and say that the names of the feilds aforementioned were the auncient names of the said feilds, but time hath worne out those names and given them new names onely some of the auncient names remaine at this day, viz: the Hoghtons, the Kerre, the Mitthom. Which Hoghtons were of late yeares divided into divers closes, and so the ancient longitude and latitude of them doth not in any one feild continue at this day."

² Land Rev., M. B. 220, ff. 27-58.

determine the area of the open field relative to the area enclosed. The five freeholders, who controlled 29 cottages and gardens had, it appears, 15 enclosed acres, while, the land attached to the forty-four customary messuages amounted to about 270 enclosed acres (mainly small parcels of pasture) and about 287 acres of common field, of which at least 107 acres were common meadow. Of open-field arable there were, therefore, not more than 180 acres, or about one-fourth of the cultivated land of the township. No rights of pasturage over the arable are mentioned, most tenants having "cattle gates" in Lyndeth Marsh or in le Inges, and sometimes common pasture for sheep on Warton Crag and Warton Marsh.

For two Lancashire townships there are fifteenth-century terriers testifying to the existence of open arable fields.[1] One recites a grant to Penwortham priory of lands at Farrington, a hamlet southeast of Preston. By it were transferred, along with a messuage, 7½ and 11 acres of arable. Of the 7½ acres, 5 were in a field which bore the name of the adjacent hamlet of Clayton (*in quodam campo vocato Claghtonfelde*), and the rest in three parcels lay respectively in Brockforlong, Stainfeldmore, and " ex parte boreali le Heghgate." The 11 acres lay, we know not how divided, " in Longestainfeld, Brokeforlong, Shortstainfeld, et le Orchards, et Catcroft medowe."[2] To judge from this grant, the subdivisions of Farrington field were few in number, scarcely more than a half-dozen. A like simplicity of field division is apparent in the other terrier, despite its greater length. This specifies the parcels which were subtracted from three bovates and three acres of arable, and from

[1] It would seem at first sight as if there were useful information in a long fifteenth-century survey of the lands of Sir Peter Legh at Warrington near the mouth of the Mersey, published by the Chetham Society in 1849 (William Beamont, *Warrington in 1465*). Apart from the messuages, gardens, and certain acres "in campo vocato Hollay," much of the land described lay " in magno campo vocato Arpeley," or in some part of it, as Le Wroe or Wetakyrs. A glance at the Warrington of today, however, shows that the reference is undoubtedly to the large tract of meadow land almost encircled by the Mersey and still called Arpsley meadows. We can learn little about field systems from intermixed acres of common meadow.

[2] W. A. Hulton, *Documents relating to the Priory of Penwortham* (Chetham Soc., 1853), p. 67 (22 Hen. VII).

ten acres of meadow, at Bolron near Lancaster.[1] Except for the 1¾ acres in the last three parcels, the arable lay in four parts of the field. At Bolronbroke and Bolrondale were 1½ acres in four parcels, in or under the Withins 2½ acres in four parcels, " super Bambrest " 4½ acres in four parcels, and in the Oldefalde 1¼ acres apparently together. Locations like these at Farrington and Bolron do not adapt themselves to a three-field arrangement. They suggest rather small hamlet fields subdivided into a few areas somewhat like midland furlongs. Throughout these furlongs the parcels of a tenant were scattered irregularly.

These characteristics are reproduced and emphasized in several thirteenth-century terriers referring chiefly to townships situated on the coastal plain between Preston and Lancaster. For the most part they record grants to Cockersand abbey, grants that seldom convey so much as ten acres of open-field arable, therein differing from the charters of a midland cartulary, which nearly always include some long specifications. In the brief Cockersand transfers it is possible, none the less, to discover in a measure the relative positions of the parcels conveyed. Typical in all respects is the charter relative to a messuage, garden, and 5¼ acres of arable at Sowerby.[2] Not only were the acres granted

[1] William Farrer, *Chartulary of Cockersand Abbey* (Chetham Soc., 3 vols. in 7 pts., 1898–1909), iii. pt. i. 819–820. The specifications run as follows: —
 " Robertus . . . recuperavit seisinam suam . . .
  de medietate unius acrae terrae iacente ex utraque parte de Bolronbroke . . .
  ac de una roda terrae in Bolrondale iacente inter terram . . .
  et de una alia roda terrae in Bolrondale per se
  necnon de medietate unius acrae terrae cum uno tofto cum suis pertinentibus in Bolrondale
  ac de tribus rodis terrae jacentibus subtus le Withins in Aldlancastre
  ed de una alia roda in eadem per se
  ac de tribus rodis terrae in eadem per se
  et de aliis tribus rodis terrae in eadem per se
  Et etiam de quinque rodis terrae in le Oldefalde cum sirposis clausuris
  et de una acra terrae super Bambrist iuxta le Lone in Scotforde cum quodam prato . . . scilicet, Morehous, continente duas acras et dimidiam
  Et similiter de duabus acris et medietate unius [acrae] terrae iacentibus super Bambrest per se
  ac de tribus rodis terrae super Bambrest per se
  ac de una roda terrae et prati super Bambrest
  Ac etiam de tribus rodis terrae in le Riddyng in Scotford
  ac de tribus rodis terrae iuxta le Standandstone
  et de una roda terrae super le Clyff. . . . "

[2] Ibid., i. pt. ii. 244 (c. 1230–1268). The arable comprised: —
 " Tres per[ti]catas in orientali parte de Stirap super aquam de Broc
  et unam dimidiam acram in occidentali parte de Stirap iuxta vadum de Quitakedich

few in number, but, as the locations show, they were not widely separated; all the parcels except the last were connected in one way or another with Stirap, Quitakedich, and the "aqua de Broc," which three in turn were near one another. The segregation which has been noted in Cumberland reappears. It can also be traced to some extent at Preston in the eight parcels which were transferred along with a burgage tenement and the third part of a toft.[1] Apart from the meadow, an assart, and a half-acre near the garden, the arable lay largely in Siclingmor and Platfordale, with something in Aldefeld and at Sewallesike; but how these areas were related we do not discover. Traces of segregation are discernible, once more, in one of the longest charters of the cartulary. From it we learn that the ten acres which the abbey acquired at Newton comprised many parcels, some of them described as selions.[2] Since 5¼ acres and more than half of the selions lay in Otemaste and Wodebinde furlongs, these two divisions of the open field contained more than three-fourths of the ten acres conveyed. In fact, one of them alone comprised about five acres, a predominance which would not be met with in a normal midland terrier.

Perhaps these descriptions may suffice to show that the open fields of Lancashire had characteristics similar to those of

<blockquote>
et unam acram et dimidiam in alia divisa . . . sequendo Quitakedich . . . ad selliones de Stirap . . .
et duas acras terrae in alia divisa . . . sequendo . . . usque aquam de Broc . . .
et unam dimidiam acram in alia divisa super terram de Leye."
</blockquote>

[1] Farrer, *Chartulary of Cockersand Abbey*, iii. pt. i. 217 (c. 1230–1255). The parcels are described as follows: —

<blockquote>
"Totam terram in assarto meo . . .
et quatuor partes terrae super Siclingmor [three parcels, each between the lands of other men]
et unam dimidiam acram super Aldefeld . . .
et tres per[ti]catas terrae in Platfordale . . .
et unam dimidiam acram in Platfordale . . .
et dimidiam acram . . . iuxta orreum meum
et totam terram meam ex utraque parte de Sewallesike
et totum pratum meum inter pratum Adae albi et commune Karrum."
</blockquote>

[2] Ibid., 175 (1262–1268). Except when otherwise specified, the following areas are in acres: —

<blockquote>
"Super Otemaste furlong," ½, ½, 1, 1, ½, ½, ½, ½, 3 selions
"super Wodebinde furlong," 1, 1, 4 half-selions
"in superiori parte viae quae ducit ad Singilton," 5 half-selions
"in inferiori parte viae de Singilton," 2 "Tungas"
"super le holderthe," 1 butt, 2 half-selions, ½ selion
"super Karfurlong," 2 "Tungas."
</blockquote>

Cumberland. By the eighteenth century they had, like the fields to the north, become largely enclosed, though certain glebe terriers of the seventeenth century indicate that the intermixture of parcels persisted in a few localities and on a considerable scale. In these and in earlier terriers of the fifteenth and thirteenth centuries the nature of the open fields becomes apparent. Nowhere were the parcels grouped systematically in the midland manner. On the contrary, they are described as lying irregularly, in areas variously named and sometimes called furlongs, while not infrequently they were segregated. Regarding the method employed in tilling the open fields no information is at hand. Since such characteristics as we know about, however, are manifestations of Celtic runrig, it seems permissible to join Lancashire with Cumberland, and assign both counties to the region within which English agriculture was affected by Celtic custom.

Cheshire

OF the counties on the Welsh border, Cheshire is most closely joined with that part of Wales to which considerable attention has been given. Since Chester is only some ten miles down the valley of the Dee from Wrexham, we shall expect to find round about this county town common fields not unlike those of eastern Denbighshire.

Late documents, however, do not tell much of common arable fields in Cheshire. The reporter to the Board of Agriculture in 1794 estimated that they probably did not amount to 1000 acres in a county of 676,000 acres, nine-tenths of which was improved land.[1] Descriptions of all the tenants' holdings at Davenham and Great Budworth in 1650 assure us that nothing but closes were to be found.[2] A great survey of Macclesfield manor and forest made in 9 James I gives minute details for some sixteen townships;[3] but throughout the entire survey there is scarcely

---

[1] Thomas Wedge, *General View of the Agriculture of the County of Cheshire*, London, 1794.

[2] Parliamentary Surveys, Cheshire, No. 11.

[3] Land Rev., M. B. 200, f. 239 (the survey comprises folios 147-357). A typical holding is described as follows: —

"Jasper Worth, esquire, claymeth to hold to him and his heyres by copie of court roll . . . Item One other tenement in the tenure of John Latham, viz:

a suggestion of open common field, except perhaps in the mention of a few unusual "parcells" of arable at Bollington.[1] Apart from these, the entire manor lay in small closes, containing for the most part from one to two acres and consisting largely of arable.

None the less, there is seventeenth-century evidence that open common fields existed in Cheshire. In 1649 the messuages and lands of the dean and chapter at Chester were surveyed. After an enumeration of several closes "situate without Northgate," the account describes a series of "parcells," mainly arable and usually of from one to two acres in extent. Though most of these are not said to be in open field, a few at the beginning of the list are so described: "In Chester Town Feild, One parcell of Ground, called Long hedge Acre . . . is in Estimacion 2 acres. . . . One parcell of ground more in Chester Town Field, near Dee Bank, called Grange Acre . . . [is] in Estimacion 1 acre, 2 roods. . . . One parcell of Arrable ground in the Lower Town Feild . . . commonly caled Burtons Acre . . . containeth by Estimacion 2 acres."[2] At least we are assured of the continued existence at Chester, in the middle of the seventeenth century, of a "town field" the constituents of which were small parcels of arable.

Not much more informing is an account of the "rectory lands" at Bowdon, dated 1654. This glebe was then leased to eight under-tenants, each with a messuage, though two were cot-

One dwelling howse and the outhowses thereunto belonginge	
One close Arr[able] called the Layefield by estimation	3 acres
One close called the Meadow place by estimation	2 "
One other Arr[able] called the Hugh close by estimation	1 acre
One other called the good crofte by estimation	1 "
One other called the Goosie Meadowe by estimation	1 "
One other Arr[able] called the Symentley Knowle by estimation	3 acres
One other Arr[able] called Symentley by estimation	1 acre
One other called the litle Meadow by estimation	½ "
One other Arr[able] called the Calfe crofts by estimation	3 acres
One other Arr[able] called the Bancks by estimation	3 " "

[1] One holding, for instance (Land Rev., M. B. 200, f. 321), includes: —
"One parcell of Arrable in the towne field .... 1½ roods
one other parcell of Arrable in the Neather or towne field . 49 yards by 5 yards
one other parcell of Arrable called the Butt in Page Croft . 20 " " 4 " "

[2] Henry Fishwick, *Lancashire and Cheshire Church Surveys* (Lanc. and Chesh. Rec. Soc., 1879), pp. 226–227.

tagers. The six held for the most part a series of closes, but four had also "lands" or strips lying intermixed in places called Eyebrookes, Church field, and Hall field.[1] These strips formed less than one-third of each leasehold, the ratios in acres being 21 to 75¼, 6½ to 85¼, 12 to 38½, 10 to 33¼. The Eyebrookes was a close, and may have been a close of glebe shared by the four tenants. Since Hall field and Church field were situated " upon the Downes," they suggest areas recently improved and subdivided. Such constituents do not go to the making of normal open arable fields.

Less vague is a survey of 1650 relative to the manor of Handbridge, just outside Chester.[2] Twenty-two of the tenants had each a messuage, a garden (never larger than half an acre), and common of pasture in Saltney Marsh. In addition, each had from one to six " lounds in the Towne feild." Always there were from one to three " lounds " in " Longefeild in the Townefeild," and eleven tenants had also a strip or two apiece " at the lower end of the Bottom in the Town feild." There were besides two parcels " within the Gullett in the Townfeild," four at " Lowhill in the Townefields," and two at Crossflatts. At times the " lound " is said to have contained one acre, and on this basis the total area of all of them would have been about sixty acres. At length we have discovered a town field, small, to be sure, but one which had its subdivisions and one in which many tenants had intermixed parcels.

In sixteenth-century terriers similar open fields are discernible in the same neighborhood. At Chester, in 2 Edward VI, the college of St. John the Baptist had several tenants, the holdings

[1] Fishwick, *Lancashire and Cheshire Church Surveys*, pp. 176–184. The items are as follows:—
> " Nyne lands in the Close called the Eye brookes, Conteyneing by estimacion 8 acres. . . .
>> Seaven Lands in the Churchfeild and eight lands in the Hall hill, conteyneing by estimacion 13 acres. . . .
>> Two lands and one head land in a Close Called the Eyebrookes, by estimacion 2 acres, 2 roods. . . . Three lands in the Church-feild and one land in the Hall feild, both upon the Downes, by estimacion 4 acres. . . .
>> Seaven lands in the Eyebrookes, by estimacion 6 acres. One land in the long acres, by estimacion 1 acre. . . . Two lands, two headlands in the Churchfeild, and three lands in the Hall feild or hall hill, by estimacion 5 acres. . . .
>> Seaven lands in the Eye brookes, by estimacion 5 acres, 2 roods. . . . Fower lands in the Church feild, with a small Cottage, by estimacion 4 acres, 2 roods."

[2] Parliamentary Surveys, Cheshire, No. 13 A.

of some twenty of whom included selions in different fields near the city.[1] These selions, which are also called riggs or lands (*terre*), are never rated in acres; they even at times serve as units of measure for the butts. No tenant had more than eleven and one-half of them, the usual holding being five. Since the selion probably contained not more than an acre, the average share in the common field did not exceed five acres and the total extent of common arable cannot have been great. Of the three fields named, one is simply the field of Chester; Spyttel field and Banke field might seem to be subdivisions of this, were it not that in certain instances each is made coördinate with it. Usually the selions of a holding lay in a single field. Only four times are they assigned to two fields, and only once to three, the division of acres in the last case being unequal (2, 2½, 4). There is, therefore, no reason for concluding that a three-field system was known to the common field or fields of Chester in the middle of the sixteenth century.

A terrier, contemporary with this from Chester and declaring itself a " bylle of the lands of Sir phylyppe Egertons," describes a holding in Tilston, a parish only about three miles across the Dee from the Denbighshire open fields of Issacoed and Pickhill. Most of the butts are assigned to the town field of Horton, itself one of the hamlets of the parish of Tilston;[2] but whether

[1] Rents. and Survs., Portf. 6/24, ff. 6–9. The list is as follows, separation by semicolons indicating different holdings: —

"In communi campo Cestrie or in campo Cestrie .... { unum pratum; iii seliones; iii seliones; iv seliones; viii seliones; iii seliones

In Spyttelfelde .... { vi terre arabiles in orientali parte; v seliones; xi seliones et dimidia; vi terre arabiles in orientali parte; v seliones

In lez Bankefelde .... { iv seliones; quinque butts continentes ii seliones et dimidiam; iv seliones; quinque butts continentes ii seliones et dimidiam

In Spyttelfelde et .... In Chesterfelde .... { ii seliones et iiii seliones; iii seliones et ii seliones; ii seliones et iiii seliones et ii seliones terre et dimidia cum uno hadlonde in Bankefeld

In Bankefelde et .... In Chesterfelde .... { v seliones et ii seliones; i selio et dimidia et iii seliones."

[2] Rents. and Survs., Portf. 1/4, No. 9. The specifications run as follows: —

"In the fylde of humfre hansons there be thow buttys . . .
In the same fylde be thow . . . [elsewhere] Another butte . . .
a hadlant lyeng in horton towne fylde . . .
In the same fylde . . four . . .

there was any grouping of butts within this field we do not learn.

Sixteenth-century arrangements at Tilston and at Chester thus seem to have been like those of the Denbighshire hamlets round Wrexham.[1] Selions in the possession of any tenant were few — seldom more than a half-dozen — and were located without any indication of grouping by fields. Often the entire open arable area was undifferentiated, being merely assigned to a hamlet. Such a "town field" must have been small and situated near the hamlet or village. Though one cannot in Cheshire, as in Denbighshire, compare total areas of townships with the areas of their open fields, the sixteenth- and seventeenth-century surveys of the former county, so far as they are extant, show fully as much enclosure as do those of the latter. In the character and extent of its open field the Dee valley was at that time a unit.

Thirteenth-century testimony regarding unenclosed fields in this part of Cheshire is not wanting. It is to be found largely in the cartulary of St. Werburgh, written soon after 1300,[2] and it accords with the sixteenth-century evidence. To sharpen our conception of a somewhat puzzling field system, it may be well to summarize and illustrate the features that appear in the charters.

The grants were usually made in selions, "lands," or butts, the areas of which were not estimated in acres,[3] a procedure

> In the same fylde . . . the clere pyett'
> In the same fylde Another butt . . .
> In the same fylde other thow . . .
> Another in the same fylde
> A butt lyeing in a fylde called the newe close . . . a hadland another butt . . .
> another butt lyeing in a fylde called unerbroke . . .
> other thow butt' lyeng in a fylde called the longe fylde . . .
> in the same fylde . . . a hadland . . . a roughst."

[1] Cf. above, pp. 179-182. [2] Harl. MS. 2062.
[3] Cf. Add. Chars. 50008, 50040, 50304, cited below. In one instance, however, a lay transaction of 1322 refers to ten acres in Aston [iuxta Mondrum], which lay " in le quytenacres, le oldefeld, Ruycedyche, Aldecrofte et in le Wallefeld " (Add. Char. 49805). Once also Abbot Simon of St. Werburgh exchanged two messuages, two crofts, two lands, and two butts in "le hedfeld" for "iii acras et i rodam iacentes inter landas suas et unum assartum continens v acras et unam rodam" (Harl. MS. 2062, f. 22b).

that emphasizes the importance of the selion as an agrarian unit. Strips described in this way were often located by furlongs, as in a midland terrier. A specification of ten of them at Claverton in the time of Edward I illustrates both characteristics.[1] Some selions lay in furlongs, others in fields, while still others, after a fashions prevalent in northwestern England, lay in areas named " Ulvesdale " and the " croft of Claverton."

Another feature of thirteenth-century Cheshire charters, more striking than either of those just mentioned, becomes apparent in a description of lands transferred at Newton-near-Chester. Twenty-one and one-half selions " in campis eiusdem villae " are characterized as follows: —

" Tres seliones que vocantur le Cleylondes
tres seliones que vocantur le styweylondes
duas seliones que vocantur le Schouelebradlondes
unam selionem que vocatur le longhevedlond
dimidiam selionem que vocatur le Cleyhalflond
unam selionem que vocatur le Brocstanlond
unam selionem que vocatur le Ioustynghevedlond
unam selionem que vocatur le Cleyhevedlond
unam dimidiam landam Iacentem iuxta eandem Cleyhevedlond
duas seliones que vocantur le Putlondes
duas seliones que vocantur le Bradelakelondes
et tres dimidias seliones in Fregrene
unam selionem que vocatur le styweylond
unam selionem que vocatur Edmundislond
unam selionem que vocatur le Schoterdichehevedlond

[1] Add. Char. 50008. The specification runs as follows: —
" Decim seliones terre iacentes in campis de Claverton, viz:
duas dimidias seyliones In Ulvesdale
et unam seylionem iacentem super le stonihulle
et unam seylionem iacentem super le Lowe
et unam seylionem extra le Lowe Iacentem in Brerifurlong
et unam seylionem Iacentem in crofto de Claverton
et duas seyliones Iacentes in le Cruftinge
et duas dimidias seliones Iacentes in le Wythines
et duas dimidias seliones Iacentes supra le Leefeld iuxta campum de Ekleston
et unam dimidiam seylonem Iacentem In Longefurlong
et unam dimidiam seylonem Iacentem iuxta Swartingesfeld."

et totam illam terram que vocatur le Bruches . . . inter terram . . . et terram.[1]

The noteworthy peculiarity here is the naming of the selions. In the case of the four headlands the use of individual appellations is, of course, not unusual. Specifications, however, do not stop with them, since the entire list is similarly distinguished. One can see why three adjacent selions, perhaps a small furlong, should be denominated Cleylondes, but it is different with the selion called Brocstanlond and the half-selion called Cleyhalflond. Since most selions of the charter were named, the usage must have prevailed throughout the common field of Newton. If so, this cannot have been very great in extent. No midland township designated separately each of its two or three thousand selions, finding it task enough to name the furlongs. The nomenclature at Newton thus points to an open arable field of restricted area, one in which individual selions might assume importance.

Still another characteristic of thirteenth-century Cheshire charters is the brevity of their descriptions of open field. The terriers cited above are exceptional in length, few others enumerating as many as six selions.[2] To be sure, the selions were often not accompanied by messuages, and hence may not have been complete holdings. At times, however, the house is mentioned, as when St. Werburgh acquired at Chester half a burgage tenement to which were attached a selion and two butts,[3] or when at Coddington a messuage was accompanied by five " half-lands " and a half-acre of meadow.[4] Small grants to monasteries are,

[1] Add. Char. 50040, *temp*. Edw. I.
[2] A typical grant to St. Werburgh is as follows (Harl. MS. 2062, f. 17): —
"vi seliones in Elton, scilicet,
　unam selionem et dimidiam in campo qui dicitur Brom
　unam selionem in campo qui dicitur Bothum
　unam selionem que . . . extenditur usque ad magnam viam
　et unam selionem que vocatur Naylont
　et unam selionem que vocatur crongeflont
　et dimidiam selionem que iacet versus metam de ẏuis."

[3] " Dimidiam burgagiam extra portam acquilonarem Cestrie et unam selionem, scilicet tertiam, a fossa iuxta viam que tendit versus flokeresbroc et ii bottas [*buttas* in the margin] iacentes inter terram suam et . . . " (ibid., f. 16b).

[4] " Mesuagium cum una dimidia Landa iacenti inter terram . . . et terram . . ., et unam alteram dimidiam Landam iuxta Le Ladeway, et unam dimidiam

of course, numerous in midland cartularies. Yet longer enumerations are nearly always to be found in them, and the absence of such in the cartulary of St. Werburgh tends once more to show that large holdings in the fields were unusual.

Not only is it possible to infer from thirteenth-century charters that Cheshire open fields were small, but these documents give no indication that the selions were ever grouped by fields. The nearest approach to such a suggestion is the location of three acres " in campo de Aston quarum una iacet super longum galewon et alia super le middilfeld et tertia in campo versus trente." [1] The first of these field names, however, together with the small area transferred, does not argue strongly for a three-field arrangement. Such fields as occasionally appear in other terriers are likely to be coördinate with furlongs or with areas variously named. Nowhere did two or three larger fields like those of the midlands gather within their bounds the selions which were conveyed. Chiefly for this reason, as in the case of the counties already discussed in this chapter, we are justified in concluding that the midland system had no hold upon the borderland of the river Dee.

It is possible, further, to discern in the charters of St. Werburgh that even in the thirteenth century the abbots were busy exchanging and consolidating parcels. Sometimes lands newly given to them lay near those which they already held. At Manley, for example, the two and one-half selions given by Robert Fitz Roger lay "in asponesfurlong, quarum una iacet iuxta sellionem que vocatur Aleyneshevedlond et alia iuxta sellionem quam henricus frater eius dedit dicto abbati aule propinquiorem et dimidia sellio [est] propinquior terre dicti abbatis in eodem campo." [2]

Elsewhere the abbots made exchanges. At Leese, Abbot Simon (1265–1289) gave in exchange for " iii acras et i rodam iacentes inter vi landas suas et unum assartum " two messuages

Landam proximam Le Ladeway, et duas dimidias Landas extendentes usque . . . Westmere cum una dimidia acra prati " (Add. Char. 50290).

[1] Harl. MS. 2062, f. 6b.
[2] Ibid., f. 21 (1265–1289).

and two crofts " cum ii landis et ii buttis in le hedfeld." [1] At Bromborough, Abbot Radulphus (1141-c. 1157) exchanged on different occasions " unam sellionem et unam Buttam . . . in Ranesfeld . . . pro una sellione iacenti in campo qui vocatur le Churchcroft; duas dimidias selliones que vocantur suchacresendes . . . pro una sellione et dimidia iacentibus in le chirchecroft . . . ; unam dimidiam sellionem in manislawefeld . . . pro una sellione et dimidia iacenti in le chirchecroft." [2] Sometimes it is evident that the exchanges looked toward consolidation for both parties. The same abbot exchanged with " Henricus filius heyle," at Weston, " pro iii selionibus in tachemedwe . . . sub crofto dicti Henrici . . . iii seliones in cliues . . . iuxta culturam abbatis et unam foreram super morshul et dimidiam rodam super pastmeslande in territorio de Aston. . . . " [3]

Exchanges like these indicate a field system which was not rigid but which easily inclined to consolidation and enclosure. At Lawton a holding given to the abbey in the thirteenth century was already a compact area, comprising a messuage and garden " cum iiii buttis ex una parte dicti gardini et aliis iiii ex altera iacentibus." [4] There is no reason why the open-field strips of a tenement, inconsiderable at best, should not have undergone a process of consolidation; they were inclined toward it both by their small number and by the absence of any grouping of the selions by fields. Since consolidation was so brief a process and was opposed by no inflexible field arrangements, one need not be surprised that it was initiated before the end of the Middle Ages.

Chester thus allies itself more closely with Wales than with the territory to the east. It appears as a county largely enclosed in the sixteenth century and almost entirely so in the eighteenth. Vestiges of open common field in Tudor surveys, however, suggest that at an earlier time most hamlets probably had a certain amount of it, and the thirteenth-century testimony, particularly that from the region near Chester, supports such a belief. This evidence reveals holdings that seldom comprised

---

[1] Harl. MS. 2062, f. 22b.
[2] Ibid., f. 20b.
[3] Ibid., f. 8.
[4] Ibid., f. 24.

so many as a dozen intermixed selions, and township fields in which the strips were so few that at times each of them could attain the dignity of a special designation. Nowhere was there a grouping of strips by fields as in the midlands, and nowhere is found mention of rights of pasture over a fallow field. The arrangement was like that which in Scotland, Wales, and Ireland was called runrig. Since in Cheshire there is no trace of continued or recurrent division of holdings among heirs, some early allotment of the common lands before the time of written records must have been final. In the twelfth and thirteenth century exchanges were being made and the first steps toward consolidation were already taking place. To the flexibility of Celtic open-field arrangements, therefore, is probably to be attributed the early enclosure of the arable in the county, so far as enclosure did not take place directly from the forest state. Such an explanation is further substantiated by the small size of most closes, as seen, for example, in the survey of Macclesfield manor.[1] To some extent, then, the seventeenth-century appearance of the fields of the county is traceable to the early existence of runrig.

Devon and Cornwall

THERE are several Devonshire surveys dating from the late sixteenth or the early seventeenth century, but too frequently they omit exact information about the condition of the fields. A survey of Topsham, for instance, though usually explaining that the " parcelle " were closes and sometimes adding that they were arable, in about one-fourth of the instances leaves them undescribed.[2] Since these undescribed parcels were relatively large, we may infer that the usual designation " clausa "

---

[1] Cf. above, p. 250.

[2] Rents. and Survs., Ro. 169 (1611). A typical holding is that of Helena Havile, widow, who had a house and stable with garden containing 2 acres; closes of arable called Butt parke and Sandell, containing 5 acres and one acre; other closes called Whittwell, Greenland, and Longland, each containing 2 acres; a parcel called the half-acre; a parcel of marsh containing 8 acres called Idons; and pasture in the marsh for twenty sheep.

was carelessly omitted.[1] One feels on safer ground in surveys like that of Sherford, in which all parcels are carefully labelled orchards, closes, or "parkes."[2] The application of the term "park" to a close of arable is characteristic of Devonshire, and its constant employment in the survey of Vielstone and Kingdon indicates the enclosed character of these townships.[3] Porlock, too, a Somersetshire township on the edge of Exmoor, resembled its Devonshire neighbors in being entirely enclosed.[4]

Despite the testimony of most sixteenth-century documents to the enclosure of Devonshire fields, there is an occasional hint that unenclosed arable might still be found. Of the manors of the marchioness of Dorset, which were surveyed in 15 Henry VIII, most lay in Somerset, but some were in Cornwall and Devon.[5] Although none of the surveys of the Devon manors are very explicit about the condition of the arable, it appears from the description of Brixham that many of the tenants held each one "furlong,"[6] comprising twenty acres of pasture and ten of arable, and that appurtenant to each furlong was "communia in communibus campis" for sixty sheep, two cows, and one horse.[7] The "communes campi" here pretty clearly bespeak open arable field, for the phrase was almost never applied to the common waste, and where it occurs elsewhere in this group of surveys it refers to certain townships in Somerset which lay in open-field neighborhoods. Inasmuch as no similar remark about common fields is vouchsafed regarding the other five Devon and Cornish townships, these by implication were enclosed. Each of them contained more pasture than arable, but

[1] To be sure, some six parcels were in Rushmore, but they were too large (7, 4, 5, 2, 5, and 3 acres respectively) to suggest open-field strips. The first three were arable, another was a close of pasture, while two are not described.

[2] Add. MS. 21605, ff. 36-43 (1606). The same volume contains (ff. 18-24) another survey of Sherford written in a hand earlier by a generation; but this one neglects to say whether its "farthings" were open or enclosed.

[3] Exch. Aug. Of., M. B. 358, ff. 64-74, 6 Jas. I. The designations, for example, are North park, Lea park, Temsty park, Wall park.

[4] Exch. Aug. Of., M. B. 385, ff. 97-106, 17 Hen. VIII.

[5] Ibid., ff. 112-208. Most important were Brixham, Woodford, and Shewte in Devon, and Trewerdreth, Trelawne, and Wadfast in Cornwall.

[6] I. e. "ferling," for the meaning of which cf. below, pp. 264-266.

[7] Exch. Aug. Of., M. B. 385, f. 200 *sq.*

arable and pasture differed little in annual value. Both were rated at from 12 d. to 16 d. the acre, whereas the arable of an open-field midland township was seldom worth more than 6 d. the acre.

A survey of the extensive Cornish and Devon estates of Lord Dynham was made in 1566,[1] describing in considerable detail more than twenty manors, among others the great manor of Hartland on the northwestern coast. Holdings here are located in large areas or by hamlets, and the parcels of which they were composed are described as closes. Such was the case with the typical holding printed by Mr. Chope, that of Agnes Dayman, situated at the hamlet of "Cheristawe." At Cheristawe were five similar tenements with areas of 21, 11¾, 21, 14½, and 25 acres, a total of 123 acres for the hamlet.[2] No township of the manor had in it more holdings than this, and usually there were fewer. So far as can be seen, the manor of Hartland consisted of hamlets the fields of which were small and enclosed.

A few phrases used in other of the Dynham surveys, however, demand attention. At Ilsington, William Prowse held "without copy one holding with a garden and one ferling of land, containing by estimation 30 acres, but he does not know where they are because they lie among the lands of the lords and of George Fourde, esq. [lord of the other half of the manor]."[3]

[1] The MS. was in 1902 in the possession of C. D. Heathcote, of Porlock, and has been described by R. P. Chope in two papers published in the *Transactions* of the Devonshire Association for the Advancement of Science, etc., vols. xxxiv (1902) and xliii (1911). In the second paper, on "The Lord Dynham's Lands," Mr. Chope sketches summarily the surveys of most of the Devonshire estates; but in the first one, entitled "The Early History of the Manor of Hartland," he transcribes all details and illustrates locations by a valuable map.

[2] Ibid., xxxiv. 438: "Agnes Dayman, widow, . . . holds by copy dated . . . 13 Henry VIII . . . a half-ferling and one clawe of land, with their appurtenances, in Cheristawe, . . . to which belong
    1 house . . ., 1 barn, 1 garden, and 1 orchard containing 1 rood
    2 closes called the Crosse parkes containing 4 acres
    1 close called Swetenham containing 2 acres
    1 close called the Hill parke containing 6 acres
    1 close bewest the towne containing 9 acres
    1 close called ye Brodewey parke containing 3 acres
    1 close called the Higher parke containing 2 acres
    1 close called ye Lower parke containing 3 acres
    and in the meadow ½ acre."
[3] Ibid., xliii. 278.

In the account of another holding of this manor occur similar statements about intermixed parcels. The customary tenement of Agnes Orchard included " divers parcels of land called *lez Shotes*, lying in the common about the bounds called *lez londscores* with the lands of William Dyggen, customary tenant of this manor, containing in all 30 acres of land in the common of Idetordowne [Haytor Down]." [1] Perhaps the translation should run, " divers parcels . . . lying in common " (if the original is *in communia*). However that be, the significant item, apart from the assertion that certain lands were intermixed, is implicit in the phrase " lez londscores." In the same manor Hugh Dyggen also held " divers parcels of land lying together about the Londscore next Idetordowne, containing in all 60 acres." [2]

Explanation of the meaning of the phrase " lez londscores " is to be had from an item relative to the Dynham manor of Woodhuish in Brixham. Here, the survey notes, " the landes . . . for the most parte lyeth by londes score in twoe commen feldes." The holdings were rated in ferlings, to each of which were assigned some 27 acres of " arable land lying at large in the fields and *lez Breches*." Altogether there were 652 acres.[3] These statements point clearly to open common fields in which parcels lay intermixed, or " by londescore." The use of the latter phrase at Ilsington, therefore, accords with the declaration that Agnes Orchard's lands lay intermixed with those of William Dyggen. Upon two of the twenty-five manors or estates of Lord Dynham which were situated in Devon and Cornwall we are thus assured of the existence of common fields.[4] At Woodhuish in Brixham they were extensive, and Brixham, it will be remembered, was that one of the Devonshire manors of the marchioness of Dorset in which common fields have already been discerned. Brixham and Woodhuish are adjacent townships lying on the southern coast at the mouth of the river Dart.

[1] Devon. Assoc., etc., *Trans.*, xliii. 279–280.
[2] Ibid.   [3] Ibid., 281.
[4] There were also " common meadows," as at Wilmington (ibid., 274). Nearly all the Dynham manors comprised wastes upon which the tenants had rights of common.

At Ilsington the intermixed lands lay on the edge of a common waste called Haytor Down. Equally unusual in situation, though in a different way, is the " Great Field " at Braunton. This Slater has described,[1] but a fuller account is available.[2] Braunton is a village in northwestern Devon, near Barnstaple, lying a little inland from the estuary of the river Taw. Bordering the river and the sea are marsh lands known as Braunton Burrows. Between the marshes and the village lies the " Great Field." " Its surface," runs the local account, " is a dead flat rising but little above the level of the marshes, and the soil is doubtless by origin a natural reclamation from the bed of the estuary. The whole field is under arable cultivation in small unenclosed plots." The Braunton rate book of 1889 states that its area was then $354\frac{1}{2}$ acres, occupied by 56 proprietors and lying in 491 strips. The strips, each containing from one-half an acre to two acres, were gathered into sixteen shots,[3] and those of each proprietor were non-contiguous. All holdings were " unqualified freehold, subject to no seigniorial rights or claims." The lord of the manor had in 1875 owned a considerable portion of the Great Field in seventy-three plots containing each about an acre, but he afterward sold them. Slater says that there are no common rights over the field.

The peculiarities manifested in this description give a possible clue to the origin of the Great Field. Its position on the map and its low-lying character suggest that it is land at some time reclaimed from the marshes; the two other manors in Braunton not adjacent to the marshes have no open field. Further, the tenure by which the field is held points in the same direction: only newly-reclaimed lands would be likely to be freehold, subject to no seigniorial rights. Probably in lieu of such rights the lord of the manor had received some fraction of the parcels. The extensive scattering of the strips may have been due to the gradual reclamation of the area, each furlong having been subdivided by lord and freeholders as it was improved.

[1] *English Peasantry*, p. 250.
[2] Devon. Assoc., etc., *Trans.*, xxi. 201 (1889).
[3] Lime tree, Harditch, Renpit, Long Hedge Lands, Broadpath, Lane end, Cuttaburrow, Higher Thorn, etc.

Fertile alluvial land would need little fallowing, and continuous cropping would leave no opportunity for the exercise of the right of pasturage during a fallow year. If these conjectures be correct, Braunton Great Field was of relatively recent origin. Perhaps the "londscores" near the common at Ilsington were also recently improved lands, in this instance taken from the waste.

In Cornwall, as in Devon, the Jacobean surveys tell of enclosed townships, occasionally hinting at the existence of common arable fields. A long account of the manor of Launceston, which describes leaseholds in many hamlets, always refers to the parcels of the tenements as closes, sometimes adding that they were meadow or pasture.[1] In a companion survey, however, a significant statement is made relative to Leigh Durrant. "Some parte of this Mannor," the surveyor explains, " lieth in Common fields which is hardly founde in any Mannor of his highness eels in Cornewall;"[2] but no description of these common fields is vouchsafed. We come upon others in a survey of Carnanton,[3] where, in 4 James I, 70 of the 960 acres accounted for still lay in some seven "common fields," of which at least three were closes. Down close contained 12 acres, held by four persons; Furze close, 8¾ acres in the hands of three tenants; and New close, 5 acres with a single occupant. The remaining " common fields " were West, North, South, and Churchway, each having an area of from 5 to 20 acres.[4] Five times the acres in common field or common close are said to be " in stichmeale," a phrase pointing to the intermixture of tenants' parcels. If we inquire into the origin of this situation, the names of the common closes at once suggest appropriation from the waste. Other items in the survey indicate that a " Downe " had recently been allotted and improved. Twelve times there is reference to acres of common pasture " in le Downs " or " in communi campo vocato

[1] Land Rev., M. B. 207, ff. 149–213, 5 Jas. I.
[2] Ibid., f. 42b.
[3] Exch. Aug. Of., M. B. 388, ff. 135–171.
[4] In West field six tenants had 19½ acres; in North field seven had 15¾ acres; in Churchway four had 3½ acres; in South field two had 5¼ acres. A few acres lay simply " in communi campo."

le Downes "; and occasionally this "communis pastura" changes into " terra arabilis et pastura," from which it is only a step to the " Downe close " with its holdings always arable.

This description, joined with that of Braunton Great Field and that of the landscores at Ilsington, seems ground for believing, not only that the common fields of Devon and Cornwall in the sixteenth century were few, but that some of them were not of ancient origin. About the antiquity of certain of the fields at Carnanton, and of the still larger ones at Brixham and Woodhuish, we know little. Nor have we information about the distribution of parcels in these Devon and Cornish fields, save that given by the nineteenth-century appearance of Braunton Great Field. This had by no means a two- or three-field aspect, the tenants' parcels being apparently distributed throughout it with the same irregularity as prevailed in the counties of the northwest.

Turning to the earlier Devon and Cornish evidence, we find two local units much in evidence, the " ferling " and the Cornish " acre." In general utility the ferling corresponded with the midland virgate, replacing it as the fourth part of a larger unit. The larger unit itself was sometimes called a virgate; in one of the fines, from a total of six virgates at Dene there were subtracted two ferlings and two and one-half acres,[1] while near Exeter we hear of the transfer of a half-virgate and a ferling.[2] In Cornwall, according to an early fine which carefully states that the sum of half an acre and two ferlings equalled an acre, the ferling was the fourth part of a Cornish acre.[3] Its area of course varied as did that of the unit of which it was the fourth part. At Brixham, as we have seen, it contained 30 acres;[4] and at the end of the sixteenth century this was its size at Woodbrooke, at Allerton, and at Sherford.[5] In a Devon fine of 22 Henry III three ferlings equalled 43 acres.[6] In Cornwall, in 1337,

[1] Ped Fin., 40-9-164 (12 John).
[2] Cott. MSS., Vitel. D IX, f. 168b (a fourteenth-century cartulary).
[3] Ped. Fin., 31-2-20.
[4] Cf. above, p. 259.
[5] Rents. and Survs., Portf. 6/61; Add. MS. 21605, ff. 19, 24.
[6] Ped. Fin., 40-12-226.

the ferling was said to contain from 4 to 5 acres, the Cornish acre being only four times as great;[1] but in a rental of 6 James I the Cornish acre was larger, three-fourths of it containing 70 English acres.[2] Thus, at different times and in different places the ferling varied in extent between 4 and 30 English acres.

Whatever may have been its size, the important question as regards field systems is whether it was a compact area or was composed of scattered strips. The best evidence on this point is from certain descriptions contained in a fifteenth-century cartulary of Torre abbey. At one time we discover that a half-ferling of unknown size is completely bounded as one block;[3] again a ferling is said to lie " propinquior ad orientem terre predictum canonicorum "; elsewhere a half-ferling lies " in hocrigge," and another half-ferling " in parte orientali de Chinrigge iuxta aquam ";[4] finally we hear of a half-ferling " unde una clawa [close] vocatur Dodemmannesland et alia clawa vocatur Wluesland."[5] In an early fine twelve ferlings of the manor of Coombe are so described as to imply that they were blocks in different parts of the village floor, and that with them were transferred the resident villein households. In Coleford there was a ferling and a half, at Tocumbe a ferling and a half, at Fostefelde two ferlings, at Haldestanc four, at Fishull one, at Blakewille one, at la Grutte one.[6] In Limerick, in 22 Henry III, two ferlings were " in Lange furlang " and " in Sholdedune."[7] Nowhere

[1] Sir John Maclean, *The Parochial and Family History of the Deanery of Trigg Minor* (3 vols., London, 1873-79), iii. 45 sq.

[2] Rents. and Survs., Portf. 2/33.

[3] Exch. K. R., M. B. 19, f. 25b: " Illud dimidium ferlingum terre . . . que se extendit a fossato . . . usque magnum iter . . . quod ducit versus Teyngnewike . . . et iacet iuxta terram ecclesie de Hanok et se extendit usque regale iter quod ducit versus hywis et iacet iuxta terram W. de Ferndon . . . et iterum iuxta pratum sub Asselonde . . . et iterum iuxta cornerium curtillagii ubi facte sunt divise."

[4] Ibid., f. 57b.

[5] Ibid., f. 33b.

[6] Joseph Hunter, *Fines sive Pedes Finium* (Record Com., 2 vols., 1835-44), ii. 46 (10 Rich. I). Coombe and Coleford are two adjacent Devonshire hamlets, but the other names are not applied to hamlets in this neighborhood.

[7] Ped. Fin., 40-12-239.

in the early fines and charters is there anything to indicate that the ferling was composed of acre or half-acre or quarter-acre strips.[1]

Devon and Cornwall thus assume in the thirteenth- as well as in the sixteenth-century documents the appearance of counties the arable lands of which were very largely enclosed. The feet of fines and the charters from this corner of England are in marked contrast with those from the midlands and even with those from the northwest. From all other English counties (except perhaps from Kent and Essex) a considerable number of fines and charters disclose on examination at least a few which record each its list of small non-adjacent parcels of arable land. The exceptional character of the Devon and Cornish documents would lead us to believe that even runrig was unknown in these two counties, were it not for the testimony of the sixteenth-century surveys.[2] How uncertain is this testimony relative to the extent and antiquity of open arable fields we have seen, but about the existence of intermixed strips it is clear. It suggests that Devon and Cornwall more closely resembled Cheshire and Wales than any other region thus far examined. In the valley of the Dee were townships which had common open arable fields, small in extent, like those of the southwest. So far as the latter were really ancient, a characteristic possibly attributable to those of Brixham, it is perhaps allowable to call them Celtic in their affinities and to assume that enclosure occurred early, as it did in most parts of Wales. With these inferences, the most probable that we can draw in view of the perplexing evidence, Devon and Cornwall take their place along with the other counties of western or northern England which in their field arrangements were subject to Celtic influence.

[1] A parcel of land in a suburb of Exeter was once designated " unum sullonem " (Cott. MS., Vitel. D IX, f. 138), but it may not have been part of a ferling or have lain in open field.

[2] It will also be noticed that in the phrases quoted in the preceding paragraph the term " rigge " was used to designate a furlong.

Conclusion

A SUMMARY of the results of the preceding examination of field arrangements in the counties of the Celtic border is now possible. It will be remembered that Scottish, Irish, and Welsh fields, differing as they might in some respects, yet had in common that which makes it possible to speak of a Celtic field system. Although this system was, without doubt, originally one of open fields, the absence of enclosure did not constitute its distinctive characteristic. Non-Celtic fields were often open, and Celtic fields, even after enclosure, sometimes bore traces of their origin. More noteworthy than the absence of enclosure was the size of the Celtic township or townland, the continued subdivision of it among co-heirs or co-tenants, the distribution throughout it of the parcels of the tenants' holdings, and the method by which it was tilled. In the counties considered in this chapter certain of these characteristics appear more clearly than others.

The small township with its hamlet settlement we have seen, behind various disguises, revealed in Cumberland documents. Since other enumerations manifest a tendency to be similarly obscure, it is difficult to determine from them alone the region characterized by this form of occupation. In the long seventeenth-century survey of the Lancashire manor of Rochdale, for instance, the hamlets themselves were so complex as to contain within their somewhat spacious boundaries several nuclei of settlement.[1]

[1] Henry Fishwick, *Survey of the Manor of Rochdale* (Chetham Soc., 1913). One division of the manor, known as Spotland, contained six hamlets and " Spotlande towne," the areas being specified as follows (pp. 163 *sq*.): —

Hamlet	Free Tenants	Acres	Copyhold Tenants	Acres	Lease-holders	Acres	Acres of Common Waste
Falinge	13	255	0	0	0	0	0
Chadwick	12	495	17	121	0	0	114
Spotland (towne)	29	1,266	40	312	1	19	672
Wolstenholme	12	796	40	558	1	33	823
Healey	11	430	9	108	1	17	240
Whitworth	6	2,588	47	726	5	3	515
Rossendale	1	2,382	6	160	0	0	0

The units of settlement named on the modern map as lying within the above areas number some fifty.

In view of the deceptive brevity of written documents, it is best, unless in each instance it be possible to investigate the stated areas, to take the less specific evidence which is furnished by the modern map. From an examination of this we shall have no hesitation in pronouncing that the counties examined in the foregoing chapter, except Northumberland, were characterized by the hamlet type of settlement. Indeed, we shall have to include other counties as well, a circumstance that leads to a further distinction.

Although the hamlet was typical of Celtic settlement, its appearance was not necessarily accompanied by a Celtic field system. Two counties of the Welsh border, Herefordshire and Shropshire, have already illustrated the divergence. On the map they are dotted with tiny groups of houses, which, though often bearing English names, are typical hamlets, while an analysis of the parish of Marden has shown us several of these grouped into a larger unit. Yet the tillage of Herefordshire and Shropshire hamlet fields was similar to that of the midlands; and, though irregularities soon arose in these fields and the decay of the midland system occurred earlier than it did farther east, the situation in the two counties assures us that hamlet settlements with inconsiderable fields did not necessarily imply Celtic runrig.

A second characteristic of the Celtic field system was its readiness to subdivide holdings, farms, or townlands among co-heirs or co-tenants in such a way that each received a share in every quality of the soil and held his arable strips under a form of intermixed occupancy known as runrig or rundale. In Scotland and Ireland such subdivision continued throughout the eighteenth century; in Wales the co-tenancy of the fourteenth century was abandoned in the sixteenth. In northern and western England little evidence is as yet available to demonstrate the prevalence there of the transmission of land to co-heirs; scholars have merely noted that the custom of certain sokes or manors in Shropshire, Herefordshire, and Monmouthshire at a relatively late time prescribed transmission by gavelkind.[1] Until further investigation

[1] T. Robinson, *The Common Law of Kent, or the Customs of Gavelkind* (5th ed., by C. I. Elton and H. J. H. Mackay, London, 1897), p. 33.

has determined the localities in which such a usage prevailed and the degree to which Celtic influence is responsible therefor,[1] no generalizations are possible. From what evidence we have it would seem that in most districts of the north and west the subdivision of socage and villein holdings, if it ever prevailed, early gave place to impartible succession, the custom which from the thirteenth century at least was usual in the midlands.

Of greater assistance in estimating Celtic influence upon the field system of English counties is a third trait of Celtic agriculture. This is the irregular disposition of the scattered parcels of the tenants' holdings throughout the cultivated area of a township; for, if it be assumed that an early subdivision of the land among co-heirs became permanent in the Anglicized border counties, such disposition would be for us the only reminder of the earlier field-history of the region. If it chanced that the dispersed parcels were in certain places reconsolidated (or if in some townships a division had never taken place), we should expect to find there enclosed areas. When, consequently, Devon, Cornwall, and Cheshire appear as counties largely enclosed in the sixteenth century, this phenomenon is explicable as a normal manifestation of the Celtic system. If some traces of arable common fields still remained within their bounds, these too are normal phenomena. Although in the southwest some such fields may have been due to improvement of marsh or down-land, other tracts are not easily so explained. At Brixham in Devon, as well as at several places in Cheshire, there seem to have been ancient arable fields that had long been characterized by intermixed holdings. Cheshire terriers of the thirteenth century give details which enable us to see that these fields were not of the midland type. In structure they were, on the contrary, like

[1] Two vague passages in the laws of Cnut which may imply that partible succession was the Anglo-Saxon usage of the eleventh century are as follows: " [If a man die intestate] Ac beo be his dihte seo aeht gescyft swyðe rihte wife 7 cildum 7 nehmagum, aelcum be þaere maeðe, þe him to gebyrige. . . . And se man, þe on þam fyrdunge aetforan his hlaforde fealle, . . . fon þa erfenuman to lande 7 to aehtan 7 scyftan hit swyðe rihte " (Cnut II 70, 78 [1027-1034], Liebermann, *Gesetze*, i. 356, 364). Chapter 34 of the *Leis Willelme* (1090-1135) is of similar purport: " Si home mort senz devise, si departent les enfans l'erité entre sei per uwel " (ibid. i, 514). Cf. Maitland, *Domesday Book and Beyond*, p. 145.

those of Lancashire and Cumberland, regions in which open fields survived longer and are more fully described.

If we turn to these two northern counties, in neither do we find such a grouping of scattered parcels as the two- or three-field system imposed. In them the strips of the holdings lay, to be sure, dispersed throughout the arable area, but the arrangement can properly be called nothing more than runrig, since nowhere is there any grouping by fields, whether two or four, three or six. In Cumberland it is even possible that the strips of a holding were at times segregated within one part of the township's arable. Whatever may have been the usual juxtaposition of a tenant's arable strips in all these western counties (and about this there is still considerable doubt), the absence of the midland alignment is a characteristic common to the field arrangements of Cumberland and Lancashire, to those of Cornwall, Devon, and Cheshire, and to those of Scotland, Ireland, and Wales. Further emphasis is put upon this characteristic by the absence in terriers and surveys of any intimation that the villagers desired to have a *continuous* stretch of their intermixed arable lying fallow at one time, as was the practice under the midland system. Although one hears much about rights of pasture over common, moor, and fell, such rights are never specified relative to a fallow field. Thus pasturage arrangements in the counties under consideration do not point to midland usages any more than do the relative positions assumed by the strips of the customary holdings.

If in both these respects the counties of the northwest and the southwest show Celtic rather than midland fields, of a final characteristic of Scottish agriculture — namely, the temporary tillage of parcels of the waste or outfield during a series of years, followed by an abandonment of the same parcels for a corresponding period of time — they furnish little evidence. To find unmistakable traces of such a custom in England it is necessary to turn to Northumberland, where its existence is established by two or three brief descriptions. Without much doubt the same practice prevailed in Cumberland, since sixteenth-century surveys record there the subdivision among tenants of areas newly improved from the waste. It seems likely, furthermore, that the custom fell into disuse much earlier in the English counties than in Scotland. We should, perhaps, think of the two regions as practicing the same system at first but developing it differently. A Scottish township continued to treat its outfield in the primitive manner, but also set aside a small infield, which by the use of manure was kept under continuous tillage; a township of the English border counties set aside no infield, but tilled in a uniform manner all land which at any time came under the plough. In England, however, a developing agriculture, since it did not create an infield, began, we may suppose, to demand that the periods of productivity of the improved furlongs be prolonged at the expense of the periods of fallow. In due course as much as two-thirds of the available arable may have been brought under yearly cultivation. If this were achieved, it would become easy to shift the location of the fallow furlongs so as to bring them together into a compact fallow field. Thereby the township would practically adopt the three-field system, a transformation which may at times have taken place in Northumberland. If this was the case, the county is to be looked upon as transitional in its field arrangements, marking the passage from the Celtic to the midland system.

Whatever may be the value of this hypothesis, it seems pretty clear that the Celtic system made its influence felt in one way or another throughout all the counties discussed in this chapter, and in all probability throughout Monmouth, Westmorland, and western Yorkshire as well. Generally speaking, then, the counties of the northwest and southwest, none of them far removed from Celtic lands, constitute that part of England which came within the sphere of influence of the Celtic field system.

CHAPTER VII

THE KENTISH SYSTEM

IT will be convenient to begin our examination of the field arrangements of the southeast of England with a study of Kent. Doubt has been expressed whether this county was ever in open field. Meitzen, with an eye upon its scattered farmhouses, which contrast with the nucleated villages of open-field districts, suggested a field system of Celtic origin, similar to that which, he thought, prevailed between the Rhine and the Weser and was largely one of enclosures.[1] Slater found no parliamentary acts for the enclosure of arable in Kent;[2] and in 1794 Boys was able to report to the Board of Agriculture, " There are no common fields in this county, and but few common pastures in this part of it [the east]."[3]

As early as the sixteenth century, indeed, Kent is referred to as one of the counties "wheare most Inclosures be,"[4] a statement that may be verified by several manorial surveys from the end of that century and the beginning of the next. A "measurement" of three manors in the parishes of Cranbrook, Goudhurst, and Hawkhurst describes large demesnes and " fermes " apparently all enclosed.[5] Similarly enclosed were the manors of Nether Bilsington (near Romney Marsh and consisting largely of marsh and woodland), Neates Court (in pasture), and Sondrisshe.[6] Throughout a hundred pages of sixteenth-century sur-

---

[1] *Siedelung und Agrarwesen*, ii. 122, 54.   [2] *English Peasantry*, p. 230.

[3] J. Boys, *General View of the Agriculture of the County of Kent* (London,.1794), p. 44. Eighteenth-century references to open-field parcels are rare, although they do occur. In 1770, for example, a Mr. Holmes at Henhurst owed tithes from 6½ acres of fallow, which was "part of [a] common field" (*Archaeologia Cantiana*, xxvii. 124).

[4] John Hales, *A Discourse of the Common Weal of this Realm of England* (1549, ed. E. Lamond, Cambridge, 1893), p. 49.

[5] Rawl. MS., B 341, ff. 31 *sq.* (1587).

[6] Add. MS. 37019 (1567); Rents. and Survs., Portf. 9/43; 6 Jas. I; Land Rev., M. B. 258, ff. 154-164, 1 Mary.

veys relating to Kentish manors or townships and collected in one volume, the tale, except for an occasional item, is one of enclosures.[1] A long survey of Northbourn likewise speaks almost entirely of closes.[2] All this, however, does not necessarily imply that sixteenth-century closes in Kent formed compact estates, as one might at first infer.

All Souls College, as owner of several Kentish estates, had maps of them made in the years 1589–1593. On the maps of the properties which lay in and about Romney Marsh, the parcels, both large and small, appear as plats rather then open-field strips and for the most part were consolidated.[3] With two manors which were situated near the mouth of the Medway in the northern part of the county the case was different. The manor of Horsham, in the villages of Upchurch, Alteram, and Ham, lay to some extent intermixed with other properties, and this characteristic is pronounced in the plan of a manor at Newington,[4] reproduced in the accompanying sketch.

Such lack of contiguity between the parcels of a holding as is shown in these instances suggests an earlier system not characterized by consolidation. That sixteenth-century closes were sometimes of recent origin is clear from an account of the manor of Westcourt or Sibertswold, which, it is said, consisted of demesnes and services. "The demesnes lye contiguously to another and they are all now in Enclosures"; the services were due from some 420 acres in sixteen closes, "for the most part lately made."[5] Although, according to a long survey, the manor of Eltham consisted largely of closes, there are references to an East field in which seven tenants still held parcels containing from ½ acre to 2 acres each.[6]

[1] Land Rev., M. B. 196, 6 Jas. I. The exceptional items tell us, for example, that at Faversham were 8 acres of arable "in communi campo vocato le Abbey wrongs," and at Shoreham 1½ acres of arable "in communi campo vocato Shorham hill" (ff. 116, 117b).
[2] Stowe MS. 858, 6 Jas. I. There is, however, mention of one acre and ten poles "in communi campo vocato Ashley field" (f. 39).
[3] All Souls Typus Collegii, iii, maps 8–14.
[4] Ibid., maps 1 (Horsham), 4 (Newington).
[5] Exch. K. R., M. B. 40, f. 7 (1616).
[6] Exch. K. R., M. B. 44, ff. 40b–50b (1605).

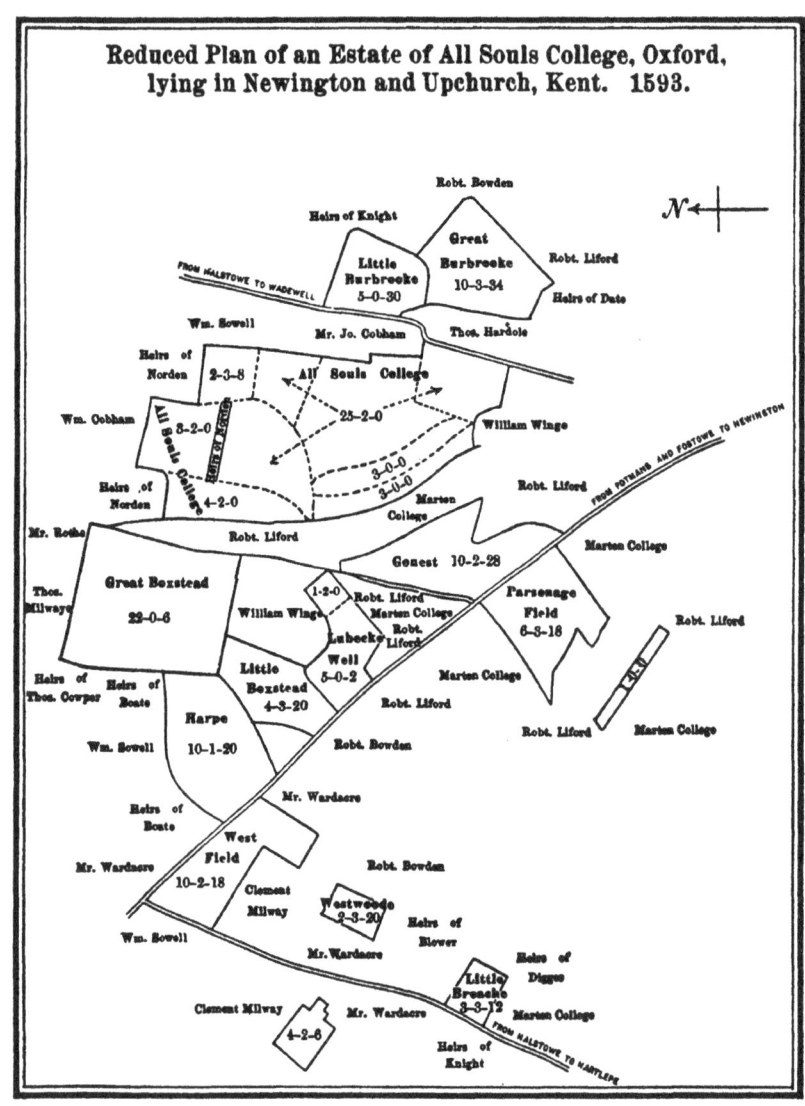

Map XII

Full and convincing testimony to the existence of open-field arable in Kent appears in certain early seventeenth-century surveys relating to St. Margaret at Cliffe, Guston, Deal, and Sutton, all situated in the southeastern part of the county. A terrier of the glebe of St. Margaret at Cliffe, dated 1645, describes it as lying in 31 furlongs, in each of which were from two to five small parcels, each parcel lying between the lands of other proprietors.[1] In Limvine furlong, for example, were four separate strips of glebe containing respectively 1 acre and 22 perches, 3 roods and 10 perches, 1 rood and 3 perches, 1½ acres and 15 perches. An earlier account of 1616 explains that the same glebe "lyeth in severall Shotts or furlongs of land . . . lying intermixed with the lands of the Tenants of the mannor of Reach."[2] Elsewhere we learn that the "manor of Reach doth consist of demesnes and services and lyeth in the parish of St. Margaretts at Cliffe neere Dover. . . . The demesnes are of three sorts — Inclosed Lands, Outlands or Downe Lands, and Commons. . . . These open fields and downe do incompasse the inclosed lands and mansion house. . . ."[3] Clearly the so-called "outlands" consisted of intermixed arable strips lying in open field.

In 1616 the demesne lands of the manor of Guston near Dover were of "two sortes, Inclosed or lying in parcells in open fielde. The inclosed lands some ly contiguously one to another and the rest lye severed amongst other mens lands." The contiguous enclosed demesne comprised 96 acres; the severed but enclosed, 16¾ acres; the unenclosed, 38 acres in 18 parcels. Of the tenants' holdings 54 acres were enclosed (whether in contiguous parcels is not stated) and 63 acres lay in 72 parcels in open field, the open fields bearing such names as the Chequer End, the Butts, Church field, and "Le Shott sive Furlong iuxta Banke."[4] In contrast with this estate, another "reputed" manor called Frith in the same parish was "onely in Demesne . . . and the whole demesne lands lye all together in an oblique lyne and no man hath any lands intermixed with the lands of the same manour."[5]

[1] Rents. and Survs., Portf. 9/55.
[2] Exch. K. R., M. B. 40, f. 47.
[3] Ibid., f. 6 (1616).   [4] Ibid, ff. 2 sq.   [5] Ibid., f. 14.

Full detail for all tenants' holdings is given in the survey of the manor of Dale, or Court Ashe, in the parish of Deal.[1] These lay almost entirely in open field. The first tenant had 56½ acres in 23 parcels lying in 16 fields,[2] the second 18 acres in 20 parcels in 13 fields, the fourth 25½ acres in 18 parcels in 14 fields, and so on. Typical among the field names were Scotten Tyght, Le Chequer, Long Tyght, Woo furlong, West furlong super le Downe, Keetwheet, Goldfrid, and Upland.

In the neighboring township of Sutton, the archbishop of Canterbury's land consisted both of parcels in enclosed fields lying among other men's lands and of parcels in open fields. The open fields were named Pising field, Barley Downe, Chequer end, the Butts, the North end, and the East hill. In them rent-paying tenants also had parcels.[3]

These four parishes, St. Margaret, Guston, Deal, and Sutton, all lie in the high down-lands of southeastern Kent, downs which today are still largely open. The surveys are relatively late, dating from the early seventeenth century. An apparently reasonable inference, then, would be that we are here dealing with stretches of common land somewhat recently improved and distributed with an attempt at equity among the several tenants. Yet why subdivide so minutely and separate so persistently? The glebe in Limvine furlong at St. Margaret might as well have been one four-acre parcel as four smaller ones, and an eighteenth-century division of a common would have made it such. The actual situation in the survey bespeaks the type of mind which subdivided the fields of the midlands, and suggests that the arrangement in Kent was not altogether recent when the surveys were made.

This inference is not without the support of earlier documents. In 4 Richard II, Thomas Menesse of Dale (in the parish of Deal) granted land as follows: —

In Dale: " una roda terre iacet in campo vocato longetheghe tres rode iacent in loco vocato Dodeham

[1] Exch. K. R., M. B. 40, ff. 8-11, 14 Jas. I.
[2] One parcel contained 18 acres, one 5½; the others were small.
[3] Exch. K. R., M. B. 40, ff. 1-2 (1616).

una roda et dimidia atte Berwhe
una roda et dimidia iacent in campo vocato Dodeham."

In Sholdon: "una roda et dimidia iacent in loco vocato Ketewode
tres rode et dimidia iacent particulariter in campo vocato Scholdonesfeld
una acra et una roda iacent in duabus parcellis apud lyden
et predicta pastura pro quatuor vaccis iacet in marisco vocato Collosschepemerch. . . ."[1]

Since the field names at Dale also occur in the survey of 14 James I as Long Tyght, Dodham, and Beere Tyght, the conditions of the seventeenth century seem to be carried back to the fourteenth.

It is desirable, however, to secure testimony from parts of Kent which do not consist mainly of downs. In a hand of Henry VIII is written a survey of Sutton at Hone, one of the archbishop's manors in the northwestern part of the county, some twenty miles east of London and five miles from the Thames.[2] The demesne, which comprised 642 acres, was enclosed, but of the 424 acres held by the freehold tenants at least 93 lay intermixed in 49 parcels in a half-dozen fields. These fields bore the names Church Down, Southfeld, Northfeld, Battesdene, Bradfeld, Jordanes Croft. Each of the last three was shared by only two tenants, but elsewhere the subdivision was more complex. Southfeld was divided among five tenants holding respectively in acres $2\frac{1}{4}$ in 3 parcels, $6\frac{3}{4}$ in 2 parcels, $4\frac{1}{4}$ in 3 parcels, 14 in 2 parcels, $29\frac{1}{2}$ in 12 parcels (the 12 once having been attached to as many as four tenements); Northfeld had four tenants, holding 1, $\frac{1}{2}$, $\frac{3}{4}$, $1\frac{1}{4}$ acres (the last in 3 parcels); Church Down fell to three tenants, whose acres numbered $2\frac{1}{2}$, $4\frac{1}{2}$ in 2 parcels, and $10\frac{1}{2}$ in 9 parcels.

This distribution of a tenant's parcels throughout fields appears quite as noticeably in an early sixteenth-century terrier of the lands of the heirs of William Hexstall, Kt., at Hoo St. Mary's, a

[1] Rawl. MS., B 335 (Reg. Hosp. St. Barth. Dovorie).
[2] Treas. of Receipt, M. B. 172.

township between the mouth of the Thames and that of the Medway. The 36 acres in question comprised 14 acres adjacent to the messuage, 3 acres of meadow in a neighboring township, and 24 small parcels in 18 places which look much like open field.[1]

Such descriptions from two northern townships suggest that there was at times in this part of the county considerable subdivision of certain holdings into small scattered parcels, and this aspect of the situation recurs in terriers and surveys of the fourteenth and fifteenth centuries. At Lewisham, near Greenwich, there is twice recorded the transfer of parcels that lay in the same field.[2] At St. Mary's Cray and Orpington, also near Greenwich, a conveyance of 40 acres, dated 26 Edward III, enumerates $11\frac{1}{4}$ acres lying in crofts, $11\frac{1}{4}$ acres " in diversis particulis in campo vocato Burfeld," and the remaining acres in ten other places, several of which are called *campi*.[3] Two instructive charters relative to lands in or near Thanet are recorded in a

[1] The distribution was as follows (Rents. and Survs., Portf. 1/4): —

	Acres	Roods (Virgae)	Day's-works (= $\frac{1}{10}$ Roods)
In Leyfeld...............	1	..	9
" Cuffeld...............	..	..	15
" Trinite................	$1\frac{1}{2}$	..	..
" Wadishawe...........	2	1	5
" Halles.................	..	1	8
" Sporadis..............	3	..	..
" Perfeld................	$\frac{1}{2}$	..	3
" Ryfeld................	1	..	.. in 3 parcels
" Barnefeld.............	..	3	..
" Newlond..............	..	1	5
" le Skeme..............	..	1	8
" Padpole...............	1	1	.. in 2 parcels
" Mershefeld............	..	..	$8\frac{1}{2}$
" Clerkyncroft..........	1	..	$8\frac{1}{2}$
" Fedelers...............	..	3	$5\frac{1}{2}$
" Greydon..............	..	..	$6\frac{1}{2}$
" Skwalmeston..........	$2\frac{1}{2}$	1	$6\frac{1}{2}$ in 3 parcels
Iuxta Waterlokestret......	..	3	

[2] In the first instance there are five acres, of which
   " una pars iacet in campo vocato Chatefeld in uno loco
   et alia pars in eodem campo in alio loco
   et tertia pars iacet in campo qui vocatur hethefeld."
The second grant relates to
   " una dimidia acra terre in campo qui vocatur Estclune ... ate myddelheg
   cum tota terra quam habet in eodem campo Ate gore " (Cott. MS., Otho B
   XIV, ff. 79, 81).

[3] MSS. of the Dean and Chapter of Christchurch, Canterbury, Lib. B, f. 18.

fourteenth-century cartulary of St. Augustine's abbey. In the parish of St. Peter's 19½ acres lay in 9 parcels, each located between the lands of persons other than the owner (e. g., "due acre iacent inter terram . . . et terram . . ."). At Chislett 4 acres in 3 parcels were similarly bounded, and other 4 in 4 parcels were assigned to different fields.[1] Finally, an extended terrier of lands just outside of Canterbury, inserted in a fourteenth-century hand in another register of St. Augustine's, describes sixteen furlongs which constituted the "Tenura de Estber' infra libertatem."[2] In each furlong are parcels, usually of ½ acre to 4 acres, held by various tenants of the archbishop of Canterbury, the prior of Dover, and the abbot of St. Augustine's. There can be little doubt here about the existence of intermixed parcels, or, considering the grouping by furlongs, about the existence of open fields.

This fourteenth-century evidence finds its prototype in thirteenth-century feet of fines. One of these, dated 21 Henry III, so describes four acres and four roods in Iwade, near the mouth of the Medway, as to give the impression that they were

[1] Cott. MS., Claud. D X, ff. 104*b*, 134*b*, 162*b*. The last enumeration is as follows: —

"Due acre et una perticata iacent in campo qui vocatur Herste
et tres perticate iacent in campo qui dicitur Meredale
et tres perticate iacent in loco qui dicitur Calespotle
et una perticata iacet in campo de teghe."

[2] Cott. MS., Faust. A I, ff. 101*b*–106. The following is the allotment of the acres of the first six furlongs (A. A. = abbot of St. Augustine's, P. D. = prior of Dover, A. C. = archbishop of Canterbury): —

	In forlango besouth ðeswey	In forlando bynorthe þesweye	In foriando qui dicitur Sharpnesse	In forlando qui dicitur Uxeastecrouche	In forlando bynorthe Lytildene Drove	In Clyfforlang
Thomas Rotyng . . .	4 A.A.	. . . .	. . . .	. . . .	. . . .	. . . .
John de Gustone . .	3	1½ A.A.	2¾ P.D.	1½ A.A.	½+1+1 P.D.	1 P.D., 3 A.A.
	. . . .	2½ A.A.	3 A.A.	. . . .	. . . .	¾+¾ A.C.
Richard Polderne . .	4 P.D.	. . . .	. . . .	4 past. A.C.	6 past., 1¾ A.A.	. . . .
	8 past. P.D.	1 P.D.	2½ past. P.D.	1 P.D.	1½ P.D.	2 P.D.
John de Ber	2 P.D.	2 P.D.	2½ P.D.	1 A.C.	3¾ A.A.	2+1 P.D.
	. . . .	1½ A.C.	3½ P.D.	1+1 P.D.	3 past. P.D.	. . . .
Quikeman de Ber . .	1½ A.C.	. . . .	. . . .	2 past. A.C.	. . . .	¾ A.C.
Simon Danyel	1½ A.C.	. . . .	1 A.C.	½+3 P.D.	¾ P.D.	1½ P.D.
	. . . .	. . . .	. . . .	. . . .	¾ A.C.	¾ A.C.
Stephen Swanton . .	. . . .	. . . .	. . . .	¾+½ P.D.	¾ P.D.	. . . .
	. . . .	. . . .	. . . .	½ A.A.	. . . .	2 P.D.

scattered parcels.¹ More instructive, however, is a fine of 20 Henry III from Barfreston, in eastern Kent, by which one-half of a carucate, excepting such land as was held in dower by two women, was transferred.² Several characteristics of this fine are

[1] " . . unam acram et dimidiam in Sweynesam iuxta terram predicti Rogeri
unam acram in Longeham iuxta terram Ricardi de Cheteneye
unam acram in Clakslond iuxta terram predicti Wyberti
dimidiam acram inter domum dicti Rogeri et domum predicti Roberti
tres perticatas inter terram predicti Wyberti et Rogeri (Ped. Fin., case 96, no. 335).

[2] Transfer of a messuage and a half-carucate, which comprised
" unam acram et tres perticatas terre que iacent sub dicto mesuagio versus austrum
et unam acram et unam perticatam que iacent inter terram ysaac de Sanwyc et terram Johannis Pent
et quatuor acras et unam perticatam terre et dimidiam versus austrum in campo qui vocatur Bestedune
et unam acram et quatuordecim sulcos qui iacent in medio campo qui vocatur Bynorthewde
et quattuor acras terre in medio pasture que vocatur northdune
et unam acram et unam perticatam et quinque pedes terre in medio pasture que iacet versus Borialem partem de Haggedale
et quinque perticatas et sex sulcos terre in campo sub Haggedale versus austrum
et duas acras et unam perticatam et septem pedes terre in pastura versus occidentem
et quatuor acras et quinque sulcos terre versus austrum in campo qui vocatur Bisuthewode
et tres acras et unam perticatam terre versus austrum in pastura de Litledune
et tres acras et unam perticatam et dimidiam et duos sulcos terre in campo qui vocatur Bromueld quarum una medietas iacet sub pastura de Litledun et altera medietas in medio campo de Bromfelde
et unam acram terre que iacet in medio pasture sub bosco de Berefrestone versus occidentem
et tres acras et unam perticatam et dimidiam et quattuor sulcos terre versus aquilonem in campo qui vocatur Langetighe
et unam acram et unam perticatam et quattuor sulcos terre in campo qui iacet versus partem orientalem de Langetigh
et unam acram et dimidiam perticatam terre versus austrum in campo qui vocatur Steuerlonde
et unam acram et sex sulcos terre versus austrum in Suthfeld
et duas perticatas et dimidiam et quattuor sulcos terre versus orientem de via que vocatur Drove
et unam acram et tres perticatas et tres sulcos terre in medio campo qui iacet iuxta curiam versus austrum

noteworthy. The parcels were not large, ranging from one to four acres and averaging about two acres. Their areas are accurately estimated even to " sulci " (furrows). They lay for the most part " in campo," " in medio campi," " in medio culture " (once), " in medio pasture."[1] The sulci occur in connection with campi, an intimation that the latter were arable; and, when in one instance it was necessary to estimate a small piece of pasture, this was done in " pedes." The campi can hardly have been conterminous with the parcels, else why " in medio campi " ? To remove doubt, the fine shows that the 3½ acres " in campo qui vocatur Bromueld " were in two parcels, one of them being " in medio campo de Bromfelde." The parcels of arable, then, lay within larger fields, and, although these are not expressly said to have been unenclosed, they probably were so. The terrier establishes the existence in northern Kent, in the thirteenth century, of holdings constituted in part of small non-adjacent parcels of arable.

The descriptive detail of the fine just quoted emphasizes what has already appeared in other terriers and surveys as a Kentish characteristic — the location of the parcels of a holding in a bewildering number of field divisions bearing local names and giv-

> et quattuor acras et unam perticatam terre in medio campi qui vocatur Osmundesteghe
> et unam acram et dimidiam perticatam et quinque sulcos terre versus aquilonem in campo qui vocatur Reteghe
> et unam acram et tres perticatas terre in medio pasture de Potynberegh
> et duas acras terre que iacent versus Potynberegh cum situ unius molendini
> et unam perticatam et dimidiam et tres sulcos terre in medio culture qui vocatur Shortestiche
> et dimidiam acram et six sulcos terre versus occidentem hortfurlong
> et dimidiam acram et decem sulcos terre versus orientem in valle sub Knolle
> et unam perticatam terre in crofta extra portum versus campum qui iacet sub Chimyno qui vocatur Drove
> et tres acras bosci de Berfreston qui iacet versus aquilonem
> et quintam bestiam cum bestiis predictarum [two women holding in dotem] in pastura de forestal ante magnam portam curie de Berefreston " (Ped. Fin., case 96, no. 276).

[1] The parcels " in medio pasture " are puzzling. It may be that old arable campi were at the time used as pastures, or it may be that pastures had been allotted among the tenants and that the latter could utilize their parcels by means

ing little clue to the husbandry employed. Except that the parcels were usually small and lay to some extent intermixed with those of other tenants, Kentish arrangements were in contrast with those of the midlands. There was no grouping of parcels into two or three, four or six, fields, with the total areas approximately equal. Nor can it be readily discovered from the surveys thus far noticed whether the parcels of a holding lay to any extent grouped in one part of the unenclosed arable area or whether they were scattered conscientiously throughout it.

Most valuable for determining this point is a survey of Gillingham, made in 26 Henry VI and preserved at the British Museum in an incomplete nineteenth-century copy.[1] Gillingham, on the lower Medway near Rochester, has today assumed an industrial character and has lost its early fields. The old survey, however, arranges the tenants' holdings in iuga, and for each of these gives boundaries and area. Since this is almost the only survey which describes iuga by bounding them, a transcription of the first thirteen boundaries is pertinent: —

" Iugum Foghell incipit ad communem viam ducentem inter Renham et Gyllyngham versus South
 ad terram de Renham et ad salsum Mariscum de Gyllyngham versus North
 et ad terram domine Alicie Passhele versus West
 et continet illud Iugum xxiii acras.
Iugum Cherlman incipit ad terram heredum Adamari Digges vocatam Wynelyng versus West
 ad Regiam stratam ducentem inter Gyllingham et Renham versus South
 ad mariscum vocatum Thomas Innyng versus North
 et ad communem viam ducentem de Berwescrosse ad Twidelswelle versus East
 et continet illud Iugum xxiiii acras iii rodas iii day.
Iugum Fissher incipit ad campum vocatum Bradefeld versus South

---

of wattles or other temporary enclosures. The phrase "in medio campo" may possibly be a variant of "in medio campi."

[1] Add. MS. 33902.

ad communem stratam vocatam Twidolestrete ducentem
ad Twidoleswell versus West
ad mariscum salsum versus North
et ad metas inter Renham et Gyllingham et terram heredum
Thome Gillingham . . . in Iugo Foghell versus East. . . .
Duo Iuga Coole incipiunt ad Iugum Fisher ex parte boriali de
Bradfeld versus North
ad metas de Renham versus East
ad Regiam Stratam vocatam Twidolestrete versus West
et ad venellam vocatam Cokkeslane . . . versus South. . . .
Fraunceis ferthyng incipit in venella vocata Cokkeslane versus
North
ad croftum Ricardi Mauncer et ad metas de Renham versus East
ad Twydolestreete versus West
et ad Iugum Hood versus South. . . .
Iugum Hood incipit ad Frounceys ferthyng versus North
ad Iugum Edweker versus South
ad metas de Renham versus East
et ad Twidolestrete versus West. . . .
Iugum Edweker incipit ad terram de Renham versus East
ad Twidolstrete versus West
ad iugum Hood versus North
et ad iugum vocatum Raynold versus South. . . .
Iugum Raynold incipit ad metas inter Renham et Gillingham
versus East
ad Twydolestrete versus West
ad iugum Edweker versus North
et ad iugum Gilnoth versus South. . . .
Iugum Gilnoth incipit ad metas de Renham versus East
ad regiam viam ducentem inter Roffam et Cantuariam
versus South
ad Twydolestrete versus West
et ad jugum Raynold versus North. . . .
Iugum Gate incipit ad Twidolestrete versus East
ad communem Stratam ducentem inter Eastcourt et Berwescrosse versus North

ad communem viam ducentem ad ecclesiam de Gillingham
et ad venellam vocatam Snorehelle lane versus South
et ad mesuagium de East court et Scottyssoole versus West.
Iugum Petri incipit ad communem Stratam vocatam Ioocesstrete versus West
ad Twydolestrete versus East
ad communem viam ducentem inter Twydolestrete et
Ioocesstrete versus North
et ad iugum Simstan et ad iugum Alreed versus South. . . .
Iugum Simstan incipit ad Ioocestrete versus West
ad iugum Alreed versus Est
ad iugum Petri versus North
et extendit ad Watescroft in Iugo Alreed versus South. . . .
Iugum Alreed incipit ad iugum Petri versus North
ad Twiddolestrete versus Est
ad Iugum Stimston versus West
ad . . . [blank] . . . versus South. . . ."

The iuga here described were clearly rectangular areas. From "iugum Fissher" to "iugum Gilnoth" they formed a series running from north to south, bounded on the east by the parish of Renham and on the west by Twidolestrete. The next four constituted a similar series lying to the west of Twidolestrete. The iugum here and elsewhere in the survey usually contained about 24 acres, although this number might drop to 5 or rise to 132. Sometimes a double iugum occurred, as in the case of " duo iuga Coole," and the fourth part of a iugum might appear as a " ferthing." As to the tenants, the distribution and areas of the parcels of their holdings relative to the first fourteen iuga may be tabulated as shown on the following page.

Few as are these fourteen iuga in proportion to the entire number, the preceding tabulation shows in a measure their relation to the holdings of the tenants. The lands of any person or estate tended to lie in neighboring iuga, whether in few or in many. The estate of the heirs of John Beausitz appears in each of the above fourteen, continues through the following eight, disappears for a long time, but reappears toward the end of the survey in some half-dozen field divisions, two of these being iuga.

THE KENTISH SYSTEM

The heirs of Thomas Gillingham fare in much the same way. Richard Bamme, who is not introduced until we reach the tenth iugum, continues to have an interest in upwards of twenty-five iuga and other areas, his total estate being large. John Digon,

Name of Iugum	Foghell	Cherlman	Fissher	Duo Iuga Coole	Frauncis Ferthyng	Hood	Edweker
Area of Iugum[1]	23 0 0	24 3 3	24 2 2	58 3 0	6 0 2	24 0 0	24 0 5
Heredes Thome de Gillyngham	1 0 7 } 2 3 4 } 8 0 6 }	8 3 8[2] 13 0 4	5 2 0 3 0 1	....	....	....	2 2 0
Heredes Johannis Beausitz	9 3 3	2 3 1	14 0 1 } 2 0 0 }	48 3 7	5 0 2	22 1 5	19 3 5
John Grenested	....	....	....	0 1 0	....	....	....
Alicia Hunte	....	....	....	1 1 0[2] } 2 0 }	1 0 0	1 2 5	1 3 0
Richard Mauncer	....	....	....	7 3 3[2]	....	....	....
Name of Iugum	Rayold	Gilnoth	Gate	Petri	Simstan	Alreed	Pilgrym
Area of Iugum	27 0 0	26 2 9	24 2 2[3]	24 0 1½[3]	24 0 3½[3]	41 3 3[3]	29 2 3[3]
Heredes Thome de Gillyngham	....	16 0 0	1 0 9½	3 0 0 } 6 0 0 }	6 0 5	2 0 0	13 1 2[2]
Heredes Johannis Beausitz	21 0 7	10 2 9	{ 1 2 7½ 3 4[2] 2 14 11 3 3 }	13 1½ 2 6 1 3 6	1 0 1 1 1 0	2 2 6 } 6 1 8 }	3 2 1
John Grenested	....	....	....	....	3 0 0 5 3 0	....	
Alicia Hunte	....	....	....	....	2 0 0 0 7[2]	....	
Richard Mauncer	2 3 3	....	....	....	3 0 4 2 4 3 0 0	....	
Domina Alicia Passhele	3 0 0	....	....	2 3 8 } 1 0 6 }	....	....	
John Lacy	....	....	1 0 2½	....	2 0	....	....
Richard Bamme	....	....	5 0 2½	....	1 0 4 1 1 0 2 0 ½	1 2 0	0 6[4] } 1 0 6 }
Johanna Jooce	....	....	1 3 6	1 2 1 2 3 8½	1 3 3 1 2 4	3 3	....
Heredes Johannis Bowes	....	....	1 8[2]	0 7	1 0 0 1 5	2 3 1 0 0	....
John Colman, Sen.	....	....	....	2 2 1½[2]	1 0 0	2 0 1 3 0	
John York	....	....	....	3 ½ 6	3 4	2 2 0 0 9 1 0 5 1 0 0	....
Nupton	....	....	....	1 0 0	....	....	6 3 3 } 1 0 6 1 0 8 }
Heredes Johannis Naste	....	....	....	....	1 0	....	
W. Mille	....	....	....	1 0	....	....	
Thomas Langle	....	....	....	....	1 3 3	....	
John Digon	....	....	....	....	1 1 2	3 8	
Heredes Ademari Diggs	....	....	....	....	....	....	2 1 0
John Harvey	....	....	....	....	....	....	1 0[2]
William Zoolay	....	....	....	....	....	....	0 4[2]
John Ram	....	....	....	....	....	....	1 0[2]
Heredes Johannis Coleman	....	....	....	....	1 0 0	2 0	....

[1] All areas are in acres, roods, and day's works, a day's work being equivalent to one-fourth of a rood.

[2] Includes a messuage.

[3] The areas assigned to the last five iuga differ slightly from the sums of the areas of the parcels in each.   [4] A garden.

at first inconspicuous, comes to have in three neighboring iuga 24 acres in 24 parcels and a croft of 5 acres. Most of the numerous tenants, however, were like John Colman or John York: they had small parcels in two or three or four neighboring iuga, but none elsewhere. It is just possible, of course, that such parcels were contiguous and formed a compact holding artificially divided among arbitrarily drawn iuga. This, however, is unlikely, and in any case the larger holdings have to be thought of as composed of parcels to some extent non-contiguous. The survey thus establishes the fact, hitherto obscure, that the parcels of a Kentish holding were not scattered throughout the expanse of the village arable, as under the midland system, but were to some extent segregated in one locality. They did not, however, entirely cohere. The field system of Gillingham may, then, be described as one of non-contiguous, yet of not widely scattered, parcels.

Over it all rested the network of the iuga, for the rectangular appearance of which this survey is our best source. Whether blocks of such a shape, regularly disposed and of uniform size, served or ever had served an agricultural end, is not explained in this abbreviated document. Other Kentish surveys, however, amplify our knowledge, and an extract from one of the best of them is printed in Appendix V. The description of units, tenants, parcels, and rents shows the same completeness which must have characterized the Gillingham original. Beginning and end are wanting, but the hand is of the early fifteenth century. The townships referred to are Newchurch, Bilsington, and Romney Marsh. Iuga are not mentioned, the units here being " dolae " and " tenementa." The accounts of three dolae (" dola Godewini," " dola Storni," " dola de Kyngessnothe ") and of three half-dolae (" dimidia dola Mawgeri," " alia dimidia dola Mawgeri," " dimidia dola de Westbrege ") have been transcribed.

Several characteristics already discerned at Gillingham reappear. The dolae are described as abutting upon or lying on both sides of certain highways, a circumstance which implies that they were compact areas; indeed, a statement is sometimes added to the effect that their acres lay " coniunctim." Their size was uniform, varying only between 40 and 46 acres. Several tenants shared in each, except in " dola de Kyngessnothe," the whole of which was held by the heirs of Jacob de Kyngessnothe. A tenant sometimes had parcels in successive dolae or half-dolae. Adam Osbarn had five parcels in one dola, four in another; the heirs of Richard Pundherst had three acres in one, three and one-half in another, and nine in a third — four parcels in all. So far as the incomplete survey permits us to judge, the parcels of a tenant did not lie in many widely separated dolae, but in a few adjacent ones.

Additional information may be got from the extracts printed in Appendix V. The parcels within each dola are named and their areas given. Names usually differed, except that a few parcels are said to have lain "in Holland"; the areas, too, of the parcels were not such as to suggest open-field strips. Both circumstances point to the absence of unenclosed arable, an inference which is the more probable since the region in question is in or near Romney Marsh. The system of iuga or dolae was therefore consistent with one of plats, whether arable, marsh, or pasture, and hence with one of enclosures. Yet if the parcels were plats or enclosures, they were none the less, in the Newchurch survey, often non-adjacent. So, indeed, the Newington plan, which has already been reproduced, shows the plats there to have been. Conditions of the late sixteenth century thus find a parallel some two hundred years earlier.

Other peculiarities of the Kentish system we can best discover by noting in connection with the Newchurch extracts the implications of another survey, perhaps the most satisfactory that we have.[1] It was made in the early fifteenth century for Thomas Ludlow, abbot of Battle, and refers to the large manor of Wye, situated in the center of the county. Except for its omission of the boundaries and locations of the iuga, it is superior to the Gillingham transcript, and it is more nearly complete and more complex than the Newchurch record.

The description, as usual, proceeds by iuga. At Wye these units varied considerably in area, comprising from 37 to 187

[1] Exch. Aug. Of., M. B. 56, ff. 108–188.

acres, the fluctuations being due to the inclusion of greater or smaller quantities of pasture or down-land. The average size of a iugum was from 60 to 70 acres, in contrast with the 24-acre iugum of Gillingham and the 40-acre dola of Newchurch. A description of two of the smaller iuga will be of assistance in drawing general conclusions: —

"Dimidium Iugum de foghelchilde

Heredes Johanis Dod tenent in campo vocato Wolvenfeld ii acras i rodam dimidiam

Agnes Broman tenet in eodem i acram iii rodas et in Strongelonde i acram dimidiam

Thomas Elyndenne tenet de iure uxoris sue in Wolvenfelde ii acras dimidiam

Heredes Johanis Selke de Broke tenent in Strongelonde iii acras iii rodas. In crofto vocato Jannescroft de Wy i acram terre

Heredes Stephani Tur tenent in Strongelond unam acram dimidiam. In Doucerede iii acras et in longecrofte i acram dimidiam

Stephanus Dod tenet in Wolvenfeld dimidiam acram dimidiam rodam

Heredes Johanis Selke iunioris tenent in Strongelond ii acras i rodam. In longorcrofte i acram et in mesuagio eorum apud Silkenstrete in capite orientali dicti mesuagii iii rodas

Johanes Petycourt tenet apud Gerardesteghile i acram terre

Andrus Martyn tenet in mesuagio suo et in Strongelond i acram iii rodas

Heredes Simonis German tanner tenent in Clerkescroft i acram iii rodas

Thomas Alayn tenet in Selkenbroke iiii acras dimidiam

Thomas Kempe tenet in Melcompesmede iii rodas prati que fuerunt Agnetis Danyel

Summa xxxiii acre i roda dimidia

[Rent and services one-half as great as for the following iugum]."

" Iugum de Clyt et Forwerde
> Hamo German tenet in East walewaye ii acras que fuerunt Thome Scot
>> Item i acram ibidem que fuit Johanis Laghame
>> Item in eodem iii acras que fuerunt Thome de Chitynden de Willielmo Barrok
>> Item in eodem i acram dimidiam que fuerunt Simonis de Tongge
>> Item in eodem iii rodas terre que fuerunt predicti Simonis
>> Item in eodem ii acras que fuerunt Johanis Hortone vocati Cukkow
>> Item in eodem i acram terre . . . que fuit Johanis Westbeche
>> Item in eodem i acram terre . . . que fuit Thome German bocher
>> Item in eodem v acras i rodam terre iacentes in longitudine ad predictam acram terre predicti Johanis Westbeche
>> Item in eodem i acram dimidiam que fuerunt Stephani Tur
>> Summa xix acre
>
> Heredes Willielmi Mellere tenent in East waleweye sub le lynche i acram dimidiam
>
> Heredes Stephani Tur tenent in Gretefeld i acram dimidiam
>
> Johanes Peticourt tenet in Eastbrettegh ii acras. Et in foldenge i acram i rodam
>
> Heredes Roberti Man tenent in Eastbrettegh iiii acras
>
> Thomas Baldewyne tenet in Westgretefeld ii acras dimidiam
>
> Heredes Johanis Selke tenent in Eastbretteghe i acram dimidiam. In Gretefelde ii acras. In Westgrettfelde dimidiam acram terre
>
> Gilbertus Baldewyne tenet in Eastbrettegh i acram dimidiam terre
>
> Summa xxxvii acre i roda

Unde de redditu per annum cum viii d. ob. pro tenemento Gilberti Forderede ix s. vi d. [at six terms] . . . ad Nativitatem domini i gallum ii gallinas Et [ad] Pascham xx ova. Et tenentes predicti debent pro predicto iugo omnes consuetudines et servicia sicut Johannes de Garde debet de proprio iugo suo de Cokelescoumbe predicto." [1]

The concluding phrases, such as are confessedly omitted in the Gillingham copy but occur in connection with each dola at Newchurch, disclose the financial aspect of the iugum. Vary as they might in area, the Wye iuga were alike in the obligations which rested upon them. Rents of assize and services were the same for all. The former were set at 8 $s.$ $9\frac{1}{2}$ $d.$ the iugum, with one-half as much for the half-iugum and one-fourth as much for the virgata. The services were not burdensome, comprising little more than the ploughing, sowing, mowing, and reaping of two or three acres yearly by each iugum.[2] At Newchurch the value of the rent of assize and of the services due from each dola was eighteen shillings. These heavier obligations imposed upon units which were usually smaller than those at Wye may have been due to a better quality of soil. At any rate, it can no longer be doubted that at the beginning of the fifteenth century the iuga and dolae were primarily financial, not agricultural, units. Whereas several tenants shared in each of them and few tenants were limited to any one of them, they did have financial unity and stability. Their midland correspondents were not the furlongs, as the boundaries at Gillingham might suggest, but rather the virgates or yard-lands upon which, as units, rents and services were always imposed in the midlands. Differ as they might in the distribution of their constituent parcels, Kentish iuga and midland virgates were alike to the rent-collector.

[1] Exch. Aug. Of., M. B. 56, ff. 116, 114b.
[2] " Et debet arare ad frumentum i acram et dimidiam acram terre domini cum facto proprio, petere semen ad granariam in manerio domini de Wy, seminare et herciare predictam acram et dimidiam Et debet arare, seminare semine domini ut supra, et herciare i acram et dimidiam acram terre domini ad sementem ordei et petere semen ut supra. Et debet falcare, spargere, vertere, cumulare, cariare in manerium domini et ad tassum furcare unam acram prati de prato domini. Et debet metere, ligare, et coppare in autumpno unam acram et dimidiam acram de

Unfortunately, as was noticed above, the iuga of Wye are not bounded in the survey. To discover whether they were compact areas, as were those at Gillingham and Newchurch, we have to attend to the field-names used, and even then we can draw conclusions only when the names are not too diverse. In the " iugum de Clyt et Forwerde " more than one-half of the parcels and of the area lay in East Waleweye; the remainder was pretty well accounted for by East Brettegh, Gretefeld, and West Gretefeld. Although no statement records that these four areas were contiguous, it is probable that they formed a block not unlike a iugum at Gillingham. Sometimes the place-names within a iugum at Wye were numerous, certain of them referring to parcels or crofts which fell to individual tenants.[1] But this circumstance need not conflict with the conception of the iugum as a compact area; it merely implies that such an area was divided into many parts.

As at Gillingham and Newchurch, the tenants of each iugum at Wye were likely to be several in number; the first iugum only was in the hands of a single proprietor. Furthermore, as in the other surveys, a tenant usually had parcels in two or three consecutive iuga; but in its greater concentration of the parcels of a holding Wye resembled Newchurch rather than Gillingham. Typical was the holding of John Baldewyne, whose parcels were situated in three iuga as follows: —

In the half-iugum Mastall,
   in Weyberghe 2 acres, 1½ roods
   in Tommestowne 3 roods, 8 day's-works
   in le Berghe 2 acres, 1½ roods;
In the one and one-half iuga Chilcheborne,
   in Chilchebournesfelde ½ acre, ½ day's-work
   in piriteghe 2½ acres
   in Watertoune et in parvo gardino 1½ rood;

frumento domini. Et debit xvii averagia et tertiam partem unius averagii per annum. Et debet inde de relevio cum acciderit xl d. Et debet sectam per annum " (ibid., f. 109b).

[1] Ten place-names, as we have seen, were connected with the half-iugum of Foghelchilde. Four tenants shared in Wolvenfeld and four in Strangelonde, but the eight remaining areas, four of which were crofts, fell to individual tenants.

In the half-iugum Ammyng [specification is only by quality of land],

olim Thome Chiterenden, terra optimi precii, 1 acre, 3 roods; terra medii precii, 1 rood

que fuerunt Simmonis de Tonge, terra optimi precii, 3½ acres; terra medii precii, 1 acre; pastura, 25⅝ acres

quondam Gilberti Mogge, terra optimi precii, 3 roods; terra medii precii, ½ acre; pastura, 1½ acres

que fuerunt Stephani Mogge, terra optimi precii, ½ acre; terra medii precii, ½ acre; pastura, 2⅝ acres

que fuerunt heredum Johanis Mogge et Hamonis Mogge, terra optimi precii, 1½ acres; terra medii precii, 1 acre; pastura, 4⅝ acres.

The "former tenants" here mentioned should be compared with those who appear in the first part of the description of "iugum de Clyt et Forwerde."[1] One-half of this iugum had come into the hands of Hamo German, his ten parcels having formerly been held by nine tenants, only one of whom bore the name of German. Since the names of the former tenants are clearly remembered, the accumulation seems to have come about somewhat recently. From phenomena like this we discover that the situation pictured in the survey was not one of long standing, and perceive that a iugum might come to be transformed from an area shared by many tenants into one held by two or three. Enclosure would readily attend upon consolidation, and by the sixteenth or seventeenth century a county largely enclosed would be a natural result.

If tendencies toward consolidation are discernible in the "iugum de Clyt et Forwerde," none the less were there tendencies in the opposite direction. Four of the eight tenants were not individuals but groups, and the group in each case consisted of the heirs of a former tenant. Even if this man's parcel had been not larger than an acre and a half, it passed to his heirs jointly. In the half-iugum Foghelchilde five of the twelve tenants were groups of heirs. Since the situation was not different with the other iuga at Wye, the actual tenants there must have been

[1] Cf. above, p. 289.

far more numerous than are the entries of the survey. At Gillingham and Newchurch, too, the heirs of a defunct tenant frequently appear as his successor. In the half-dola of Westbrege the only tenants were three such groups. The custom of transmitting a holding to all the heirs of a tenant rather than to one of them is of course the distinguishing feature of the Kentish tenure known as gavelkind. This usage was in marked contrast with that of the midlands, where the virgate or fractional virgate of a customary tenant passed intact to a single heir or to a new tenant. The antiquity of this Kentish custom, so unusual and so frequently perceptible in the fifteenth century surveys, now deserves consideration.

In tracing the earlier history of the iugum much depends upon nomenclature. One of our most valuable documents at this point, since it admits of comparison with the fifteenth-century survey, is a Wye rental of 5 Edward II, which records the tenants of all iuga and the rents accruing from them.[1] On examination we discover that the names of the iuga are practically the same as in the survey of 150 years later, and that the surnames of tenants are frequently unchanged. In the half-iugum Coklescumbe the heirs of John Hughelyn are tenants in the earlier rental, Simon Hoghelyn and Johanna Hoghelyn in the later survey. While surnames often thus persist in the same iugum, there is also a tendency for them to shift from one iugum to another. In the rental, " Gilbertus Dod et Simon Dod tenent iugum de Eastchilton pro Willelmo de Chiltone "; in the survey, tenants by the name of Dod are found in half-iugum Coklescumbe, in iugum Waleweye, and in half-iugum Foghelchilde.

A peculiar difference between rental and survey lies in the fact that in the former the tenants, whether few or several, hold " for " (*pro*) one or more persons. The two Dods held " for " William de Chiltone. The one-and-one-half iugum Chelcheborne is held " pro " Hugo Mogge and " Walter de Chelcheborne et socii " by Richard de Coumbe, Richard de Broke, Stephen Renaud, the heirs of William de Chilcheborne, Gilbert de Chilcheborne, Stephen Baldewyne, William Mogge, Richard Mogge,

[1] Exch. Aug. Of., M. B. 57, ff. 95–105.

Adam Mogge, and Robert "filius Alain Mogge." This is an elaborate process of sub-letting, both tenants and sub-tenants being many. In the fifteenth-century survey only one group of tenants, presumably the old sub-tenant group, is referred to.

The significant feature of the rental, however, is the similarity between the names of the iuga and the names of the tenants (not sub-tenants). The tenant of the iugum East Chilton is William de Chiltone. Similarly, " Hamo pistor tenet iugum de Wytherstone pro Gilberto de Wytherstone," and " Thomas de Bronesford tenet iugum quod vocatur Aula de Bronesford." The one-and-one-half iugum Cukelescumbe is held by several sub-tenants for a group composed of Walter de Cukelescumbe, the heirs of Hugo de Cukelescumbe, and the heirs of "Radulfus molendinarius et socii." Thus, so far as names are concerned, the tenants of the iuga stand at times in intimate relations with the iuga themselves. With the sub-tenants' names, naturally, such is not the case. In the later Wye survey, too, where we are dealing only with actual holders, who correspond to the former sub-tenants, little similarity between surnames and the names of iuga is to be expected; and, indeed, the matter of nomenclature is there of little importance.

The identity between surnames of immediate tenants and names of iuga in the rental of Edward II has noteworthy implications. It might seem that the tenant at times got his name from the iugum rather than the iugum from the tenant. Hugo de Cukelescumbe, Walter de Chilcheborne, Gilbert de Witherstone, William de Chiltone, are tenants of the iuga whose names they bear. Yet even in these instances it is probable that the personal name was originally derived from some place-name older than that of a iugum at Wye. On the other hand, there can be no doubt that several of the names of the Wye iuga were derived from personal names, and in one instance we can see this happening. Two sub-tenants confer their names upon the iuga which they hold. In the rental Gilbert Dod and Simon Dod held the iugum of East Chilton for William de Chiltone; in the survey this iugum appears as one of the " duo iuga de Doddes que vocantur Gerold et East Chilton." Other iuga in both rental and survey are obviously designated by personal names. Such are "iugum Willelmi de lewte et Stephani de Chiltone," "dimidium iugum Orgari pistoris," "iugum [Stephani] Bard et Neel," "dimidium iugum de Bisshop," "dimidium iugum de Knottes et Someres." In the Newchurch survey the dolae bore the names of Godewin, Mawger, and Storn. Since iuga and dolae therefore sometimes came to bear personal names, we may at once inquire what this implies.

If a person gives his name to a certain area of land or even derives his name from it, that land is presumably his own in some more or less intimate sense. The dola named from Godwin and the iugum named from Stephen Bard must at one time have been in their occupation or ownership. A clue to the interpretation of the Wye evidence from this point of view is to be had in a still earlier Wye rental, one from the days of Edward I.[1] Although it is merely an enumeration of payments and of the persons responsible for them, the names of the latter, we discover, are names later borne by the iuga.[2] Most significant is the frequency with which the persons answerable for rents appear as groups of heirs. Where documents of the fourteenth and fifteenth centuries speak of "iugum de Clyt et Forwerde," the rental states that the "heredes Cliteres" and the "heredes Forwerd" pay $7\frac{1}{2}d$. Where in the later records we have been wont to hear of the half-iugum Foghelchilde, we learn from the early rental that the "heredes Foghel" pay $3\frac{3}{4}d$. In short, the rental transports us to a time when most of the iuga were, to be sure, not in the hands of their eponymous tenants, but in the hands of the heirs of such tenants. It is a period antecedent to the two stages pictured by the later rental and the still later survey. Back of it is yet an earlier period, the existence of which seems guaranteed by the use of the term " heirs "; for, if the heirs of a tenant hold a parcel of land, the tenant in question must once have held it either for himself or as representative of his family group.

[1] S. R. Scargill-Bird, *Custumals of Battle Abbey* (Camden Soc., 1887), pp. 101–136.

[2] In a half-dozen instances, iuga, not persons, are named as responsible for the payments.

Four stages in the history of the iuga at Wye thus emerge. At a date still undetermined each iugum or half-iugum was attributable to a single tenant, who either gave it his name or, if it already bore a topographical designation, possibly took his name from it. By the end of the thirteenth century it had passed to his heirs, who held it as a group of co-tenants. In the first half of the next century these heirs were distinguished from another group of tenants who held for (*pro*) them, but whose members had to only a slight extent the same surnames as the group of heirs. By the fifteenth century one group of tenants alone remained, and their names have scarcely any connection with the names once traditional in the iugum. To how small a degree the interests of these last tenants were bound up with the iugum is shown by their acquisition of parcels in other iuga as well. The history of the iugum was therefore one of continuous subdivision and reapportionment, largely due to the practice of transmitting landed property to groups of heirs, who in turn at times sublet it.

The effect of such a history upon the appearance of the iugum can be conjectured. Division among co-heirs probably involved giving to each his share of the several qualities of land within the iugum. In the fifteenth century the half-iugum Ammyng gave no names to its parcels, but grouped them as " terra optimii precii, terra medii precii, et pastura." [1] Since allotments of different quality must frequently have been non-contiguous, the tenants of a subdivided iugum would find their holdings consisting of scattered parcels. But neither in this condition nor as a compact block before subdivision can the iugum have been fitted into the framework of the midland system. Had the arable of the township been divided into two or three large fields, the iugum as a compact area would in a particular year have been either entirely fallow or entirely sown. That fifteenth- or even thirteenth-century Kentish holdings consisted of scattered parcels does not, therefore, imply midland husbandry. One must remember, too, that the parcels of the iugum might be meadow or pasture as well as arable, they might be open or enclosed. Only

[1] Cf. p. 292, above.

one approach to the aspect of the midland system do Kentish postulates allow. If in the case of the arable co-heirs or co-tenants at times devised some system of coöperative ploughing, there may have arisen within a iugum something resembling a midland furlong. But such a furlong did not combine with other similar ones to form two or three large fields.

With the key to Kentish field arrangements above given, the interpretation of early charters becomes simple. The scattering of parcels is explicable; it was, indeed, normal. The multiplicity of names arises from a reference not only to the field divisions of one iugum, but in all probability to those of two or three iuga. The varying areas of the parcels are appropriate to Kent as they would not be to the midlands, and the small size of most of them was a natural outcome of more or less frequent subdivision. Parcels might or might not assume the appearance of arable strips, according to the tenants' attitude toward coöperative ploughing. Apparently they did practice this on the down-lands of the southeast. In general, then, a fourteenth- or fifteenth-century map would show the parcels of a holding as a network of non-contiguous plats or strips often considerably segregated in one part of the township's area. In its primary methods and results the Kentish system was not unlike the Scottish or the Irish; transmission to co-heirs or co-tenants wrought similar effects in each case. The difference lay in the original units. In the Celtic countries it was the entire township which was first subjected to subdivision; in Kent, it was the smaller iugum or dola.

No conjecture has yet been hazarded as to when the iugum or dola was in the hands of the tenant with whose name it came to be connected. Since iuga are Domesday units, they must have antedated the Conquest. Yet most of the names which they bear are later. At Gillingham the personal names Fissher, Hood, Pilgrym, have no flavor of antiquity. At Wye, in the earliest rental, many of the list seem to be from Norman England. Such are Gilbertus de Wythereston, Willielmus de Pirye, Roger et Juliana de Rengesdon, Radulphus molendinarius, Richardus Besant, and many others. A few names, however, suggest

Saxon or Danish connections, and these are significant. There was a "iugum Wlstani" and a "iugum Orgari"; at Newchurch the dolae bore the names of Godewin, Mawger, and Storn. Apparently iuga and dolae had once been in the hands of Saxons or Danes but had been largely appropriated by victorious Normans, who thereupon affixed to them Norman names and transmitted them by tenure of gavelkind. What we should like to discover is whether, before the Conquest, the units had been thus transmitted and whether therefore there had been several tenants of a iugum or dola. To judge from nomenclature alone, one might assume that there had not, that Wlstan, Orgar, Storn each held an undivided tenement. But we know too little about the transmission of socage holdings in pre-Norman days to maintain that one son inherited to the exclusion of his brothers.[1] The individual Saxon name attached to each iugum or dola may have been that of the head of a family group and the group may have held the tenement collectively without formally dividing it — a parcel of folk-land. We must thus be content with consigning the Kentish iugum, as a compact rectangular area of from 25 to 60 acres, into the hands of a single Saxon or of a Saxon family, by whom it was cultivated without thought of any two- or three-field system.

In order that the system may the better be traced outside the borders of the county, a word should be added regarding other Kentish units of land-holding. In the first place, the iugum had its subdivisions, the fourth part having already made its appearance at Gillingham under the name of "ferthing." At Wye the same fraction was called a virgate, and a " virgatam Throfte, vocatam Throfteyerde, continentem L acras" was bounded as one block. Farther on we learn that it paid only one-fourth of the usual relief, " sicut de quarta parte unius iugi."[2] At Sceldwike there is mention of a " dimidia virgata terre " from which $5\frac{3}{4}$ acres are granted away, and at Selling there was an agreement " de tribus iugis terre et una virgata."[3] At Eltham

[1] Cf. below, p. 304, n. 1.
[2] Exch. Aug. Of., M. B. 56, f. 136.
[3] *Archaeologia Cantiana*, i. 262 (Ped. Fin., 10 Rich. I), iv. 308 (8 John).

virgates entirely superseded iuga as rent-paying units, the villein tenements amounting to 28¾ virgates of 7½ acres each.[1]

This connotation of the term *virgata*, a word often found in descriptions of Kentish land, is, however, unusual. Elsewhere in Kentish documents virgata meant a rood, or the fourth part of an acre. Three virgatae, for instance, equalled " una dimidia acra . . . et una virgata,"[2] and seven virgatae equalled " una acra et tres virgatae."[3] Such is nearly always the significance of the term in charters and surveys.[4]

In some places the larger " sulung " persisted as the rent-paying unit, without any reference to iuga. So late as 30 Henry VI the arrangement in Thanet was that 50 tenants held Mergate " swyllung " of 210 acres, while 13, 10, 20, and 15 tenants held respectively the other sulungs of Savlyng, Westgate, Syankesdon, and Hertesdowne.[5] All sulungs contained 210 acres except the last, which comprised only 146. At Estrey details are given about the four " sullinga " and the services due from each in a roll written in a hand of the thirteenth century and said to be " de novo compositus sed ab antiquo rotulo abstractus."[6] Of these sulungs, each containing about 200 acres, the first is described as follows: " In sullunga de Ruberghe sunt ccv acre unde Willielmus de Ruberghe tenet xlv acras et Willielmus Iuvenis et socii sui lx acras et Ricardus cyece et socii sui lx acras et Stephanus filius Normanni et socii sui xl acras." From this it seems likely that the sulung consisted of contiguous blocks of land, and it is clear that one man was often put at the head of his " socii " as responsible for some 60 acres. Other Christchurch manors at the end of the thirteenth century were assessed by sulungs. In the manor of Ickham we find four of them.[7] Again, "apud Monkton sunt xvii swolung de Gavilykind," each

[1] *Archaeologia Cantiana*, iv. 311 (Inq. p. Mort., 47 Hen. III).
[2] Rawl. MS., B 335, f. 54, 9 Edw. II.
[3] Cott. MS., Claud. D X, f. 123.
[4] E. g., Ped. Fin., case 95, no. 133, 11 Hen. III: " De dimidia acra terre et una virgata et dimidia."
[5] MSS. of the Dean and Chapter of Christchurch, Canterbury, Reg. St. Augustine's E xix. f. 182*b*.
[6] Ibid., Roll E, 184*a*.
[7] Ibid., Lib. J, f. 10*b*.

rendering yearly 20 s. " de mala " and 1½ d. the acre " de gablo." [1] At Adisham there was a tendency to use the term generally — to speak, for example, of 36 acres " de Swylenglonde " in Pedding (a field name) held by six tenants.[2] Certain manors of St. Augustine's were in the thirteenth century divided into sulungs. At Chislet, "Hec sunt consuetudines . . . scil. de quinque sulinges et dimidia"; and a Littlebourn rental begins, " Apud Sircham habetur dimidia sullung de c acris et debet de qualibet acra 1 d. de gablo." [3]

Alongside these definite units, the sulung, the dola, the iugum, and the quarter-iugum called virgata or ferthing, there is one less definite, the " tenementum." It appears in the Wye survey. Following the list of iuga are many tenementa, each containing from 60 to 70 acres, each paying a considerable rent (e. g., 15 s. 9 d. from 70¾ acres) and doing ploughing, reaping, and mowing services much as did the iuga. Iuga and virgatae sometimes appear among them, and like the iuga they comprised parcels held by different men. It seems natural to infer that this second part of the rental describes lands improved or assessed more recently than were the old iuga, and at a time when there was a tendency to abandon the ancient term for a new one. This conjecture is strengthened by a fourteenth-century custumal of Eastry,[4] which is detailed enough in its field names to admit of comparison with the earlier one of the thirteenth century. [5] In the later roll the term sulung, the basis of assessment in the first, does not appear, and the same lands are grouped under new and smaller units called tenementa, for each of which there are many tenants. In Eastry the sulung seems to have broken directly into tenementa without reference to iuga; in Wye the tenements took their places beside the iuga.

One other unit, more distinctively Kentish, should be noticed. This is the "day's work," often abbreviated to "day" or "dai." In the survey of Wye nearly all parcels are given in acres, roods,

[1] MSS. of the Dean and Chapter of Christchurch, Canterbury, Reg. St. Augustine's E xix. f. 1b.
[2] Ibid., f. 21b.   [3] Cott. MS., Faust. A I, ff. 56, 120.
[4] MSS. of the Dean and Chapter of Christchurch, Canterbury, Roll E, 188.
[5] Cf. above, p. 299.

and days. The days seldom exceed ten, and a comparison of items leads to the conclusion that ten day's-works constituted one rood. The smallest Kentish unit of superficial land measure was thus not the pole or perch, but the equivalent of four poles.

Of the five units above described three were certainly not widespread in England, at least under these names. The sulung, the iugum, and the day's-work are not often mentioned outside of Kent, and hence may without great inaccuracy be called Kentish. The appearance of any of these names elsewhere will suggest the Kentish system.[1] The terms tenement and virgate are of course common, but the connotation which they had in Kent is, for the virgate at least, distinctive.

In default of accessible documents, the methods of tillage employed by manorial tenants in the open fields of mediaeval Kent are not easy to ascertain. There is, however, no reason to think that the demesne may not at times have lain intermixed with the tenants' land,[2] as it often did in the midlands. If such were the case, the record which we have concerning the tillage of demesne lands may to some extent be representative of methods of tillage in general. In any event, it will disclose the fact that in fourteenth-century Kent certain arable lands were tilled more continuously and with better results than were similar lands in most other parts of England.

Evidence regarding the tillage of the demesne can be drawn, as heretofore, from the extents contained in the inquisitions post mortem, especially in those of the late thirteenth and early fourteenth centuries.[3] These documents make it clear that in the midlands the average annual value of an acre of arable which was left fallow every second or third year was from 4 d. to 6 d., with an occasional drop to 3 d. and a rise to 8 d. In Kent the percentage of the arable left fallow and the annual value of an acre did not at times greatly differ from this. At Hothfield, in 12 Edward III, 80 acres from a total of 200 were untilled, and 6 d. is stated to have been the average annual value of an acre.[4] Elsewhere, while

---

[1] This is less true of day's work (cf. p. 228, n. 2).
[2] Cf. above, p. 275.   [3] Cf. above, p. 46.
[4] C. Inq. p. Mort., Edw. III, F. 56 (1), 17 July, 12 Edw. III, Hothfeld: " Sunt ibidem cc acre terre arabilis que valent per annum c s. praetium acre vi d. De

the area sown was about two-thirds of the total, the part left fallow had a distinct value as pasture, presumably because it was enclosed. At Throwley, for example, the 110 acres that were sown from a total of 160 were each valued at 6 d., but the remainder bore pasturage worth 2 d. the acre.[1] Similarly, at Brabourn the demesne arable when sown was worth 6 d., but when unsown the pasturage of each acre was worth 3 d. from Easter to All Saints.[2]

Most interesting and most significant, however, are the somewhat numerous Kentish manors on which in the middle of the fourteenth century all the acres of the demesne were sown yearly, — "possunt seminari quolibet anno." They were so sown on nearly all the manors of Giles de Badlesmere in 12 Edward III. Under these conditions the value of an acre often became 12 d.[3] Occasionally it did not rise above 6 d. or 8 d., but this was in the down-country of the east.[4] Such annual tillage of the entire demesne, with the resultant high valuation per acre, is a circumstance very unusual for the fourteenth century. It was seldom to be met with outside of Kent, but was there a normal concomitant of the flexible field system which the surveys of the county have shown us. In a region in which it was generally known that much land could be sown yearly, and in which there

quibus seminabantur hoc anno cxx acre ante mortem dicti Egidii et residuum iacet ad warectam."

[1] C. Inq. p. Mort., F. 65 (11), 21 May, 15 Edw. III, Throwley: "Sunt ibidem clx acre terre arabilis que valent per annum quando seminantur iiii li. pretium acre vi d. et quando non seminantur pastura cuiuslibet acre valet ii d. De quibus seminabantur ante mortem predicti Willielmi de semine yemali et quadragesimali cx acre."

[2] Ibid., F. 45 (24), 11 Edw. III, Brabourn: "Sunt ibidem cccxlii acre terre arabilis in dominico que valent . . . quum seminantur pretium acre vi d. Et quum non seminantur pastura cuiuslibet acre valet a festo Pasche usque festum Omnium Sanctorum iii d. Et a predicto festo Omnium Sanctorum usque festum Pasche pastura earundem nihil valet quia nihil vendi potest."

[3] Ibid., F. 56 (1), 3 July, 12 Edw. III, Badlesmere: "Sunt ibidem ccc acre terre arrabilis que valent per annum xv li. pretium acre xii d. et possunt quolibet anno seminari et seminabantur hoc anno ante mortem predicti Egidii." So too, with difference of areas, was it at Chatham, Kingston, Tong, Sibton, Wilderton. At Erith there were 243 acres of "terra arabilis in marisco" worth 3 s. the acre, and 68 other acres of arable worth 20 d. the acre.

[4] E. g., at Chilham 8 d., at Ringwold 6 d., at Whitstable 6 d. (ibid.).

was no system of two or three open fields, agriculture appears to have advanced more rapidly than elsewhere in England.

We are at length in a position to summarize the characteristics of the Kentish field system. In part they are negative. The arable fields of a township were not divided into two or three large areas in each of which all virgate or bovate holders had strips and one of which was usually left fallow. On the contrary, all the improved land of the township was marked off into more or less rectangular areas called iuga, dolae, or tenementa, all serving as units for the assessment of rents and services. If an actual fifteenth-century holding be considered, it appears that the constituent parcels did not lie consolidated within any iugum. Instead, they were likely to be scattered throughout several iuga, but through those which lay mainly in one section of the township. This situation seems not to have been the original one, but to have arisen from the subdivision of a once compact holding among co-heirs or co-lessees. The acquisition by many of these new tenants of parcels in other iuga gave rise to the discreteness which the fifteenth century knew. The parcels at that time were arable, meadow, or pasture; and so far as they were arable and were ploughed by a large coöperative plough they may well have been strips like those of the midlands. On the downs in the southeastern part of the county such have been discerned. The rotation of crops was variable, sometimes resembling that of the midlands, but frequently tending toward an unbroken succession. The absence of a three-course rotation, and especially of a large compact fallow field, made easily possible the reconsolidation of scattered parcels as soon as the tide turned in that direction. It apparently did so turn from the fifteenth century, and hence Kent early became characterized by the consolidation and enclosure of its farms.[1] Toward this enclosure the flexible field system contributed in no negligible degree.

How ancient was the custom of subdividing holdings among heirs is not altogether clear. It was observed at Wye in the time

[1] The persistence of a heavy four-horse plough did not prevent enclosure, since Boys in his report to the Board of Agriculture notes both phenomena (*General View of the Agriculture of the County of Kent*, pp. 21, 41, 44, 70).

of Edward I, when the heirs appear in most cases as descendants of Normans. Yet a few iuga there derived their names from Anglo-Saxons or Danes — an indication that in some sense the iuga were before the Conquest connected with individuals. Whether the Anglo-Saxon tenant held for himself or for his family group must be left undetermined. If the latter relationship was the existent one, the custom of gavelkind is carried back to pre-Conquest days.[1] However this be, the Kentish system, in the subdivision and reconsolidation of its holdings, was not unlike the Celtic. It was in the size and shape of their respective units that the two systems differed. The iugum of the one was rectangular and relatively small, the townland of the other irregular in shape and larger. An explanation of these facts and of the origin of the Kentish system will be hazarded in a concluding summary and synthesis.

[1] Maitland remarks that there is no reason for assigning the body of Kentish custom characterized by tenure of gavelkind to a period earlier than the Conquest. Elsewhere he notes that the Kentish *villani* of Domesday Book seem not particularly distinguished from those of other counties among whom a system of impartible successions may at that time have prevailed. (Pollock and Maitland, *History of English Law*, 2d ed., Cambridge, 1898, i. 187; ii. 272, 263).

CHAPTER VIII

THE EAST ANGLIAN SYSTEM

EIGHTEENTH-CENTURY survivals of open field in East Anglia offer suggestions about the character of the field system that once prevailed there, and this information is considerably amplified by sixteenth-century surveys. With the key thus secured earlier and less detailed data can be interpreted.

Enclosure awards from Norfolk drawn up after 1750 show little surviving open arable field, and those from Suffolk almost none. The plan and appendix prepared by Slater to illustrate parliamentary enclosures in the northern county convey a wrong impression.[1] Relying as he did upon the acts which authorized the awards, he failed to perceive a peculiarity of Norfolk procedure. For it came to be customary in the county, even when there was but little open arable field within a township, to ask for a nominal re-allotment of the entire township in order to obviate any persisting common rights and to establish authoritative titles to ownership. This procedure comes to light through a comparison of certain enclosure awards with nearly contemporary maps and surveys of the same townships made by W. J. Dugmore in 1778.[2] Relative to Weasenham All Saints, Weasenham St. Peter, and Wellingham, the enclosure award of 1809 declares that " all lands and grounds in the said several parishes . . . do contain by measure 4406 acres," and this amount is forthwith allotted. One might conclude that all or much of the area in question was open field, were it not that the earlier Dugmore map reveals at Weasenham only 421 acres of open arable field (in 337 parcels) and at Wellingham only about 50 acres. The Sparham and Bil-

---

[1] *English Peasantry*, pp. 197, 215, 290. He remarks that after 1793 the acts fail to make mention of areas.

[2] The Dugmore maps, which were drawn for Thomas W. Coke, Esq., are among the Holkham MSS., and the awards referred to are either in the same collection or at the Shire Hall in Norwich.

lingford enclosure award of 1809 allots almost all of the 3533 acres which constitute the two townships, but the Dugmore map shows that not more than one-sixth of this area was in open field.[1] Although the entire parish of Longham containing 1286 acres was the subject of the enclosure award of 1814, the map of 1778 shows it already enclosed except for some twenty-one strips of open field which contained less than 50 acres. Finally, the Warham award of 1813, which announces that the entire parish of 2303 acres is to be divided and allotted, should be interpreted in connection with a map of 1712, in which one-half of the parish is seen to be already enclosed.[2] Because of this aspect of parliamentary enclosure in Norfolk, any inference as to the amount of arable open field existent there after 1750, so far as it is based upon parliamentary petitions and acts, is untrustworthy.

Nor, indeed, is it possible to get from the awards themselves very accurate information on this subject. The plans, which alone are useful, are so intent upon the new allotments as only occasionally to indicate by fine lines what the old arrangements were. In a general way, however, the existing enclosures left unchanged in an award can usually be distinguished by their irregular, more or less quadrilateral shape. To judge from them, it appears that at times hardly any open-field strips remained, the award evidently having been made in order to abolish certain rights of common which might still be claimed over enclosed land.[3] The phraseology of other awards makes it difficult, if not impossible, to determine how much of the area actually affected by them was waste and how much was arable.[4] Under these cir-

[1] About 125 acres in Sparham and from 400 to 500 acres in Billingford.
[2] The Warham map of 1712 is at Holkham Hall.
[3] Among such awards are those relating to Great Walsingham, Little Walsingham, and Houghton (1812), Mileham and Beeston (1814), Gresham and Sustead (1828), all at the Shire Hall, Norwich. A similar award relative to Winforton, Herefordshire, has been noted above, p. 140.
[4] The preamble of the Wootton award of 1813 declares that it is concerned with 394 acres of "open and common fields, fens, commons and waste lands." Each allotment, however, refers more accurately to "commons, fens, and waste lands," while the plan shows that the area in question was unimproved land on the outskirts of an enclosed township. The Wymondham award of 1810 notes that the "lands and grounds directed to be divided, allotted, and inclosed contain 2285

cumstances an examination of all the available awards and plans, such as has been made for Oxfordshire and Herefordshire and such as here too would be the only safe basis for a generalization, promises an unsatisfactory return for a great expenditure of labor. If undertaken, it would, we may safely surmise, show few townships with so much as one-third of their improved area still in open field, and in most townships the fraction would be less than one-tenth.[1] William Marshall's description, written in 1787, supports this view. It runs as follows: "Some remnants of common fields still remain; but in general they are not larger than well-sized inclosures. Upon the whole, East Norfolk at large may be said to be a very old-inclosed country. . . . [The] few common fields [left] . . . are in general very small; ten, twenty or thirty acres; cut into patches and shreds of two or three acres, down to half an acre, or, perhaps, a rood each. . . . Towards the north coast some pretty extensive common-fields remain open; and some few in the southern Hundreds." [2]

If it be true that Norfolk arable fields were very largely enclosed without the aid of parliamentary act, the period at which the process took place most rapidly becomes a matter of interest. The subject cannot here be adequately discussed, but the testimony of one or two groups of documents may be noted. Sixteenth- and early seventeenth-century surveys, of which there are several from this county, concur in representing the

acres." A schedule denominates 1934 acres as "Total com. Allots." and 351 acres as "Total Field Allots." Only from the excellent map do we discover that the 1934 acres were the old enclosures, and that the 351 acres were parcels on the outskirts of the township looking very much like waste land. Both these awards are at the Shire Hall, Norwich.

[1] The common field at Fellbrigg before its enclosure in 1771 was, according to the reporter to the Board of Agriculture, unusually extensive. He remarks that the township had remained time out of mind in the following state: 400 acres inclosed, 400 common field, 100 woodland, 400 common heath (Nathaniel Kent, *General View of the Agriculture of the County of Norfolk*, London, 1794, p. 23). Relatively large was the expanse of open arable field at Ormesby, where in 1845 it amounted to 700 acres, in contrast with 1464 acres of old enclosures and 521 acres of roads and commons. At Weasenham and Wellingham about 500 acres were in 1809 unenclosed in townships comprising an area of 4406 acres.

[2] *The Rural Economy of Norfolk, comprising the Management of Landed Estates and the Present Practice of Husbandry in that County* (London, 1787), i. 4, 8.

open fields as still mainly intact.[1] Considerable enclosing therefore took place at some time between 1600 and 1750. How much of it occurred before 1714 may be in part discerned from a summary of the plans of nineteen estates lying in a somewhat larger number of townships in the central and eastern part of the county and belonging to St. Helen's, the Boys', and other hospitals in Norwich.[2]

It will be seen that about eighty per cent of the total area of these estates was enclosed when the plans were made; but whether most closes had arisen through encroachments upon the open fields may be doubted, since in nine estates all or a part of them bordered upon the respective common wastes. In those estates, however, in which enclosures from the waste play a smaller part, the open and enclosed areas nearly balance each other. In the Trowse and Bixley property of 58 acres the area of the scattered enclosed plats was 34 acres, that of the open-field strips 24 acres; in the Buxton estate there were 20 acres enclosed, 18 open; at Shropham, the open-field strips seem to have contained about the same number of acres as the scattered closes, neglecting enclosures from the common; at Snitterton unenclosed strips predominated. In general, then, it is possible that not more than one-half of the open arable fields in the region round Norwich had been enclosed before 1714.

The method followed in bringing about enclosure before this date was the piecemeal one described by Nathaniel Kent at the end of the eighteenth century and still employed at that time.[3]

[1] Certain of these surveys are referred to below (p. 313 *sq.*) in the discussion of field arrangements. Corbett, describing the open fields of six Norfolk villages, infers that in four instances enclosure had affected not more than one-half of the arable and in the two others much less than this ("Elizabethan Village Surveys," p. 87).

[2] Cf. p. 309. These plans, which are among the Norwich records kept in the castle, were made known to me through the kindness of J. S. Tingey, Esq.

[3] "There is still a considerable deal of common field land in Norfolk, though a much less proportion than in many other counties; for notwithstanding common rights for great cattle exist in all of them and even sheep walk privileges in many, yet the natural industry of the people is such, that wherever a person can get four or five acres together, he plants a white thorn hedge round it, and sets an oak at every rod distance, which is consented to by a kind of general courtesy from one neighbor to another" (*General View of the Agriculture of Norfolk*, p. 22). William

THE EAST ANGLIAN SYSTEM 309

Township	Enclosed			Open Field	
	Area in Acres	No. of Parcels	How Situated	Area in Acres	No. of Parcels
Heathell..............	176	22	A compact area near the common.	..	..
Trowse	19	8	Non-contiguous parcels.	8	7
Swanton Morley	73	13	A compact though elongated area.	..	..
Burston..............	181	17	A compact area.	..	..
Forncett	83	14	A compact area near the common.	..	..
Bixley	85	12	A compact area.	..	..
Cringleford	45	7	One group of four closes, another of three.	..	..
Alborow	139	20	An elongated area twice broken by tongues of land.	..	..
Trowse and Bixley......	34	10	Six non-contiguous parcels.	24	20
Hellington	164	24	A compact area adjoining the common.	3	2
Calthrope and Woolverton	94	19	Four non-adjacent parcels.	3	1
Sprowston	13	2	Two non-adjacent closes.	8	4
Barton	97	18	A compact area adjoining the common.	1½	2
Shipham	70	..	Three non-adjacent blocks adjoining the common.	5	3
Snitterton	18	6	Four closes in the common and two home closes.	27	18
Buxton, Haveringham, Straton	20	7	Two detached closes in the field and five home closes.	18	20
Great Melton	71	11	Three non-adjacent closes of 11 acres in the field, the remainder in two large blocks bordering the common.	9	8
Sallows and Wroxham ..	119	..	A compact area bordering the fen and marshes.	26	23
Shropham	186	23	Many non-contiguous parcels, and 89 acres of common.	62	40

Plans of the Norwich hospital estates often show single strips enclosed as long rectangular "pightles." At Shropham there were seventeen such, most of them non-adjacent and with a total area of 38 acres; one containing seven acres is labelled "formerly several pightles." At Great Melton there was in Bow field an enclosed piece of one acre and another strip of one and one-half acres "partly inclosed"; at Shipham two separate acre strips were partly enclosed. In these plans, too, is discernible another characteristic of Norfolk fields which was conducive to piecemeal enclosure. At Shropham a half-acre strip is labelled "a piece in Clark's Close," and near by two other strips together containing one acre are "pieces in another close." In these cases the enclosure of a furlong had preceded consolidation of ownership. It is not, however, by any means certain that such conditions arose only after the sixteenth century; for early charters sometimes refer to non-contiguous strips in the same croft, as, for instance, " tres pecie divise in Lyckemillecroft." [1] Whether the phenomenon be early or late, it undoubtedly contributed to informal enclosure.

If we turn from the enclosure of Norfolk open fields to consider the aspect of such of them as did persist into the early years of the eighteenth century, we find the plans of the estates of the Norwich hospitals still instructive. Although, as we have seen, about one-third of these estates were enclosed and another third had in each case only three or four detached strips of land, the remaining third retained considerable open field. It is the situation of these open-field strips that for the moment is of interest. An estate of 38 acres in Buxton, pictured in the accompanying plan, extended into two adjacent parishes. Near the farmhouse were five closes containing together 12 acres, while at a distance was a detached close of 6 acres. The remaining 20

Marshall notes the inconvenience arising from such procedure. " But another species of intermixture, much more disagreeable to the occupier, is here singularly prevalent. It is very common for an inclosure, lying, perhaps, in the center of an otherwise entire farm, to be cut in two by a slip of glebe or other land lying in it; and still more common for small inclosures to be similarly situated" (*Rural Economy of Norfolk*, i. 8).

[1] P. R. O. Ancient Charter A 3138, *temp*. Edw. I.

acres lay in twenty-one strips, all non-contiguous, except that three abutted upon others. The parcels, however, were not widely scattered. All lay in the same part of the township not far from the farmstead, and some were obviously either in the same furlong or in adjacent furlongs. Another estate of 255

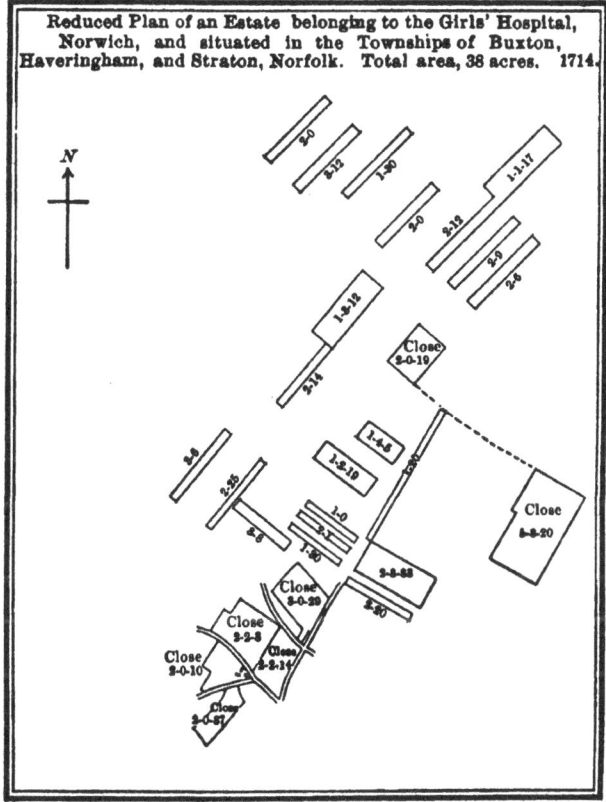

MAP XIII

acres in Shropham lay on the edge of the common, from which 90 acres of " Breck Lands " had been appropriated.[1] Eighteen other enclosures, which together contained about 100 acres, were for the most part detached from one another, and several had evidently been strips of open field. The remaining 65 acres still lay in open field divided into forty-two strips, some of which were in the same furlong (here called " field "). Although the parcels

[1] Cf. plate, p. 312.

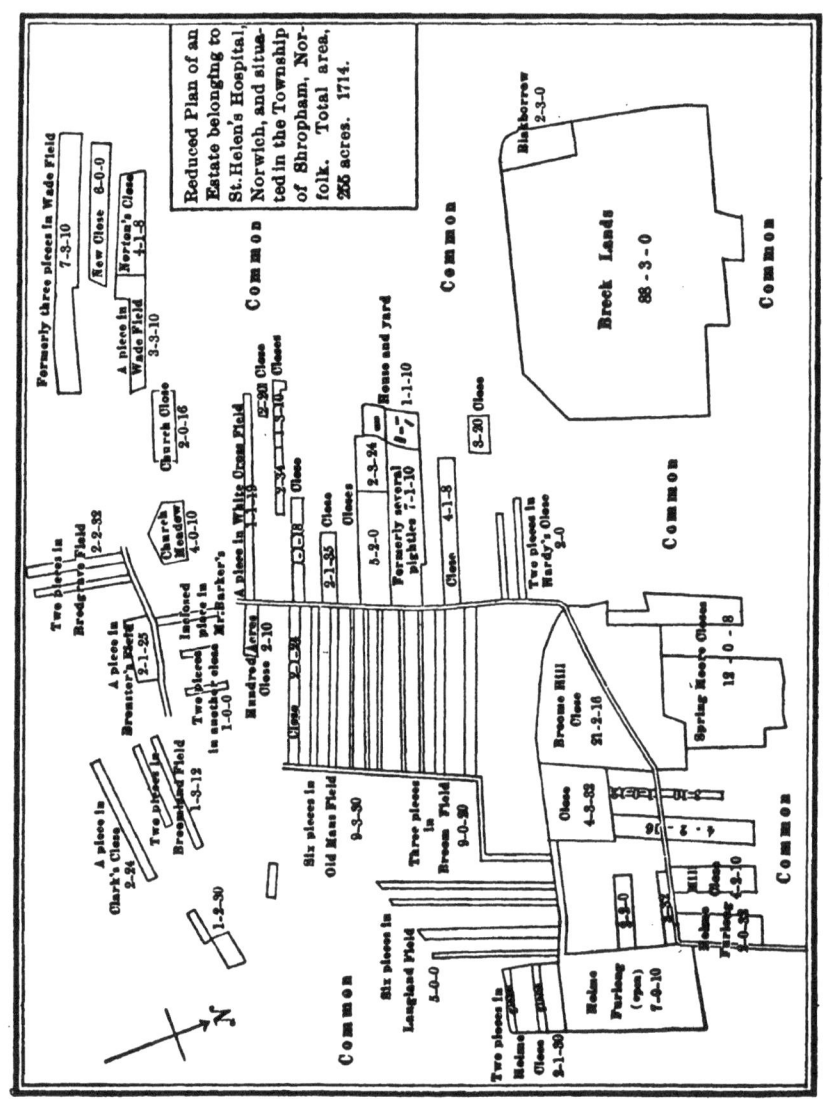

MAP XIV

of this estate, both open and enclosed, were numerous and disparate, they did not lie scattered throughout the township, but were near the farmhouse. As at Buxton, the farm was one of non-contiguous parcels lying in the same section of the village area. The four other hospital estates which retained most open field show similar characteristics.[1]

These field arrangements can in 1714 scarcely have been recent. If they were, they would have to be explained as representing a transitional stage between widely-scattered strips on the one hand and enclosures on the other. Apart from the cumbersomeness of a procedure that would eventually necessitate another exchange, the uniformity of the evidence tells against an original wide dissemination of strips. Although in some cases tentative exchanges may have occurred, such a process can hardly have gone on systematically and to the same degree in all the properties before us. Not one of the six is an estate with parcels scattered throughout the village area. If, on the other hand, it be true that the arrangements of 1714 are an inheritance from a considerably earlier period, we have a contrast to the midland system, the essence of which lay in a wider and more nearly uniform distribution of parcels.

For the sixteenth century, admirable data regarding field arrangements are furnished by surveys and maps such as were made to describe many of the earl of Leicester's estates.[2] In these documents there is usually no subdivision by " fields " in the technical midland sense,[3] but the enumeration proceeds by furlongs, frequently called *stadia* or *quarentinae*. Sometimes, happily, these furlongs are so grouped that we can tell in what part of the village area certain of them lay. They are referred to " precincts," divisions formed usually by the highways that traverse the township.[4] East Carleton and Hethilde were each

[1] These estates were at Snitterton, Great Melton, Trowse and Bixley, Sallows and Wroxham.

[2] These are among the well-arranged records at Holkham Hall, for access to which I am indebted to the Rt. Hon. the Earl of Leicester, K. G.

[3] Corbett, in his study of certain Norfolk surveys (" Elizabethan Village Surveys," p. 70, remarks upon the unimportance of the fields.

[4] Miss Davenport found mention of only precincts in the Forncett records (*Economic Development of a Norfolk Manor*, Cambridge, 1906, p. 1).

thus subdivided into five precincts;[1] Burnham Sutton had three, one being considerably smaller than the others;[2] Weasenham was cut by the Massingham road into a North and a South precinct, the two being approximately equal in extent.[3] When "fields" do occur in the surveys they are as inconsequent as the precincts, being determined by the topography of the parish, the relative position of its highways, or the points of the compass. In the same region there thus arose two, three, four, or five fields. Castle Acre had three, West, Middle, and East, divided by highways and of approximately the same size;[4] not far away, Warham had five unequal fields.[5]

Among such haphazard fields or precincts we should hardly expect to find an equal distribution of the parcels of the various holdings. From the accompanying tabulation of a few of those at Castle Acre, which has most the semblance of a three-field parish, it can be seen how indifferent to the "fields" was the distribution of acres.[6] In the first holding nearly three-fourths of the acres lay in West field, in contrast with one-twelfth of them in Middle field; in the second holding three-fourths, again, lay in West field, with the remainder in Middle field and none whatever in East field. The third holding redresses the balance by assigning to East field nearly 70 per cent of its acres and to West field less than 10 per cent. Still other holdings lay largely in Middle field, like that of Domina Bell, 80 per cent of whose land was there. Such arrangements are, of course, inconsistent with the midland adaptation of fields to a three-course rotation of crops.

At West Lexham the departure from the midland system took another form. As a map of 1575 shows,[7] there was no division

[1] Stowe MS. 870 (field-book of 13 Eliz.).
[2] Rawl. MS., B 390 (field-book of 38 Eliz.).
[3] Holkham Records, field-book of 42 Eliz., and map of the same date (cf. below, p. 327).
[4] Holkham Maps, No. 18. The fields lay to the north of the village, whence two highways extend to the north and northwest. Middle field lay between these highways, the other fields to the west and east respectively.
[5] Holkham Deeds 182, 30 Eliz.
[6] Holkham Deeds 57 (field-book of 25 Eliz.).
[7] It is among the Holkham Records and is sketched in the accompanying cut.

DISTRIBUTION OF THE ACRES OF SELECTED HOLDINGS AT CASTLE ACRE, NORFOLK

Tenants	Messuage	Enclosed				West Field				Middle Field				East Field			
		Demesne		Free	Native	Demesne		Free	Native	Demesne		Free	Native	Demesne		Free	Native
		of the Prior	of the Earl			of the Prior	of the Earl			of the Prior	of the Earl			of the Prior	of the Earl		
Roger Jent	3	...	...	2¼	1½	3	3	28¾	...	4	...	¼	...	10	...	...	...
Hugo Wood	3	14¼	...	¼	...	4½	8½	12½	...	...	...	9½	...	...	...	...	...
John Wingfeild	5	...	1¾	75½	...	...	15½	20	...	...	1¾	95½	...	1	10	263¾	1½
John Gaye	4	...	...	2½	1½	¾	...	38½	1¼	...	...	5	1¼	...	...	...	...
John Mower...............	1[1]	15½	...	...	...	6	8½	...	...	19	...	...	...	2¼	...	...	...
John Warde	1	1¼	...	1½	...	2½	2	...	...	15	...	...	...	1¼	...	5½	...
Margeria Lackford	1	...	...	1½	...	...	...	48¼	...	...	...	5¼	...	...	...	...	...
Domina Bell	2	...	...	41	...	2¼	2¼	17¼	...	...	...	140¾	...	...	...	11¾	...
John Alcock	1	3	...	½	...	¾	4	...	...	½	...	1½	...	1¼	...	...	...
Richard Ellven	1	...	...	...	16¾	...	...	...	...	...	...	...	...	5	...	...	...

[1] And mill.

by fields or precincts, but the two largest holdings lay in strips in the northern two-thirds of the township, while the small holdings were for the most part thrown together in intermixed strips south and southeast of the village. Neither in the township as a whole nor in this southern part of it is there trace of tripartite division.

Of all sixteenth-century Norfolk records those of Weasenham give the most satisfactory idea of the management of the open field in the northwestern part of the county. Particularly useful are (1) a large map of 1600, in two parts, giving the names of many of the open-field furlongs, with an accompanying field-book recording the areas and locations of the constituent parcels of the tenants' holdings; and (2) the note-book of a Weasenham farmer, George Elmdon, describing the sowing of his lands in 1583, 1584, 1588, and 1589. From this group of documents we can discover how the tenants' holdings were distributed among the precincts and how the parcels of Elmdon's holding were grouped for tillage. The map is here outlined, selected holdings from the field-book are summarized, and the information from the note-book is both tabulated and interpreted topographically.

The map is slightly incomplete in that it does not give the whole of one sheep pasture.[1] A later plan shows this pasture or "fold-course," which is in the northwestern part of the township, to have been at least twice as large as the other fold-course, which is represented. Apart from these two fold-courses, some small commons, and a few enclosures, all lands were open arable field. This open field, constituting about two-thirds of the township's area, was cut into two nearly equal parts by the Massingham road, which divided the Northern from the Southern precinct. In the Southern lay the hamlet and parish of Weasenham St. Mary with its church; in the Northern, the hamlet and parish of Weasenham St. Peter, the church here being just south of the road. If we turn to the field-book to discover the relation of the tenants' holdings to the precincts, we find that the larger holdings were nearly always unequally divided between precincts and that the smaller ones frequently lay wholly within one

---

[1] The outline of the map is on p. 322.

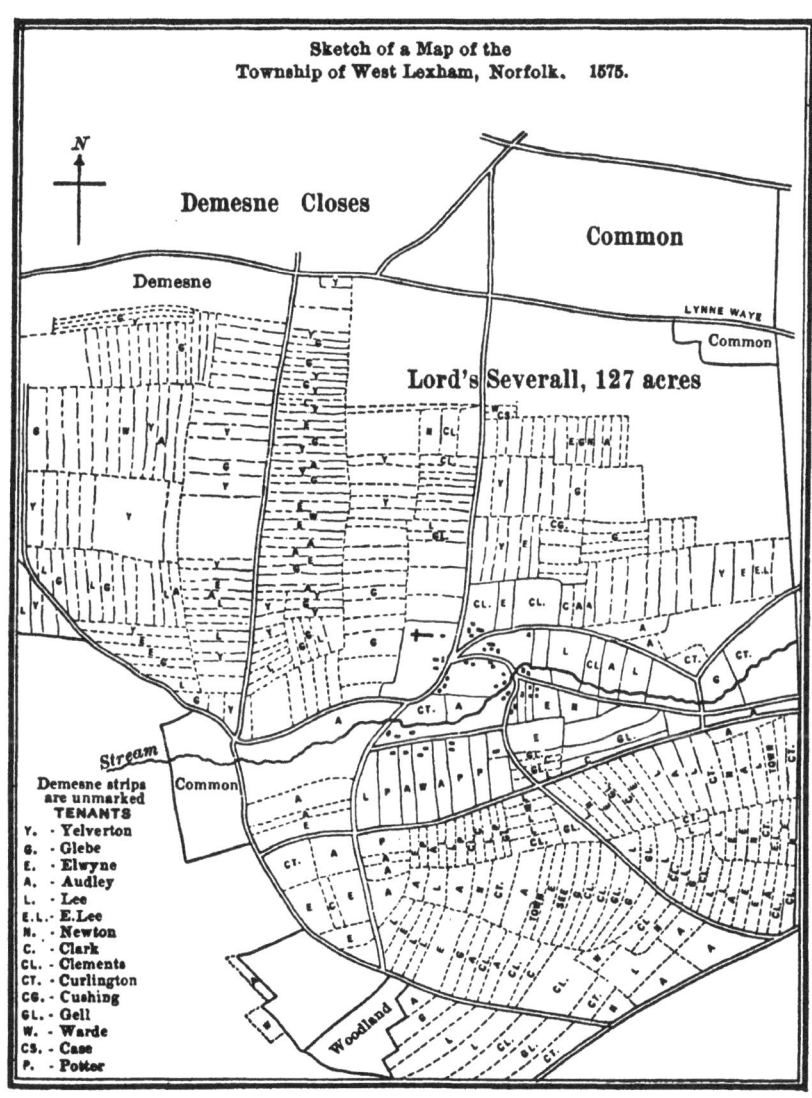

MAP XV

precinct.[1] This arrangement precludes the possibility that Weasenham was at the end of the sixteenth century cultivated as a two-field township.

One of these late sixteenth-century holdings about which we are well informed is that of George Elmdon, who, as noted above, in four separate years (1583, 1584, 1588, 1589) made careful record of how each parcel of his arable was sown. The account, which is unique and valuable, shows that much of the holding was leasehold and varied somewhat in amount from year to year. When in 1588 Elmdon drew up a list of the lands in his possession, he had 71 acres of enclosed arable and meadow, 12 acres of open meadow, and 199 acres of open arable field. The notes that describe the sowing of crops account for from 120 to 160 acres yearly, but in the record of any one year a part of the open-field arable usually fails to appear, because laid down for the time in grass. An idea of the character of these notes will be got from the following transcript, relative to the year 1584: —

"Wynter corne stubble to sowe barlye on pro anno 1585

In dritelakesmere	xi^a
super Overgate versus boream	xii^a iii^r di.
Dowespitt	iii^r
Roysdike als Rushdike di. et ii^a yt John Burges had wheat on	ii^a di.
Rougham deale	i^a di.
Ildemere furlonge et howlond	iii^a et di.
Netherdotslands iiii pec'	iii^a di. rod.
Saleyard	i^r di.
Lawell furlonge	ii^a i^r
Nether blacklonds	i^r di.
Nether calgrave	iii^r
Abb[utting] super marketstie versus austrum	i^a
Item in the Southfeild iii^r pease ex occidente de horswonge, yt was barlie the last yere	iii^r
[Total, 40¾ acres.]	

---

[1] Holkham Records, field-book of 42 Eliz. The following cases are typical: —

Tenant	Messuages	Southern Precinct	Northern Precinct
Manor of Northall	1	183½	111½
Manor of Easthall	1	205	117½
Edward Coke, Esq.	6	103½	327
Anthonie Brovner, Esq.	1	130½	18¼
John Burgess	1	16½	2¼
Hillarie Forby	2	3	29
John Barton	1	30½	17½
John Billings	1	5	1½
Thos. Linge	1	....	3½
Thos. Wright	1	....	15

" Ollands broken upp befoer Xrmas 1584 to sowe barlie on pro an. 1585 by gods grace

In Langmere furlonge linge togeather ex latere australi v^r, v^r, i^a, i^a, i^a di. old olland	vi^a i^r
Itm a gorie three rods prope le wyndmill in bastard Sommerley at Baylies request for ease in drivinge cattle	iii^r
In Burnhowe deale of olland very latelie layd	iii^r
In stadio abb[utting] super northall milhill versus occidentem	i^a
In newdikemere of olland latelie layd	i^a
In newbie prope Mr. Yelvertons	i^a di.
Itm my haye close	i^a di.
Itm Cookes close	i^a
[Total, 13¾ acres.] Summa	lv^a i^r

Inde pease viii^a di. rod, Otes iii^r, Barlye xlvi^a i^r di.

" A breif of all my somerlies pro anno 1585

In Westgate feild voc[atum] the Southfeild, viz., ulvescroft, horsewong, brokback, Brenwonge, et Newbie	xx^a di.
In Mildele et Longlond	i^a iii^r
In the newe broken feild	xi^a i^r di.
In Raisdele	i^a di.
In Blackland feild	xiiii^a i^r di.
[Total, 48 acres.]¹	

" Wynter Cornes god prosper it sowen at Mich[aelmas] 1584

Barkers croft in una pecia	v^a
Hallonge furlong in xiv peciis	ix^a iii^r
Langmere " " iiii "	vi^a iii^r
Endike " " iii "	vi^a di.
Howland " " i pecia	i^a i^r
Abutt[ing] super millhill bothom versus occidentem in iii peciis	ii^a iii^r
Burnhowedele in i pecia	i^a di.
In stadio ad finem borialem de Shortlond furlong in i pecia	iii^r di.
In stadio ad finem orientalem ex latere australi de Auppitt furlong in i pecia	i^r di.
Micklecrofte in una pecia	i^a
Powlesfeild in iiii peciis	ii^a i^r di.

" Wynter corne god blesse et prosper it sowen As[cension] 1584

Wheat in pawles feilde nere Auppitfurlong and belowe the wyndmill	viii^a iii^r di.
and above the wyndmill ex parte australi de Massingham wey sowen before this vth of october, 1584	ix^a iii^r di.

¹ In the margin are written in Arabic numerals other areas, the sum of which is 53½ acres.

"Messylen sowen at michelmes 1584

The upper end of i^a di. in Burnhowe bothome
i^r di. in howlond furlong
di. acra ex austro de le wyndmill
the v small lands in Hallonge furlong .... ii^a i^r
Summa messilen iv^a ii^r di.
[Total winter corn and "messylen," 61½ acres.]"[1]

This account, it will be seen, begins with an itemization of 40¾ acres, which in the autumn of 1584 were winter-corn stubble. Since alone they did not suffice for the barley crop of the next year, some 14 acres of "ollands" were broken up before Christmas to be added to them. These ollands were parcels which for a longer or shorter period ("old olland," or "olland very latelie layd") had been in grass, the term being applied to land enclosed, but particularly to strips in open field. Most of the 14 acres were of the latter character, and nearly one-half of them lay in Langmere furlong near the wheat stubble. Similar strips of open-field arable under grass have been met with in Leicestershire and in Durham in the early seventeenth century,[2] where in a measure their presence betokened the decay of the open field. Here in Norfolk they were a reserve upon which the tenant could draw at any time to increase his allotment for a particular crop. In 1588 Elmdon was tenant of some 27 acres in Ildemere furlong and Westlongland furlong, but he ploughed only 14 of them; the rest were probably in ollands. To judge from the divergence between his total open-field arable in that year and the portion which he ploughed, his ollands must have amounted to about one-fifth of his open field (38 acres out of 199).

What further appears from the enumeration is that practically the same areas of open field were set apart for winter corn, for spring corn, and for "somerlie" or fallow; and this is true not only for the year 1584 but for 1583 and 1588, as the following summary shows: —

"1583.	Acres
Sowen wynter corne at Michaelmas	41½
Wheat Stubble at Michaelmas	41¼
Pease and Barlie Stubble at Michaelmas	[38 +]

[1] Holkham Deeds, 2d series, 231. [2] Cf. above, pp. 35, 106.

1584.
Wynter cornes sowen at Michaelmas and at Ascension 56⅞ ⎫ 61½
Messylen sowen at Michaelmas .... 4⅜ ⎭
Wynter corne stubble to sowe barlye on .... [40¾] ⎫ 54½
Ollands broken upp befoer Xrmas to sowe barlie on.. [13¾] ⎭
Somerlie pro anno 1585 .... [48]
1588.
    Wynter Corne growing in Maye 55⅛
    Barlie, otes, pease, and fetches god bless them sowen in Maye 54⅛
    [Somerlie not given]
1589 (July 15).
    Wynter corne nowe growinge .... [32]
    Barlie, pease, otes, sowen '89 .... 64⅞
    Somerlie pro anno 89 et pro siligne 90 .... 48 "

The only divergence from symmetry here occurred in the year 1589, when the area under spring corn was increased at the expense of the winter-corn crop. So relatively exact was the division of the other years as strongly to suggest a three-course husbandry, and the suggestion becomes a certainty if the parcels sown together at any time be followed from year to year. Although the group which, for example, was under winter corn in 1584 was not precisely the same group that was under spring corn in 1585 and again in 1588, it was nearly the same. Perhaps one-third or one-fourth of the parcels changed during the period; but enough remained constant to establish the important fact that a three-course husbandry was to a large extent employed on tenants' land in Norfolk open fields at the end of the sixteenth century.

If it were true that a three-course husbandry implied a three-field system, we should at this point declare Norfolk fields akin to those of the midlands. Since, however, the one does not necessarily involve the other,[1] and since certain features about the Weasenham descriptions are unusual, it is desirable to locate, if possible, the three groups of parcels into which Elmdon's open-field arable was roughly but pretty continuously divided. This we may do approximately by comparing the names and descriptions of his parcels with the excellent map of 1600. The result is shown on the accompanying plan.[2]

---

[1] Cf. above, p. 45.
[2] Although most of the furlongs of the original map are named, certain of them are not, a fact which renders the exact location of a few of the parcels problematic.

It will be noticed that at least two-thirds of Elmdon's parcels lay in the Southern precinct — that is, to the south of Massingham way — and that within this precinct they tended to concentrate near "Overgate" and "Milstye." None were east of

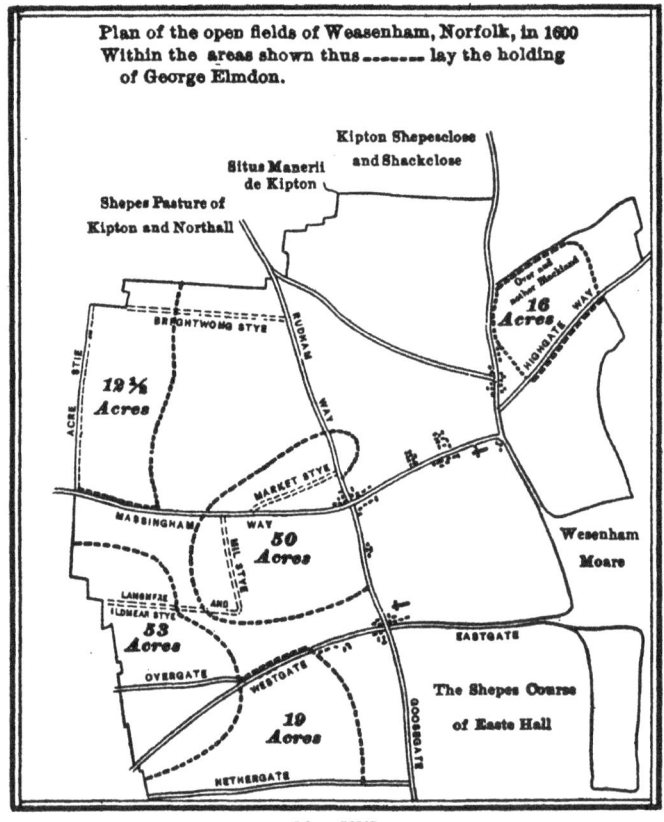

MAP XVI

the church, and apparently there were none near Goosegate, which ran to the south. A few of the Milstye group extended into the Northern precinct, on the other side of Massingham way. Elsewhere in this precinct, which was larger than the Southern, Elmdon had only two small groups of parcels, one toward the west, the other toward the northeast. In another way we thus arrive at the conclusion which has already been reached from a

The names of the highways and field paths are of assistance, however, and the rough grouping of the sketch is reasonably accurate.

consideration of the totals of the field-books of Weasenham and Castle Acre, the conclusion namely that in Norfolk a tenant's arable acres were likely to be concentrated within a particular precinct or field of the township.

Holdings so constituted can be reconciled with the existence of a two- or three-field system only on the assumption that a township had groups of two or three fields and that the parcels concentrated lay in one of these groups. Of Elmdon's three groups of parcels that were assigned successively to winter corn, spring corn, and fallow, two, those near Overgate and Milstye, were each distinctly segregated; and the two may conceivably be thought of as having lain in two compact fields. These fields would, however, have constituted only about one-fourth of the total unenclosed arable, an excessive concentration implying that there were five or six other similar fields. Apart from the attribution of so large a number of fields to Weasenham in the sixteenth century, a difficulty arises regarding Elmdon's third group of parcels. This, instead of being compact, broke into three sub-groups, one lying in the southwest of the township near the Overgate group, the others in the Northern precinct widely removed one from the other. What we actually have, then, is a concentration of two groups of parcels and part of the third within a relatively circumscribed portion of the township's arable. Such locations preclude a six-field arrangement, and one of eight fields does not comport with three-course husbandry. Despite the three-course rotation of crops at Weasenham, therefore, the distribution of Elmdon's parcels conflicts with the assumption that the township was one of three or of six fields.

Other features of Elmdon's notes emphasize this conclusion. The writer never says, as he so easily might have done had the system been simple, " Sown with winter corn, all parcels in X field." On the contrary, he nearly always assigns his strips to furlongs, with only an occasional mention of fields. The area round Westgate is at times, to be sure, vaguely referred to as Westgate field or South field, but in it lay parcels devoted to different crops. In it, too, lay at least two-thirds of Elmdon's holding. Since for these reasons it cannot be thought of as functioning like a midland field, the designation without doubt had merely a topographical connotation. Again, we hear of Kipton field, so named from its proximity to the site of the manor of Kipton. Here lay 12½ acres of Elmdon's land; but inasmuch as these same acres are elsewhere referred to as in the " newe-broken field " and were situated not far from the "Shepes pasture of Kipton and Northall," the " field " in question may have been a newly-improved tract of arable.

As was just noted, parcels within the same field were at times not under the same crops as neighboring parcels. Although, in 1583, 13 acres of Blackland field lay in wheat stubble, 4¼ acres of the same field were at the same time in pease stubble. One of the parcels in the extract quoted had been under barley for one season and was to be so during the next year.[1] In the furlong called Newbie there were in 1589 some parcels fallow and some sown with winter corn. All this implies considerable flexibility in the utilization of the open field. The existence of ollands, or strips of grass in the midst of winter and spring corn, testifies further to the same characteristic and helped make it possible.

This flexibility appears most strikingly in the sowing of Elmdon's open field in 1589. In this year, it will be remembered, the acres hitherto equally divided between winter corn, spring corn, and fallow were unequally apportioned. In preceding years, to each crop and to the " somerlie " had been allotted about 40 acres (1583) or about 55 (1584, 1588). In 1589, the total acreage accounted for was 145 acres. As usual, one-third of this, 48 acres, was somerlie; of the remainder only 32 acres were devoted to wheat, while upon 65 acres spring grains were sown. Such expansion and contraction of the acreage assigned to a particular crop would have been possible under a three-field system only if all tenants had agreed to shift for the year the boundaries of the three fields. In Elmdon's note-book there is no hint that his dispositions rest upon communal arrangements of this sort.

As a result of the implications of Elmdon's notes, we are led to conclude that a three-course rotation of crops in the open

[1] Cf. above, p. 318.

fields of Norfolk was not necessarily indicative of a three-field system. On the contrary, it proved feasible to till parcels concentrated in one part of the open field in such a manner as to allot one-third of them to winter corn, one-third to spring corn, and to leave one-third fallow. Under these circumstances each third naturally consisted, as far as possible, of neighboring parcels. It may seem, however, that the introduction in the same year of two crops and a fallow within a limited portion of the township's arable was a retrogression from the principles of the midland system; that the obvious convenience of the large and simple divisions of that arrangement was sacrificed, inasmuch as a compact fallow field for the pasturage of sheep and cattle thus became impossible in East Anglia. Such questionings are pertinent and bring us to a new aspect of the subject, namely, the provision made for the pasture of cattle and sheep in East Anglian fields.

Certain items regarding Weasenham may serve as introduction. The map of 1600, and still better that of 1726-28, show two large "sheep's courses" distinct from the open fields, one appertaining to the manors of Kipton and Northall, the other to the manor of Easthall.[1] Relative to the open strips themselves the schedule accompanying the later map gives information. Apart from 717 acres of "break" (the former sheep's course of Northall and Kipton), the largest of Sir Thomas Coke's farms comprised arable land described as follows: —

"Subject to its own flock and including whole year grounds, 265 acres,
  Subject to the flock of Easthall, 56 acres,
    and to Lord Townshends [flock] in Little Raynham and Martin Raynham, 42 acres."

From this it appears that the sheep within a township fell into flocks, each manor having its own flock. Any particular parcel of open-field arable was "subject" to a certain flock, perhaps not to that of the proprietor of the land in question. Pasturage arrangements were not devised with a view to the township as a whole, as in a midland village, where rights of pasturage over

[1] Cf. above, p. 322.

the commons and fallow field inhered in the community and were jointly exercised by all its land-holding members. In Norfolk pasturage rights over certain pastures and certain portions of the fallow field (together called " fold-courses ") appertained only to particular proprietors, other land-holders being excluded. Since this was the practice, it would have served no end had George Elmdon's acres been distributed among three fields. When one-third of them lay fallow, they would not have been open to all the sheep of the township, but would have been reserved for a particular flock. All arable which in any year lay fallow in the township did not form one common pasture, but had to be subdivided in accordance with the claims of the several flocks.

If it be thought that the Weasenham evidence in this matter is insufficient, there is fuller information relative to Holkham, near by. A map of this township, dated 1590, discloses its pasturage arrangements,[1] which are further explained by the report of a special royal commission sent in 1584 to ascertain the queen's rights in the " common wastes."[2] The map shows a large South field equal in size to both of the other fields, which were known as Church and Stathe. The marshes to the north next the sea constitute one common, the Lyng on the southeast another. Across fields and commons are traced the boundaries of four fold-courses, each comprising about one-fourth of the township's arable and common waste,[3] but the boundaries nowhere correspond with those of the fields. Three of these fold-courses represent the three manors of the township. The fourth " is fed with the sheepe of one Edmund Newgate and others the Inhabitunts and house holders there. But whether Newgate's be taken as a folde corse or no we [the jurors] knowe not."[4] This arrangement

[1] Holkham Maps, No. 1, sketched in the accompanying cut.

[2] Duchy of Lancaster, Special Commission, No. 350. The commission and a part of the return have been printed by Hubert Hall, *A Formula Book of English Official Historical Documents* (2 pts., Cambridge, 1908–09), ii. 17.

[3] They are known as North course, Caldowe course, Wheatley's course (also called Grigg's), and Newgate's course. The first includes one-half of the marshes, all of Church field, and a part of South field; the second, a large part of South field and one-half of the Lyng; the third, the remainder of the Lyng, with parts of South field and Stathe field; the fourth, a part of Stathe field and one-half of the marshes.

[4] Duchy of Lancaster, Special Commission, No. 350. The jury continues: " No

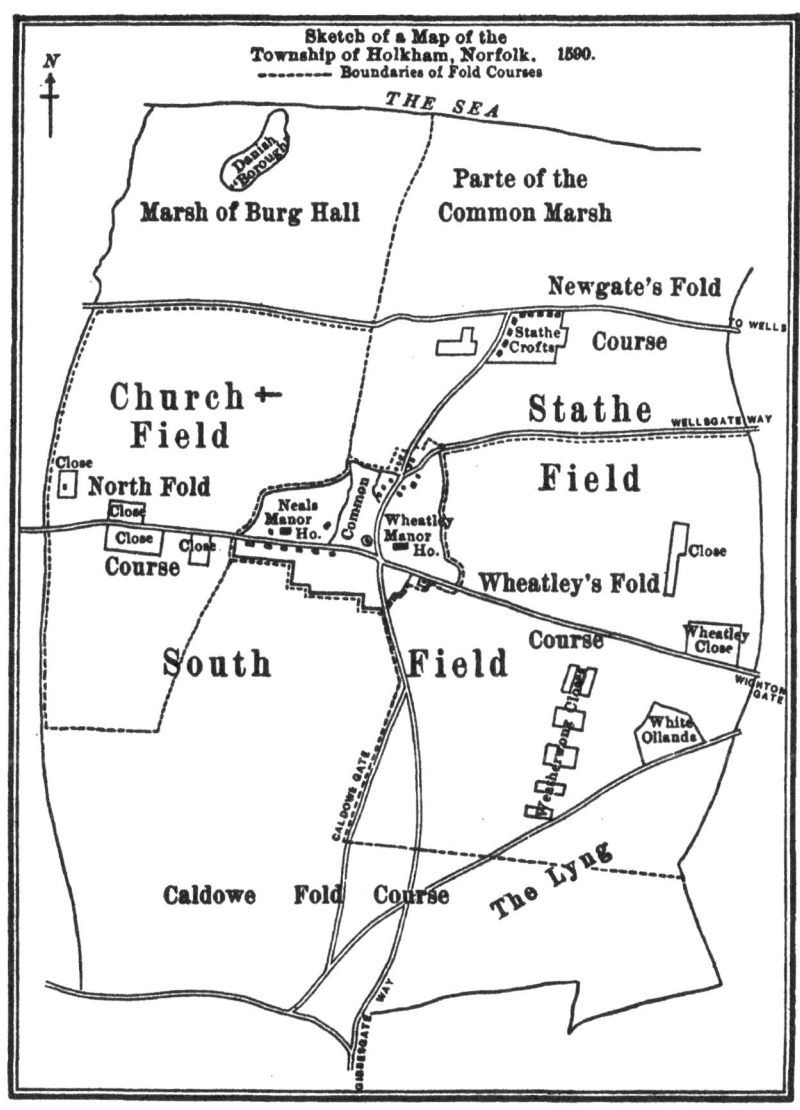

Map XVII

can only imply that each of the four flocks of sheep always had at its disposal about one-fourth of the unimproved land of the township, and after harvest time could also be pastured over such of the arable as lay within the bounds of its fold-course. If we should try to picture the arable of any course as comprised within one of the three fields, we should at once see that such an arrangement would not have permitted the flock of the course in question to fare the same in winter-wheat years as in fallow ones. At times all of the course would have been under crops, at times all of it fallow. If stubble fodder was always to be available for each of several flocks of a township, a system different from the typical three-field one must have been evolved. Within each fold-course some parcels of land must have been under winter crops, others under spring crops, and others fallow.

The actual situation is disclosed in an indenture of 26 Elizabeth that conveyed the Holkham manor called Nealds, alias Lucas.[1] This manor, we are told, consisted of 25 acres which formed the site of the manor-house, 234 acres in South field, 67 in Church field, and 88 in Stathe field. Appurtenant to these lands were certain common rights of pasture, viz.: —

(a) " Item a Liberty of Fould course and Fouldage and shacke with shepe in the southe fielde of Holkham [but, as the map shows, by no means over the entire South field].

(b) " Item a common of pasture . . . for horse, neate, and sheepe at all tymes in the year in fourteen score acres lyinge in the southe parte of Holkham Common Lynge. [This area and the preceding comprised " Caldowe fold course " on the map.]

(c) " Item another common of pasture . . . in all tymes of the year for horse, neate, and swyne in all the commons of Holkham aforesayde.

(d) " Item another common of pasture . . . for horse, neate, and swyne [but not for sheep] uppon all the feilds, grounds, and marshes within Holkham aforesaid lyinge freshe and unsowne

man ought to kepe or mainteyne any folde corse within the marshe. But of late there is one Edmund Newgate taketh upon [him] to kepe five hundred shepe there whereas before tyme his Grandfather and others Kepte not above two hundred yet there upon theire privat marshe."

[1] Holkham Records, uncatalogued.

yearly from the feaste of St. Mychael the archeAngell or the ende of harveste until the annunciation of our Ladye or untill suche tyme before the sayde feaste . . . as the said feilds and grounds be sowen agayne."

From this it is clear that all the village cattle ranged over the entire common waste throughout the year and over the unsown fields from October to February. From February to October they had no access to the fallow arable, which was reserved for the various flocks of sheep. Each flock of sheep, furthermore, never passed beyond the bounds of its fold-course; within this course it was presumably folded from day to day over the fallow acres. Since in all probability wattles were used, no inconvenience arose if sown and fallow acres lay side by side. Hence came the flexible, particularist, more modern system that was employed in the days of George Elmdon. It was an arrangement far better for the soil than was that of the midlands, since by it each parcel of arable was assured of fertilization during the fallow season. Some of the thriftless convenience of the midland system may have been sacrificed, but superior agricultural method and profitable sheep-raising were compensations.

Touching the subjects discussed above — the distribution of the parcels of a holding throughout the arable area of the township and the rotation of crops practiced upon them — we should like testimony from an earlier time than the end of the sixteenth century, and we happily find it in various items that carry back a little the regime of insignificant fields and three-course husbandry. Most numerous are data relating to Holkham. This township, it will be remembered, revealed on the map of 1591 a large South field and two smaller fields to the north next the sea, called Church or West field and Stathe or North field. That the distribution of a tenant's acres among these three fields had for a long time been unequal, becomes apparent from an examination of several earlier terriers. In 26 Elizabeth the manor of " Nealds " allotted its arable to the three in the proportion of 233, 66, and 87 acres respectively,[1] while in 3 Edward VI the apportionment of the lands of Edward Newgate was 13, 7, and

---

[1] Holkham Records, uncatalogued.

55 acres.[1] In a terrier of "Pomffrett" lands drawn up in 30 Henry VIII the subdivision was 16, 2½, and 6 acres.[2] In 2 Henry VII Sir Thomas Briggs's manor of Hillhall was so situated that 132 acres lay in North field, 45 in Church field, and 390 in South field.[3] The same manor was smaller in 17 Edward IV, and of its parcels 128 acres were in North field, with 68 in Stathe field, 33 in Church field, and 105 in South field.[4] The nearest approach to an equal division of acres in the fifteenth-century Holkham terriers is the assignment of 19 acres of Galfridus Porter's holding to South field and 17 to North field;[5] but in a contemporary conveyance of 39 Henry VI the 50 acres which John Newgate transferred to his son Thomas lay almost entirely in the northeastern part of the village arable area.[6] The distribution of the acres of fifteenth- and sixteenth-century holdings among the fields of Holkham thus seems to have been chronically irregular.

For discovering what rotation of crops was favored in fifteenth-century Norfolk a Holkham charter of 20 September, 16 Richard II, is of value.[7] In an exchange of lands, John and Isabell Lyng "invenerunt terras seisonatas ut inferius patet, vidz.:

i acra in crofta de terra frista semel arata non compostata,
iiii acre dimidia in campo australi semel arate de terra firista non compostate [fallow, probably to be sown with wheat the next year]
iiii acre dimidia in eodem campo que fuerunt cum ordeo anno elapso [seminate] semel arate non compostate [barley stubble, probably to lie fallow for a year]
iiii acre in eodem compostate anno elapso non arate [probably sown with wheat in the year past and to be sown with barley in the coming season]."

Here not only were the 12 acres all in the same field, but they were apparently under a three-course rotation and were manured

---

[1] Holkham Records, uncatalogued.   [2] Ibid.
[3] Holkham Field-book, 75.   [4] Ibid.
[5] Holkham Records, uncatalogued.
[6] Holkham Deeds, 2d series, 29.   [7] Ibid., 77.

at least once every three years. The sixteenth-century system is carried back to the end of the fourteenth, and assurance is given regarding the method of fertilization about which later surveys are silent.[1]

Other East Anglian townships furnish early evidence not unlike that from Holkham. In the time of Henry VIII a manor at Scratby is described as comprising, besides 9 acres enclosed, 29 acres in twelve parcels in South field and 16 acres in nine parcels in North field.[2] A Great Massingham terrier of the late fifteenth century enumerates the parcels of three holdings in four unnamed fields.[3] The respective areas in acres were 8½, 8⅞, 3¾, 7½; 9, 6⅞, 3¾, 6⅞; 2¾, 2½, 1½, 1½: an irregularity in apportionment not to be remedied in any instance by a combination of the last two areas. In 8 Henry V a manor at Ormesby was held by several tenants whose parcels lay principally in North, South, and Little fields, but so unevenly distributed as often to be almost entirely within one field.[4] At Rockland in 23 Henry VI the manor of Kyrkhall Moynes had its small parcels principally in South and West fields, although there was something in North and East fields.[5] Obviously, the distribution of the acres of fifteenth-century holdings among fields was as capricious as in the sixteenth century.

Relative to the matter of tillage, a lease of the manor of Beddingfield, Suffolk, dated 19 Richard II, instructs us as to how certain lands there were sown. Eleven acres were in wheat, of which two were "compostate," two were fallow, six sown with barley, eight with peas, thirteen with oats.[6] Whether these acres were open or enclosed we do not learn, but the small number left fallow points to a husbandry almost as far advanced for

[1] The few other Holkham transfers of the fourteenth century which take the trouble to mention fields are regardless of exact division. Such is the case with 3 acres which in 4 Edward III were in two pieces in South field, and with 6 acres of which in 2 Edward II at least 4¾ were in South field (Holkham Deeds, 2d series, 37, 24).

[2] Rents. and Survs., Portf. 12/59.
[3] Ibid., 22/46.
[4] Ibid., 22/54.
[5] Add. MS. 33228.
[6] Add. Char. A 3338: "Terra seminata cum frumento xi acre unde ii acre compostate; item terra warecta ii acre; item cum ordeo vi acre; item cum pisa viii acre; item cum avena xiii acre."

the fourteenth century as was the contemporary Kentish tillage which cultivated all arable acres yearly.[1]

The foregoing testimony of surveys and terriers tends in general to show that Norfolk "fields" had from the beginning of the fifteenth century no agricultural significance, and that, although a three-course rotation usually prevailed, it was not dependent upon a three-field system. It may be objected, however, that an earlier system was then in decay, one in which names of fields had more than topographical connotation. Without doubt an old system was in decay in the fifteenth century, but scarcely in the sense intimated. What this earlier situation was must now be explained.

With regard to tillage, the custom in East Anglian fields before the sixteenth century was not unlike that practiced by George Elmdon in the days of the Armada. Information on this subject is to be had from extents of Norfolk demesne lands contained in inquisitions post mortem of the first half of the fourteenth century.[2] Sometimes these contain statements which, with change of areas, are like the following from East Bradenham: "Sunt c acre terre arabilis per minus centum de quibus possunt seminari per annum lx et seminate fuerunt ante mortem predicti Rogeri [inquisition dated 16 June, 11 Edward III] et valent per annum xx solidos, pretium acre iiii d. Et totum residuum nihil valet per annum quia iacet ad warectam et in communi."[3] The significant information here is that one-third of the demesne land was fallow throughout every third year, and then was of no value, since it lay in common. Occasionally the phrase is "in communi campo," leaving no doubt that the demesne was open common field.[4] The townships about which this could be said lay in eastern as well as in western Norfolk, a fact that fixes the custom upon the entire county.[5] Although similar remarks about

[1] Cf. above, p. 302. [2] Cf. above, p. 46.
[3] C. Inq. p. Mort., Edw. III, F. 51 (11).
[4] "Sunt cxx acre terre arabilis . . . de quibus iiii^{xx} acre terre seminabantur hoc anno . . . et residuum iacet ad warectam et in communi campo" (ibid., 51 (10), Newton).
[5] Ibid., 46 (3) Gayton, 45 (18) Rainham and Islington, all three in the west of the county; 51 (10) Caistor and Hellesdon, in the center and east.

Suffolk townships are less numerous,[1] they occur often enough to show that a three-course rotation upon open-field demesne could at this time be found throughout East Anglia.[2]

Testimony like this, if from midland counties, has been cited as quasi evidence for the existence of a three-field system there.[3] Without doubt the presumption is that arable land which was fallow and common every third year lay in three (or six) common fields. Where other evidence, therefore, points to the prevalence of the three-field system, as it does in the midlands, a statement like that quoted above may be looked upon as credible testimony that a particular township had three compact open fields. With East Anglia, however, the case is different. There we have seen in the sixteenth century a three-course rotation of crops upon open fields divorced from a three-field system, and a similar situation in the fourteenth century is not improbable. The implication of the phrases in the extents will therefore depend upon other contemporary evidence touching the location of holdings in the open fields. To interpret such evidence we shall have to determine the nature of East Anglian units of villein tenure. Inasmuch as no reference to a virgate, the unit prevalent in the midlands and the north, has thus far been found here, it may be that the omission points to peculiarities of field arrangements, just as the character of the Kentish iugum lay at the base of a unique field system. If so, the nature of the early unit of villein tenure in East Anglia assumes increased importance and demands attention.

During the sixteenth century and even at a much later period certain parcels of an East Anglian holding were often said to

[1] C. Inq. p. Mort., Edw. III, F. 41 (1), Thurston; 41 (19), Monewden, Badmondisfield, and Lidgate, the last two on the Cambridgeshire border.

[2] But it was not universal. At Kettleburgh, in Suffolk, the soil was of poor quality (*debilis*), " et quolibet anno medietas iacet frisca et iacet in communi per totum annum " (ibid., 51 (2)). More noteworthy was the case of a tenement of sixty acres at Wymondham: " Sunt ibidem lx acre terre arabilis ... unde seminabantur hoc anno semine yemali ante mortem predicte Alicie [inquisition, 2 June, 15 Edw. III] xx acre et semine quadragesimali xxx acre. ... Et x acre terre predicte iacent in communi per totum annum quum non seminantur " (ibid., 65 (13)). Thus it was sometimes customary to fallow only one-sixth of the arable. [3] Cf. above, p. 46.

belong to one or another "tenementum"—to Smith's tenement, for example, or to Bunting's. In the fifteenth century all the parcels of a holding could at times be assigned to the tenementa of which they had once formed portions. A tabulation of the acres of two holdings described in a survey of Bawdsey, Suffolk, dated 16 Henry VI, will make this clear, and will show incidentally the insignificant part played by field divisions at that time.[1]

Sitting tenant	Names of the Tenementa to which the Parcels belonged	En-closed	In le Melle-feld	In Lyllond	In le Estfeld	In Dal	Miscel. and Unspec.
Thomas Mekilborgh	Alex. Frebner..	..	..	..	..	1	..
	Ston..........	..	..	..	1	..	1
	Chouks.......	1 [2]	..	$\frac{1}{4}$	..	$\frac{1}{4}$	$1\frac{3}{4}$
	Fenkell.......	..	..	..	$1\frac{1}{8}$	$\frac{1}{4}$	$\frac{1}{4}$
	Filles	..	..	..	$1\frac{1}{4}$	..	..
	Chapman.....	..	..	..	..	..	$\frac{1}{4}$
	Gellys........	..	..	$\frac{3}{4}$	..	..	..
	Hailles........	..	..	..	1	..	..
John Godwyn	Frebner	$\frac{1}{16}$ [3]	1	..	..	..	..
	Ipris	..	$2\frac{1}{2}$	..	..	..	..
	[?]	4	..	..	..	..	..
	Godewyn¯.....	..	..	..	..	$\frac{3}{8}$	..
	Crour'	2	..	..	..	..	..
	Crunnok	$\frac{1}{16}$ [2]	..	..	1	..	..
	Hobard	$1\frac{1}{8}$	..	..	..	..	..

Surveys of East Anglian manors dating from the late fourteenth or from the fifteenth century are likely to be cast in a mould much like that of Bawdsey. They point, it is clear, to a still earlier time in which the tenementa were the principal agrarian units of the township, instead of merely serving, as they did in the fifteenth century, for the apportionment of rents and services.

In surveys and rentals of the early fourteenth or, better still, of the thirteenth century, the tenementa assume their earlier prominence. Surveys were not then drawn up, as at Bawdsey, under the names of the contemporary tenants who had parcels in different tenementa, but they were arranged according to the tenementa themselves, in each of which the parcels were assigned

---

[1] Add. MS. 23948. [2] With toft. [3] With cottage.

to the tenants of the hour. Nearly always a tenementum was held by several men, who usually had parcels in other tenementa as well. In rentals of the time, too, brief and parsimonious of names as they are, one often finds reference to such items as the tenementum of John Smith "et parcenarii sui" or "et participes sui."

One of the most instructive of these earlier surveys is that of Martham, Norfolk, giving as it does a detailed account of field arrangements.[1] Situated near the east coast, this manor of Norwich priory was surveyed in 1291, the fourth year of Prior Henry of Lakenham. The villein holdings are described with reference to the tenants who formerly held them, and are then assigned with minute specification to the contemporary tenants. A typical description of one such holding is as follows: —

" Thomas Knight tenuit quondam xii acras terre de villenagio que vocatur i eriung[2] et reddit inde . . . [services and rents follow]. Sciendum est quod xii acre de villenagio vocantur unum Eriung. Et quilibet tenens unum Eriung faciet in omnibus sicud predictum est de tenemento Thome Knight. Et habentur in Martham xxii Eriung et iii acre de villenagio et omnes isti herciabunt totam terram Ville exceptis terris quesitis ad siliginem, avenam, et falihes.

De quibus xii acris terre nunc sunt xii tenentes, viz., Martha Knight tenet iii acras et dimidiam de quibus
    dimidia acra iacet in campo de Martham qui vocatur Estfeld . . .
    Item dimidia acra iacet in eodem campo . . .
    " i roda et dimidia iacent in campo qui vocatur Monechyn . . .
    " dimidia acra iacet in Damiottoftes . . .
    " " acra " " Fendrovetoftes . . .
    " " Roda " " Morgrave . . .
    " i Roda et dimidia iacent in Tofto suo cum mesuagio . . .
    " dimidia Roda iacet in Monechyn . . .
    " xxx perticate iacent in eodem campo . . .
    " dimidia Roda iacet in Westfeld . . .
    " xxx perticate iacent in Monechyn . . .

[1] Stowe MS. 936, ff. 37–115.
[2] An eriung is the Anglo-Saxon term for a ploughing or plough-land.

Johanes Knyght tenet i acram dimidiam et i Rodam terre
iacentes in campo de Martham de quibus
dimidia acra iacet in Westfeld . . .
Item dimidia Roda iacet in Fendrove . . .
" i Roda et dimidia iacent in Estfeld . . .
" i " " " " " eodem campo . . .
" i " iacet in Estfeld . . .
Andreas Knyght tenet i acram et dimidiam et i Rodam terre
iacentes in campo de Martham. De quibus
dimidia acra iacet in Estfeld . . .
Item i Roda iacet in Fendrovetoftes . . .
" i " et dimidia iacent in Tofto suo cum mesuagio . . .
" i " " " " " Monechyn . . .
" i " iacet in Westfeld . . .
Willielmus Anneys tenet i acram de qua
dimidia acra iacet in Estfeld . . .
Item dimidia acra iacet in Estfeld . . .
Beatrix Knight tenet i acram terre de qua
dimidia acra iacet in Estfeld . . .
Item dimidia acra iacet in eodem campo . . .
Robertus Aleyn tenet i rodam iacentem in campo de Martham qui vocatur Estfeld . . .
Robertus Wuc tenet dimidiam acram In Estfeld . . .
Hugo Balle " " " " " . . .
Robertus filius Roberti mercatoris tenet i acram terre de qua
dimidia acra iacet in Westfeld . . .
Item dimidia Roda iacet in eodem campo . . .
Item i Roda et dimidia iacent in Fendrovetoftes . . .
Willielmus Folpe tenet i Rodam et dimidiam iacentes in
Estfeld . . .
Willielmus Godrich tenet i Rodam et dimidiam iacentes in
Estfeld . . .
Et omnes isti tenentes reddunt servicia pro pleno Eriung sicut fecit Thomas Knight in tempore suo."[1]

Relative to a holder of "mulelond" the record runs,[2] "William Hereman tenuit quondam vi acras terre que vocantur Mulelond

---

[1] Stowe MS. 936, f. 39b. [2] Ibid., f. 70b.

pro xvi d. de Redditu [obligations follow] . . . De quibus nunc sunt ix tenentes," whose holdings are detailed as before. There seems to have been no unit of mulelond, since the holdings of the "former tenants" contained a variable number of acres. Similarly, the socage land is referred to "former tenants" in varying amounts, and these holdings too had been parcelled out among contemporary tenants.

In the case of the villein eriung of Thomas Knight it is easy to see that the existing situation had come about through a subdivision of the twelve-acre holding among heirs.[1] Four tenants still bore the name of Knight and had the largest shares in the eriung, together retaining seven and three-fourths acres of it. They may have been three or four generations removed from the ancestor who gave his name to the holding. If so, Thomas Knight lived in the late twelfth or early thirteenth century.

After this introduction to a thirteenth-century Norfolk survey, we may proceed in our inquiry regarding field systems. If the foregoing references to the East and West fields of Martham suggest that a two-field system prevailed there at the end of the thirteenth century, its existence should be revealed in the distribution between these fields of the acres of the old units. Since at Martham the villein eriungs were the holdings most invariable in size, being always in theory twelve acres, they should be looked upon as the standard units and most likely to be evenly divided between fields. The condition of a few typical holdings, villein and other, is pictured in the following table: —

Former Tenant ("tenuit quondam")	Tenure	Number of Present Tenants	Number of Messuages	Area of Closes in Acres	West Field		East Field		Miscel. Areas	Total Area
					Number of Parcels	Area in Acres	Number of Parcels	Area in Acres		
Thomas Knight	villein	12	2	¾	6	1½	15	6½	3¼	12
Thomas Knight	socage	3	0	0	0	0	8	3	¾	3¾
Humfridus de Sco ..	socage	10	7	10¼	5	¾	3	1½	2¼	14¾
Humfridus de Sco ..	villein	4	0	0	8	1½	0	0	¾	2¼
Syware filius Galfridi	villein	18	0	3½	24	7½	1	¼	½	11¾
Syware filius Galfridi	socage	0	0	½	7	2¼	3	¾	½	4
Stephanus Byl	villein	3	1	½	0	0	9	3½	1	5
Nicholas Haral	mulelond	3	1	3	0	0	9	3	½	6½

[1] That socage and villein holdings in East Anglia were ever subject to partible transmission seems to have escaped the notice of legal historians.

From the table it is clear that the holding of no "former tenant" was divided equally between the two fields, and this is true whether land of all tenures be considered or, as is more to the point, whether attention be confined to villein land. Even in the midlands free land was not always very evenly apportioned among fields. That the Norfolk villein eriung, however, the unit which corresponded with the evenly-divided midland virgate, should show an indifference to equal division between "fields," and an inclination to lie largely in one of them, is significant. It implies that the East and West fields had no agrarian importance at the time when the eriung took form.

To know just how the parcels of an eriung lay in relation to one another would be information well worth having. Unfortunately, they are described in the Martham survey as they had come to appear in the hands of the numerous tenants of 1291. How many there were and how related when the "former tenants" held them we are left to puzzle out from the incomplete boundaries that are given. The description of one of the half-eriungs which lay entirely in the West field is substantially as follows: —

Wm Godhey tenuit quondam vi acras de vilenage pro dimidio Eriung. De quibus sunt nunc xi tenentes, viz.:[1]

Robert Koc, messuage and two tofts; 1½ roods next Osbertus Harald on the north; 1½, Simon Koc N; 2, Walter de Scoutone W; 2, Thomas Mome W.

Robert de Hyl	1½, Alan de Syk W.
Richard Mercator	1 , John clericus E; 1, Thos. Mogge W; ½, Robert Koc S.
Simon Koc	½, Roger Mercator S.
Osbertus Harald	2 , Robert Koc S; 25 poles, Rob't Koc S.
Beatrix Harald	25 poles, Osbertus Harald S.
Simon Cok	1 , Robert Koc S.
Walter de Scoutone	2½, Robert de Hill E.

[1] The neighbor on one side of each parcel is specified, as, for example, "iuxta Osbertus Harald ex parte aquilonari," but the parcel is not completely bounded. The abbreviated locations which follow should all be read in this way, the areas being in roods unless otherwise specified.

Alan de Syk   1½, Robert de Hill E.
Alexander de Sco   2 , Robert Koc E.
Robert filius Willielmi
   Webbester   10 poles, Robert Koc S.

From this description it is pretty clear that many of these parcels lay not only in the same field but also side by side. Such were the parcels of Robert Koc and Osbertus Harald, of Robert Koc and Simon Cok, of Robert de Hyl and Alan de Syk. One begins to suspect that the six acres of William Godhey's halferiung were after all not necessarily much separated one from another. Since the Martham descriptions are somewhat inconclusive in this respect, we turn to another survey that furnishes what is perhaps our best evidence relative to the appearance of the original East Anglian villein tenementum.

This survey, which is incomplete at beginning and end, relates to Wymondham, a township southwest from Norwich, and is written in a hand of the time of Henry VII.[1] Although many descriptions are not detailed and others break off with the statement that the residue of the tenementum is in the hands of the lord, certain of them are, none the less, instructive. The tenementa were by no means so subdivided in the fifteenth century as were those at Martham in the thirteenth. If this should lead one to suspect that no great period of time had elapsed since they were in the hands of their original tenant, the suspicion would be dispelled by the discovery that not one of the existing tenants of a tenementum bore its name as his surname, after the manner of the Knights who continued to share in Thomas Knight's eriung at Martham. The tenementum here, as there, probably goes back at least to the thirteenth century. More ancient than that it can scarcely have been, if we may judge from such names as Toly, Crisping, Caly, and Davys.

The novel feature about these Wymondham tenementa is that they can in some instances be shown to have been nearly compact areas:—

"Tenementum Toly iuxta Grishaugh continet i mesuagium, xi acras, iv Rodas terre . . . Unde

---

[1] Land. Rev., M. B. 206, ff. 188-215.

Thomas Knyght alias Kette tenet dictum mesuagium, ix acras, iii Rodas terre . . . iacentes iuxta Grishaugh

Ricardus Deynes tenet ii acras pasture inclausas . . . abuttantes . . . super Grishaugh versus austrum." . . . [1]

Here the entire tenement bordered upon " Grishaugh " and cannot have been in more than two parcels at most.

" Tenementum Havercroft continet xiiii acras terre et bosci cum mesuagio vacante, Unde

Thomas Caly tenet totum tenementum iacens in partrikefeld . . .

Edwardus Groote tenet inde ii acras terre in Cobaldisfeld." [2]

Six-sevenths of this tenement lay in " partrikefeld," a feature in which it resembled one of its neighbors.

" Tenementum Ricardi Aleyn continet i mesuagium edificatum et xxxii acras et dimidiam terre, Unde

Galfridus Symond tenet dictum mesuagium ac xxvi acras et dimidiam terre, pasture, et subbosci in parkrykefeld . . .

Johanis Caly tenet v acras terre . . . in eodem campo . . .

Thomas Cooke alias Blexter tenet unum inclusum infra mesuagium suum vocatum Benecroft . . . et continet i acram iii Rodas

Item tenet unam Rodam dicti tenementi iuxta Benecroft." [3]

Practically all of this tenement lay in " partrikefeld." Finally a small holding is briefly dismissed as follows: —

" Tenementum Pering continens iiii acras terre in una pecia restat in manu domini." [4]

These four illustrations, which are particularly comprehensible since subdivision is slight and locations are traceable, make it certain that the early tenementum was at times a nearly compact area. Three of the above tenementa were relatively large, and two of these lay almost entirely in a single " field." Some non-adjacent parcels there may, of course, have been in this field, as descriptions of other tenementa imply was at times the case. The twenty-four acres of " Tenementum Cobalds," for example, all of which except two acres were held by Thomas Neker, lay " in

---

[1] Land Rev., M. B. 206, f. 208.   [2] Ibid., f. 209.
[3] Ibid., f. 210.   [4] Ibid.

diversis peciis in Cabaldisfeld, partrikfeld, et Domiham hallfeld."
In general, however, the details of this survey reinforce the impression got from the half-eriung of William Godhey at Martham. So far as we can ascertain the appearance of the original tenementum in Norfolk, it seems to have been either a compact area or a group of not widely separated parcels.

After examining above the appearance of a sixteenth-century Norfolk holding, we proceeded to inquire into the pasturage arrangements of that date and found them based upon so-called fold-courses.[1] A division of each township was set off as the fold-course for a certain flock, and over the part of this which lay fallow in any year the flock was folded from February to October. Since the thirteenth-century tenementa were quite as regardless of a three-field disposition of their parcels as were the sixteenth-century holdings, we shall expect to find in early documents pasturage arrangements not unlike those which later prevailed.

Useful information touching this point is given in a series of extents and custumals drawn up in 1278 and referring to the manors of the bishop of Ely, several of which were in Norfolk and Suffolk. Three items in particular relate to methods of tillage. First of all, it appears that the tenant of a full villein holding was bound to carry manure for the lord and spread it upon the fields. Sometimes he carried for a half-day, sometimes he drew five or six cartloads, and once, it is estimated, the labor occupied all the tenants for a week.[2] Evidently stabling of stock and manuring of fields were to some degree practiced.

Such a device, however, was not the chief reliance for maintaining the fertility of the soil. As in the sixteenth century, this

---

[1] Cf. above, p. 325 *sq.*
[2] " Item iste cariabit fimum domini per dimidiam diem semel in anno. . . . Et quotiens opus fuerit sparget fimum a mane usque ad horam nonam . . . " (Cott. MS., Claud. C XI, f. 221, Derham, Norfolk). " Et debent cariare quindecim muncellos composti in quoscunque campos dominus voluerit pro uno opere. Unde duo moncelli vel tres facient unam carectatam " (ibid., f. 259, Glemsford, Suffolk). " Et iste et omnes pares sui cariabunt totum compostum domini per unam septimam ad festum Sancti Michaelis. . . . Et quod cariaverint debent spargere " (ibid., f. 243*b*, Bridgham, Norfolk).

end was achieved by the folding of sheep and cattle upon arable fallow, a usage likewise revealed by the custumals. On many manors the tenant of villein land had to make wattles and carry them about. Often he furnished five with ten supports and moved them at least once a year.[1] Such procedure, it may be said, refers merely to the demesne acres upon which the sheep of tenant and lord were folded together. In certain instances it is indeed specifically declared that the tenants' sheep shall lie " in falda domini " throughout the year and the cattle from Pentecost to Michaelmas.[2] Had this always been the case, an enclosed demesne might account for the requisitions, and we need assume no unusual field system. It is the third item of the extents that forces us to believe that the system was unique.

This item specifies the payment of "faldagium." " Et dabit de faldagio ad Gulam Augusti per annum pro quolibet bove unum denarium. Pro qualibet vacca sterili unum denarium. Pro qualibet vacca cum vitulo duos denarios. Et pro qualibet iuvenca duorum annorum vel pro quolibet Bovecto eiusdem etatis unum obulum. Et pro quinque ovibus unum denarium. Et ideo nec oves sue nec averia sua iacere debent in falda domini."[3] The payment of foldage according to this scale exempted the tenant's sheep and cattle from being folded with those of his lord over the demesne acres. Upon several manors, especially in Suffolk, the tenant had no obligation either to fold sheep or to pay if he did not, the custom being that " oves sue non iacebunt in falda domini."[4] In the Ramsey cartulary the same privilege is recorded in slightly different phrase. That the villein " habet suam faldam," or at least had it during a part of

[1] " Et dabit decem palos et quinque cleyas falde sine cibo. . . . Et portabit quinque cleyas falde domini et totidem palos semel in anno de uno campo in alium sine cibo . . ." (Cott. MS., Claud. C XI, f. 243b, Bridgham).

[2] " Oves sue iacebunt in falda domini per totum annum preter oves matrices tempore agnilis. . . . Et omnia alia averia sua iacebunt in falda domini a pentecoste usque ad festum Sancti Martini preter vaccas. . . . Et boves similiter iacebunt in falda domini inter pentecostem et festum omnium sanctorum si non dederit cupam pro eis ut supradictum est " (ibid.).

[3] Ibid., f. 221b, Derham.

[4] Ibid., f. 259b, Glemsford; f. 265, "Herthirst"; f. 272, Rattlesden; f. 279b, Hitcham; f. 288b, Barking; f. 296b, Wetheringsett; f. 303, Brandon.

the year, was a custom there in the twelfth as well as in the thirteenth century.[1]

How, we may now ask, could a tenant's privilege of having his own fold be realized ? Under a system of enclosures there would have been no difficulty, but in the twelfth and thirteenth centuries East Anglian fields were largely open. Assume now that the arable of a township was divided into two or three (four or six) compact divisions cultivated like those of the midlands. There it was the practice for sheep and cattle to roam over the entire field which lay fallow, the lord's acres (if in open field) and the tenants' acres sharing alike. If under such a system tenant or lord were to have had " sua falda," he would have been obliged to hedge about his parcels with wattles, thereby sacrificing the prime advantage secured by the compact fallow field — the freedom from attending much to the wandering sheep and cattle. Since one aim of the midland system was to attain this convenience, we do not hear about the use of wattles in midland open fields or about any tenant having " sua falda."

Apply again the privilege of " sua falda " to such a field system as was practiced in Norfolk in the sixteenth century. There the flock of each manor had in the township a definite area, apart from unploughed pasture and waste, over which it had rights. Beyond this area it did not pass, and within it some parcels were fallow and some were sown each year. To protect the growing corn wattles must have been necessary. Since the lord's flock had to be kept from the cultivated acres and folded upon the parcels of fallow until harvest time, the complexity would in no wise have been increased if the tenant were to employ the same procedure relative to his acres. He, too, like his lord, might well have had some of his parcels under crops, and others fallow with his sheep folded upon them. The villein's privilege of having " sua falda," recorded in the Ely cartulary, thus accords entirely with the Norfolk method of pasturing sheep, but not at all with that of the midlands. That it is noted in the twelfth- and thirteenth-century documents argues for the early existence

[1] *Cartularium Monasterii de Rameseia* (ed. W. H. Hart, Rolls Series, 3 vols., 1884–93), i. 423, iii. 261, 262, 264.

of the East Anglian system, and the case is strengthened by the divergence manifested in the customs of the manors of Ramsey abbey. No one of the long list of the midland possessions of that abbey possessed the privilege of independent foldage. Yet, as we have seen,[1] the two Norfolk manors had it, and the selection of them for such a favor suggests that they were in a condition to take advantage of it as the others were not.

Pasturage arrangements adopted in East Anglia thus concur with the disposition of the parcels of a tenementum relative to the fields, in pointing to a unique field system. Such descriptions of this system as have so far been utilized are, except for certain items in regard to foldage, not earlier than the late thirteenth century. It remains to inquire whether it is possible to discover at what time the tenementa took form.

If we were to judge from names alone, we should not assign them to a period earlier than the thirteenth century. It is easy to see that the land which, in the Martham survey of 1291, Thomas Knight is said to have held (*quondam tenuit*) would soon be known as Knight's tenementum, that the socage land would become Knight's free tenementum, and the eriung Knight's villein or bond tenementum. Thomas Knight, himself, as we have seen, must have lived either in the early thirteenth century or at the end of the twelfth. The names which attached themselves to the tenementa at Wymondham and at Baudsey often included surnames, as in the case of "tenementum Ricardi Aleyn, or "tenementum Alexandri Frebnere." Since villeins seldom bore surnames before the thirteenth century, the nomenclature of the surveys would seem to assign the tenementum to a period not much earlier than this.

Even if the names of the tenementa did not much antedate 1200, there is reason for thinking that the unit itself was older, though not always, to be sure, under the name "tenementum." This term became usual only in the fourteenth century, and Thomas Knight's holding, though referred to as a tenementum, properly bore the infrequent Anglo-Saxon designation

[1] Cf. above, p. 342.

eriung.¹ In the thirteenth century the units of villein tenure often assumed other names. " Plena terra " was much in favor. The excellent series of Ely extents already quoted frequently employs this phrase and attaches to it, as to a unit, the enumeration of villein services. Its area was uniform within the same manor. At Walpole it contained 30 acres; at Walton, 24; at Feltwell, 20; at Northwold, 48; at Terrington, 24.² Sometimes no name at all was given to the full villein holding. The Ely manor of Emneth leaves unnamed its unit of 23 acres,³ and the Ramsey cartulary finds no term to apply to holdings " in lancectagio." ⁴

At this point it will be of assistance to note the way in which Norfolk manors are treated by this cartulary in its two series of extents, one from the middle of the twelfth century, the other from the middle of the thirteenth.⁵ Ramsey had only two considerable manors in Norfolk — Brancaster and Ringstead — whereas in the midlands she had many. In the latter the villein holdings were always denominated virgates, and the enumeration of virgates is usually lengthy. At Ringstead, however, as we learn, " Non sunt ibi hydae, vel virgatae terrae. Aestimantur, tamen, quod ibi sint quinque hydae terrae praeter dominicum." At Brancaster, " Ibidem sunt decem hydae. Nescitur, quot virgatae faciunt hydam, nec quot acrae faciunt virgatam." ⁶ Farther on we are told that three of the Brancaster hides were villein land. The extents which thus deny the existence of virgates

¹ The word occurs in an important passage in the Ramsey cartulary. Cf. below, p. 348.
² Cott. MS., Claud. C XI, ff. 192, 199, 254, 258b, 182.
³ Ibid., f. 206.
⁴ In this cartulary such is the usual designation for villein land. " Gilbertus Potekyn . . . recognovit viginti quatuor acras terrae, quas tenet de domino Abbate, esse lancectagium Abbatis, et quod debent omnes consuetudines serviles, salvo corpore suo " (court roll of 1239, *Cartulary of Ramsey Abbey*, i. 424).
⁵ In the first series we find that " Eadwinus de Depedale tenuit in diebus Regis Henrici, et *nunc tenet* . . . " (ibid., iii. 261); many extents of the second series are dated 1250–1252. Unlike the tenants in the second series, those in the earlier one usually have no surnames, and their names have a more archaic Saxon or Danish character than was usual a century later.
⁶ Ibid., i. 405, 413.

and are somewhat vague about hides date from about 1240 to 1250.[1]

From the absence of the term virgate, however, it does not follow that units of villein holding were non-existent. On the contrary, the uniformity in size which characterized holdings " in lansectagio " both at Brancaster and at Ringstead points conclusively to a recognition of such units. At Brancaster the three villein hides in the thirteenth-century extent were constituted as follows: 38 holdings of 12 acres each, 17 of 24 acres, two of 60 acres, two of 30 acres, and four of 15 acres. In the extent of a century earlier we find 39 holdings of 12 acres each, one of 32 acres, three of 16 acres, and two of 18 acres. Obviously at both periods the unit was 12 or 24 acres. At Ringstead the holdings were less symmetrical. In the thirteenth century there were 13 holdings of 10 acres, two of 14, and single holdings of 28, 22, 12, 8, and 7 acres. In the twelfth century there were ten eight-acre holdings, with one of 12 and one of 11 acres. The unit seems to have shifted from eight to ten acres and the total villein land to have increased considerably. Mr. Hudson notes the existence of similar unnamed units of villein land in two extents which he publishes. In the manor of Banham in 1281, out of 32 customary tenants who together held 244 acres of arable, seven had 7 acres each and five others had multiples of 7; in the manor of Bradcar in Shropham six customary tenants in 1298 had 8 acres each and the seventh had 6 acres.[2]

If Norfolk units of villein tenure, even though unnamed, seem to have existed in the twelfth and thirteenth centuries, it may appear fanciful to insist upon the absence of the term virgate in descriptions of them. They might well enough, it will be said, have been called virgates or half-virgates. By midland extent-makers, indeed, the terms were sometimes applied to the Nor-

[1] Neither extent is dated, but none in the series bears a date later than 1252. That of Ringstead is followed by a court roll of 1240, which seems to be later than the extent, for in it Stephanus Clericus recognizes that he holds his land " in lanseagio," a dependence which has not been admitted in the extent (*Cartulary of Ramsey Abbey*, i. 411).

[2] William Hudson, " Three Manorial Extents of the Thirteenth Century," Norfolk and Norwich Archaeol. Soc., *Norfolk Archaeology*, xiv. 11, 8.

folk unit as they were not by the resident Norfolk population. Accustomed as men of the midlands were to calling the full villein holding a virgate, they not unnaturally persisted in the usage when they came to speak of East Anglia. There are several instances in the Ely cartulary.[1] Usually it is made clear that the term is merely a substitute for the " plena terra," which turns into a virgate under our eyes. At Derham, as at several other places, the customary tenants who hold plenae terrae (" de operariis plenas terras tenentibus ") are forthwith called virgate-holders.[2]

That this use of " virgate " was, however, imported rather than native seems conclusive from the usage of two large groups of early documents, records which, drawn up within the county, furnish most of our information regarding early units of land-holding. These are the feet of fines and the Domesday returns. In the fines of midland counties the virgate constantly recurs. In Norfolk and Suffolk, however, an examination of several hundred of the earliest fines reveals the term only in connection with one village, Walsoken, which, situated on the Cambridgeshire border in the fen country, was organized by virgates, like its midland neighbors.[3] The Domesday usage is the same: in connection with no East Anglian manor except Walsoken is the term virgate used to designate a villein holding.[4]

[1] The Ramsey cartulary also once uses the term virgate in connection with the two Norfolk manors, but this happens in a brief summary of all the manors of the abbey in which the attribution of hides and virgates brooks no interruption. Since this summary is contemporary with the detailed thirteenth-century extents which explicitly declare that virgates are unknown in Brancaster and Ringstead, it is obvious that the virgates crept in through hasty cataloguing. " At Brancaster 40 acres make a virgate, 4 virgates make a hide; at Ringstead 30 acres make a virgate, 4 virgates make a hide " (*Cartulary of Ramsey Abbey*, iii. 213).

[2] Cott. MS., Claud. C XI, ff. 221 (Derham), 233b (Shipham), 209 (Pelham), 248 (Bridgham).

[3] " De dimidia virgata terre et de tertia parte dimidie virgate terre " (Pedes Finium, Case 154, no. 180, 4 John).

[4] As printed by the Pipe Roll Society, two other fines mention virgates: one from Riston is concerned " de duabus virgatis terrae et dimidia et tribus bovatis terrae (*Feet of Fines*, xvii. 22); the other, from Upton, relates to a dispute between Stephanus de Ludington and Robert le Wile " de i virgata terrae " (ibid., 35). These fines are in all probability wrongly assigned by the Public Record Office cataloguer to Norfolk. They date from the first year of Richard's reign, when

This peculiarity of nomenclature, this avoidance of the name appropriate to the midland unit, is thus at once early and persistent. It points to some fundamental difference between the East Anglian and the midland servile holding, a difference that can hardly have lain in the nature of the services exacted from the respective tenants; for, although East Anglian obligations seem in general to have been lighter than those of the midlands, they were similar in kind. May not divergent field systems have been reflected in the usage? Just as the Kentish unit avoided the midland name because the iugum was essentially unlike the virgate, may not the East Anglian eriung, plena terra, or tenementum have done so for the same reason?

Besides emphasizing the early distinction between midland and East Anglian field systems, the above excursus into nomenclature has disclosed something about the earliest appearance of the East Anglian unit. A villein holding, the area of which was uniform in a given township, is revealed in the Ely extents of the thirteenth century, where, too, it is nearly always named. It is discernible, though unnamed, in the Ramsey extents of the twelfth century. In the same century, however, the unit sometimes assumed the name by which it was later designated at Martham; for in an extent of Stephen's time there is record of a holding of three "ariunges," our earliest specific reference to an East Anglian unit of villein tenure.[1] For Domesday is noncommittal. Frequently as it speaks of iuga or virgates in other counties, in the description of Norfolk and Suffolk (except at Walsoken) it carefully avoids reference to any units except hides and acres. Since the acres of the survey are never parcelled out to the villeins on a manor, we cannot tell whether there existed in 1086 the unnamed units which had taken form at Ringstead and Brancaster some seventy-five years later.

the name of the county is often missing from the fine, as is the case in both these instances. There is another Riston in Yorkshire (a land of bovates) and there are several other Uptons. The Upton in question was probably not far from Luddington, with which Stephen was connected. Of the three Luddingtons in England not one is in or near Norfolk. On the other hand, Luddington in Lincolnshire is only some twenty miles distant from an Upton in the same county.

[1] *Cartulary of Ramsey Abbey*, iii. 285.

At this point our evidence fails, leaving us in the twelfth century with an East Anglian unit of villein tenure which did not exactly resemble either the midland virgate or the Kentish iugum. It was not, like the former, a group of small arable strips divided evenly between two or three fields; nor is it certain that it was always, like the latter, a compact area. At Wymondham a few tenementa were more or less compact, and at Martham several of the strips of an eriung seem to have been not far distant from one another. Yet, as shown by the thirteenth-century survey of the latter township, the large number of strips in the eriung and the probable disparateness of some of them make us hesitate to believe that as a rule the eriung assumed the form of an undivided parcel of land. Probably it was sometimes compact, sometimes a group of not widely-scattered parcels. At times it resembled the Kentish iugum; at other times it was such a holding as a Kentish tenant would have had after the subdivision of iuga had begun, many of his parcels still lying in the ancestral iugum, while others, which had been acquired, were dispersed throughout neighboring iuga.

In what way can such an aspect of the East Anglian eriung or tenementum be explained ? Was this unit affiliated more with the virgate of the midland system or with the iugum of the Kentish system ? Before answering this question, we must give attention to the intimate connection which existed between the location of the parcels of the tenementum and the pasturage arrangements prevalent in East Anglia. The early custumals, we have noticed, usually record whether a tenant had or had not his own fold (*sua falda*), whether he might or might not pasture his sheep upon his own fallow acres. It may be that the attention which they give to this matter points to a greater development of sheep-raising in East Anglia than elsewhere in England; it is more likely, however, that it signifies a superiority in agriculture. Arable fallow was naturally better fertilized when sheep were folded regularly upon it than when the township herd and flock wandered aimlessly over it every second or third year, as they did in the midlands.

But to comprehend East Anglian pasturage arrangements one has to consider another factor than agricultural method, namely, the manor. Throughout the midlands, as Maitland pointed out, manor and township tended to coincide.[1] Even if there chanced to be two or more manors in a township, they all adapted themselves to the two- or three-field system precisely as did a single comprehensive manor: demesne horses, cattle, and sheep roamed over the waste and over the fallow field along with the beasts of the tenants. In East Anglia, however, the existence of several manors within a township was the rule rather than the exception, a rule, indeed, which tended to be almost universal.[2] Furthermore, as we have seen, the manors of a township insisted upon individuality in pasturage arrangements. Except during the autumn and winter seasons, the flock of sheep which each maintained was not allowed to range over the unsown lands with the flocks belonging to the other manors of the township; it was restricted to its own fold-course, where it enjoyed exclusive privileges. Such particularism, antagonistic as it was to action by the whole township, proved irreconcilable with the practice of the two- and three-field system of tillage.[3]

It thus appears that pasturage arrangements in East Anglia, so far as they had to do with fold-courses, were bound up with the co-existence of two or more manors within a township. If we may assume that fold-courses were as ancient as the manors to which they appertained, it becomes possible to form conjectures about the time of their origin. The petty manors of East Anglia are everywhere apparent in Domesday Book.[4] In that record, too, Norfolk and Suffolk boast of many " commended " (i. e. slightly attached) freemen, to whom may naturally be

[1] *Domesday Book and Beyond*, pp. 22, 129.

[2] Ibid., p. 23. Miss Davenport notes that in 1086 in the hundred of Depwade, Norfolk, every township with possibly one exception was held of more than one lord (*Norfolk Manor*, p. 7).

[3] Whether this particularism in pasturage had any connection with the determination of what constituted a manor in East Anglia cannot be here discussed, but in view of the vexed state of this latter question the consideration of such a possibility is not unworthy of attention.

[4] Of the 659 Domesday manors of Suffolk, 294 are rated at less than one carucate and only 70 at five or more carucates. Cf. *Victoria History of Suffolk*, i. 369.

referred the other feature peculiar to East Anglian pasturage arrangements — the privilege, namely, of independent foldage. From the character of the Domesday record, therefore, it seems possible to infer that the fold-courses of petty manors and the particularist foldage of certain tenants may have been existent prior to 1086.

We may now return to the question of the origin and affiliation of the eriung or tenementum. The foregoing digression relative to the pasturage arrangements of East Anglia has served to suggest a connection between the agrarian system there developed and the small manors and numerous freemen of pre-Domesday times. May there not also have been a connection between these same manors and the East Anglian unit of villein tenure? The hypothesis deserves consideration, despite the difficulties which it at once encounters. For two views are current regarding the general relationship to the manor of the Anglo-Saxon unit of villein tenure. In the opinion of some writers this unit antedated the manor and represented the original holding of one of the households of a free village community; when the manor was imposed upon this community, the holdings suffered change of status, not change of form.[1] The contrary opinion is that the persistent uniformity in the size of these holdings within a township points to a landlord's activity.[2] Without discussing this question in its wider bearings, or accepting the latter opinion in the form in which it was stated by Seebohm, we may here note that a fusion of the two views offers a tenable hypothesis relative to the origin of the East Anglian tenementum.

This unit, as has been explained, often in the thirteenth century assumed the appearance which a Kentish holding took on at some time after the disintegration of the iuga had set in. Assume, now, that there were once in East Anglia units like the Kentish iuga. Assume that they were divided among heirs and that some of the new tenants acquired parcels in other iuga, as they did in Kent. Assume, finally, that while the new holdings were in this condition a manorial system was imposed upon them.

[1] Maitland, *Domesday Book and Beyond*, pp. 337–338.
[2] Seebohm, *English Village Community*, pp. 176–178, 419.

It would be natural for the new lords to desire uniformity of size in the units from which rents and services would henceforth be due. What more natural, then, than that they should discard the antiquated and perhaps forgotten iuga and assess their tenants on the basis of actual holdings ? To equalize the areas of these holdings so as to make them full units or half-units it would only be necessary to shift a few parcels here and there. Some holdings may have been found compact and may have been left so. The outcome of such a readjustment would be tenementa and eriungs like those met with in the thirteenth century. Conjectural as this hypothesis is, it explains more simply than any other the aspect and characteristics of the East Anglian unit of villein tenure. If it be accepted, the tenementum becomes a derivative of the Kentish iugum, the result of an arrest in its disintegration and the making permanent for a time of the stage of decline then reached.

There remains the question whether any unusual event in East Anglian history may have contributed to the break-up of an ancient iugum and perhaps have had something to do with the formation of the manorial system which, in accordance with the foregoing hypothesis, created the new units and the new pasturage arrangements. For answer there must be further resort to conjecture. Domesday Book, as has been noted, shows us that the petty manors and numerous freeholds of East Anglia were in existence earlier than 1086. That these features are in no wise to be attributed to the Norman Conquest is apparent from the assumption of the survey that the conditions which it describes go back, in general at least, to the time of the Confessor. Before this date the most pronounced social revolution which Anglo-Saxon East Anglia experienced was the Danish invasion. That the Danes came in sufficient numbers to make permanent settlements is proved by the place-names of the region. To the Danes also is probably to be attributed the larger free element in the population which in 1086 still persisted here, as elsewhere in the Danelaw.

In a well-settled area, such as East Anglia undoubtedly was before the coming of the Danes, the intrusion of a considerable number of new settlers, who were also conquerors, must have wrought agrarian changes. Foremost among the problems which would naturally arise was that of providing the new-comers with land. One readily surmises that the humbler among the invaders became small freeholders, and that the more powerful came into control of many acres along with the tenants already settled thereupon. From the latter appropriation arose the petty manor. Upon the new lords — Danes, or perhaps at times Anglo-Saxons who had profited by disturbed conditions — fell the task of rating the holdings of their new tenants with an eye to uniformity of size within each manor. To them, in short, was due the creation of East Anglian tenementa and eriungs.

One naturally asks why incoming Danes brought into existence in East Anglia a unit different in aspect from the virgate and bovate found elsewhere within the Danelaw. The reply is that the midland system of Lincolnshire and Northamptonshire was much the same system as prevailed in Scandinavian countries.[1] Danes and Anglo-Saxons agreed in their method of tilling township fields. Hence when the Danes settled in northeastern England there was no need of a readjustment, either on the part of freemen or on the part of conquerors who may have developed into manorial lords. No difference, therefore, would in the future be perceptible between the field system of the northern Danelaw and that of Wessex. In East Anglia, however, the Danes probably found a field system divergent, then as later, from that of the midlands. To this they adapted themselves, being without doubt the minority of the population. It was, like their own, a system of open fields, and at the time of their arrival had become one of scattered parcels. In temperament and customs they were not hostile to the process of subdivision and dispersion, and they may even have contributed to the disintegration which after the re-rating once more set in throughout East Anglia. But how far the responsibility for this later movement rests with them is uncertain and does not particularly affect the hypothesis sketched above. According to that hypothesis, to state it once more in taking leave of the subject,

[1] Meitzen, *Siedelung und Agrarwesen*, i. 22.

the East Anglian field system was in origin similar to the Kentish, but was so modified before the Norman Conquest through the settlement of the Danes and the formation of the manorial system that by the thirteenth century it had developed pasturage arrangements and a unit of villein tenure peculiar to itself.

CHAPTER IX

THE LOWER THAMES BASIN

THE four counties which lie between East Anglia, Kent, and the circuit of the midland system, together forming what may be called the basin of the lower Thames east of the Chilterns, are Surrey, Middlesex, Hertfordshire, and Essex. To the north this basin is drained by the small rivers Colne, Lea, Roding, and by the coastal streams of Essex; to the south by the Wey and the Mole. For the most part it is sharply bounded by high hills. The Surrey downs stretch from Croyden southwest to Aldershot, while high-lying heath and the forest of Windsor extend northward to the Thames. On the northwest and north the Chiltern hills and foothills continue the boundary to the corner of Essex, whence it is no longer upland but the river Stour flowing down to the sea. Although the basin of the lower Thames is not strictly conterminous with the four counties mentioned, it is nearly so. The exclusion of a strip of Surrey on the south and of the edge of Hertfordshire on the north is compensated for by the inclusion of southern Buckinghamshire and a patch of Bedfordshire. The region is practically that which must have been occupied by the East Saxons and Middle Saxons in the sixth century.

In its field systems this area differed somewhat from the Kentish, East Anglian, and midland districts, but borrowed characteristics from each. The unit of villein tenure was in general not the iugum or the tenementum (although there are interesting exceptions), but the virgate. This midland feature, however, was such in name rather than in reality. The virgates here did not consist, as they did in the midlands, of parcels equally distributed between two or three fields; instead their parcels lay irregularly throughout several furlongs, shots, fields, or crofts. In this irregularity they approximated to fifteenth-century Kentish holdings and East Anglian tenementa. Although the region was more or less homogeneous in these respects, it will be best to present the evidence county by county and then attempt to make certain generalizations.

Surrey

ARABLE open field in Surrey persisted until the period of parliamentary enclosure, the reporters to the Board of Agriculture in 1794 estimating the total at some 12,000 acres. The largest amount in a single township was the 800 acres at Epsom, while the townships which had more than 350 acres apiece numbered only about a dozen.[1] Although the open-field arable thus constituted no large fraction of a township, parliamentary awards often refer as well to considerable stretches of waste.[2] At Ewell, for instance, Slater, following the act of 1801, reports that 1200 acres were enclosed.[3] The award and map, enrolled in 1803, show the enclosing of down on the south and of common on the north (Chessington common) amounting to some 350 acres, but the arable allotted was not more than 600 acres.[4] The map fails, as enclosure maps so often do, to indicate the old field names or arrangements. Even were these given, we should be disinclined to accept them as representative of an early field system, since the arable constituted so small a fraction of the township's area.

The reporters to the Board make statements which seem to ally Surrey tillage with that of the midlands. In a general way

[1] W. James and J. Malcolm, *General View of the Agriculture of the County of Surrey* (London, 1794), pp. 45–50. The reporters' list is as follows: from Carshalton to Sutton and Cheam, 3000 acres; Ewell, 600–700; Epsom, 800; Ashted, 700; Fetcham, 150; Bookham, 450; East and West Clandon, 300; Merve and Horsehil, 510; Egham, 300; Hythefields, 250; Thorp, 350; Mortlake, Putney, Wandsworth, and Battersea, 1340; Runnymead, 160; Yard Mead and Long Mead, 100; Weybridge and Walton meadows, 350; Send Common Broad Meadow, 365; Scotches Common Broad Meadow in Send parish, 50; Send Little Mead, 70.

[2] The award and map for Croydon are among the few that have been printed: J. C. Anderson, *Plan and Award of the Commissioners appointed to Enclose the Commons of Croydon*, Croydon, 1889.

[3] *English Peasantry*, p. 301.

[4] The map reveals the northern half of the township entirely enclosed, while its open field lay compactly in the southern half, stretching toward the downs. Where arable and downs met on the east, another enclosed area of some 200 to 300 acres was marked off as North Loo Farm. The award is in the Public Record Office.

they remark upon the similarity between the open fields of the county and those of other counties, describe the three-course rotation, and even mention the tripartite division.[1] The conclusion of their account, however, shows that they were not at the moment describing what they saw. Mankind has at length, they say, become "more thoughtful and more enlightened," and has "changed somewhat of the mode" of cultivation. The description is intended to be historical and general, the reporters assuming that the three-field system, which in their day they still saw farther up the Thames, had once prevailed in Surrey. This natural assumption we should likewise make were the earlier evidence in accord with it. Since, however, the testimony of surveys and terriers conflicts with the conjectural but seemingly straightforward account of the reporters, it will have to be given in some detail.

Somewhat voluminous is the careful transcription of numerous Surrey terriers drawn up in 1–2 Edward VI and probably relating to monastic lands.[2] Regarding many townships we learn of little more than the existence of common fields, the specification being that so many acres lay " in communi campo " or " in communibus campis."[3] In the longer terriers no holding is evenly divided between two or three comprehensive fields, as would surely have happened several times in the description of an equal

[1] "According to the common field husbandry of this county [which is similar to that of other counties] ... very little or no variation could take place; and therefore wheat, barley, and oats have been the uniform routine, and their chief aim has been to get the wheat crop round, be the ground rich or poor, shallow or deep. The custom of each manor in the arable lands for the most part was to lay them in three common fields; and in so doing they were enabled to pursue a course of wheat, barley or oats, and the third remained in fallow.... But as mankind became more thoughtful and more enlightened, finding the bad effects of this sort of husbandry, and being precluded the advantage of winter crops; seeing also the absurdity of fallowing, they wisely made an agreement among themselves (wherever they could possibly effect it) and changed somewhat of the mode by the introduction of the artificial grasses" (*General View*, etc., p. 38).

[2] Land Rev., M. B. 190; Treas. of Receipt, M. B. 168, 169.

[3] Of this nature are terriers relating to West Cheam, West Molesey, East Molesey, Esher, Waddington (in Coulsdon), Malden, Witley, Claygate, Pirbright, Lambeth, Ashstead, Eashing, Shalford (Land Rev., M. B. 190, ff. 107, 38*b*, 48, 40*b* and 117, 68*b*, 170, 130*b*, 136*b*, 138, 156, 189, 225*b*, 226).

number of midland holdings.   Parcels are, to be sure, sometimes located in fields, but this is done in the most incidental manner. Several typical terriers in which " fields " are distinguished from furlongs make clear this characteristic.[1]

[1] At Kingston a leasehold is described as a toft and three roods, " cum vi acris terre arabilis, viz.,
> iii acre iacent in quodam campo vocato le Combefeld in diversis parcellis
> et i alia acra iacet in quodam campo vocato le litlefeld in brokefurlong
> et i alia acra iacet in eodem campo apud le Chapelstyle
> et sexta acra iacet in eodem campo in tribus partibus " (Land Rev., M. B. 190, f. 163).

At Sutton there was a freehold " tenementum cum xiii acris terre arabilis eidem tenemento pertinentibus iacentibus in communi campo in diversis locis, viz.,
> iiii acre insimul iacent subtus le halle
> et iii acre terre insimul iacent in le Russhemede . . .
> et iii acre [et] dimidia insimul iacent in australi campo de Sutton apud Suttonsplott
> et ii acre [et] dimidia iacent apud le Fowleslowe " (ibid., f. 62b).

At Ewell the 6½ acres which are appurtenant to a tenement and garden " iacent in diversis particulis in campo vocato Southefeld, unde
> due acre iacent ex parte occidentali vie ducentis de Ewell versus Bansted
> et alia acra iacet ex parte orientali vie ducentis de Ewell versus Reigate iuxta Chillisbusshe
> alia acra iacet in Estmarkefurlong inter terram . . . et terram . . .
> alia acra iacet in Southelong inter terram . . . et terram . . .
> dimidia acra iacet apud Balardespit
> sexta acra iacet inter viam regiam ducentem de Ewell versus Reygate . . . et aliam viam ducentem versus Walton " (ibid., f. 91).

At Walton-upon-Thames a leasehold of 2 Edward VI consisted of 15¼ acres of pasture and woodland and 34¼ acres of arable.   Two-thirds of the arable lay in closes, but 10½ acres were " in diversis parcellis in lakefield."   A freehold in the same township consisted of a tenement at Payneshill, 7 acres in crofts, 5½ acres of meadow, and " in communi campo vocato Lakefeld
> iiii acre et dimidia apud Hokebusshe
> alie iiii acre
> i alia acra iuxta le Lake
> iii acre et dimidia apud Guldford corner . . .
> ii alie acre ibidem
> iii rode abuttantes super Stonyhill " (ibid., f. 73b).

At Worplesdon a freehold comprised a messuage with 1½ acres adjoining, and
> " ii acras terre ibidem iacentes in communi campo vocato le Greate Worthe . . .
> ac dimidiam acram terre in dicto communi campo . . .
> ac unam acram et dimidiam in communi campo ibidem vocato le Litleworth " (ibid., f. 170b).

Another freehold in Worplesdon had " xi acras terre arabilis divisim iacentes in ii^bus communibus campis ibidem vocatis le Greateworth et le litleworth."   In the first lay 10½ acres in five parcels, in the second one half-acre parcel.

At Weybridge three fields recur in two copyholds.   In the first we hear of a messuage with 5½ " acras terre arabilis divisim iacentes in communibus campis, unde
> iiii acre iacent in Wodhawfeld
> et i acra dimidia iacent in le Townefelde
> ac etiam iiii acre terre iacent in Pircrofte
> iiii acre terre et pasture in Townegaston
> ac ii acre et dimidia prati " (ibid., f. 32).

Only from Kingston have we a terrier which gives such a description of acres as would imply a two- or three-field system; and even here nothing but chance and the smallness of the holding are responsible, since other Kingston terriers mention other fields. At Sutton a South field, it appears, figured beside three equally important nondescripts. At Walton all the acres lay in a single field, Lake field, and at Ewell all were similarly in South field. Worplesdon had, besides its Great Worth, where were most of the acres, its "Litleworth", contributing to each holding a parcel or two. Even if, by accepting the doubtful Pyrcroft, we posit three fields in the Weybridge terriers, we find the acres apportioned with no symmetry, Townefeld having fewest in one holding and most in the other. At West Clandon, where again there were three fields, East field received 4½ acres in contrast with 3 acres assigned to Tonge field, 1¾ acres to West field, and 1¼ acres to miscellaneous areas.

It may be objected that, since all the holdings just mentioned were leasehold and freehold, or, if copyhold, were not estimated in virgates, irregularity in field arrangements might naturally be expected to appear. The objection would be valid were it possible to discover in Surrey any instances of symmetrical arrangement against which, as against a background, the foregoing

Here only the first two parcels seem to lie in the common field, but the other copyhold suggests that Pircroft is to some extent common. Appurtenant to a tenement called Hudnetts one finds a small close and

"ii acras terre arabilis iacentes in Wodhawfeld
iii acras et dimidiam terre arabilis iacentes in campo vocato Pyrcroft
iiii acras terre arabilis divisim iacentes in Townefeld" (ibid., f. 35b).

At West Clandon a freehold tenement has appurtenant three gardens, a croft, and " xi acre terre arabilis quarum

iii acre iacent in quodam campo vocato Tongefeld
dimidia acra in Northehill
dimidia acra in Southehill
una roda terre in Westfeld
una roda in Bassettehawe apud Hordunstile
due acre et dimidia in Estfeld super culturam vocatam Northefore
dimidia acra in eodem campo vocato Estfeld super culturam vocatam longrowe
dimidia acra in eodem campo super culturam vocatam Shuldmere
dimidia acra in eodem super culturam vocatam Shelfegate
dimidia acra in eodem campo super culturam vocatam Pyrrewe
dimidia acra in Westfeld super culturam vocatam Litledean
dimidia acra in eodem campo vocato Westfeld super culturam vocatam threyerden
dimidia acra in eodem campo super culturam vocatam Westlongland " (ibid., f. 14b).

terriers stand out as exceptions. Not only do such instances fail to occur in this series of surveys, but four detailed descriptions of virgates confirm the evidence just given.

Dating from the fifteenth and sixteenth centuries, these descriptions are as satisfactory illustrations of the appearance of Surrey holdings as can be desired.[1] Each virgate was appurte-

[1] At Epsom the virgate which was held with the tenement called "Synotes" by copy of 1 Henry VIII comprised, along with two parcels which may have been added: —

> " i croftam terre vocatam marters
> et iiii acras terre in Chotley
> unam acram terre apud Whiteweshill
> ii acras terre apud Wersdeynsknoll
> et iii acras terre apud Hadbrought
> et iiii acras terre in communi campo
> unam acram iacentem apud Churchefurlong
> et i acram et dimidiam iacentes in Gorybroke
> et i acram [et] dimidiam iacentes in Middlefurlong
> et dimidiam acram apud Werisdenknoll nuper Johanis Hellowes
> et ii acras terre iacentes in Mysden pertinentes ad nuper officium Coquinari " (Land Rev., M. B. 190, f. 61).

At Battersey in 1 Edward VI a tenement had appurtenant " dimidiam virgatam terre et prati, unde

> una acra terre iacet apud Tyethbournehawe . . .
> due acre contigue iacent in Longstrete . . .
> et alia acra terre iacet in Croche . . .
> una acra iacet in Stonyland . . .
> dimidia acra terre iacet in le Grotton . . .
> ii acre terre iacent separatim in medmeney . . .
> una roda prati iacet in Sladonditch " (ibid , f. 16).

In 31 Henry VI the abbot of Chertsey granted to William Frydey at Chobham " unum mesuagium, unum curtilagium cum dimidia virgata terre vocata Eyrey . . . in villenagio . . . unde

> due acre terre iacent in campo vocato Burifeld inter terram . . . et terram . . .
> et due acre iacent in campo vocato Beanlonde inter terram . . . et terram . . .
> et iiii acre iacent in campo vocato Gretestene
> et due acre iacent in campo vocato lytilstene [two acre-parcels]
> et una pecia terre continens ii acras iacet apud Estonlanende iuxta campum vocatum Gretestene
> Et unum pratum . . . et unum pratum . . . [no areas] " (Exch. K. R., M. B. 25, f. 264b).

In 33 Henry VI the abbot of the same monastery conveyed at Chertsey "unum cotagium et unum curtilagium cum dimidia virgata terre cum suis pertinentibus vocata proutfotes . . . unde

> predictum cotagium cum curtilagio iacet ibidem in vico vocato Eststrete
> et tres acre pariter iacent in campo vocato Estfelde in cultura vocata Syllyn . . .
> et dimidia acra terre iacet ibidem iuxta Coppedeheg' . . .
> et due acre terre pariter iacent in campo vocato Myllershe . . .
> et una acra iacet in campo vocato yonder Estworthe . . .
> et dimidia acra terre iacet in campo vocato heder Estworthe . . .
> et una acra prati iacet in prato vocato Estmede . . .
> et una alia acra prati iacet ibidem " (ibid., f. 221b).

nant to a messuage, and hence maintained a household; two were situated in river townships, two in the center of the county; all were in open field, and in all the parcels were largely arable. Yet in none was there any grouping of the strips by fields. At Epsom a "common field" appears once, but without further specification and on a par with hills and furlongs. At Battersey no fields or even furlongs are mentioned, while the field names are curious. At Chobham we get trace of four "fields," Burifeld, Beanlonde, Gretestene, and Lytilstene; but one of them, Gretestene, with an adjacent parcel contained half of the holding. At Chertsey the arable was divided among three fields, Estfeld, Myllershe, and Estworth, but in the proportion $3\frac{1}{2}$, 2, and $1\frac{1}{2}$. No one of the terriers therefore pictures a three-field system.

In two grants made by the abbot of Chertsey at East Clandon in 11 Henry IV we are able not only to follow the description of a virgate but to compare with it the account of a fourth-virgate, or "ferlingata."[1] The virgate was large, containing $13\frac{1}{2}$ acres

[1] Exch. K. R., M. B. 25, f. 284. The respective descriptions are as follows: —

"Unum mesuagium et unam virgatam terre vocatam Crouchers continentem Triginta quatuor acras et dimidiam terre que William atte Crouche quondam . . . tenuit in villenagio in Estclendone in diversis locis, unde

mesuagium et curtilagium iacent per viam Regiam ducentem usque Shure versus Ripplee
una crofta et quatuor acre et dimidia terre vocate Clausland iacent . . .
due crofte vocate puricroftes continentes tres acras terre iacent . . .
due crofte vocate northecroftes continentes sex acras terre iacent . . .
una roda terre vocata Shamelondesbutte iacet iuxta mesuagium predicti tenementi . . .
una acra terre iacet in Penstede . . .
dimidia acra terre iacet in Penstede predicta . . .
una alia dimidia acra terre iacet in eadem Penstede . . .
due acre terre vocate Swythecroftes iacent . . . inter . . . halvyncroft . . .
alia dimidia acra terre iacet in le Haluyngcroft predicto . . .
tres rode terre iacent in halewyngcroft predicto . . .
alia dimidia acra terre iacet in le Overshorebrude . . .
una acra terre iacet in le Overshorebrude . . .
una dimidia acra terre iacet in le Overshorebrude inter terram communem vocatam le Doune . . .
una alia dimidia acra terre iacet in le Nethereshorebrude . . .
alia dimidia acra terre iacet in quodam forlango vocato horeslowe . . .
alia dimidia acra terre iacet apud Lytelhegge . . .
una acra terre vocata Cowshoteacre iacet . . .
tres rode terre iacent apud Longedenesende . . .
una acra terre iacet apud Coppedthorn . . .
dimidia acra terre iacet sub le Coppedthorn . . .
una roda terre vocata le houstedell iacet inter terram vocatam Stonycroft . . .
una alia roda terre iacet in Scoldmere . . .
una alia roda terre vocata Rokeyerdmele . . .

in crofts and 21 acres of common land lying in thirty-one parcels. A few of the parcels were in the same field areas — three in Penstede, three in Overshorebrude, two in Halvyngcroft — but most were disparate and without inclination to group themselves by fields. The ferling, which was about one-fourth as large as the virgate, appears relatively less enclosed. Noteworthy about its parcels was their location in field divisions that are not once named in the description of the virgate. The symmetry and uniformity which might be expected under three-field arrangements were thus entirely wanting.

Longest of the Chertsey terriers, four of which have just been quoted, are those relative to Egham, which lies on the Thames just east of Windsor Park. In them two virgates and three half-virgates granted by the abbot in 2 Richard III are described in full.[1] The length and breadth of the strips are often given, e. g., " i pecia terre in Northcrofte . . . longitudine xxvi perticas, latitudine in utraque parte iii perticas, continens dimidiam acram, dimidiam rodam, viii perticas." More comprehensible than a transcript of the holdings is a tabulation showing the parcels arranged by fields and permitting a comparison of the virgates. Each holding includes a messuage, a curtilage,

> dimidia acra terre iacet apud Thistelford . . .
> una parva butta terre vocata Pilchebutt' iacet . . .
> dimidia acra terre iacet supra Bradvor . . .
> una acra terre iacet in le Stonycroft . . .
> dimidia acra terre iacet super bradvor . . .
> tres rode terre iacent super Wowefor . . .
> dimidia acra terre iacet super le Westhulne . . .
> una acra terre iacet super le Westhulne . . .
> una parva pecia terre continent' [sic] unam acram et dimidiam vocatam le Sturtes iacet . . .
> dimidia acra terre iacet super le Northfor . . .
> tres rode terre iacent apud Godhurne. . . . "

> " Unum toftum et unam ferlingatam terre . . . in villenagio . . . continentem septem
>   acras terre et unam rodam unde
> predictum Toftum continet unam acram dimidiam terre . . .
> una alia acra terre iacet in Rogersdene . . .
> una alia acra terre iacet in le Shorebrude . . .
> una alia acra terre iacet super le Westeshulne . . .
> una alia acra terre iacet apud le Merk . . .
> una roda terre iacet apud le Merk predictum . . .
> dimidia acra terre iacet super le Inlond . . .
> alia dimidia acra terre iacet super le Middelfor . . .
> una alia acra terre iacet super longworthe."

[1] Exch. K. R., M. B. 25, f. 238b.

and a group of strips the areas of which in acres are as follows: —

Field Divisions	Virgata Gilberti et Johanis Morcok	Virgata Matildis de Bakeham	Dimidia Virgata Williemi at Well de Inglefeld	Dimidia Virgata Agnetis at Well	Dimidia Virgata Johannis at Well
Ermersh	1, $\frac{1}{2}$, $\frac{1}{4}$, $\frac{1}{8}$, $\frac{1}{4}$, $\frac{1}{8}$, $\frac{1}{2}$, $\frac{1}{2}$, $\frac{1}{4}$ 1 pecia prati 1 pecia terre arabilis	1, 1	$\frac{3}{8}$, $\frac{3}{8}$	$\frac{3}{8}$, $\frac{1}{4}$	$\frac{3}{8}$, $\frac{3}{8}$
Southcroft ..	1, $\frac{3}{4}$, $2\frac{1}{2}$ (enclosed) $\frac{1}{2}$, $\frac{1}{4}$, 1, 1, 1 pecia	....	$1\frac{1}{2}$	....	$1\frac{1}{8}$
Northcroft ..	$\frac{3}{4}$, $\frac{3}{8}$	....	$\frac{1}{4}$	$\{\frac{1}{4}, \frac{3}{4}, 1\frac{5}{8}, \frac{3}{8}, \frac{3}{8}\}$	$\frac{1}{4}$
Burgcroft ...	....	....	$1\frac{1}{2}$	....	$1\frac{1}{2}$
Hillarshe	1 pecia	....	$4\frac{3}{4}$	$4\frac{3}{8}$	$4\frac{3}{4}$
Miscellaneous	$\frac{1}{2}$ apud Sidenehill $1\frac{1}{2}$ apud le Swell 1 pecia iuxta grangiam $1\frac{1}{4}$, $\frac{1}{4}$ unspecified	$\{\frac{1}{2}$ Medilgrove $\frac{3}{4}$ Langfur $\frac{1}{2}$ Shortfer$\}$	....	....	2 "prepresture"
Crofta inclusa	$\{1\frac{3}{8}, 1\frac{1}{4}, \frac{7}{8}, 1\frac{1}{2}, \frac{1}{2}$ gravetta$\}$	$4\frac{1}{4}$, $3\frac{1}{4}$	$\{$1, 2 (terra et bruerium) $4\frac{7}{8}$ (terra et bruerium) $1\frac{3}{8}$ (terra et bruerium)$\}$	$1\frac{3}{4}$, 1	$\{\frac{1}{2}$, 1, $6\frac{1}{8}$, $\frac{5}{8}$ (pastura)$\}$
Pratum and "more"	1 pratum, 3 more	....	$\frac{7}{8}$	....	$1\frac{1}{4}$, $\frac{5}{8}$

Every holding, it thus appears, comprised several enclosures, but the largest part of each still consisted of parcels intermixed with those of other men. Although the township lies near the Thames and the name Ermersh may suggest marsh land, most of the unenclosed parcels seem to have been arable. Those of the several holdings were very unevenly distributed among the field divisions. The half-virgates had large parcels in Hillarshe, but not so the virgates; one holding had eight parcels in Southcroft, another had five in Northcroft, the others not more than a parcel or two apiece in these areas; Burgcroft appears but twice; the first holding had ten parcels in Ermersh, the others only two each. The location of many parcels in " crofts " suggests a subdivision of old enclosures rather than normal open field. Considered together, therefore, the Egham terriers not only fail to evince any trace of a three-field system, but even seem to be prohibitive of such an arrangement.

Since complete surveys are always more convincing than terriers, any comprehensive evidence of this sort available from Surrey is important. Somewhat late, to be sure, is the description of Banstead, which in 1680 pictures little of the township unenclosed.[1] Five of the tenants were then possessed of a few acres vaguely assigned to " the common field," the area of which proves to have been only about 24 acres. The fields may to some extent have been reduced in size, since we find mention of " Upper, Middle, and Lower Common field Closes "; but it is not clear that they had ever been large.

Another survey, earlier by three-quarters of a century, records all the holdings at Byfleet and Bisley.[2] Byfleet, a township adjacent to the river Wey, was then entirely enclosed. At Bisley there is note of a few small common fields, the combined area of which was about 100 acres and nearly all of which fell within a dozen copyholds.[3] Most acres were in Neltrow and Widcroft, a few being in Burcroft. Since in nearly all of the copyholds the enclosed area exceeded in amount that which was open, from an agricultural point of view it mattered little that

---

[1] H. C. M. Lambert, *History of Banstead in Surrey* (Oxford, 1912), pp. 194–216.
[2] Land Rev., M. B. 203, ff. 80–133.
[3] Their areas in acres may be tabulated as follows: —

Customary Tenants	Enclosed	Neltrow	Northill	Widcroft	Southash	Burcroft	Unspecified	Common Meadow
Robt Cobbett, m..	½	7	..	1	..	..	..	2½
Edm. Bonsey	...	..	..	..	..	..	9	
Martha Lusher, 2 m.	11¾	3	..	2½	2	1	..	2½
Jo. Symons, m. ...	21½	2	..	2	..	1		
Robt. Cobbett ...	7½	..	3	2½	..	1	..	1
Wm. Farnham, m..	10¾	9 acres in these three			..			
Jo. Hone, m.	11½	2	..	2	..	2	..	2
Henry Lee, m. ...	16½	3	..	5	2	2	..	1
Henry Lee, m. ...	...	2	..	5	1	2	..	1½
Joseph Hone, m. ...	8	4	2	3	1	1	..	1
Henry Rutter, m..	14½	..	..	..	..	..	10	3

there was no symmetrical distribution of acres among the field divisions. The common fields of Bisley were at this time probably similar to those of many Surrey townships when the formal enclosure of these took place in the eighteenth or nineteenth century. Nor had the Bisley fields changed much during the two centuries preceding the survey, if we may judge from the mention of three of them in an indenture of 6 Henry IV.[1] At Bisley, as at Egham and East Clandon, irregular field arrangements thus antedated the sixteenth century.

Not all sixteenth-century common fields in Surrey were so meagre as those of Banstead or Bisley. A field-book written in a hand of about 1600 describes furlong by furlong all the parcels of open field in "Keyo and West Sheen alias Richmond," the total being some 650 acres. By tabulating and summarizing the information there given, we get what is perhaps our best view of relatively extensive open fields in the county at the period in question. All holdings larger than five acres are noted in a schedule in Appendix VI.

The smaller holdings, which averaged about 1½ acres and consisted of from one to three parcels, numbered nearly thirty. Each of them lay in one division of the township's arable, a characteristic not indicative of a midland field system. Nor for the larger holdings was there a general arrangement by fields, the furlongs instead having a substantial importance. After being told about Kew field and Kew heath, we come upon the "lower field," in which there were at least two shots, and possibly more. Thereafter we are guided upward and southward only by furlongs, since "East field" was no more than a shot. To discover any simple field system governing the distribution of acres is difficult. Kew field was of interest to only three tenants, one of whom had nothing in the Richmond furlongs, and a second but little. If we disregard Kew field and try to arrange the remaining furlongs in

---

[1] Exch. K. R., M. B. 25, f. 264. Of the five acres of arable from which tithes were owed by a certain John Willere,

"una acra iacet in Campo vocato Northull
et una acra et dimidia pariter iacent in Campo vocato Wydecroft
et una acra iacet in Campo vocato Eltrowe
et una acra et dimidia pariter iacent in Campo vocato Westeworth."

three groups, the following combination is probably the most feasible: —

	The Lower-field furlongs (three)	"The shott butting upon the park west" and Church furlong	Upper Dunstable to Maybush shot inclusive
Sir Henry Portman	0	3¼	8¼
Will. Portman, Esq.	[16½]	24	28¼
John Burd, iure uxoris	[7¼]	2	2¾
Stephen Pierce, Gent.	9¾	19¾	41¾
Vincan Jones, Gent.	28	30½	24¾
— Payne, Gent, iure uxoris	24	29¾	33½
Robt. Clarke, Esq.	14½	10	13½
Geo. Charley, Gent.	5	8¾	10
Mary Crome, vidua	6	12¼	11
Lady Wright	5	3¼	6½
Barth. Smith, iure uxoris	2	3½	6¾
Lott Peerce	5	1	2
The Church land	[2]	2	2¾
Thos. Smith	0	4½	5¾

In the case of three tenants, Jones, Payne, and Clarke, this grouping would make a three-field system not altogether impossible, but elsewhere the misfit is complete. Any other arrangement of furlongs, whether by three or four fields, is equally futile. The irregularity of the Richmond field system at the end of the sixteenth century seems pretty clearly demonstrated.

From a survey of 1522, earlier by three-quarters of a century than that of Bisley or than the Richmond field-book, we have the items which relate to the manor of Merstham.[1] Although the holdings here were tending to accumulate in the hands of a few men, they are still differentiated in the survey. Usually at least half of each lay in open field. When, however, we begin to examine the location of the constituent acres we at once encounter difficulties. For there were no comprehensive fields. The parcels of the larger holdings lay in as many as twenty field areas, often called furlongs, the amounts assigned to each being usually from one-half acre to three acres;[2] and no grouping of these furlongs to form any kind

---

[1] The extracts were copied in 1710 from a "Rentall of the Lordshipps of Merstham and Charlewood," and have been printed in *Surrey Archaeological Collections*, 1907, xx. 94–114.

[2] The following holding, though not of the longest, is typical: —

of system can be other than highly conjectural and inconclusive. It is evident that at Merstham no emphasis was put upon the midland combination of furlongs into three large fields; the furlongs possessed rather an independence and flexibility which admitted of any arrangement desired.

If we turn now to our earliest sources of information regarding Surrey fields, the fines and charters of the thirteenth century, we shall get plenty of evidence that open fields were usual, but none pointing to the existence of a regular field system. Various fines dating from 10 Richard I to 19 Henry III locate the small parcels of the transferred lands in such way as to leave no doubt that open-field strips were in question.[1]

In some terriers, furthermore, the location of strips is instructive. At Walworth a lease of 17 Edward II enumerates twenty parcels containing $20\frac{1}{4}$ acres of arable which lay in Ellenebussh, Lolipette, Longewygheth, Fowes, and Southcroft.[2] In four instances, indeed, it is possible to discover in a measure how the parcels which constituted a virgate were disposed. At Mitcham a half virgate and six acres are defined as " illam medietatem que ubique iacet in campis de Inlond, Bery, Battesworth, Burforlang, Spirihey, Westbroc, versus umbram . . ."[3] At Carshalton ten acres taken from a virgate comprised two acres in Hodicumbe and

"William Holman for a tenement, garden, and croft on the backside called Barkleyes containing by estimation 2 acres and a halfe . . .
    for half an acre in North Deane in the common Feild . . .
    And for one yard in Swynk Furlong . . .
    And halfe an acre in North Worth . . .
    And for one yard in Towneman Meade . . .
    And 3 yards in Tottbury Bush shott . . .
    And for 2 acres in a peece in Crooked Land . . .
    And for one acre in Heyforlong . . .
    And halfe an acre there . . .
    And for halfe an acre there in Ashtedd . . .
    And for another halfe acre there . . .
    And halfe an acre in Tottbury Hill Furlong . . .
    And for halfe an acre in Tottbury Bush Shott . . .
    and for halfe an acre in Little Bosefeild Shott . . .
    and for one acre in the upper shott in Quarrepittden . . .
    And for one acre in Great Oate Croft . . ." (ibid., 106).

[1] Such open-field strips are attributable to Camberwell, " Bechom," " Maudon," Kingston, and Thorp: Ped. Fin., 225-1-44 (10 Rich. I); 225-2-2 (1 John); 225-3-44 (14 John); 225-4-13 (3 Hen. III); 225-4-21 (3 Hen. III).

[2] MSS. of the Dean and Chapter of Christchurch, Canterbury, Lib. B, f. 35b.

[3] Ped. Fin., 225-9-30, 19 Henry III.

three in Hegecroft, while the others were situated at Stikelehelde, Twiseledeweie, Westhehe, Cwernherst, and Netherathe.¹ At Polestede, in a transfer of two virgates, one was described as " aliam virgatam . . . eiusdem Phillippi, scilicet,

duas acras terre et dimidiam que iacent versus austrum subtus viam que est inter Polestede et losne
et duas acras in Westden et in Coster... una
v acras dimidiam in bromhell versus boreal'
unam acram et iii acras in Estden et in Melherse
tertiam partem unius acre et i acram prati et tertiam partem prati subtus polested et capitale mesuagium . . ." ²

Such curious and varied descriptions of the parcels of a virgate indicate that in the thirteenth as well as in the seventeenth century the open fields of Surrey between the downs and the Thames were not divided into two or three or four large fields among which the acres of a holding were equally distributed. The midland system was not in vogue, and the reminiscent history of the reporters to the Board of Agriculture is not sustained by contemporary evidence. The fields were numerous, were curiously named, sometimes being called furlongs, and the distribution of the acres of a holding among them was irregular. What the affiliations of this unsymmetrical system were can best be discussed after a study of neighboring counties has been made.

Before we leave Surrey it should be noted that on the Kentish border a virgate in the early documents does not resemble one which lay in the plain to the north and west of the downs. In the high rolling country between Croydon and Reigate a virgate often seems to have been a more or less compact parcel of land, with no scattering of the acres. At Banstead the 24 acres which were granted from a virgate lay " in Snithescroft." ³ At Sanderstead the fourth part of a virgate was " unum campum terre . . . et quinque acre in hadfeld quas Ricardus filius Swein essartavit." ⁴ At Gatton the half of two virgates may perhaps have been slightly more disparate, comprising as it did

[1] Ped. Fin., 225-3-4, 5 John.
[2] Ibid., 225-1-41, 10 Rich. I.
[3] Ibid., 225-2-8, 1 John.
[4] Ibid., 225-2-15, 1 John.

" septem acras terre in Neweland
   et duas acras terre et dimidiam ad Horscroft
   et duas acras terre et dimidiam ad Kinerod
   et dimidiam acram ad Wudappeltre
   et medietatem curtilagii . . . in eadem villa quod vocatur
      Chapmanhag." [1]

The existence of compact virgates in this region need not imply that there were no common fields. A Coulsdon rental of 11 Henry VII specifies several crofts, "cum aliis terris in communibus campis de Wentworth, Churchden, prestyslond." [2] Other documents furnish a clue to the nature of these fields, and any one who has seen the bald chalk downs of the neighborhood can surmise what was their character. In a lease of 9 Henry VI there is mention of " viginti acre terre pariter iacentes in campo vocato Wenteworthe "; [3] and in a Coulsdon charter of 18 Edward II 48½ acres of arable are described as lying " in communi campo in loco qui vocatur Toldene iuxta ferthyngdoune." [4] The common fields were, it seems, nothing less than the slopes of the downs, in which parcels were likely to be large; and the very frequency with which the fields were named " dene " points to the same conclusion. Fields of such a character in eastern Surrey help to explain the tendency of virgates in that region to lie in a few parcels, or even in a single parcel. The virgates did, indeed, begin to take on somewhat the aspect of Kentish iuga, with which, as we shall see, they were probably allied.

Hertfordshire

WHAT has to the modern student become the typical three-field township of England, is, with no inconsiderable irony, located in a county not characterized by the three-field system. Had Seebohm gone ten miles to the south or to the east he would have found no field arrangements like those of Hitchin. For it happens that the long northwestern boundary of the county falls within the midland area and just beyond the hills that bound

---

[1] Ped. Fin., 225-5-25, 8 John.
[2] Exch. K. R., M. B. 25, f. 330.
[3] Ibid., f. 347.
[4] Ibid., f. 336.

the valley of the lower Thames. In consequence there was in Hertfordshire a fringe of townships, of which Hitchin was one, as typically three-field or six-field as anything to the north.

In this region it was that the reporter to the Board of Agriculture in 1794 noted the persistence of open fields. " The land [of the county] " he says, " is generally inclosed, though there are many small common fields . . . which are cultivated nearly in the same way as inclosed lands; the larger common fields lie toward Cambridgeshire." Almost all of Ashwell he found unenclosed.[1] This township, along with Hinxworth and one or two other places, is a projection of Hertfordshire between Bedfordshire and Cambridgeshire, belonging topographically with the latter counties and like Hitchin falling within the midland area.[2] Adjacent to Ashwell is Kelshall, a plan of which, made at the end of the eighteenth century, apparently for purposes of enclosure, shows six large open fields stretching northward from the village to the heath, which lay on the Cambridgeshire border.[3] Not far away, on the northern slope of the hills, lay the manor of Lannock in the parish of Weston. Here, too, an early seventeenth-century description of the demesne divides it among three fields in the midland manner.[4]

[1] D. Walker, *General View of the Agriculture of the County of Hertfordshire* (London, 1794), pp. 48, 52.

[2] A Hinxworth terrier in an early seventeenth-century hand, describing the lands " which Bray holds to Calldecott farme," enumerates them as follows (Add. MS. 33575, ff. 46–48): —

In the Windmill Field or Clay field	3¾ acres in	6 parcels
In Waller field	11 "	" 22 "
In Bennill field	1 "	" 1 "
In Saltmore field	12¼ "	" 19 "
In Blackland field	26¼ "	" 52 "

This enumeration does not indicate clearly the character of the field system. The township may originally have had two fields, one of which is here represented by Blackland field; or the terrier may be incomplete, since it begins very abruptly; or, once more, the farm which Bray held may have had enclosed lands which permitted the irregular distribution of parcels throughout the fields.

[3] Add. MS. 37055. The fields were Baldock Way, Crouch Hill, Stump Cross, Sibbern Hill, East Little, and Beacon.

[4] After specifying the *scitus manerii* of 17 acres and a woodland of 35 acres, the account continues relative to the arable: " Que quidem terra arabilis dividitur in tribus Seysonibus [the *culture* being designated]. . . . In illo campo quod iacet iuxta

Another projection of Hertfordshire, comprising the townships of Long Marston, Puttenham, Wilston, and Tring, runs into the midland area between Bedfordshire and Buckinghamshire. The open fields of all these townships except Puttenham were enclosed by an award of 1799.[1] According to the enclosure map, Long Marston had three fields (Langdale, Mill, and Lolymead); Wilston had three, somewhat subdivided, to be sure, but still clearly discernible (Near East and Far East, Bennell, Lince Hill with Blackmoor and Moor Hill); and Tring had seven, apparently grouped as Dunsley and Parkhill, Hazely and Gamnill, Hawkwell and Hitchin and Gold.

Turning now from the three-field edge of Hertfordshire to the body of the county, what field system do we find? Evidently one which is irregular in much the same way as that of Surrey. In so far as the holdings of a township lay in open field, the fields were many and there was no symmetry in the distribution of parcels among them. Indeed, at a fairly early date certain townships contained no open field whatever. In 28 Charles II a survey was made of the manor of Hemel Hempstead with its members, Flanden, Eastbrook, Boxhamstead and Bovingdon, a large area in the Chiltern region of the west.[2] All the parcels are expressly stated to have been closes, and there is no trace of open field. In an earlier survey of 1607 relating to Berkhamstead and its neighbor Northchurch most of the parcels appear as closes, though a half-dozen common fields are mentioned. Of these the two most often named probably lay in Northchurch, an indication that even at this early date Berkhamstead was almost, if not quite, enclosed.[3]

If we pass northeastward along the southern slope of the Hertfordshire hills, we come successively to Little Ayott, Kings Walden, Weston, Clothall, and Ardeley. The appearance of

viam que ducit de Sheneffeld versus Wylie et abuttat super boscum de Langenoke . . . [are] ccxxi acre iii Rode et vi acre de novo adquisite. In Seysona de Gravelefeld . . . ciiii^{xx} et v acre et Roda et x perticatas. In Seysona de Duxwellefeld . . . cclx acre preter iii perticatas " (Add. MS. 33575, ff. 57-58).

[1] K. B. Plea Ro., 45 Geo. III, Mich.
[2] Land Rev., M. B. 216, ff. 39-70.
[3] Ibid., M. B. 365, ff. 1-25.

each of these townships is described in sixteenth- or early seventeenth-century documents, all of them implying the existence of open fields but never the existence of a two- or three-field system. At times the amount of open field was very slight. In 1636 " a coppie of the Survey of Little Ayott " (probably relating to demesne) took account of 83 acres of woodland, 64 acres of park land, and twenty-three closes containing together 329 acres, while in Church field there were but 19 acres in twelve parcels and in Nellwyn field but 24 acres in nineteen parcels. Only one-ninth of the tillable lands here still lay open.[1]

From Kings Walden the field detail contained in three terriers is far more explicit. The latest, dated 1654, rehearses the " particulers of the landes liing in the Common Feildes belonging to the Berry and Parsonage Farme taken out of former notes with some additions ";[2] another of 1568 relates to all the "copyehold londes of John Camfyld holden of the manor of Kings Waldon ";[3] the third is a valuation, in an Elizabethan hand, of the possessions of Sir William Burgh, knight.[4] The Burgh estate comprised a manor house, 266 acres of enclosed land, 80 acres of woodland, and " clxx Acres of Arable lande lieing in sondrie peeces in divers fieldes of Kinges Walden, Powles Walden, and Polletts." These open-field acres (except the ten in Powles Walden and Polletts), together with the parcels of the two other terriers, are shown in the table on the next page. No uniformity is perceptible in these terriers, except in the two larger holdings, which show a preponderance of acres in Mill field. Since in both holdings the acres in question comprised more than one-third of the total but less than one-half, Mill field can hardly have been one of three fields. Especially would its slight representation in Camfyld's holding tell against such an hypothesis, while the location of one-third of this copyhold in Howcroft once more precludes any simple three-field arrangement.

On the top of the Hertfordshire hills is situated the parish of Weston, the northern slope of which, constituting Lannock

[1] Add. MS. 33575, f. 23.
[2] Ibid., f. 242.
[3] Ibid., f. 141.
[4] Ibid., f. 4.

Location of Common-field Parcels	Berry and Parsonage Farm	Acres of John Camfyld's Copyhold	Sir William Burgh's Estate, 170 Acres
In Hadden field	21 acres in 2 parcels	½, ½, ½, ½	20⅛ acres
" Hadden dell shott	....	½, ½, ½	
" Mylle field............	66 acres in 17 parcels	½	73 "
" Royden field	30 " " 11 "	½, ½, ½, ½, 1	30¼ "
" Legates field	9½ " " 10 "	....	22 "
" Hanger field...........	2¾ " " 3 "	....	5 "
" Fogman field..........	....	....	8 "
" Fogman downe	11¼ " " 5 "	....	6½ "
" Floxmoore field........	....	....	1 acre
" Wind mill field........	10¾ " " 11 "		
" Wooden	4½ " " 5 "		
" Breadcroft	2¼ " " 2 "	½	
" Landmead downe	6 " " 4 "		
" How croft	....	½, ½, 9¼	
" Astyge	....	1, 1	
" Woodden valye	....	2	
" Woodden hyll	....	½	
" Sandy shote...........	....	¾, 1, ½, ½, 1	
		Two crofts of 3 and 2 acres	

manor, we have seen disposed in three fields. Weston itself, however, appears to have departed from this arrangement. A mid-seventeenth-century terrier of two tenements of " Mr. Faireclough his land in Weston " names eleven closes of 74 acres at the beginning of the list, seven more of 140 acres at the end, with the following items between: —

	Acres	Roods	Perches	
Frontley Feild:				
A hedge Row	0	2	0	44–1–7
In the upper shott	1	0	0	
a peice next weevers mead	13	1	17	
peace tree pightell	1	2	33	
another close more there	3	3	36	
Woodgate shott	23	3	1	
Ye corner [5 parcels]				13–2–34
Lince Feild:				
Most part of whitehill furlong	13	0	1	73–0–15
in the second furlonge..............	3	0	14	
a peice more there	5	2	0	
another great peice more there	44	0	0	
[7 other pieces]	7	2	0	
Fitks grove shott, the upper shott, neitherdown shott .. and walkerne shott [= 102½ acres] with 4 other pieces				115–3–30[1]

[1] Add. MS. 33575, f. 317.

Despite the bracketing, this arrangement scarcely bespeaks three fields. Even if " ye corner " be added to Frontley Feild, the sum of the acres is not half so great as the area of the last group. Two other items from Weston seem decisive. In a rental of 11 Charles II there is mention *inter alia* of

"Dimidium unius virgate terre vocate Bondsland in a feild called Lince Feild [and]

Unum aliud dimidium virgate knowne by the name of Daviesland iacentis in severall parcells liing in Lince Feilde aforesaid." [1]

Since half-virgates in three-field townships do not usually lie entirely in one field, we must conclude that the township of Weston marks the transition from the midland area, as represented by Lannock manor, to the region of irregular fields.

Adjacent to Weston are Clothall and Ardeley, the latter lying farther down on the southern slope. A terrier of the manors of Kingswodebury and Mundens, situated in these two parishes, is dated 5 Edward VI.[2] The demesne, which lay very largely in Clothall, consisted of 452 acres of "londs, medowez, Fedyngs, wods, and pasture," together with

" erable londes in Sheldon felde in the
    parishe of Ciothall............ 28 acres in 8 parcels
erable acres of lond in the Westfield
    of Clothall"................. 38½ acres in 13 parcels.

The manors comprised also "londs in the occupation of Thomas hawez," which with the exception of some 20 acres lay in the parish of Ardeley. The Ardeley acres were located indifferently in fields or furlongs, the terms being apparently interchangeable.[3] At Clothall and Ardeley, as in the other townships along the south-

[1] Add. MS. 33575, f. 315.
[2] Add. MS. 33582, ff. 4-9.
[3] Netherwykdane furlong, ½, ½, ½, ½, ½ acres; Snaylesdel, ½; Lodley felde, ¼, ½, ½, 1, 1, 3½, ½, one pightell; Brokefeld (or furlong), ½, 2, one little pightell; Dockelond, ¾, ½; Scalfurlong, ½, ½; Depewellshott, ¾, ½, ½; Holmshot furlong, ½, ½, ½, ½, ½; Newellfeld, ½; Hoggyswelfurlong, ½, ½, ½; Asthill (furlong or field), 1, ½, ½, ½, ¼ (meadow); Banbery feld, 1½, ½, 1½; Kybwellfeld, 1, ¾, 1; Rybrade, 1 (a headland); Scotesden furlong, 1; Colecrofte, [?].

ern slope of the Hertfordshire hills, there is thus in the terriers no trace of a three-field grouping.

Leaving the hilly townships of the north and passing to the more level region of the southeast, we come upon the large manor of Ware, about which much information is given in a sixteenth-century book of manorial jottings. If from this collection of court rolls, rentals, and incomplete surveys we select three terriers of typical copyholds,[1] we shall find that the three had a common interest in Wykfeld only. The first had two acres there, the last four; but the other parcels of their open field lay disparate. Most suggestive of the terriers is the second, in which all the acres except two were in Breckelfeld, a proximity reminiscent of the East Anglian system.

[1] "Robertus Forde . . . tenet sibi et heredibus suis . . . per copiam datam . . . Anno regni Regis Henrici VIII^{vi} xxxvi^{to}

> unum croftum terre vocatum Gallocroft continens . . . quinque acras
> unam actam terre iacentem in Warefeld inter le borne et salmonscroft
> et duas acras terre in Abelyingstok
> et duas acras terre iacentes in Wykfeld subtus wolkechen hedge
> et unam acram terre iacentem apud waringehowgate
> et unam acram terre inclusam apud le gravel pytts."

"Christoferus wright tenet per copiam curie datam xiii^{tio} die Junii anno secundo Regis Edwardi vi^{ti}

> unum tenementum vocatum Hills iacens et existens in Baldockstrete . . .
> unam acram terre in Brekelfeld iuxta le Claypitt
> unam aliam acram terre in eodem campo iuxta Dunsell erosse
> duas acras terre in eodem campo nuper Thome Peerse . . .
> tres acras terre in eodem campo abuttantes super le Peertre
> et duas acras terre in campo sive clausura iuxta wyken lane."

The third terrier details the surrender, in 20 Henry VIII, into the hands of the lady of the manor, of

> "unum tenementum cum gardino adiacenti apud werengo hill,
> decem et octo acras terre arabilis particulariter in diversis locis iacentes, unde
> quinque Acre iacent in duobus peciis in West feild
> et quatuor acre terre in wykfeld subtus parcum domine
> et duo crofta insimul iacent et inclusa tenemento predicto Annexata continentia novem acras terre insimul
> cum tribus acris prati in parco de ware predicto
> ac tres acras terre in pippeswell field inter terram . . . et terram . . .
> et unam acram prati in parke meade quondam Johanis Ostwyke . . .
> unum pratum continens tres acras terre vocatum Sondlese. . . .
> Ac unam peciam terre continentem quinque acras . . . iacentem apud Goodyerefeld inter Ripariam . . . et terram domine . . .
> et quinque acras terre vocatas Lokeholme . . . iacentes inter Ripariam et Riponam . . .
> Necnon decem acras terre subtus parcum domine in Dymershott
> ac etiam tres acras terre iacentes in wykfeild vocatas Ladymere" (Add. MS. 27976, ff. 67b, 89b, 47).

Still farther in the direction of London is Cheshunt, a parish of the Lea valley adjacent to Waltham abbey. Relative to two manors lying mainly in this parish but extending beyond it up and down the river, we have surveys of 19 James I.[1] In the manor of Periers and Beaumond, reaching into Wormley and Tunford on the north, the larger holdings were leaseholds, and for the most part consisted of parcels of pasture and meadow with a few acres in the common meadows. In the other manors of Theobalds, Crosbrook, and Collins, lying towards Waltham, there was still much open field. Here, too, the larger holdings were held by lease, the copyholders having only messuages with at best bits of land attached. The most important leaseholds are summarized in Appendix VI in a schedule which accounts for all the Cheshunt open field. No fewer than 250 arable acres of the manor were common and unenclosed, while the irregularity of field arrangements is perceptible at a glance. As at Edmonton, which was just down the valley of the Lea in Middlesex, a tenant usually had parcels in two or three fields;[2] but the two or three were seldom the same. No one of them had the prominence which " le Hyde " had at Edmonton, though Holbrooke was favored. Such surveys well illustrate the field situation which was likely to be found just to the north of London at the end of the sixteenth century.

If we turn from sixteenth- and seventeenth-century evidence to that of the thirteenth and fourteenth centuries, we shall find the same line of demarcation in the county. One of the townships north of the hills in the Bedfordshire plain is Hexton, where an early charter of Walter de la Ponde bestowed 23 acres of land upon St. Albans abbey. After describing eight parcels which contained 12 acres, the charter gathers under the rubric " Et in alio campo " the remaining twelve or thirteen parcels.[3] At the time, therefore, Hexton was in two fields, as were several townships in this part of Bedfordshire.[4] Another Hertfordshire

[1] Land Rev., M. B. 216, ff. 16–38.
[2] Cf. below, p. 381, and Appendix VI.
[3] Cott. MS., Otho D III, f. 152b. At the end of the charter the manuscript is injured.
[4] Cf. below, Appendix II.

village situated in the midland plain appears to have had three fields. In 1297 the demesne lands at Norton were so grouped for a three-course husbandry that the totals by "seasons" amounted to 64, 68, and 66 acres;[1] and the assignment of this demesne to such areas as Westfeld, Neitherwestfeld, and Stokefeld renders it likely that open field is referred to.

At Kensworth, however, which lies on the crest of the hills, the existence of a similar situation at an early time is less credible. A lease from there, dated 1152, provides, to be sure, that on its expiration the lessee " reddit eis [canonicis Sancti Pauli] totum bladum lxx acrarum de hiemali blado seminatorum et similiter totum bladum lxx acrarum de vernali blado seminatorum et quatuor xx acras warectetas."[2] There is, however, nothing to show that these acres, which were probably demesne,[3] lay in three common fields. Indeed, certain later evidence tells against the existence of such fields at Kensworth; for terriers of several freeholds and copyholds of the time of Henry VII, one of which describes a half-virgate, show no three-field grouping of parcels.[4]

[1] " Ad seysinam unam pertinent in campo qui vocatur Neitherwestfeld xxiv
   acre terre arabilis et in bokefeld continentur xl acre
      ad secundam seysinam pertinent in stokefeld xxxvii acre, in Cellenelond
      xv acre, in Sondishot xiv acre, in Lepescroft ii acre
      ad tertiam seysinam pertinent in Westfeld xxiii acre, xxxiv acre, i acra, et
      viii acre " (MSS. of the Dean and Chapter of St. Paul's, Lib. I, f. 150).
[2] Ibid., Lib. L, W. D. 4, f. 35.
[3] At so early a date only demesne acres would be leased.
[4] Of the six larger copyholds three are summarized in the following table (from MSS. of St. Paul's, Press A, Box 62). The fourth description shows that the freeholds were not dissimilar to the copyholds: —

" Thomas Albright senior tenet unum mesuagium et xii acras unam rodam terre ac libertatem bosci pertinentem ad dimidiam virgatam terre,
    iii acras, scilicet, iacentes eidem mesuagio iuxta Basse Croft
    iii acras abuttantes in dictum Wodgrove
    iii acras iacentes super Stokkyng hille
    i acram iacentem apud longemere
    iii rodas iacentes super le lynches super Stokynghill
    i acram extendentem ad Croucheway
        dimidiam acram in eadem cultura abuttantem super eandem viam.

" John Ellam tenet unum toftum et xxiii acras terre arabilis, [scilicet],
    x acre iacentes inter terram Thomam Albright et terram . . .
    vii acras super Blakehill . . .
    vi acras iuxta le Kensworth down.

Of three Hertfordshire townships to the north of the summit of the hills, one thus lay in two fields in the thirteenth century, another probably in three, while the Kensworth evidence stops short with showing us a three-course rotation upon demesne lands.

South of the crest of the hills the early evidence changes, two fractional virgates being so described as to imply that parcelling out of virgates among fields was not there usual. A fine of 9 John relative to Wheathamstead transfers " illam quartam partem unius virgate terre cum pertinentibus que iacet versus austrum et occidentem a via que tendit de molendino de B[atford] usque ad mesuagium ipsius Rogeri."[1] If it be urged that this quarter-virgate may have been demesne, the same objection will not apply to the description of a half-virgate at Tewin. The account is contained in a charter, copied into an early fifteenth-century cartulary of St. Albans, by which " Adam filius Walteri, parsone de Aete, [conveys to] Sanson de tebreg pro homagio et servicio suo dimidiam virgatam terre in thebreg, illam, videlicet, dimidiam virgatam terre que iacet inter terram predicti Sansonis et terram Roberti de thebreg' sub parco de Aiete versus le su . . . Reddendo inde annuatim . . . duos solidos."[2] Homage, service, and rent imply customary land, and the whole description, like the one preceding, suggests that the plots of land in question were undivided.

It should by no means be assumed that all Hertfordshire virgates or fractions thereof were at an early time compact areas.

" Thomas Affrith tenet x acras terre arabilis divisim iacentes in campis de Kenesworth, [scilicet],

    iv acras apud le Greneway ducens a Kenesworth versus le Down
    i acram . . . apud Gatepath
    i acram et dimidiam . . . super Stokynghill . . . super terras Willielmi Flyndey
    i acram super easdem terras
    dimidiam acram iacentem iuxta terram domini la Zouche
    ii acras . . . iuxta terram Johannis Movell
    dimidiam acram in Crouchedene et est forera et extendit super Croucheway.

" George Ingleton tenet [libere] xi et dimidiam acras terre divisim iacentes, unde

    v acre similiter iacent iuxta le Spitleway
    iv acre iacent super Aldreht
    ii acre similiter iacent ibidem
    dimidia acra iacet in le Galowfurlong."

[1] Ped. Fin., 84-7-18.   [2] Cott. MS., Jul. D III, f. 64.

On the contrary, one may see how exceptional were the two just described by referring to several early terriers from the Essex border, all of which testify to the existence of intermixed parcels in open field. One of these, from Alswick, describing a half-virgate given in the thirteenth century to the priory of Dunmow, enumerates the acres as follows: —

" iv acras iacentes in scalchedelle
  et dimidiam acram et dimidiam Rodam que iacent [iuxta] terram Stephani decani in alio campo adversus gravam decani
  et duas acras et dimidiam Rodam que iacent in eodem campo in duas partes
  et acram et dimidiam et unam Rodam per se usque ad viam
  et ii acras dimidiam rodam minus [sic] inter Alwardeshei et Siwinesho
  iii rodas sub domo ẏvonis clerici iuxta terram sparche
  et unam acram in puse crofth
  et ii acras quae iacent in campo adversus pucheleshei inter terras brici et sparche
  et iii Rodas que appellantur Hevodacher
  et ii acras terre quae iacent circa dellam." [1]

Unmistakable characteristics of open fields are here visible: $17\frac{1}{2}$ acres were divided into ten parts, many of them small; two parcels were in the same field, another lay *per se*, and another formed a head acre.

The parcels of this terrier had, however, been subjected to no two- or three-field arrangement. Although evidence of such a plan might escape us in a single charter, it would scarcely fail to appear in the numerous terriers that are available. Such records the feet of fines, the cartulary of Waltham abbey, and that of St. John Baptist at Colchester supply in no grudging measure.[2]

[1] Harl. MS. 662, f. 67b.
[2] Typical illustrations from each of these sources follow: —
At Barkway, in the northeastern hills, Abbot Adam of St. John Baptist's, Colchester, conceded to Robert le Moine, about 1200,

" mesagium cum una acra terre et dimidiam [sic] iuxta fontem qui uocatur Bedewelle
  et aliud mesagium . . . cum dimidia acra
  et tres acras cum prato eiusdem latitudinis extra forum de Berqueia
  et unam acram et dimidiam penes Bruneslawe
  et duas acras et dimidiam iuxta Suineslauue
  et septem acras in Hocfeld
  et unam acram et dimidiam in campo de Ried
  et unam acram iuxta Tieuesstrate
  et octo acras que adjacent ad Tieuesstrate
  et unam acram et dimidiam super Malmhelle
  et tres acras que extendunt super Holeuueie
  et quinque acras in campo de Ried . . " (*Cartularium Monasterii Sancti Johannis Baptiste Colecestria*, ed. S. A. Moore, Roxburghe Club, 2 vols., 1897, ii. 630).

Unfortunately, however, virgates are seldom described so fully as they are in the instance just quoted, although the statement that a messuage accompanies the parcels often shows that tenants' holdings are in question. The characteristics of open field — small parcels in several localities and sometimes two parcels in the same furlong — are usually apparent in these terriers. Practically all of them, however, fail to group parcels according to two or three fields. The only item suggesting such a grouping is from Stanstead, where in the thirteenth century thirty arable acres were divided among three fields as follows: —

> "In campo qui vocatur Alfladesfelde quindecem acras
> et in campo qui vocatur Kyngesfelde decem acras
> et in campo qui vocatur Bokkeberwefeld quinque acras." [1]

Even these acres, it should be noted, were unaccompanied by a messuage and were unequally divided among the fields. Since this is only one of a long series of Stanstead grants, and since all the others apportion their acres unequally among fields, there is

At Munden a fine of 15 John enumerates six acres as follows: —
> "i acram in Netherlee iuxta essartum
> et ad chevicias illius acre unam rodam et dimidiam
> et Infra essartum duas acras et i rodam
> et in Hertwelleschote dimidiam acram
> et in campo qui vocatur Pucheleslee unam acram et dimidiam . . .
> et in eodem campo unam acram et dimidiam
> et in Buttes unam rodam et dimidiam
> et in Bradecroft duas acras et dimidiam
> et ad chevicias de Sewanesfeld unam rodam
> et in eodem campo de Sewanesfeld tres rodas cum forera
> et unam acram bosci " (Ped. Fin., 84-7-29).

A grant at Stanstead, typical of many that occur in the Waltham cartulary, runs as follows: —
> "Concessi viginti acras terre mee arabilis in villa de Stanstede, scilicet,
> duas acras terre et dimidiam cum mesuagio que sunt ex opposito molendini de Stanstede
> et tres acras et dimidiam super Ketteshell
> et unam et dimidiam ulterius in eodem campo
> et ex opposito de Ketteshell ex parte altera duas acras et unam rodam . . .
> et quinque rodas terre et dimidiam ultra le Newestrate . . .
> et duas acras et unam rodam ad quercum . . .
> et duas acras ulterius in eodem campo
> et unam croftam cum sepibus et fossatis continentem quatuor acras iuxta mesuagium Jordani Partryke . . .
> et unum mesuagium cum crofta quod vocatur hosgodeshamstall quod continet tres rodas . . .
> et tres acras prati mei et dimidiam " (Harl. MS. 4809, f. 146).

[1] Harl. MS. 4809, f. 147b.

no reason for interpreting the passage as evidence of the existence of a three-field system.

Hertfordshire testimony of the thirteenth century thus concurs with that of the sixteenth. Although from the first the county probably had numerous enclosures and considerable woodland, there was doubtless in the more open regions an abundance of open field. This field was, however, irregular in character, the parcels of arable, so far as can be seen, not being grouped by furlongs into two or three large areas. On the contrary, the fields were, as in Surrey, numerous, often curiously named, and presumably small. The origin and affiliation of such a field system can best be discussed in connection with similar questions regarding the other counties of the lower Thames valley.

Middlesex and the Chilterns

THE western half of Middlesex retained much open field until the period of parliamentary enclosure. Slater's list of acts includes the names of nearly all the townships of this part of the county, and considerable of the area tabulated must have been arable. From the eastern half, however, only two townships figure in his enumeration, Enfield and Edmonton. To the latter the enclosure act assigns 1231 acres, but a Jacobean survey makes it clear that not more than 500 of them can have been arable common field.[1] The 3540 acres mentioned in the Enfield act undoubtedly comprised a certain amount of arable, since the reporter to the Board of Agriculture in 1793 bewails the existence of "a large tract of common field land watered by the New River, at present condemned to lie fallow every third year."[2]

The Jacobean survey of Edmonton just mentioned illustrates well the irregular field system of eastern Middlesex. The village lies halfway between London and Waltham abbey in the valley of the Lea, not far from the point where the three counties of Hertfordshire, Middlesex, and Essex meet. Most of the numerous tenants held a few acres of customary land, although the

[1] Land Rev., M. B. 220, ff. 110–185.
[2] T. Baird, *General View of the Agriculture of the County of Middlesex* (London, 1793), p. 36.

messuages were often freeholds leased by them.[1] Fully half the township lay in closes, usually pasture land, and many tenants had parcels in the common marsh. Over and above the pasture and marsh there was considerable unenclosed arable divided into many fields, in most of which three or four tenants had parcels. These fields, numbering about a dozen, seldom contained more than twenty acres apiece. Typical of them were Langhedge, Okefeld, Hegfeeld, Dedfeeld, all of which appear in the holdings reproduced in Appendix VI. Only one open-field area, that called " le Hyde," was large and shared in by many tenants. It was quite normal for a tenant to have, along with his enclosed pasture, arable acres in le Hyde and in perhaps one other area, an irregular arrangement which was of course incompatible with a two- or three-field system.

This being the situation in eastern Middlesex, it only remains to inquire whether conditions were similar in the rest of the county. In the west the nearest approach to a three-field arrangement appears at Feltham. This township, situated in the plain of the Thames, on the highway from Staines to Hampton, is described in a survey of 2 James I.[2] At that time the fields were three, with names reminiscent of the midlands (Further field, Middle field, Home field), and the copyholds were divided with more or less equality among them. It may well be that this was a township cultivated in the midland manner.[3]

Elsewhere the evidence tells against the creeping of midland habits down the Thames. Cold Kennington, the village that gave its name to the manor which embraced Feltham, had not three fields but two, and in the holdings that are specifically described (four of the six are not) the division of acres between these

[1] A dozen of the copyholds have been summarized in Appendix VI.

[2] Summaries of the most important copyholds are given in Appendix VI.

[3] Slater's intimation that three fields were enclosed by the Cowley and Hillingdon enclosure act (*English Peasantry*, p. 287) should not mislead us. The petition for this act asks that " certain Common Fields called Cowley Field, Church Adcroft, and Sudcrofts " be divided between the two parishes as well as apportioned anew to the tenants and enclosed. These fields were, therefore, not those of a three-field township, but fields that chanced to be common to two townships. Nor are their names the usual ones for three important fields. Cf. *Journal of the House of Commons*, 21 Jan., 35 Geo. III.

fields was irregular.[1] An Elizabethan terrier of Harleston farm, which was the property of All Souls College and had its open-field lands in Willesdon, after enumerating the closes, proceeds with the open-field parcels, which it locates in seven fields or shots and in two meadows.[2]

From the thirteenth century, also, the specifications of Middlesex virgates fail to suggest two or three fields. At East Greenford a fine of 4 John describes a half-virgate as comprising 9 acres in Lukemere and 3 in the field "toward Bramte," while East field and West field receive only 1½ acres each.[3] At Laleham the fourth of a virgate is assigned in a thirteenth-century transfer to seven localities, four of which were furlongs.[4] To be sure, the Laleham enclosure map of 1803 dubs the small open field above the village North field and the large one below it South field;[5]

1 The survey is combined with that of Feltham, just referred to (Land Rev., M. B. 220, ff. 97–98). William Newmann had a messuage, a close of one acre, and 2½ acres of arable in Court field, 15¼ in West field; Anthony Taylor had a messuage, a close of four acres, and 2½ acres in Court field, 13 in West field.

2 All Souls MSS., Terrier 32 (1593). The distribution of open-field acres is as follows: —

Hungerhill, ¾, ¼, ¾, ¾, ½, ¼, ¼, ¼, ½, ½, ⅛, ½, ¼, ⅛, ¾, ⅛, ½, 1¼ [= 8⅜ acres]
Knowles shoot, ½, ½, ½, ¼ [= 1¾ acres]
Blacklands, ¾, ½, ½, ¼, ⅛ [= 2⅛ acres]
Fortune feild, ½, ½, ¼, 1, ½, 2, 1¼ [= 6 acres]
The great marshe mead, ¼, ¼, ¼, ¼, ⅜, ⅜, 2, 1, ⅛, ⅛, ⅛, ⅝, ⅛, ⅛, ¼, ⅜, 1½ [= 9⅜ acres]
The little marshe mead, 1, 1, 3, ½, ½ [= 6 acres]
Brontfeild, ½, 1, ¾, 1, ¾, 1, ½, 1, 1, 1, ½, ¼, ¼, ¼, ¼, 1¼, ⅛, ⅛, ¾, ¾, ½, ½, ¼, ½, ⅛, ¼, ⅛, ⅛ [= 15¾ acres]
Meesdonn field, 1, ½ [= 1½ acres]
Little manicrofts, ¼, ⅜, ⅜ [= 1 acre]

There is also a map showing the closes and open-field strips of the farm (Typus Collegii, ii, maps 18–22).

3 "vi acras terre in Lukemere versus occidentem
et in eodem campo tres acras versus orientem
et in campo qui se extendit versus bramte [?] duas acras versus astrum
et in eodem campo versus ecclesiam iiii particas terre pro i acra versus aquilonem
et in Estfeld i acram et dimidiam versus Horsendune
et in Westfeld i acram et dimidiam versus eundem Horsendune" (Ped. Fin., 146–2–24).

4 Ped. Fin., 146–3–22. The distribution is as follows: —
"In Langfurland tres perticatas terre ex parte Occidentali
et in Middelfurlang dimidiam acram ex parte orientali
et in Brocfurlang dimidiam acram ex parte Occidentali
et in Retherford dimidiam acram ex parte orientali
et in Brache unam perticatam ex parte orientali
et in Shelpe dimidiam acram prati ex parte Occidentali
et in Bottefurlang unam perticatam prati ex parte Occidentali."

5 C. P. Rec. Ro., 43 Geo. III, Trin.

but these names must have been merely topographical, since, apart from the divergence in the size of the fields, a township in this fertile region could hardly at any time have been tilled under a two-course rotation. Indeed, we know that in the near-by township of Sutton, held by the canons of St. Paul's, the three-course rotation usual on the demesnes of their manors was early employed.[1] It is probable that the Sutton demesne was unenclosed,[2] although the names of the divisions in which it lay, as well as the area assigned to each, preclude the midland system of two or three large fields.[3] Thus the earlier Middlesex evidence is in conflict with that of the Jacobean survey of Feltham, the three seventeenth-century fields of which township must have been exceptional.

If it be true that the midland field system did appear in the Middlesex plain, there is no doubt that the manifestations of it there were isolated from the midland area by the interposition of a different system, one which followed the Chilterns to the Thames and crossed it east of Reading. For the evidence from this Chiltern region regarding irregular fields is full and convincing. If we follow the river up from Windsor into the midland plain, we shall in so doing have an opportunity to observe Buckinghamshire and Oxfordshire townships on the one hand and those of Berkshire on the other.

The Buckinghamshire reporters to the Board of Agriculture stated that the occupiers of the common fields of Horton (500 acres), Wraysbury (200 acres), Dachet (750 acres), Upton (1500 acres), Eton (300 acres), and Dorney (600 acres), all townships lying near Eton, " have exploded entirely the old usage of two crops and a fallow and have a crop every year."[4] May not this deviation in 1794 from the three-course rotation which prevailed

[1] A lease of 1283 specifies that 44 acres were sown with corn and 18 with rye or mixtilion, 60 with oats and 12 with barley, while 64 lay fallow (MSS. of the Dean and Chapter of St. Paul's, Lib. I, f. 24).

[2] A measurement of 1299 attributes its acres to various quarentenes (ibid., f. 33*b*).

[3] There were in all 90 acres in Suthfeld, 47 in Breche, 9 in Hamstal, 36 in Estfeld, 9 in Northfeld, 66 in Westfeld, 22 in Eldefeld (ibid., f. 33).

[4] W. James and J. Malcolm, *General View of the Agriculture of the County of Buckingham* (London, 1794), p. 27.

elsewhere in the county have been facilitated by the existence of fields already irregular ? The reporters do not tell us when the change took place, but it may have been long before they wrote.

Most instructive in showing the character of Buckinghamshire open fields in this region is a survey of Farnham Royal made in 6 James I. Although by far the greater part of the manor was enclosed, some 250 acres of unenclosed field are described.[1] While there were four recurring common fields, the system was by no means a four-field one, nor was it even reminiscent of two fields. The acres were unequally apportioned, Hawthorne field receiving most of them and any field being liable to neglect. If in the eighteenth century the fields of all the townships round Eton were like those of Farnham two centuries before, transition from a three-course husbandry was invited by the location of the tenants' acres.

Ascending the Thames past Henley, we come into the small plain about Reading, where the valley widens just east of the main ridge of the Chilterns. Regarding sixteenth-century fields here we are instructed by two useful surveys of townships within five miles of Reading — those relating to Sonning, Berkshire, and Caversham, Oxfordshire.

Since the Sonning survey is arranged according to three tithings, the tenants and fields of which differ, we have in it, so far as tillage is concerned, the record of three independent townships.[2] The tithing of Okingham was practically enclosed.[3] A yard-land there usually consisted of some twenty enclosed acres, for the most part arable. At times there were from two to four acres in an open field, but such fields are too insignificant to merit attention. More open was the tithing of Wynnershe, several of the copyholds of which have been summarized in Appendix VI. Occasionally yard-lands were here enclosed (e. g., those of Agnes Astell and Robert Phillipps), but most of them had considerable arable and some meadow in the fields. This arable,

---

[1] The holdings which contain most of it are transcribed in Appendix VI.
[2] The tenants of the several tithings have rights of pasture in the same commons.
[3] Land Rev., M. B. 202, ff. 74–82.

apart from a few outlying acres, was disposed within a group of three fields (Demys, Whetershe, Benhams) or, if not in these, within another group of four (Stony, Goswell, Old Orchard, Rudges). Whichever group, however, a yard-land favored, among the fields themselves there was no equal apportionment of acres. If a three-field system employing six fields had ever been in force, here it had fallen into decay, a supposition which the presence of numerous enclosures renders not incredible. The third tithing, the one called by the parish name of Sonning, was not unlike Wynnershe.[1] Although at times a holding there was enclosed (e. g. the half yard-land of John Gregory), most of the cultivated land lay in open field. Of the four fields which most often recur, to the one called Bulmershe there was seldom assigned more than an acre, whereas Charfielde frequently received a greater number of acres than did all the remaining fields together. A three-field system can hardly have been constructed on such foundations.

Across the river from Sonning is Caversham, which the survey of 5 Edward VI pictures as already largely enclosed. The virgate holdings in this township show that frequently not more than between one-fifth and one-tenth of a tenant's land lay in the open fields.[1] Yet the fields were numerous, a dozen of them being mentioned and a half-dozen often recurring. Usually a holding had its acres in only three or four of them, and then with no regularity. Small fields which, like these, played so slight a part in the economy of a township could easily depart from any systematic cultivation without inconvenience to their tenants, and apparently those at Caversham had done so. If a three-field or a six-field arrangement ever existed there, it had disappeared before the middle of the sixteenth century.

Passing farther up the Thames, we reach the outposts of the region of irregular fields. These lay in Oxfordshire, either on the northwestern slopes of the Chilterns or in the bottom lands below. Watlington and Ewelm represent the former, Warborough and Bensington the latter. Typical holdings from Jacobean surveys of each of these four townships, which are situated near

[1] Illustrative holdings are given in Appendix VI.

one another, are summarized in Appendix VI. In all four open field largely predominated. The number of fields, however, varied from township to township, and the acres held by individual tenants were nowhere evenly divided among the fields. Ewelm perhaps approached most closely to the midland arrangement. Three open fields, Grove, Middle, and Church, frequently recur, and in three or four instances the tripartite division of the open-field acres of a holding was nearly achieved. In many cases, however, one or more of these fields are disregarded, while as many as a dozen others are mentioned. At both Watlington and Bensington there were about a dozen fields, in from one to seven of which the acres of a tenant might lie. No group of three fields stood prominently forth in either township, nor can a six-field arrangement be discovered. The fourth of the townships, Warborough, did to be sure, have six fields; but here too, if we try to combine any three with any other three, we shall get such improbable apportionments of tenants' acres as 4, $1\frac{1}{2}$, $8\frac{1}{2}$. Since there are several such inequalities for each equitable division, we are forced to consider the open fields intractable like those of the other townships. It may be added that nineteenth-century enclosure plans and awards concerned with these four places evince no regularity in field arrangements. To judge, then, from all the instances noticed above, it seems probable that the irregular fields of Surrey and Middlesex extended into the Chiltern region of the three counties to the west, and came to an end only when they reached the plain of southeastern Oxfordshire.

Essex

THE early field system of few English counties is so difficult to describe as that of Essex. At the time when records of it were first made, much of the county was already enclosed. The earliest evidence thus assumes peculiar importance, but since it is of a fragmentary nature it forbids any but tentative conclusions.

Like Kent, Essex was referred to in the sixteenth century as one of those counties " wheare most Inclosures be."[1] A descriptive

---

[1] John Hales, *A Discourse of the Common Weal of this Realm of England* (1549, ed. E. Lamond, Cambridge, 1893), p. 49.

rental of St. Paul's manor of Heybridge in 1675, and a plan of New College's manor of Homechurch Hall in 1662, show closes only,[1] and so do various accounts of late sixteenth-century conditions. A two-hundred-page survey of Westham, a short one of the manor of Lawford Hall, an excellent one of Eastwood Bury, a plan of four " tenements " in Woodham Ferrers, describe enclosures.[2] From the times of Henry VIII and Edward IV we hear principally of crofts in a detailed rental of Rivenhall, in the fragment of a survey of Sandon, in extracts from the court rolls of Crepinghall, in a description of tenants' holdings at Newhall in Boreham, and in a full account of the manor of Wikes.[3] Finally, fourteenth- and fifteenth-century terriers of the lands of various chantries at Colchester seem to be concerned mainly if not altogether with enclosures.[4]

Such evidence might raise the question, as it did in Kent, whether common arable fields ever existed, and the topography of the county might suggest that Essex was isolated from its western neighbors by stretches of forest through which open-field usages never found their way. It is true in a measure that the western boundary of the county was reinforced by tracts of forest. Toward Hertfordshire lay Hatfield Chase, toward Middlesex the wider reach of Epping. These forests, however, seem not to have acted as barriers to colonization or communication. To judge from the frequency with which Domesday hamlets were scattered throughout them, their settlement was not long delayed;[5] and the numerous possessions of Waltham abbey within the bounds of Epping at an early period indicate that communication with the home manor to the west cannot have been difficult. There is thus no topographical reason why western Essex should

[1] MSS. of the Dean and Chapter of St. Paul's, Press A, Box 62; Rawl. MS. B 311.
[2] Exch. Aug. Of., M. B. 425, ff. 1-113, 3 Jas. I; Add. MS. 34649, 1 Jas.; Rawl. MS. B 308, 8 Eliz.; Harl. MS. 6697, ff. 20-24, 21 Eliz.
[3] Rents. and Survs., Portf. 2/44, 7/47; ibid., Ro. 196, 3 Edw. IV; Treas. of Receipt, M. B. 163, f. 47.
[4] Philip Morant, *History and Antiquities of the County of Essex* (2 vols., Chelmsford, 1816), i. 150-158.
[5] See the Domesday map in *Victoria History of Essex*, i. 426-427; also W. R. Fisher, *The Forest of Essex*, London, 1887.

have been isolated from Hertfordshire and Middlesex in its field systems. Nor does it seem to have been. The dividing line in field usages, passed through the county rather than along the border, and set apart the northwest as a region indistinguishable from Hertfordshire in the aspect of its open fields. In the central and southeastern part of the county, however, different arrangements and possibly Kentish affinities are perceptible.

The northwest is a continuation of the Hertfordshire highlands, that here form part of the boundary of the midland plain. In Essex the river Cam, flowing northward, issues from the hills, which are noticeably lower than in Hertfordshire. From both the valley and the hill region we have several terriers that agree in demonstrating the prevalence of open fields in this part of the county. In every way these fields were similar to those of Hertfordshire, and especially noteworthy is the fact that the numerous parcels of a holding were never grouped as if lying in two or three large arable areas.

In some terriers the parcels were seldom larger than an acre and were widely dispersed throughout the fields. At Wenden, for example, the six acres which in 8 John formed part of a virgate were located in twelve places.[1] At other times several parcels fell within one of the open-field divisions, the names of which were of the most varied sort, often being reminiscent of hill and woodland. Fifteen acres at Arkesden which were given to the monks at Walden early in the fourteenth century are illustrative of conditions in the district. The twenty parcels were located as follows, the areas being in acres: —

"in campo qui vocatur Newey, 1, ¼
"in campo qui vocatur Mapeldeneswell, 1¼, ⅞
"in campo qui vocatur Apostolgrove, 2, ⅞, ½
"in campo qui vocatur Witedune, ¾
"in campo qui vocatur Blakedune, ¾, ¾, ½, ½, ½, ⅞
"in campo qui vocatur Stockyng, 1
"in campo qui vocatur Burgatesshot, 1
"in campo qui vocatur Sevenacres, ¾
"in campo qui vocatur Langeland, ¾, ½
"in campo qui vocatur Wyndemelnessot, ¼."[2]

[1] *Feet of Fines for Essex* (ed. R. E. G. Kirk, Essex Archaeol. Soc., 1899, etc.), i. 37 (no. 197).   [2] Harl. MS. 3697, f. 143b.

Elsewhere the parcels were larger, and might seem to have been closes were it not that the field names and the assignment of more than one parcel to the same field division reassure us. A list of such parcels in the following terrier is illustrative of some sixty pages of similar matter in the Walden cartulary, and establishes the existence of open fields at Saffron Walden, in the upper valley of the Cam: —

> " due acre in Benepistel
> sex acre et una roda quarum unum caput extendit super manlond . . .
> due acre super Sortegravehill . . .
> due acre super Sortegravehill que vocantur le Gorey . . .
> una acra et tres rode in eodem campo
> tres acre et una roda super Putemannesdole
> tres acre et una roda et quarta pars unius rode in Goredlond
> septem acre et tres rode in Middelsot
> dimidia acra ex opposito Eustachii de Broc
> una acra . . . in eodem campo." [1]

The Dunmow cartulary takes us a little farther toward the center of the county and fixes the probable limit of open field. There can, for instance, be no doubt about its existence at Henham, where ten acres were disposed in seven parcels among various field divisions.[2] At Henham, too, we hear of " totam dalam terre . . . que iacet apud le' 'helz de Rokey inter terram Radulfi Rafur et terram Galfridi Dolling. Et dalam illam in eodem campo inter terram Arnulfi et Rogeri le hog." [3] These dales recall the open fields of northern England, and in the guise of " doles " recur elsewhere in this region. [4] Just to the west of Henham a fine of 40 Henry III locates sixteen acres at Manewden in nine parcels.[5] Other instances of scattered parcels hereabouts are available, but perhaps enough have been adduced to show

[1] Harl. MS. 3697, f. 89b.
[2] Harl. MS. 662, f. 59b. The acres were distributed as follows: —
    3½ in Bermvelehe   ½ in Hofeld
    ½ in viclande   3 in Coperdenefeld
    1½ in Cockesdenefeld   1 in crofto meo
[3] Ibid., f. 58b.
[4] Land in Alsewic, Herts, is specified as seven doles, each separately described (ibid., f. 116). At Middleton one of the parcels in a holding was in the fourteenth century described as one " dole terre continens i acram que vocatur Sheppelond in Sturfeld (Rents. and Survs., Portf. 25/17).
[5] Kirk, *Essex Fines*, i. 216 (no. 1286).

that there can be no hesitation in assigning this corner of the county to the province of open field. Its open fields, however, were not of the midland type, but resembled those of Hertfordshire. So far as can be seen, parcels were not arranged with a view to a two- or three-course husbandry accompanied by pasturage of the fallow. The terminations *dene, dune (done)*, and *lee* suggest, further, furlongs in a woodland area; and it is possible that a township's arable arose from the continued assarting of the waste, with an adaptation, but no adoption, of midland arrangements.

Throughout all of the county except the northwestern corner traces of open-field husbandry are slight. Seldom, even in the early fines or charters, do we meet with the series of small parcels which betray the presence of intermixed arable strips. Since these fines are both numerous and specific and do not fail to ascribe small parcels to the northwest,[1] their failure to record similar phenomena in the remainder of the county becomes a telling negative argument against the existence of open arable fields there.

Later Essex surveys and terriers have the objectionable habit of merely reciting the parcels of the tenants' holdings without grouping or describing them in any way. The sixteenth-century documents referred to at the beginning of this chapter are useful in that they go so far as to indicate which parcels were closes. What we should like to know, however, is the character and appearance of the primitive villein holding. Inspection of the fines and extents reveals the fact that the Essex unit was often the virgate or yardland.[2] It appears as such on three of St. Paul's manors in 1222,[3] and on many Waltham manors tenants held virgates.[4] There were at Bocking in the thirteenth century $22\frac{1}{4}$ virgates, 10 "forlands," and 7 half-forlands, each forland doing

[1] Cf., in addition to instances already cited, 6 acres in 13 parcels at Heydon, and $5\frac{1}{2}$ acres in 4 parcels at Birchanger (Kirk, *Essex Fines*, i. 41, 61, 9 John, no. 228, and 6 Hen. III, no. 91).

[2] Ibid., 9 sq. *passim*.

[3] Beauchamp, Wickham, Tidwoldington: W. H. Hale, *Domesday of St. Paul's*, Camden Soc., 1858, pp. 27, 33, 52.

[4] E. g., Woodford, Nettleswell (13th century): Cott. MS., Tiber. CIX, ff. 205b–210.

one-fourth as many days' week-work as a virgate.[1] Virgates are also found at Berneston in 13 Henry VI and at Felsted Bury in 41 Edward III.[2] Sometimes, however, the villein unit took less committal names. At Hadley there were 22½ "terre," and 6 "moneday londs," the services from each being given in detail.[3] Lalling, described in the same series, had 12½ " terre " from which villein services were due.[4] Elsewhere no name whatever is given to the unit; in the thirteenth-century enumeration of services due at Borley, for example, we find merely " de singulis xx acris terre." [5]

Although information to the effect that the virgate or a corresponding unit was the standard villein holding can be deduced from the extents and from Domesday,[6] it is impossible to discover from them or from other documents hitherto cited what was the aspect of the virgate. For that reason the later descriptions contained in a Jacobean survey are of importance. The survey in question describes the large manor of Barking in southwestern Essex, not far from London. Although for the most part it neglects to group the acres of its holdings in any way, there are happy exceptions, of which the three following descriptions are typical: —

" Idem Johanis [Trewlove] similiter tenet unam virgatam terre custmumarie et heriottalis vocatam Coryes . . . iacentem ex parte orientali de le Fyve Elmes in Daggenham cum vi denariis redditus annuatim percipiendis de una crofta terre vocata Whites continente per estimationem iii acras que nuper fuit parcella predicte virgate terre; de qua quidem virgata terre quatuor clausa continent per estimationem xii acras terre arabilis insimul iacentia inter terram . . . et tres alia clausa residuum predicte virgate . . . continent per estimationem vii acras terre arabilis

---

[1] MSS. of the Dean and Chapter of Christchurch, Canterbury, Lib. B, ff. 115b, 132b.
[2] Rents. and Survs., D. of Lanc. Portf. 2/5; ibid., Ro. 188.
[3] Add. MS. 6160, f. 68 (early 14th century).
[4] Ibid., f. 69.
[5] MSS. of the Dean and Chapter of Christchurch, Canterbury, Lib. B, f. 143.
[6] e. g., at Horndon, Liston, Creping Hall, East Donyland, etc.: *Victoria History of Essex*, i. 560–566.

et iiii acras bosci et insimul iacent super terram Jacobi Harvey. . . .

"Thomas Humfrey tenet sibi et heredibus suis per copiam . . . unam virgatam terre custumarie et heriottalis vocatam longyerd . . . continentem per estimationem vii acras terre arabilis et septem acras bosci abuttantes super Blackhethe versus boriam et venellam ducentem a le kings highway versus . . . et terram nuper Thome Pruey . . . versus Austrum et terram Josephi Haynes armigeri versus boriam. . . .

"Johannes Pragle tenet per copiam . . . unam virgatam terre . . . vocatam Beesdown ab antiquo Roughlands continentem per estimationem Novemdecim Acras terre arabilis iacentem in parochia de Dagenham abuttantem super terram liberam predicti Johanis versus occidentem et terram nuper Thome Cowper vocatam Sawgors versus boriam et terram Roberti Scott generosi et terram pertinentem le Almeshouse de Romford versus occidentem."[1]

Two of the above virgates consisted of arable and woodland, the third of arable only. While the first of the three comprised two groups of closes probably separate, the others were compact areas, and, though nothing is said about their being enclosed, such was without doubt their condition. The nature of the virgate of southwestern Essex at the end of the sixteenth century thus becomes apparent. It was sometimes, at least, a compact area usually divided into closes of arable and woodland.

The testimony of earlier documents confirms that of the Barking survey. A glebe terrier at Kelvedon declared in 1356 that the vicar should "have 62 acres of arable land whereof 52 acres lie together near the aforesaid mansion in one field called the Churchfield with the hedge adjoining, and nine acres in a field called Lyndeland as enclosed with hedges and ditches."[2] Most important of the early documents, however, are the feet of fines. After 1235, to be sure, they rarely mention virgates, but the following descriptions are informing. At Dunmow, which was near the open-field part of the county, the fourth of a virgate was

---

[1] Land Rev., M. B. 214, ff. 285, 312, 318.
[2] Essex Archaeol. Soc., *Trans.*, new series, 1911, xi. 7.

specified in 3 Henry III as one-half of a messuage, "with the field called Wudelehe, and with a moiety of Smithescroft."[1] At Laver, in 5 John, the half of a virgate consisted of "all the land which lies between Bredenewell and the wood towards the west, and 3 acres of land which lie between the road (*cheminum*) of the same town and the wood towards the north."[2] Lastly, at Havering in the southwest 50 acres were in 15 John taken from one and one-half virgates and located in such a way as to bespeak complete enclosure.[3]

Thirteenth- as well as sixteenth-century accounts of Essex virgates thus describe them as being largely consolidated; nowhere, except in the northwestern part of the county were they composed of small scattered parcels. Mention of them in the fines and charters becomes so infrequent after the first quarter of the thirteenth century as to render generalization somewhat unsafe, but the evidence at hand points unanimously and unmistakably to the largely consolidated virgate as characteristic of much of the county. The case is strengthened by the descriptions of numerous holdings which were not virgates. These, too, were composed, not of small scattered strips, but of larger areas which may have been little separated. Certainly the impression carried away from a perusal of the Essex fines is very different from that given by the fines of most other English counties. One feels that they resemble rather closely the equally unusual fines of Kent. If, whether in terrier or survey, we trust to the appearance of the virgate holdings or even to the aspect of holdings of any sort, we shall be inclined to ally the greater part of Essex with its southern neighbor in respect to its field arrangements.

[1] Kirk, *Essex Fines*, i. 52 (no. 29).
[2] Ibid., i. 32 (no. 146).
[3] Ibid., i. 46 (no. 257). The locations were as follows: —
    2 acres in the croft called Hamstall
    18 acres in the croft called Nortfeld
    5 acres in the croft called Laiacre
    11 acres in the croft called Phistelcroft
    5 acres in the croft called Brigfeld
    9 acres in the croft called La Dune.

Conclusion

BEFORE summarizing the results of this chapter we may profitably give a little attention to another group of early sources which has elsewhere been of some value in determining the character of early field systems. This is the series of extents contained in the inquisitions post mortem. By explaining whether the acres of demesne lands lay one-half or one-third fallow and in common, these extents have heretofore supplemented the evidence got by locating in the fields the parcels of the holdings.[1] Fourteenth-century records of this kind from the midland counties have frequently assured us that the demesne was thus fallow and common; others from East Anglia, while they have revealed the same three-course rotation as prevailed on common lands in the sixteenth century, have not forced us to conclude that a three-field system was existent at the earlier period any more than at the later one, when, as we know, it did not prevail. Kentish extents, on the other hand, have in our examination of them not admitted the possibility that demesne lands in Kent ever lay one-third fallow and in common. If, as occasionally happens, one-third of them are said to have lain fallow, the value put upon the pasturage of these shows that during the fallow season they were not open to common use.[2] Furthermore, we have found Kentish demesne sown yearly and valued as high as 12d. the acre, an undoubted indication of superior agriculture.

It is time now to inquire whether any information, relative either to improved tillage of the demesne or to the distribution of demesne acres between two or three common fields, is available from the extents of the counties of the lower Thames valley. Although, like the other documents from this region, these extents are annoyingly noncommittal, those of the decade 7–16 Edward III, which have hitherto been referred to, give testimony of a general character. In the first place, it is noteworthy that the

[1] Cf. above, p. 46, pp. 301–302.
[2] " Sunt ibidem ciii^{xx} acre terre arabilis que valent per annum quando seminantur iiii li. pretium acre vid. et quando non seminantur pastura cuiuslibet acre valet ii d. De quibus seminabantur ante mortem predicti Willielmi de semine yemali et quadrigesimali v^{xx}x acre " (C. Inq. p. Mort., Edw. III, F. 65 (11), Throwley).

highly-valued demesne acres of Kent nowhere appear. Valuations of the arable do not differ particularly from those of the midlands or of East Anglia, but range, like them, from 4 d. to 8 d. the acre. In the second place, statements, usual in the midland or East Anglian extents, to the effect that one-half or one-third of the demesne acres were without value because each year they lay fallow and in common, seldom occur in similar documents from the counties of the lower Thames. Surrey furnishes no such declarations in the extents of the decade referred to; Middlesex contributes the curious information that one-third of certain demesne acres were intermixed fallow and yet retained a considerable value;[1] the Hertfordshire instances, of which there are four, relate to townships near the northern border of the county;[2] and in the numerous Essex extents only once does the phrase characteristic of the midlands occur.[3] Not that there are in the extents of the decade in question no other traces of common usages or of the three-course rotation of crops. On the contrary, demesne acres are sometimes said to lie in common from the end of harvest till January,[4] and a three-course rotation was at times practiced

[1] C. Inq. p. Mort., Edw. III, F. 66 (32), Parva Greenford: " Sunt ibidem vi^{xx} acre terre pretium cuiuslibet acre de iiii^{xx} iiii d. et residue xxxx acre iacent frisce inter friscas aliorum hominum que valent per annum quelibet acra ii d."

[2] Ibid., F. 42 (18), Offley: " Et residuum predicte terre, viz., cc acre iacent in communi et valent per annum quum seminantur xxxxi s. viii d. . . . et quum non seminantur nihil valent per annum quia per totum annum iacent in communi."

Ibid., F. 52 (7), Berkhamstead: " Sunt ibidem ccc acre terre arabilis quarum duo partes seminabantur ante mortem predicte Johannis et tertia pars iacet ad warectam et in communi . . . et quando non seminatur nihil valet quia iacet in communi."

Ibid., F. 64 (20), Reed: " Sunt ibidem c acre terre arabilis que valent per annum xxx s. iiii d. . . . Et inde seminantur ante mortem predicti Thorne iii^{xx}x acre et residuum iacet in communi."

Ibid., F. 64 (20), Widdiall: " Sunt ibidem ccc acre terre quarum cc valent per annum lxvi s. viii d. . . . Et inde seminabantur hoc anno semine yemali et quadragesimali cxxxx acre. Et residuum iacet ad warectam et in communi."

[3] Ibid., F. 61 (10), Tolleshunt: " Sunt ibidem cciiii^{xx}v acre terre arabilis de quibus due partes possunt seminari per annum et tunc valet acra per annum quando seminatur iiii d. . . . et totum residuum nihil valet quia iacet ad warectam et in communi per totum annum."

[4] Ibid., F. 38 (1), Moulsey, Surrey: " Sunt lx acre terre arabilis que valent per annum xx s. . . . et non plus quia iacent in communi a festo sancti Petri ad vinculos usque ad festum Purificationis beate Marie."

upon enclosed demesne.[1] Of evidence, however, which proves the practice of three-course husbandry upon demesne acres lying in common fields there is only the brief amount just cited. Apart from the testimony of four townships in northern Hertfordshire and of one in Essex, we have from a considerable body of extents no suggestion that a three-field system prevailed in the counties of the lower Thames.

As to these exceptions, the Hertfordshire townships, lying as they do on the borderland of the midland area, may well have adopted midland husbandry without coming to be in any way abnormal phenomena. The Essex instance, however, is more difficult of explanation. Tolleshunt is situated, not in northwestern Essex on the edge of the midlands, where three fields might be expected, but in the eastern part of the county near the coast. The statement, then, that one-third of the demesne lay fallow and common there seems to import into the region the usages that lay behind similar statements in Norfolk and Suffolk extents. In those counties, as we have seen, a three-course rotation of crops on common fields did not, either in the sixteenth century or in the fourteenth, necessarily imply a three-field system. The same may have been true of Tolleshunt, and the field arrangements there may have been like those with which we have become familiar at Weasenham in Norfolk.[2]

Forms of tillage other than a three-course rotation of crops were also known in fourteenth-century Essex. At Chingford, in 12 Edward III, 240 acres from a total demesne of 260 acres were sown;[3] at Newport and at "Lachlegh" during the same decade

[1] C. Inq. p. Mort. Edw. III, F. 66 (27), Bennington, Herts: "Sunt ibidem ccc acre terre arabilis quarum due partes seminari possunt per annum. Et valent si seminantur lx s. viiii d. pretium acre iiii d. Et quando non seminantur pastura eorum duarum partium valet per annum xvi s. viii d. pretium acre i d. et non plus quia terra illa est valde petrosa et inde male herbata. Et dicunt quod due partes seminabantur ante mortem dicti Petrum. Sed tertia pars, viz., c acre de predicta terra iacent ad warectam que valet per annum viii s. iiii d. pretium acre i d."

[2] Cf. above, pp. 316–325.

[3] C. Inq. p. Mort., Edw. III, F. 56 (1): "Sunt ibidem cclx acre terre arabilis. ... De quibus seminabantur ante mortem predicti Egidii de seisona hyemali cl acre et de seisona quadragesimali iiii^{xx} x acre."

160 acres out of 220.¹ These ratios recall that which we have seen maintained at Beddingfield in Suffolk a half-century later,² and suggest that the tillage of the Tolleshunt demesne may not have been the usual Essex practice. Such a belief is fostered by the isolation of the instance. Of the forty Essex extents contained in the inquisitions of the decade 7–16 Edward III, only at Tolleshunt is the demesne described as lying at the same time one-third fallow and in common. In view of these circumstances, it is scarcely necessary to abandon the conclusion reached from a study of Essex fines, charters, and surveys — the conclusion that the field system of Essex was not that of the midlands, but resembled either the East Anglian system or the Kentish.

Having ascertained that the extents from the four counties of the lower Thames basin during a decade of the fourteenth century are almost entirely indifferent to the three-field system, we may proceed to summarize the more positive results of this chapter. The counties in question have been discussed together, not so much because of their topographical unity as because their field systems had certain characteristics which differentiated them from their neighbors on all sides. Unlike Kent and East Anglia, they designated the unit of villein tenure a virgate; unlike the midlands, they did not distribute the parcels of a virgate between two or three large arable fields.

Along with the characteristics which they had in common, however, went certain divergences that distinguished one county from another. In Hertfordshire and Middlesex there was no exception to the use of the term virgate, and the occurrence of that unit was usual at a late date. With regard to Essex neither of these generalizations is valid. Other units were there sometimes substituted for the virgate, notably the " terra " and an unnamed area of uniform size, both already met with in East Anglia. The

---

[1] C. Inq. p. Mort., Edw. III, F. 66 (33), Newport: " Sunt ibidem ccxx acre terre arabilis . . . unde seminabantur ante mortem predicte Margarete clx acre de semine yemali et quadragesimali."

Ibid., F. 56 (1), " Lachlegh ": " Sunt ibidem ccxxii acre terre arabilis. . . . De quibus seminabantur ante mortem predicti Egidii de seisona hyemali iiii^{xx}v acre et de seisona quadragesimali lxxvi acre."

[2] Cf. above, p. 331.

integrity of the virgate, moreover, was not long maintained in Essex, where the use of the term so late as the sixteenth century was unusual.

In Surrey virgates bore the midland name and continued intact; but the county furnishes one deviation from the customary nomenclature which is significant in determining the affiliation of the field system of at least a part of the region. This divergent nomenclature occurs in an extent of Ewell, undated, but at least as early as the thirteenth century.[1] In this extent the tenants' holdings are never rated in virgates but always in iuga. The first tenant held " unum iugerum [iugum] terre continens xiii acras terre," twelve others had one iugum each, one had one and one-half iuga, three had three iuga each, and fourteen had half-iuga. Although no field detail relative to the iuga is given, we are able to supply a certain amount from a field-book of 8 Henry IV.[2] In the latter document as we pass from furlong to furlong, each composed largely of acre and half-acre parcels in the hands of many tenants, we often meet with such items as " dimidia acra quam tenet Thomas Wagmore de tenemento Wowards."[3] Now, one of the tenants of a half-iugum in the extent was Rogerus Woward; and by looking closely we shall find that many of the parcels of the field-book were still attributed to tenementa which bore the names of the tenants of the extent. In the interim between the drawing up of the two documents the iuga had come to be called tenementa and the constituent parcels of each iugum had fallen into the hands of divers new tenants. The latter change is precisely that which thirteenth-century tenementa in Norfolk underwent, and the Ewell field-book in its attribution of parcels to tenementa is like a fifteenth-century Norfolk field-book.[4] How much the parcels of the Ewell tenementa were dispersed throughout the open arable area cannot be precisely ascertained, for the field-book often neglects to attribute strips to their respective tenementa. Considerable scattering there

---

[1] *Register or Memorial of Ewell, Surrey* (ed. Cecil Deedes, London, 1913), pp. 135–162. The texts printed are from a sixteenth-century transcript.
[2] Ibid., pp. 1–135.
[3] Ibid., p. 35.
[4] Cf. above, p. 334.

certainly was, since parcels belonging to the same tenement are often widely separated in the field-book's enumeration. The date at which the tenementa were in the hands of the tenants whose names they came to bear is determined by the extent. Since this document is cast in the usual thirteenth-century form, and since in 8 Henry IV a parcel of a tenement was still occasionally in the hands of a descendant of the original holder,[1] the extent undoubtedly belongs to the thirteenth century. What we see, then, at Ewell are thirteenth-century tenementa, very much like those of Norfolk, bearing the names of contemporary Kentish units. As in Kent, too, the subdivisions of the rood at Ewell were known as " day works."[2] Thus, the Ewell field arrangements, reproduced probably in many Surrey townships, become a connecting link between the East Anglian and Kentish systems.

Essex as well as Surrey shows East Anglian and Kentish analogies. Its " terra " and unnamed unit of villein tenure were East Anglian; its " day's work," a unit of measure often used was Kentish, and there were of course Kentish counterparts for the consolidated or nearly consolidated virgates of Essex. Except for the northwestern part of the county, deviation from the original Kentish system was less than in East Anglia or in Surrey. Especially are the compact holdings of Essex noteworthy. Although we have no evidence that these were rectangular blocks, as were the Gillingham iuga, nevertheless the descriptions of virgates at Barking are not unlike those of iuga at Newchurch and especially at Wye. Only the name differed; whereas at Wye the virgate was the fourth part of the iugum, in Essex it was, for purposes of estimate, the fourth part of the hide. Units so essentially alike in aspect seem to assure us that the system prevalent in Kent extended to the north of the Thames.

It is doubtful whether the northwestern corner of Essex should be included in the above generalization. Much more open field was to be found there than in the rest of the county, and the terriers of holdings are very much like those of Hertfordshire in

---

[1] " I acra quam tenet Petrus Saleman de tenemento suo." One of the iuga of the extent was held by Johannes Saleman. *Register of Ewell*, pp. 34, 141.

[2] Ibid., p. 159.

the number, location, and naming of their parcels. Indeed, it would seem that the entire hilly district extending from Essex to the Thames ought to be considered as one whole. It embraces, besides northwestern Essex, Hertfordshire and the Chiltern region of southern Buckinghamshire and southern Oxfordshire. In its early days this region must have been, even more than at present, a wooded area. Denes, dells, groves, hills, and lees, which so often recur in the terminations of the names of openfield divisions in this region, suggest the original condition of the arable field. Hills and forests, as it happens, have been features not without influence upon field systems. In a territory where woodland was relatively extensive, where it was somewhat difficult to transform waste into arable, tenants can have had no concern about a compact fallow field to supplement the pasture. If, further, an additional arable furlong were at any time to be improved from the waste, one non-adjacent to the existing arable may now and then have been selected. The most feasible spot for improvement may often have been a valley or a slope which, after being brought under cultivation, would still be surrounded by woodland. These considerations should be kept in mind when surmises about the origin of the field system of this Chiltern region are made. Under the circumstances, it would seem hazardous to posit either midland or Kentish affiliations. It is improbable that simple two- or three-field arrangements, with virgates divided between the fields, were ever existent there; yet there may have been such in the earliest days, and the later irregularities may have arisen from the addition of assarted areas. On the other hand, there is no reason to assume a Kentish origin for the system. The villein units were named differently from the Kentish, they were not compact areas, they were never rated in " day's works," they were not subdivided among cotenants. The Chiltern area should, therefore, be looked upon as a boundary region so influenced in its field system by its topography that its original affiliations cannot readily be discovered.

Middlesex remains. In the east its open fields seem to have been like those of Hertfordshire; in the west it is just possible that some of the townships of the Thames plain were in three fields. The character of a Jacobean survey of Feltham is the principal reason for admitting the latter possibility; other evidence tells for irregular fields, like the adjacent ones of Surrey. What is clear is that the plain on both sides of the Thames west of London constituted a region where the midland system and the Kentish system came into contact. In Middlesex, the former seems to have prevailed, in Surrey the latter. The outcome was a hybrid system difficult to follow in its origins; and, indeed, this difficulty pertains to the field arrangements which characterized the entire lower valley of the Thames. Scarcely any part of England is so dependent upon conjecture for the writing of its early field history. For this reason it is to be hoped that new documents may in time dissipate some of the uncertainties which this chapter leaves unsettled.

CHAPTER X

RESULTS AND CONJECTURES

IN an introductory chapter it was suggested that a study of field systems might throw light upon the history of English agriculture; it was intimated, too, that a discrimination between regions characterized by different field arrangements might be of importance for the history of English settlement. The time has come to inquire whether these predictions have been fulfilled.

The preceding chapters have, it is hoped, established certain general conclusions. The current view that the two- and three-field system was prevalent throughout England has been rejected, and it has been shown that this system was restricted to a large irregular area lying chiefly in the midlands. This central area reached northward as far as Durham and southward to the Channel; it extended from Cambridgeshire on the east to the Welsh border on the west.[1] In the counties farther toward the southeast, the southwest, and the northwest different field systems have been discovered. Whatever the dissimilarity between these, they have shown agreement in not dividing the unenclosed arable of their village fields into two or three parts to each of which one-half or one-third of every tenant's parcels were assigned.

A marking-off of central England as the precinct of the two- and three-field system is significant for the history of agriculture. The development of this art has depended primarily upon the extent to which and the manner in which the soil has been utilized for the multiplication of agricultural products. At one end of the line of development stands the unenclosed open waste, parts of it transiently improved for purposes of tillage, as in the Scottish outfields;[2] at the other end stands the modern enclosed farm, its acres cultivated in accordance with the principles of convertible husbandry. Between these termini lie two well-marked

[1] Cf. map facing the title-page.   [2] Cf. above, pp. 158 sq.

phases of progress. The first is the reduction of the waste to regular and considerable, but still open-field, tillage; the second is the enclosure of the now well-established arable fields and remaining commons, accompanied by some increase in improved pasture and by the substitution for the old fixed succession of crops and fallow of a varied rotation of grains and grasses. The first phase comprehends the development of open-field systems, the second the history of enclosure. With both these subjects the preceding chapters have been concerned, but with the latter somewhat incidentally.

What, now, does our investigation show to have been the relation between the subdivision of England according to field systems and the lines of agricultural development just indicated? Precisely this, that enclosure was earliest achieved outside the precincts of the midland system. The map which Slater has roughly constructed from the list of eighteenth- and nineteenth-century enclosure acts shows that the midland area was the one where open fields lingered longest.[1] Gay had already shown that the small enclosures of the sixteenth century took place particularly within this region, and had correctly inferred that the open fields then encroached upon lay largely within those counties in which such fields were especially to be found.[2] In most counties lying without the midland area, unenclosed arable fields, so far as existent, disappeared for the most part before the era of parliamentary enclosure. Only Surrey, Middlesex, Hertfordshire, and a part of Norfolk then retained any appreciable stretches of them. In Northumberland, Durham, and Cumberland they had vanished rapidly after the sixteenth century. Earlier still they had ceased to be characteristic of Devon and Cornwall, Cheshire and Lancashire, Suffolk, Kent and Essex.[3]

One reason for this early disappearance of unenclosed arable lies in the nature of the field systems prevalent in these counties. Since both Celtic and Kentish systems were in part determined

---

[1] *English Peasantry*, p. 73.

[2] "Inclosures in England," *Quarterly Journal of Economics*, xvii. 576, 593-594. In a footnote contemporary authorities are cited.

[3] Cf. above, chapters VI, VII, IX.

by the custom of subdividing land among heirs, some intermixture of the parcels of tenants' holdings naturally appeared wherever either system was practiced. But the Celtic system did not necessarily imply an extensive development of runrig, especially if the region were a pastoral one; and the Kentish system did not render immobile the intermixture of tenants' strips. It was possible under both systems for holdings to retain a certain degree of compactness, a fact which naturally facilitated enclosure. At any rate, no close connection between a three-course rotation of crops and three large fields ever arose. Often, too, there were in the counties in question tracts of woodland or waste, moor or down, so large that it was possible to set little store upon the use of the fallow arable for pasture, a feature which the midland system always emphasized. If it did seem desirable thus to utilize the fallow arable, as happened in Norfolk, wattles were employed. Freed in one way or another from the pasturage needs of the midlands, and disposed with none of the symmetrical arrangement there prevalent, the open-field arable acres of the non-midland counties readily yielded to enclosure at an early time. Such is the first and not the least noteworthy effect which field systems have had upon the agricultural development of England.

The midland system, on the contrary, exerted upon this development an influence which was to some extent inhibitive. It delayed enclosure. The correspondence between its precinct on the one hand and the regions of the persistent open field of the parliamentary awards on the other, shows in a general way that it was peculiarly favorable to the preservation of unenclosed arable, that it served, indeed, as a protective shell. In order to view this relationship more closely we have given somewhat careful attention to the later enclosure history of Oxfordshire. In consequence it has become apparent that those townships which longest remained open were the ones which clung most tenaciously to the old system. If this was the case in the eighteenth century, when incentives to abandon the traditional tillage were strongest, the protection afforded by the system was probably even more effective during earlier centuries, when there was less thought of change. The persistent open field of the midlands, therefore, coincides with the precincts of the two- and three-field system, because in a general way field system and unenclosed arable here stood to each other in the relation of cause to effect.

Although the midland field system was inherently adverse to enclosure, it should not be inferred that within the large area characterized by it no progress took place between Anglo-Saxon days, when a two-field system was probably in use, and the early nineteenth century, when enclosure was for the most part accomplished. For it is one of the cardinal theses of this book that, owing to changing field arrangements within the midlands, agriculture did develop there during the centuries in question. The first important movement of this sort was a transition from two-field to three-field tillage, a change which, according to our evidence, seems to have been brought about in many parts of the eastern and northern midlands during the thirteenth and fourteenth centuries. The second change was later, occurring apparently between the middle of the sixteenth century and the middle of the eighteenth. In some places it took the form of a subdivision of two fields into four, three of which were tilled annually; elsewhere it appeared as the transformation of regular fields into irregular ones, a process probably attended by improved tillage and certainly often accompanied by considerable piecemeal enclosure.

Evidence regarding the second change is the more abundant, and considerable of it has been cited.[1] Several Tudor and Jacobean surveys have established the fact that departures from the two- and three-field system were known in certain parts of England as early as the sixteenth century, especially in the counties of the western midlands from Durham to Somerset, and above all in the valley of the Severn. Typical of the disappearance of open fields in this region is the enclosure history of Herefordshire, which has been examined in some detail.

The open arable fields of this county had before the days of parliamentary enclosure so shrunken that they constituted not more than two and one-half per cent of its total area. The abandonment of communal tillage, and hence the achievement of enduring agri-

[1] Cf. above, chapters III, IV.

cultural progress, had been brought to pass, if not so promptly as in many non-midland counties, at least earlier than in the eastern midlands. For this progress the county seems to have been indebted to certain irregularities in its field arrangements, some of which were already apparent in Jacobean times. These irregularities were in turn due to divers causes. The situation of townships within fertile river valleys, which throughout midland England often proved itself an influence conducive to the appearance of irregular fields, was characteristic of a large part of Herefordshire. Another general influence, the location of townships within a forested area settled and improved relatively late, was not without effect in the county. Along with these wide-reaching causes of irregularities in field systems, irregularities which in turn were conducive to enclosure, went a special circumstance, probably operative in other counties of the western midlands as well as in Herefordshire. This was the small size of township fields. In a region characterized by hamlet settlements, as some of these western counties were, the improved arable was not great in amount and the tenants were not numerous. Departures from a regular system were easier to make than where fields were large and tenants many; and our evidence goes to show that they were frequently made. The outcome of this and of the other influences mentioned was often a multiplicity of small fields. Jacobean surveys and enclosure awards have served to illustrate these fields and have shown how they facilitated piecemeal enclosure. For piecemeal enclosure was the form of agricultural development naturally adopted by districts circumstanced like Herefordshire.

The course of events differed in Oxfordshire, a county which, because of its situation in the more eastern midlands, serves to exemplify the agricultural progress of that region. The first and dominant fact disclosed by our inquiries is that large tracts of open arable common field persisted in the county until the second half of the eighteenth century. Some thirty-seven per cent of its area then remained in this state and had to be enclosed by act of parliament. One should not infer from this that a certain amount of open arable field had failed to escape enclosure between the Middle

Ages and the period in question. Large parts of some townships and all of others had succumbed to it. Conducive thereto were certain of the causes operative in Herefordshire — situation in a river valley or in a forest area; contributory, too, was the cherished passion for country estates manifested by the new gentry of the sixteenth and seventeenth centuries. The salient feature of agricultural development in Oxfordshire before 1760, however, and presumably the one characteristic of many counties of the eastern midlands, was not the enclosure of open fields but the improvement of them as they lay unenclosed. The redland district of northern Oxfordshire is typical. Once characterized by two-field townships, it began from the end of the sixteenth century to subdivide the two fields into four and to get thereby an annual return from three-fourths of the acres rather than from one-half of them. In the eighteenth century still more of the arable was annually tilled and the rotation of crops became as complex as upon enclosed lands.

Improvement in the tillage of unenclosed fields was not confined to Oxfordshire. We have testimony to the early appearance of four fields in southwestern and northeastern England, in the valleys of the Severn and Trent. Irregular fields, too, of which we have found many throughout the midlands as early as the sixteenth century, probably reflected other forms of improved cultivation, the nature of which is not always discernible from the surveys. In so far as these irregularities did not correspond with a changing tillage of arable, they imply that the arable strips were transformed to meadow, a phase of development peculiarly suited to river valleys. As it happens, we have given most attention to such transformation in Durham open fields during the seventeenth century; but hints from other regions indicate that it was far from unknown throughout the northern midlands.

In whatever way, therefore, open arable fields underwent change before the middle of the eighteenth century, whether they submitted to a process of piecemeal enclosure with some conversion to meadow and pasture, or whether on the other hand they attained to a higher standard of tillage, remaining arable the while, the fact cannot be escaped that field systems, either as cause or as manifestation, were associated with agricultural development. For this reason the preceding chapters have a bearing upon the history of English farming.

If the influence of divergent field systems upon the progress of the enclosure of open arable fields is reasonably clear, there is more doubt about the interpretation of this diversity in relation to the history of the early settlement of England. The traditional account of the Anglo-Saxon occupation, as gleaned from Bede and the Anglo-Saxon Chronicle, has by modern scholars been brought into connection with other evidence. Most opposed to it and most suggestive is Seebohm's theory of a Roman origin of the manor, its fields, and its class distinctions.[1] Meitzen and Maitland have pointed to the contrast between nucleated and scattered settlements, with an intimation that the latter were of Celtic origin.[2] Chadwick has done much to abolish the distinction between Angles and Saxons, concluding that only Kent, the Isle of Wight, and the southern coast of Hampshire were occupied by a distinct branch of the invaders.[3] Turner, finally, has sketched a theory which discerns Roman elements in the five-hide manor, and possibly in the rod of southern England.[4]

All students of Anglo-Saxon England agree upon the dominance which the new-comers of the fifth century exercised upon institutions. Legal, military, and political organization became Germanic. The spoken language retained few Celtic words, while villages and towns assumed names which in their terminations at least are Teutonic. If any Roman or Celtic influence survived, it was in matters connected with the lowest stratum of society, the stratum engaged primarily in the cultivation of the soil. By enslaving a considerable mass of the British population, itself already Romanized, the conquerors could, it is clear, have created

[1] *English Village Community*, pp. 409 sq.
[2] Meitzen, *Siedelung und Agrarwesen*, ii. 118 sq., and Anlage 66a; Maitland, *Domesday Book and Beyond*, p. 15.
[3] H. M. Chadwick, *Origin of the English Nation*, p. 88.
[4] G. J. Turner, *Calendar of the Feet of Fines relating to the County of Huntingdon*, pp. lxx, cix.

a rural estate not unlike the later mediaeval manor. In taking over the cultivators of the soil they might also have adopted the methods of tillage already practiced by the sitting tenants. As a result, the ancient tenant-holding and its relation to the township's arable would have persisted after the Germanic conquest. On the assumption, therefore, that the Romano-Celtic population was to some extent assimilated rather than exterminated, we should expect to find in Anglo-Saxon England a sub-stratum of servile dependents whose holdings had Roman or possibly Celtic characteristics. What should appear in the extant evidence as testimony to the existence of a conquered and depressed group are Roman or Celtic agrarian usages and early traces of serfdom. This was Seebohm's thesis, and to a limited extent it is Turner's.

The subject discussed in the preceding pages is one that touches the history of settlement at just this point. The nature of field systems depends primarily upon the relation of the unit of villein tenure to the arable fields. For this reason it is pertinent to inquire in what measure the systems that have been described are Anglo-Saxon, Celtic, or Roman. So far as this point can be ascertained, additional matter will be at hand for solving a troublesome problem of English social history.

One limitation of our evidence touching field systems which seriously impairs its applicability to the problem above described is its relatively late character. Little of it antedates the thirteenth century, a period itself seven hundred years removed from the Germanic invasion. Although Domesday Book and certain twelfth-century documents refer to the unit of villein tenure, they disclose scarcely more than the names it bore, giving no descriptions that relate it to the arable fields in which it lay. More informing are the Anglo-Saxon charters, which in a few instances testify to the existence of intermixed parcels and by phrases in the boundaries hint at open-field usages. Even the assurance, however, that some form of open field existed in midland England in the tenth century is not very valuable for our purpose, partly because the information is still four centuries later than the coming of the Germans, but still more because, in view of the fact that open-field arable of one sort or another could be found throughout England in the thirteenth century, the existence of it at an earlier time is almost a postulate. What we should like to know are the varieties of open field with which the Anglo-Saxons were familiar, and of these the charters tell us nothing. It is necessary, then, to assume that distinctions which obtained in the thirteenth century are assignable to the period that saw the accomplishment of Saxon settlement, or at least to the period that followed the Danish invasions. If this assumption be not admitted, variations in field systems have nothing to tell about the settlement of England.

If it be granted, however, that field arrangements as we find them in the thirteenth century represent more ancient usages, the preceding chapters have implications. It has appeared that a large midland area was characterized by a two- and three-field system. That this system was not Celtic an examination of Scottish, Irish, and Welsh evidence has made clear. In Celtic countries we do not find the arable of a farm, township, or townland divided into two or three equal compact areas and tilled under a rotation of two crops and a fallow. This was, on the other hand, or it came to be, a custom prevalent in Germany, especially east and south of the Weser.[1] Since this is the region from which the invaders who settled midland England appear to have come,[2] it is probable that the two- and three-field arrangements of the midlands represent Germanic usage.[3] If this be true, the thorough Germanization of central England suggested by various practices is confirmed by the testimony of field systems. No Romano-Briton population remained there in numbers large enough to preserve either a Celtic or a Roman method of tilling the soil.

The westernmost territory which thus yielded to the invasion of Teutonic custom is interesting, since it did so rather grudgingly. It comprised the counties of Herefordshire and Shropshire, a fertile region early occupied by the Magonsaetan. Here,

---

[1] Meitzen, *Siedelung und Agrarwesen*, i. 33–36, 67, 169, and Atlas, Uebersichtskarte; Hanssen, *Agrarhistorische Abhandlungen*, i. 171.

[2] Chadwick, *Origin of the English Nation*, pp. 88, 91, 116; map, p. 112.

[3] Cf. Meitzen, as above, ii. 110.

as in the midlands, the invaders impressed upon their new conquest an open-field system which, according to our earliest evidence, was one of three fields. In another respect, however, they seem to have adopted the habits of their predecessors: their settlements were small and of the hamlet type. Perhaps they assimilated a part of the Briton population itself along with the Celtic type of settlement. Place-names here evince more of a commingling of Celtic and Anglo-Saxon elements than is usual in the midlands — a further indication that there was in these two counties a more equitable balance of Celtic and Germanic forces in matters of settlement and agriculture than appears elsewhere in England, unless it be in Northumberland.

Other counties of the west and north diverged more sharply from the midland model. Often townships and settlements in them were small, as in Celtic countries. In Cornwall, Devon, Cheshire, and southern Lancashire open arable fields seem never to have been numerous or, in any township, extensive. The same cannot be said of Northumberland, Cumberland, and northern Lancashire, where such fields were relatively frequent in the thirteenth century and comprised the largest part of the tilled land of each township. But whether large or small, numerous or infrequent, the open arable fields of all these counties were not of the midland type. In no instance (with perhaps a reservation relative to Northumberland) were they divided into two or three equal parts to which the strips of each holding were equitably assigned. In appearance they were more like Scottish or Irish open fields, in which the strips were said to lie in runrig. The underlying principle of runrig was the assignment to each tenant of a share in every kind of soil within a township, whenever an occasion for distribution arose. Since the several qualities of land were likely to lie in various parts of the cultivated area, a scattering of parcels was to some extent the result. Recourse to runrig, therefore, brought about either temporarily or permanently a dispersion of the parcels of a holding. Yet there was no guarantee that these would be as symmetrically located throughout the arable area as they were in two- and three-field townships. There might even occur a segregation of parcels, a feature which certain Cumberland terriers seem to reveal. In general, however, the parcels in the larger township fields which lay in runrig, especially those in Northumberland, doubtless remained widely dispersed.

Why, then, it may be asked, was the custom of allotting strips in runrig incompatible with a three-field system? The answer is that it was not necessarily incompatible, since runrig might under certain conditions develop into the system in question. To understand what these conditions were we must turn to another aspect of Celtic field arrangements still visible in eighteenth-century Scotland. This was the practice, appropriate to primitive agriculture, of improving successively different parts of the waste and allowing each part in turn to revert to fallow for a series of years. Traces of such a custom are also perceptible in documents from Cumberland, but are more apparent in others from Northumberland, a fact that has led us to formulate the hypothesis that the English border counties originally had the same field system as Scotland but developed it differently.

In both regions, we may surmise, field arrangements were based upon runrig, a device that assigned to all tenants within the township strips in any tract of waste brought under transient tillage. As agriculture advanced, however, the two regions expanded this system in different ways. Scottish husbandry turned to an intensive tillage of the arable which lay nearest the homestead, the so-called "infield," and by the aid of manure took from it an annual crop, the remaining "outfield" being treated in the old manner. In the English border counties, on the other hand, no permanent differentiation was made between infield and outfield; but, as the demand for a greater return from the soil grew, the period of fallow which had been allowed to the transiently improved parcels of waste was shortened. Eventually, we may suppose, it was reduced to an interval of one year in three, as it appears in fourteenth-century Northumberland extents. When this stage was reached, transition to a three-field system was feasible, involving only such regrouping of the parcels of the holdings as would render compact the area left fallow each year. Since the advantage of a fallow field of this sort lay in its utility for purposes of pasturage, the transition in question may have occurred in those places in which it was desirable to utilize all available pasture. Where, on the other hand, the moor and fell of the township furnished ample grazing ground, it may never have come about at all. Certain features of the Northumberland evidence, especially the marking-off of fields on sixteenth-century maps, suggest that some townships of this county may have adopted the three-field system; but the unequal division of the parcels of the tenants' holdings among the fields leaves the matter in doubt. Other Northumberland townships probably never created three equal compact fields. Nothing whatever in the evidence from Cumberland and northern Lancashire leads us to think that the three-field system ever developed there.

Cheshire, southern Lancashire, western Somerset, Devon, and Cornwall were regions in which there was either a far slighter extension of runrig in early days or a more rapid consolidation of scattered parcels than in Northumberland, Cumberland, and northern Lancashire. In explaining how runrig arose we have had occasion to point out that a farm, township, or townland might show no trace of it if the custom of joint succession had not been effective, or if the landlord had intervened to prevent subdivision, or if he had at any time exercised his authority by reconsolidating the subdivided arable; it might appear in only a modified form if the lands to be divided were meadow or pasture rather than arable and the need of marking out strips for coöperative ploughing did not arise. One or another of these factors seems to have been at work in the counties now under consideration. Traces of runrig have been found in each of them, but the intermixed strips show evident tendencies toward early disappearance. In Cornwall and Devon, furthermore, the lands so divided were at times apparently improved waste or marsh. The conclusion suggested is that the counties in question were subjected to Celtic influence in the matter of field systems, but in a different way from those to the north: the original farms, apparently like many in pastoral Wales, sometimes escaped subdivision, or at least escaped it to such a degree that reconsolidation was easy and was achieved at an early date.

Southeastern England with its divergent field systems was widely separated from the counties in which Celtic influence was manifested. Since the great midland area stretched between, there would seem to be little *a priori* probability that irregularities in southeastern fields were of Celtic origin. It is of course possible to argue that all English field systems arose on the basis of runrig, as did the Celtic. On this hypothesis the two- and three-field arrangements of the midlands would be such adaptations of runrig as have been suggested above relative to Northumberland,[1] and the systems of the south and east would be other manifestations of it. Such a theory, however, ignores the fact that the midland system was that of the Germans in their home land and was thus more than any other essentially Teutonic. Or are we to assume that the Germans, both in Germany and in England, had a genius for developing runrig into a more regular system? At all events, the hypothesis would encounter a further difficulty in the fact that the peculiarities of the fields of southeastern England were not all Celtic in type. Settlements and fields here were not small, as they were on the western border, and certain of the earliest units which are met with in the southeast had no correspondents at all in the west.

The Kentish system is at once most divergent and most comprehensible. The best-defined feature of it is the iugum, the unit of villein tenure, which, compact and rectangular in shape, had its exact counterpart nowhere else in England. If we ask whether the continent offered analogies, we are at once reminded of Roman measurements of land. The application of these, as Meitzen has shown,[2] resulted in a superficial unit of the sort actually found in fourteenth-century Kent. This similarity is of the highest importance; for, despite the centuries that have to be bridged, we are led to the inference that the Kentish field system was of Roman origin. While the Anglo-Saxons who occupied the midlands and the south established there the elements of a two- and three-field system, the Germans who occupied Kent seem to have adopted Roman arrangements and to have

[1] Cf. above, p. 225.
[2] *Siedelung und Agrarwesen*, i. 276-321, and Anlage 29.

maintained a closer agrarian tie with Roman Britain than persisted elsewhere in the island.

The field arrangements of the other southeastern counties are more difficult to interpret, and in attempting to discover their origin we advance farther into the realm of conjecture. To explain the formation of East Anglian eriungs and tenementa an hypothesis has been sketched which in brief is as follows. The peculiar pasturage arrangements of Norfolk and Suffolk, arising from the possession by individuals and small manors of the privilege of independent foldage, is suggestive of a connection between the formation of these manors and the development of the field system of the region. Inasmuch as the manors antedate Domesday Book, the foldage privileges may be looked upon as correspondingly early. A more decisive feature in the East Anglian system, however, is the aspect assumed by its unit of villein tenure when we first get descriptions of it in the thirteenth century. Its compactness in some instances and the segregation of its parcels in others reveal its similarity to the Kentish iugum; but it was usually less like the intact Kentish unit than like a Kentish holding after the iuga had for some generations been subdivided and a tenant had come to hold parcels in several neighboring iuga. This feature of the East Anglian tenementum is perhaps best explained by the supposition that a pre-Norman organization of petty manors in East Anglia arrested for a moment the disintegration of ancient iuga which were once characteristic of the region, and established as new units the holdings that we find. Such a reorganization of the agrarian situation we have tentatively attributed to the Danish invasion, since to that intrusion was due the greatest social upheaval of Anglo-Saxon days. In this way the East Anglian and Kentish field systems, originally similar, may have come to be unlike each other. Should these inferences be correct, the area within which Roman influence persisted after the invasions of the fifth century is enlarged to include, along with Kent, two other counties of the southeast.

Essex, situated as it is between Kent and East Anglia, could with difficulty have escaped falling within the same sphere of agrarian influence. Nor, as a matter of fact, was its field system of such a character as to tell against a belief that it did so. To be sure, the villein units in Essex were virgates, as they were not to the north or the south; but the virgates in a large part of the county tended to be compact areas which may well have been related to the Kentish iuga. Similarities of nomenclature, too, especially the employment of the term " day's work," emphasize the connection with Kent.

Like peculiarities tempt us to extend the Kentish system to Surrey. " Iuga " and " day's works " were once known at Ewell, and what we learn about the later field arrangements of the county is not prohibitive of an early prevalence of the Kentish system within its borders. Division of holdings among three arable fields seems never to have prevailed there; nor can the aspect of a villein holding have differed greatly from that of one in East Anglia, or from the appearance of one in Kent after the disintegration of the iuga had set in. In view of these circumstances, the most credible hypothesis relative to Surrey is the assumption that, like East Anglia and Essex, it was originally within the Roman sphere of agrarian influence; that, like these three counties, it diverged somewhat more from the norm than did Kent; and finally that, like East Anglia, it reorganized the disintegrating iugum, adopting for the new unit the name of the midland virgate, a name likewise favored in Essex.

Whether the same hypothesis should be applied to the region which constitutes Middlesex, Hertfordshire, and the Chilterns is uncertain. This is an area which, with the exception of its northern fringe and possibly the flat plain west from London, seems not to have known the three-field system. On the other hand, its field nomenclature and the absence of consolidation which the parcels of its holdings reveal apparently leave it without the sphere of Kentish or Roman influence. A factor that enters into the situation is the hilly character of the district, which was doubtless once heavily forested. Probably much of it was settled later than the plains round about, and a large part of the arable was undoubtedly improved from the forest state. Whether the tiny settlements which thus extended their tillage organized their first fields after midland or after Kentish models is a question that must be left undetermined. Since the region forms a borderland between these two spheres of influence, some settlers may have come from the midlands and others from the southeast. The only thing that is clear is the development of arable fields through the assarting of the waste in such manner that the tenants' holdings came to comprise a certain amount of unenclosed land lying in scattered parcels.

The foregoing explanation of the field systems of southeastern England, hypothetical as it is in part, does at least leave us with a generalization which, if true, is important. It implies that throughout five counties of the southeast the influence of Roman Britain in agrarian affairs persisted after the Germanic conquest of the fifth century. Either the conquerors showed extraordinary flexibility in adopting a field system with which they must have been unfamiliar, or they spared a part of the native population who, as serfs, continued to employ their own agricultural methods. Since the latter supposition is the more credible, we are led to posit a greater survival of the Romano-Celtic population in southeastern Britain than in the midlands.

Anglo-Saxon England is thus, so far as field systems are indicative of settlement, divisible into three parts. The large central area, stretching from Durham to the Channel and from Cambridgeshire to Wales, was the region throughout which Germanic usage prevailed, presumably because of the thoroughgoing nature of the fifth-century subjugation; the southeast was characterized by the persistence of Roman influence, a circumstance which implies that the conquest was less destructive there than to the north and west; the counties of the southwest, the northwest, and the north retained Celtic agrarian usages in one form or another, a retention that is readily comprehensible in view of the difficulty with which, as we know, these districts were slowly overpowered by the invaders. This subdivision of Anglo-Saxon England, together with the evidence upon which it is based, constitutes the contribution which the study of field systems is able to make to the history of pre-Norman conquest and settlement.

APPENDICES

APPENDIX I

A. EXTRACTS FROM A SURVEY OF KINGTON, WILTSHIRE

Harl. MS. 3961, ff. 40-62.   9 Henry VIII

KYNGTON

TERRARIUM Omnium Terrarum et tenementorum unacum finibus, Redditibus, et heriectis eiusdem manerii, Factum ibidem mense Marcii Anno Regni Regis Henrici Octavi Nono Et Anno Domini Ricardi Beere Abbatis vicesimo Quinto coram Fratre Thoma Sutton, Cellerario forinseco, per Sacramentum et fidelitatem Ricardi Snelle, prepositi ibidem, Johannis Tanner, Willelmi Neck, Thome Mylle, Thome Coke, Walteri Tourney, Henrici Belle de Langley, Johannis Kyngton, Ricardi Broune, Willelmi Torney, Walteri Amyett, et Thome Amyett de Kyngtone, ceterorumque tenencium Domini ibidem ad Idem Terrarium vocatorum et distincte examinatorum preter specialem perambulationem et mensuracionem factam ibidem atque probatam.

COMMUNIA

Est ibidem quedam Communa vocata Langleyhethe continens cccx acras ubi dominus et tenentes custumarii communicare possunt cum omnimodis averiis omni tempore anni.   Et ulterius Thomas Montague — heres — Baroni et Johannes Gangelle et eorum tenentes in Langley communicare possunt in eadem.

NUNDINE

Sunt ibidem Nundine in festo Sancti Michaelis Archiangeli unde Tolnetum extenditur communibus annis xvi s.

FIRMA TERRARUM DOMINICALIUM IBIDEM

Ricardus Snell, firmarius Domini, tenet Curiam Dominicalem, viz., Aulam, Cameram, Coquinam, grangiam, boveriam, Domum Columbarium, et Croftam in boriali parte Curie, continentes insimul ix acras dimidiam.

Item tenet xx acras prati in Pekyngelmed A festo Purificationis Beate Marie usque amocionem feni.

APPENDIX I

Item tenet pasturam De Ruydon continentem xxxi Acras i perticatam per redditum xiii s. iiii d. solvendo annuatim Prioresse rectricique de Kyngtone in pecuniis pro xi averiis euntibus in predicta pastura per firmarium domini ibidem viii s. vi d. ultra xiii s. iiii d. predictos.

Item tenet pasturam et subboscum de Ynwode continentes xx acras per redditum iii s. iiii d.

Item tenet cxxvi acras iii perticatas terre arabilis in duobus Campis unde

in Campo orientali

   in La hamme xiiii acras dimidiam
   In la Deene iiii acras in ii^{bus} particulis
   In Farndelle ix acras dimidiam in iiii^{or} particulis
   In Sourelond unam acram dimidiam
   In Middelfurlong xx acras
   In Manshulle iiii acras dimidiam
   In Smethyescroft x acras

[Total, 54½ acres]

Et in Campo occidentali

   in Brechefurlong iiii acras iii perticatas
   In Brodefurlong xv acras dimidiam
   In Wellemore x acras dimidiam
   In Lordeshulle xiii acras
   Item ibidem ii acras prati
   In Overlordeshulle xi acras
   et in Colroft vi acras dimidiam inclusas

[Total, 56¾ acres]

Redditus lxvii s. viii d.

Bosci

Est ibidem quidam boscus vocatus Haywode continens cccc acras bosci et subbosci unde vendi possunt quolibet anno xxv acre subbosci. si copis bene preservetur et superintendatur pretium acre xiii s. iiii d, et sic quilibet copis recrescet in xvi annis.

Liberi Tenentes ibidem

Johannes Saunders tenet unum tenementum apud Haywode in feodo quondam Thome Bolehide per redditum annuatim ii aucarum precium viii d.

Thomas filius et heres Cristoferi Troponelle tenet unam virgatam terre in feodo quondam Edwardi Basyng et nuper Thome Troponelle per redditum annuatim v s. ix d. ob. q., unde ad festum Natalis Domini xx d., Pasche xx d., Johannis Baptiste ix d. ob. q., et Sancti Michaelis Archiangeli xx d., et heres eius solvet relevium post mortem, etc. viz., duplicem redditum unius anni.

Abbas et conventus de Malmesbury tenet unum mesuagium in feodo in villa de Malmesbury quondam Willelmi de aula per redditum x d.

Priorissa De Kyngtone tenet unum mesuagium aut duo in Langleygh unde sectam facit (ut fertur) ad dies legales tentas ibidem bis per annum.

Tenentes per Consuetudines Manerii ibidem Dimidii Hidarii

Isabella Russelle vidua tenet dimidiam hidam terre in Kyngton unde messuagium cum Curtillagio continet unam acram et in duobus Clausis annexatis vii acras.

Item tenet x^lvi acras dimidiam terre arabilis in ii^{bus} Campis unde in Campo orientali

    in Wydenalle quinque acras dimidiam in iiii^{or} particulis
    In occidentali parte de Barefurlong dimidiam acram
    In Sowarlondes ii acras
    In farendelle iiii acras dimidiam
    In Hendley iiii acras in iii^{bus} particulis
    In Smalemede dimidiam acram
    apud Ellenstubbe i acram
    In Strottefurlong iiii acras in ii^{bus} particulis
    apud la naysshe iii acras in iii^{bus} particulis
    apud Culnerwall unam acram
    apud Byddelyate dimidiam acram

                          [Total, 26½ acres in 19 parcels]

Et in Campo occidentali

    in Bradfurlong iiii acras in ii^{bus} particulis
    In orchadlondes ii acras
    apud Grovelondeshegge unam acram dimidiam
    apud la heele ii acras dimidiam
    In evydeen unam acram dimidiam
    Item ibidem iiii acras in ii^{bus} particulis
    Super Cowngroveshulle ii acras in ii^{bus} particulis
    In Wynewodes ii acras dimidiam in ii^{bus} particulis

                          [Total, 20 acres in 12 parcels]

    Et cum obierit dabit domino heriectum. Finis xx s. Redditus xi s.

Willelmus Necke De langley tenet dimidiam hidam terre unde messuagium cum Curtillagio, gardino, et pomerario continet dimidiam acram, et unum Clausum annexatum continet iiii acras iii perticatas prati.

Item xiiii acras dimidiam terre prati et pasture in separali unde in australi parte tenementi sui iii acras dimidiam perticatam, In Northeclose iii acras iii perticatas, In Langcroft iii acras, In Oldelond ii acras iii perticatas, In Hetheloce ii acras.

Item tenet xliii acras i perticatam terre arabile in ii^{bus} Campis unde in Campo boriali

   In Beerefurlong unam acram in ii^{bus} particulis
   Juxta Blakebuysshe dimidiam acram
   In Clandfeld In overfurlong ii acras in iii^{bus} particulis
   In Nitherfurlong unam acram dimidiam in iii^{bus} particulis
   apud Thornewell dimidiam acram
   In Holdeen dimidiam acram
   In Farendoune iiii^{or} acras in ii^{bus} particulis
   Super Mycheldale ii acras in iiii^{or} particulis
   In Netylmede dimidiam acram
   In Burymede unam acram in ii^{bus} particulis
   In Whytelond unam acram in ii^{bus} particulis
   In Hendley unam acram dimidiam in iii^{bus} particulis
   Super Whetehullehed dimidiam acram
   In Crownefurlong dimidiam acram
   apud Childaker unam acram in ii^{bus} particulis
   In Millefurlong ii acras in iiii^{or} particulis
   In Langdowne dimidiam acram
   In Shorldowne ii acras in v particulis
   In Walfurlong i acram
   In Hendley dimidiam acram

                            [Total, 24 acres in 41 parcels]

Et in Campo occidentali
   apud Galleaker dimidiam acram
   In orientali parte de haywode unam acram dimidiam in iii^{bus} particulis
   In Thyckefurlong unam acram in ii^{bus} particulis
   apud Barnardes Shave unam perticatam
   In Strottefurlong dimidiam acram
   apud lordeshulle unam acram in ii^{bus} particulis
   apud Northstock i perticatam

apud Jacobescrosse unam acram
In la Slade unam acram in ii^{bus} particulis
apud Wecheanger dimidiam acram
apud Shortcrosse dimidiam acram
In Staperlond dimidiam acram
In la More unam acram in ii^{bus} particulis
Juxta Somerleas unam acram
In Wydefurlong dimidiam acram
In Barrefurlong dimidiam acram
In Worthy dimidiam acram
apud Blakelond dimidiam acram
In Hangyn Cliffe unam acram in ii^{bus} particulis
apud Merlynpytte dimidiam acram
Super Fursehulle i perticatam
Super Whetehull unam acram dimidiam in iii^{bus} particulis
In Odgarston i acram in ii^{bus} particulis
In Pesefurlong dimidiam acram
In la hamme i acram in ii^{bus} particulis
In Overhamme unam acram in ii^{bus} particulis

[Total, $19\frac{1}{4}$ acres in 37 parcels]

Et cum obierit dabit domino heriectum. Finis cxiii s. iiii d. Redditus xvii s. ix d. ob.

[Four other dimidii hidarii have similar holdings.]

Virgatarii

Walterus Tourney De Langley virgatarius tenet unum Mesuagium cum curtillagio continenti i perticatam et in iii clausis annexatis x acras iii perticatas.

Item tenet unum Toftum cum iii clausis vocatis Jeffryes continentibus insimil quinque acras dimidiam.

Item tenet xix acras dimidiam terre arabilis in ii^{bus} Campis unde in campo boriali

in Barrefurlong unam acram in ii^{bus} particulis
In Clanfeld unam acram dimidiam in iii^{bus} particulis
Super Mycheldale unam acram in ii^{bus} particulis
Super Hendlye i acram in ii^{bus} particulis
In Whytelond dimidiam acram
In Boriali parte de Burymede unam acram in ii^{bus} particulis
In australi parte eiusdem unam acram in ii^{bus} particulis
apud Chyldaere unam acram in ii^{bus} particulis

In Myllof dimidiam acram
apud Whetehulhed dimidiam acram
In Wallefurlong dimidiam acram
In Shortdun dimidiam acram

[Total, 10 acres in 20 parcels]

Et in Campo occidentali
apud Galleacre unam acram in ii^{bus} particulis
In orientali parte de Haywode dimidiam acram
In Lordeshulle dimidiam acram
In Thyckefurlong dimidiam acram
In Strettfurlong dimidiam acram
In Morefurlong unam acram in ii^{bus} particulis
In Stapefurlong dimidiam acram
apud Shortcrosse dimidiam acram
In Pessefurlong dimidiam acram
In la Worthy dimidiam acram
In Odgarstone dimidiam acram
In Whetehulle unam acram in ii^{bus} particulis
In la hamme unam acram dimidiam in iii^{bus} particulis
In Wecheanger dimidiam acram

[Total, 9½ acres in 19 parcels]

Et cum obierit dabit domino heriectum. Finis xl s. Redditus xiiii s. viii d. ob.

Willelmus Taylour De Langley virgatarius tenet unum mesuagium et Toftum alterius virgate terre unde Curtillagium continet insimil i perticatam et in uno Clauso annexato quinque acras i perticatam.

Item tenet vii acras terre prati et pasture in separali unde in Northe-cloce unam acram iii perticatas, In Oldlondes quinque acras i perticatam.

Item tenet xxx acras iii perticatas terre arabilis in ii^{bus} Campis unde in Campo boriali
  in Barrefurlong unam acram in ii^{bus} particulis
  In Clanfeld dimidiam acram in iii^{bus} particulis
  In Thornewelfurlong iii perticatas in ii^{bus} particulis
  In Mycheldale unam acram dimidiam in iii^{bus} particulis
  In Netylmede dimidiam acram
  In Burymede unam acram in ii^{bus} particulis
  In Child acre unam acram i perticatam in iii^{bus} particulis
  In Okeworthe dimidiam acram

APPENDIX I

In Myllefurlong iii perticatas
In Troefurlong dimidiam acram
apud Whetehulleshed unam acram dimidiam in iii^{bus} particulis
In Wallefurlong dimidiam acram
In Langdoune iii perticatas in ii^{bus} particulis
In Shortdoune dimidiam acram
Item ibidem unam acram i perticatam
et in Guardeene i acram

[Total, 13¾ acres in 28 parcels]

Et in Campo occidentali
apud Galleacre unam acram in ii^{bus} particulis
Subtus Haywode unam acram dimidiam in iii^{bus} particulis
apud Barnardshave unam perticatam
In Strettfurlong dimidiam acram
In Thyckefurlong unam acram in ii^{bus} particulis
apud Langcrosse dimidiam acram
Super Lordeshulle unam acram in ii^{bus} particulis
apud Jacobbescrosse i acram in ii^{bus} particulis
In Stapulfuriong dimidiam acram
In Somerleas ii acras in iiii^{or} particulis
In Hangyngcliffe unam acram in ii^{bus} particulis
In Merlynpytt unam acram in ii^{bus} particulis
In Whetehull unam acram dimidiam in iii^{bus} particulis
In Odgarston dimidiam acram in ii^{bus} particulis
In Wecheanger unam acram iii perticatas in iiii^{or} particulis
In Worthe i perticatam
In Pessefurlong i perticatam
In la hamme unam acram dimidiam in iii^{bus} particulis
apud Northstocke unam perticatam

[Total, 16½ acres in 38 parcels]

Et cum obierit dabit domino ii heriecta. Finis xx s. Redditus xx s.
[Fifteen other virgatarii have similar holdings.]

DIMIDII VIRGATARII

Henricus Belle tenet dimidiam virgatam terre in Langley unde mesuagium cum Curtillagio continet i perticatam et in ii^{bus} Clausis annexatis vii acras.

Item tenet xi acras dimidiam terre arrabilis in ii^{bus} Campis unde in Campo boreali
in Berfurlong dimidiam acram

apud Blakebusshe dimidiam acram
In Thornewell dimidiam acram
In Mucheldale dimidiam acram
In Whitelond unam acram in ii^{bus} particulis
In henlee i perticatam
In Burymedefurlong i perticatam
In Whetehulhed dimidiam acram
In Millefurlong i perticatam
In la Downe unam acram i perticatam in iii^{bus} particulis
In Overclanfeld dimidiam acram
       [Total, 6 acres in 14 parcels]
Et in Campo occidentali
in Lordeshulle iii perticatas
Subtus Haywode dimidiam acram
In Thickefurlong dimidiam acram
In Stretfurlong i perticatam
In Churchewaiefurlong dimidiam acram
In la More unam acram in ii^{bus} particulis
In la Hamme i perticatam
In Overhamme dimidiam acram
In Whetehull dimidiam acram
In Berfurlong i perticatam
In Hangcliff dimidiam acram
       [Total, 5½ acres in 12 parcels]
Et cum obierit dabit domino heriectum. Finis xx s. Redditus vi s. ii d. ob.

Robertus Coke De Langley dimidius virgatarius tenet unam mesuagium cum Curtillagio continentem dimidiam acram et unum Clausum annexatum continens iiii^{or} acras dimidiam.

Item tenet unum Clausum pasture apud Northcloce continens ii acras separales.

Item tenet xii acras terre arabilis in ii^{bus} Campis unde
in Campo boriali
apud Barrefurlong unam acram in ii^{bus} particulis
In Clanfelde unam acram in ii^{bus} particulis
Super Mucheldale dimidiam acram
In Henlee dimidiam acram in ii^{bus} particulis
In Burymede iii perticatas in ii^{bus} particulis
In Millefurlong dimidiam acram

In Wallefurlong dimidiam acram
In Croefurlong dimidiam acram
In Langdoune i perticatam
Subtus Langdoune i perticatam
In Shortdoune dimidiam acram

[Total, 6¼ acres in 15 parcels]

Et in Campo occidentali
 apud Galleacre dimidiam acram
 Super Lordeshulle i perticatam
 Subtus Haywode iii perticatas in ii^{bus} particulis
 apud Fordesyate dimidiam acram
 In Thyckefurlong dimidiam acram
 In Strottefurlong i perticatam
 In la More dimidiam acram
 In la hamme iii perticatas
 Super Whetehulle dimidiam acram
 In Odgarstone unam perticatam
 In pessefurlong dimidiam acram
 In Beerfurlong i perticatam
 In Furcyhulle i perticatam

[Total, 5¾ acres in 14 parcels]

Et cum obierit dabit domino heriectum. Finis xx s. Redditus vi s. ii d. ob.

[Ten other dimidii virgatarii have similar holdings.]

Robertus Hagges tenet unum Cotagium cum Curtillagio in Kyngton continenti dimidiam acram.

Item tenet Toftum i placee terre in orientali parte Cotagii continentis unam perticatam.

Item tenet iii acras terre arabilis in ii^{bus} Campis unde
in Campo orientali Super Manneshulle unam acram dimidiam
Et in Campo occidentali in Ruydone unam acram dimidiam.

Finis iii s. iiii d. Redditus ii s. vi d.

Johannes Purymane, Paynter, tenet unum Cotagium cum Curtillagio in Langley continenti dimidiam acram et i hammam prati vocatam Mullehamme continentem i perticatam dimidiam cum ii Meerys annexatis in boriali Campo continentibus iii perticatas. Item tenet iii acras dimidiam terre arrabilis in eodem Campo in v particulis in furlongo vocato Clanfeld.

Item tenet annuatim ii plaustra in bosco domini apud Haywode in emendatione tenure predicte. . . . Finis ——. Redditus v s.

B. EXTRACTS FROM A SURVEY OF HANDBOROUGH, OXFORDSHIRE

Land Revenue, Misc. Bk. 224, ff. 96–145.   4 James I

Comitatus Oxoniae.   Manerium de Hanberough.

Supervisus Manerii predicti factus die Julii Anno regni domini nostri Jacobi Dei gratia Anglie Scotie Francie et Hibernie Regis fidei defensoris, etc., viz., Anglie Francie et Hibernie Quarto et Scotie xxxix, per Henricum Lee prenobilis garterii militem, Franciscum Stonerd, militem, Johanem Herche, Arm., et Johanem Herch, Jun., gen., virtute Comissionis dicti domini Regis extra Scacarrium suum eis et aliis directe, super sacrum Tenentium ibidem, viz., Richardi slutter, Jun., Johanis Wellr, Johanis Ford, Richardi Deane, Henrici Horne, Thome Lymborough, Richardi Rowland, Galfridi Hischman, Johanis Salter, Stephani Rugs, Rogeri Brooks, Willielmi Wrighte, Richardi Fletcher, Qui dicunt super sacrum suum quod

LIBERI TENENTES[1]

Rogerus Brooke Tenet libere per copiam datam quinto die Decembris Anno Regine Elizabethe xl^{mo} certas terras iacentes iuxta silvam vocatam Pindsley Coppice, viz.,

Terram arabilem vocatam le Sarte iacentem iuxta Pinsley Coppice per estimationem iiii acras

Boscum in Pinsley Coppice vocatum a hedge acre per estimationem
i acram

Habendas prefato Rogero et heredibus suis nuper procreaturis secundum consuetudinem Manerii predicti. Redditus per annum v d. ob., relievum x d.

Henricus Salter Tenet libere per copiam datam xxiiii^{to} die Marcii Anno Regine Elizabethe xliiii^{to} unam parcellam prati vocatam Ferretts meade, unam parcellam pasture vocatam Fulwell, et unam parcellam terre vocatam le Sartes nuper Willielmi Salter, patris sui, viz.,

parcellam prati in prato vocato Ferretts Meade per estimationem
iii acras ii rodas

---
[1] In the manuscript this rubric is in the margin.

parcellam pasture vocatam Fulwell iuxta Pindsley Coppice per estimationem   iii rodas
parcellam terre arabilis vocatam le Sartes in Myllfeld per estimationem   i acram ii rodas

Habendas sibi et heredibus suis nuper procreaturis secundum consuetudinem Manerii. Redditus per annum xii d. q., relievum xii d. q.

Jacobus Woldridge Tenet libere per Copiam datam xx° die Julii Anno Regine Elizabethe xliii° unum mesuagium sive Tenementum et unam clausuram eidem adiacentem in Handboroughe cum pertinentibus ex sursumredditu Galfridi London, viz.,

    Domum mansionalem iiii spaciorum, horreum iii spaciorum, unum stabulum cum aliis le outhouse iii spaciorum, Coquinam ii spaciorum, gardinum, pomarium, et Curtilagium, cum parva Clausa adiacenti, per estimationem   i acram ii rodas

Habendas sibi et heredibus suis secundum consuetudinem manerii. Redditus per annum xvi d., relievum ii s. viii d.

Thomas Martyn Tenet libere per Copiam datam xiiii die Martii Anno Regis Regine Elizabethe xxxix° unum Cotagium et unam peciam terre vocatam a garden plott cum pertinentibus vocatas Smartes ac dimidiam acram terre arrabilis iacentem in Mylfeild ex sursumredditu Ricardi Richardson, viz.,

    Domum mansionalem v spaciorum, horreum, et stabulum, et gardinum, per estimationem   i rodam
    Terram arrabilem in Myllfeilde per estimationem   ii rodas

Habendas sibi et heredibus imperpetuum secundum consuetudinem Manerii. Redditus per Annum iii d. Relievum vi d.

Georgius Cole, gen., Tenet per Copiam datam —— die ——Anno —— unum Cotagium vocatum Pynes, viz.,

    Domum mansionalem iiii spaciorum, gardinum, pomarium, et Curtilagium, per estimationem   iii rodas
    Clausam pasture vocatam Irenmongers per estimationem   ii acras
    Duo Cotagia vocata Clarkes v spaciorum, gardinum, pomarium, et Curtilagium, per estimationem   ii rodas
    Clausam pasture adiacentem   ii rodas
    Alium Cotagium ii spaciorum, gardinum, et Curtilagium, per estimationem   ii rodas
    Clausam terre arabilis vocatam Ridinges per estimationem   vii acras

Habendas sibi et heredibus secundum consuetudinem Manerii. Redditus per annum xxiii d. ob. q. Relievum duplex redditus.

Heredes Martini Culpeper, Militis, Clamant tenere libere per Copiam non ostensam certas terras in Hanborough, viz.,

    Domum mansionalem viii spaciorum, gardinum, cum clausa adiacenti, per estimationem   ii acras

    Terram arrabilem in le Hide vocatam Hutchins Hilles per estimationem   viii acras

    Clausam arrabilem vocatam Old Close per estimationem   vi acras

    Domum mansionalem vocatam Trumplettes ii spaciorum, gardinum, per estimationem   i rodam

Habendas sibi et heredibus suis secundum consuetudinem Manerii. Redditus per Annum ii s. v d. ob.   Relievum duplex redditus.

[There are several other similar freeholds.]

Custumarii

Heredes Martini Culpeper, Militis, Tenent per copiam — datam — die — Anno regis — unum mesuagium et unam virgatam terre cum pertinentibus in Hanborowe, viz.,

    Domum mansionalem ii spaciorum, unum horreum ii spaciorum, gardinum, et Curtilagium, cum clausa adiacenti, per estimationem   iiii acras

    Clausam terre arabilis vocatam Keenes per estimationem   vi acras

    Terram arabilem in Southfeild per estimationem   x acras

    Terram arabilem in Mylfeild per estimationem   x acras

    Terram arabilem in Myddlefeild per estimationem   viii acras

    Pratum in Southmeade per estimationem   i acram

    Pratum in Nyemeade per estimationem   i acram

    Pratum in Cowmore per estimationem   i acram

    Pratum in Fenlake per estimationem   ii rodas

    Communiam pasture ut supra

Habendas sibi et heredibus secundum consuetudinem Manerii. Redditus per Annum v s.   Finis duplex redditus.   Harrietum ——. Annualis valor dimittendus ——.

Ricardus Weller Tenet per Copiam ut dicitur sed non ostensam unum mesuagium et dimidiam virgatam terre cum pertinentibus, viz.,

    Domum mansionalem iii spaciorum, unum horreum iii spaciorum, Coquinam ii spaciorum, gardinum, pomarium, et Curtilagium, per estimationem   ii rodas

    Terram arabilem in le Myddle feild per estimationem   i acram i rodam

APPENDIX I

Terram arrabilem in le South feild per estimationem
                                                             iii acras ii rodas
Terram arabilem in le Myllfield per estimationem   i rodam
Pratum in Cotland Meade per estimationem   ii rodas
Pratum in Cheny Weare Meade per estimationem   ii rodas
Communiam pasture ut supra
    Habendas sibi et heredibus secundum consuetudinem Manerii. Redditus per Annum iiii s. vi d. Finis duplex redditus. Harrietum ——. Annualis valor dimittendus ——.

    Galfridus Hitchman Tenet per Copiam datam xvii die Novembris Anno Regni Regine Elizabethe xxii° unum messuagium et dimidiam unius virgate Terre cum pertinentibus nuper Elizabethe Hitchman, viz.,

Domum mansionalem vi spaciorum, horreum iiii spaciorum, i stabulum i spacii, pomarium, gardinum, et le backside, per estimationem   i rodam
Clausam pasture domui adiacentem per estimationem   i acram
Terram arrabilem in Myddle Feild per estimationem
                                                   v acras dimidiam
Terram arrabilem in South Feild per estimationem   v acras
Terram arrabilem in Myll Feild per estimationem   v acras
Pratum vocatum Stone Acre per estimationem   i acram
Pratum in Cowmore per estimationem   i acram
Communiam pasture pro xxx ovibus
    Habendas sibi et heredibus suis secundum Consuetudinem manerii per Redditum per annum ii s. vi d. Finis v s. Herrietum una vacca. Servicium, etc. Annualis valor dimittendus ——.

    Johannes Weller Tenet per Copiam datam xii° die Septembris Anno Regni Regis nunc Jacobi Anglie Francie et Hibernie Regis iii° et Scotie xxxix° unum mesuagium sive Tenementum et duas Clausuras et dimidiam virgatam terre cum pertinentibus ex sursumreddito Martini Varney, viz.,

Domum mansionalem ii spaciorum, unum horreum ii spaciorum, gardinum, et Curtilagium, per estimationem   i rodam
Clausam pasture vocatam Peakes per estimationem
                                                      i acram ii rodas
Clausam terre arrabilis vocatam Peakes per estimationem
                                                      i acram ii rodas

Terram arrabilem in le Myddlefeild per estimationem
                 iii acras ii rodas
Terram arrabilem in le Myllefeild per estimationem iii acras
Terram arrabilem in le Southfeild per estimationem
                 iii acras iii rodas
Pratum in Southmead per estimationem   ii rodas
Pratum in Nyemead per estimationem   ii rodas
Pratum in Cowmore per estimationem   ii rodas
Pratum in Fenlake per estimationem   i swathe
Communiam pasture ut supra

Habendas sibi et heredibus secundum consuetudinem Manerii. Redditus per Annum ii s. vi d. Finis duplex redditus. Harrietum optimum averium. Annualis valor dimittendus iiii li.

Johannes Bates, Clericus, Tenet per Copiam datam xxiiii^{to} die Marcii Anno Regni Regine Elizabethe xxxix° unum mesuagium sive Tenementum et dimidiam virgatam terre cum omnibus pertinentibus ex sursumredditu Ricardi Sampson filii et heredis Roberti Sampson, viz.,

 Domum mansionalem iii spaciorum, unum horreum ii spaciorum, unum stabulum i spacii, gardinum, pomarium, et Curtilagium, per estimationem   ii rodas
 Clausam pasture domui-adiacentem   ii acras
 Terram arrabilem in Myddlefeild per estimationem
                 iii acras ii rodas
 Terram arrabilem in Southfeild per estimationem
                 iii acras ii rodas
 Terram arrabilem in Myllfeild per estimationem   iii acras
 Pratum in Southmeade per estimationem   ii rodas
 Pratum in Nyemeade per estimationem   ii rodas
 Pratum in Cowmore per estimationem   ii rodas
 Pratum in Fenlake per estimationem   i rodam
 Communiam pasture pro omnibus Averiis in omnibus Communiis, etc.

Habendas prefato Johanni Bates et heredibus suis imperpetuum secundum consuetudinem Manerii. Redditus per Annum ii s. vi d. Finis duplex redditus. Herrietum optimum averium. Annualis valor dimittendus iiii li.

Ricardus Stutter, junior, Tenet per Copiam datam xvi^{to} die Martii Anno Regni Regine Elizabethe xliii° unum Tenementum et unam vir-

APPENDIX I

gatam terre cum omnibus pratis, pascuis, et pasturis eidem pertinentibus cum pertinentibus et alium Tenementum cum dimidia virgata terre cum pratis, pascuis, et pasturis eidem pertinentibus, viz.,

    Domum mansionalem xii spaciorum, unum horreum iii spaciorum, unum stabulum iii spaciorum, unum le Shepehouse iii spaciorum, gardinum, ii pomaria, et Curtilagium, per estimationem   i acram

    Clausam pasture vocatam le Corne Close cum alia clausa adiacenti vocata Heathfield per estimationem   x acras

    Terram arabilem in le Southfeild per estimationem   x acras

    Terram arabilem in le Myddlefeild per estimationem   x acras

    Terram arabilem in le Myllfeild per estimationem   x acras

    Pratum in Southmeade et Nyemeade per estimationem   iii acras

    Pratum in Cowmore per estimationem   i acram ii rodas

    Pratum in Fenlake per estimationem   iii rodas

    Alium domum mansionalem iii spaciorum cum gardino et pomario in occupatione Johannis Holloway per estimationem   ii rodas

    Communiam pasture pro omnibus averiis in Einsham heath et Kinges Heath

Habendas prefato Ricardo et heredibus suis imperpetuum secundum consuetudinem Manerii. Redditus per Annum vii s. vi d. Harrietum optimum averium. Finis duplex redditus. Annualis valor dimittendus x li.

Rogerus Brooke Tenet per Copiam datam quinto die Decembris Anno Regni Regine Elizabethe xl^{mo} unum Mesuagium, unum horreum, unum pomarium, unum clausum, unum gardinum, cum omnibus edificiis eidem mesuagio pertinentibus, et unam virgatam terre in Handboroughe ex sursumredditu Willelmi Watson et Isabelle uoris exius, viz.,

    Domum mansionalem vi spaciorum, ii horrea vi spaciorum, coquinam iii spaciorum, stabulum i spacii, unum shepehouse iiii spaciorum, gardinum, pomarium, et Curtilagium, cum parva Clausa adiacenti, per estimationem   iii acras

    Duas clausas pasture vocatas Heath Closes per estimationem   iiii acras

    Terram arrabilem in Southfeilde per estimationem   xi acras i rodam

    Terram brueriam ibidem per estimationem   ii rodas

    Terram arrabilem in Myllfeild per estimationem   vi acras i rodam

    Terram leazuram in Myllfeild per estimationem   i acram

    Terram arrabilem in Myddlefield per estimationem   vi acras

    Pratum in Southmeade per estimationem   i acram

Pratum in Nyemeade per estimationem   i acram
Pratum in Cowmore per estimationem   i acram
Pratum in Fenlake per estimationem   ii rodas
Communiam pasture in omnibus Campis, etc.

Habendas prefato Rogero Brooke et heredibus suis imperpetuum secundum consuetudinem Manerii. Redditus per annum v s. Finis x s. Harrietum optimum averium. Annualis valor dimittendus vi li.

[Several other customary tenants hold similar virgates or half-virgates.]

Terra Dominicalis per Copiam

Rogerus Legg Tenet per copiam datam —— die —— Anno Regis —— unam acram et unam rodam terre arabilis et unum rodam prati que sunt terre dominicales et vulgariter appellantur Buryland, viz.,

Terram arabilem in Myddlefeild per estimationem dimidiam acram
Terram arabilem in Southfeild per estimationem   i rodam
Pratum in bysouth et meadhey per estimationem dimidiam acram
Terram de Leyland in le Bushyehide per estimationem i rodam
Terram de ley iuxta domum predictum per estimationem i rodam

Habendas sibi et heredibus secundum consuetudinem Manerii. Redditus per annum ii s. iii d. Finis duplex redditus. Annualis valor dimittendus ——.

Heredes Stephani Culpeper Tenent per copiam datam die —— anno Regis —— quatuor acras terre arrabilis et unam acram prati que sunt terre dominicales et vulgariter appellantur Buryland, viz.,

Terram arabilem in Southfeild per estimationem   i acram
Terram arabilem in le hide per estimationem   i acram
Pratum in Bysouth per estimationem   i acram
Pratum in Meadhay per estimationem   i acram
Terram leazuram in bushiehide per estimationem   i acram

Habendas sibi et heredibus secundum consuetudinem Manerii. Redditus per Annum ix s. Que quidem premisse similiter clamantur per heredes Martini Culpeper.

Ricardus Lous Tenet per Copiam datam xiii° die Marcii Anno Regni Regine Elizabethe xxiii° octodecim acres terre arrabilis et tres acras prati que sunt terre dominicales et vulgariter appellantur Buryland, viz.,

Terram arrabilem in le hide per estimationem   xiii acras
Terram arrabilem in le Myddlefeild per estimationem i acram

APPENDIX I 437

Terram arrabilem in le Southfeild per estimationem   iiii acras
Pratum in Meadhay per estimationem   iii acras
Pratum in Bysouth per estimationem   iii acras
Habendas sibi et heredibus secundum consuetudinem Manerii. Redditus per annum xxvii s. Finis duplex redditus. Annualis valor dimittendus xl s.

[There are several other lessees of demesne lands. At the end of the survey are the signatures of the jurors.]

C. SUMMARIES OF TUDOR AND JACOBEAN SURVEYS WHICH ILLUSTRATE NORMAL TWO- AND THREE-FIELD TOWNSHIPS

Areas are in acres unless otherwise specified. Messuages are indicated by m., virgates by virg., cottages by cott., tenements by tent., and gardens by gard.

BRAILES, UPPER AND NETHER, WARWICKSHIRE

Land Rev., M. B. 185, ff. 181–229. 5 Jas. I

Custumarii (Upper Brailes)	Enclosed Unspec.	Pasture	Arable in the Open Common Fields		Common Meadow			
			North field	South field	North field "Leis et hades"	South field "Leis et hades"	Upper Md.	Nether Md.
Thos. Baldwyn,[1] m., 2 virg.	$1\frac{1}{8}$	..	12	12	$1\frac{1}{2}$	$1\frac{1}{2}$	$\frac{3}{4}$	1
Margar. Napton, m., 1 virg.	$\frac{3}{4}$	..	5	4	$\frac{1}{4}$	1	$\frac{1}{4}$	$\frac{3}{4}$
Marion Warde, m., 2 virg.	1	$\frac{1}{16}$	14	14	$1\frac{1}{2}$	1	$1\frac{3}{8}$	$1\frac{1}{2}$
Roger Marshall, m., 1 virg.	1	..	4	4	1	$1\frac{1}{4}$	$\frac{1}{4}$	$\frac{3}{4}$
Dorothea Nicolls, m., 2 virg.	1	$\frac{3}{4}$	10	9	3	$1\frac{1}{2}$	$\frac{1}{2}$	$1\frac{1}{2}$

Custumarii (Nether Brailes)							"Lotted ground" in "le heath"	Gallow hill
Thos. Bishopp, m., 2 virg.	1	..	11	9	3	2	2	2
Ed. Walker, m., 1 virg.	1	1	6	7	4	4	1	1
Serack Ockley, m., 1 virg.	1	..	8	7	3	2	1	1
Wm. Gardner, m., 2 virg.	1	1	12	12	4	4	2	2
Ric. Rymell, m., 1 virg.	$1\frac{1}{4}$	..	7	6	$3\frac{1}{2}$	$2\frac{1}{2}$	1	1

There are many similar holdings.

[1] Baldwyn has "communia pasture in quibusdam pasturis iacentibus in upper brailes vocatis . . . [nine pastures named] pro viii averiis, v equis, iiii^{xx} ovibus." Other tenants fare proportionally.

Shipton-under-Wychwood, Oxfordshire
Land Rev., M. B. 189, ff. 96–100. 6 Edw. VI

Custumarii	Enclosed	Arable in the Open Common Fields		Common Meadow
		East field	West field	
Thos. Bradshaw, m., 2 virg.......	6	44	35	4½
Eliz. Gybbons, m., 3 virg.........	9½	40	40	11
Thos. Altofte, m., [?] virg.........	3	66	60	7½
John Whytinge, m., 2 virg........	3	46	40	3½
John Harris, m., 2 virg...........	¾	42	39	4
Michael Hucks, m., 3 nocate	2	20	19	1½
Thos. Hucks, m., 1½ virg.........	1¾	32	22	6
Henry Parrotte, m., 2 virg........	1	30	30	6
Rich. Canburye, m., 1 virg........	1	24	17	2
Rob't. Sell als. Taylor, m., ½ virg. ...	3	17	8	4½

There are four other copyholds.

Charlton Abbots, Gloucestershire (in the Cotswolds)
Exch. K. R., M. B. 39, ff. 166–170. [Edw. VI]

Copyholders	Pastura Separalis	Arable in the Open Common Fields		Common Meadow
		West field	East field	
Jac. ——, m., cott., 1 virg. ...	2 ferrundells [1]	24	24	9
Edm. Copping, m., 1 virg.....	1 ferrundell	24	24	9
Rog. Drewe, m., 1 virg.......	½ acre	24	24	9
Katerina Drewe, m., 1 virg. ...	1 ferrundell	24	24	9
Wm. Diche, m., 1 virg........	½ acre	24	24	9
Mariana Bedell, m., 1 virg. ...	½ acre, 1 ferrundell	3	3	..

This is a complete list of the copyholders. They all have stinted pasture, e.g., " pro c bidentibus, xii animalibus, ii equis."

Weston Birt, Gloucestershire (in the Cotswolds)
Rents. & Survs., Portf. 2/46, f. 150. 1 Edw. VI

Copyholders	Enclosed	Arable in the Open Common Fields		Common Meadow
		North field	Campus Australis	
Hen. Parker, m....................	1	21	21	½
Thos. Drewe, m., 1 virg..........	1	21	21	½
Jo. Redford, m., 1 virg...........	1	28½	29½	..
Thos. Cirtle, m., 1 virg...........	1	20	17	¾
Jo. Tyler, 3 m., 2 virg............	2	40	39	1¾
—— Clerk, m...................	½	8	7	¼
Wm. Holboroughe, 4 m., 2 virg.....	12	30	20	¾

A complete list.

---

[1] The ferrundells are located, e.g., " ad partem australem ecclesie," and obviously constitute the village closes.

APPENDIX I

South Stoke, Somerset

Land Rev., M. B. 225, ff. 150–159. 6 Jas. I

Copyholders	Enclosed				Arable in the Open Common Fields	
	Arab.	Md.	Past.	Unspec.	East field	West field
Wm. Hedges, m.	$2\frac{3}{4}$	1	$4\frac{3}{4}$	..	14	14
Wm. Mercer, m.	$\frac{1}{2}$	3	$5\frac{1}{2}$	$5\frac{1}{2}$	25	29
Alice Willis, m.	..	$3\frac{1}{4}$	5	..	16	16
John Dagger, m.	$2\frac{1}{4}$	1	2	$4\frac{1}{2}$	6	6
Joane Browne, m.	1	3	10	1 wood	20	20
Laurence Smythe, m.	$\frac{1}{2}$	4	$10\frac{1}{2}$	$9\frac{3}{4}$	3	$6\frac{1}{2}$
John Awburd, m.	$\frac{1}{8}$	$1\frac{1}{2}$	7	..	14	16
Editha Reade, m., 1 yearde of land	3	$3\frac{1}{2}$	$7\frac{1}{4}$	..	16	20
Thos. Hudd, m., $\frac{1}{2}$ yearde of land	7	3	$12\frac{1}{4}$	..	9	20

Gillingham, Dorset

Land Rev., M. B. 214, ff. 1–47. 6 Jas. I

Custumarii	Arab.	Enclosed Md.[1]	Past.	Arable in the Open Common Fields			Common Meadow
				South field	North or Woodhouse field	Madgeston field	
Dorothie Dirdoe, m.	$\frac{1}{2}$	..	$7\frac{1}{2}$	6	4	2	..
Jo. Mullins, m.	$\frac{1}{8}$	1	..	2	4	..	3
Wm. Helmes, m.	$\frac{1}{2}$	11	10	2	$2\frac{1}{2}$	..	..
Geo. Jukes, 2 m.	$2\frac{1}{4}$	$12\frac{1}{2}$	3	6	11	..	..
Wm. Bowles, m.	$\frac{1}{4}$	$5\frac{3}{4}$	..	7	$8\frac{1}{2}$	..	$2\frac{3}{4}$
Ric. Cooke, 2 m.	2 perches	1	..	3	5	..	1 and 4 gates
Wm. Sheppard, m.	$\frac{3}{4}$	$7\frac{1}{2}$	5	11	10	1	$1\frac{1}{4}$ and 8 gates
Joan Mountier, m.	1	3	$7\frac{1}{2}$	22	12	..	$4\frac{1}{4}$

The copyholders have "common in the forest."

[1] Some of the meadow was probably open. The usual phrase is, e.g., "one meadowe called broade meade 4 acres."

APPENDIX I

WELLOW, ISLE OF WIGHT

Exch. Aug. Of., M. B. 359, ff. 26–36. 6 Jas. I

Custumarii	Enclosed	Arable in the Open Common Fields	
		East field	West field
Elenora Herring, m.	$3\frac{1}{4}$	$5\frac{3}{4}$	$10\frac{1}{2}$
Agnes Hellier, m.	$4\frac{3}{8}$	$5\frac{1}{2}$	9
Robt. Gopard, m.	3	2	5
John Bull, m.	$5\frac{1}{4}$	$5\frac{1}{2}$	2
Thos. Gillett, cott.	$1\frac{5}{8}$	$\frac{1}{4}$	...
Jo. Goodall, m.	$2\frac{3}{4}$	6	$8\frac{1}{2}$
David Dore, m.	4	7	10
Thos. Dore, m.	$5\frac{5}{8}$	$4\frac{1}{2}$	3
Jo. Squib, m.	$4\frac{1}{4}$	7	7
Ric. Bigger, m.	3	$7\frac{1}{2}$	11
Thos. Urrye, m.	$4\frac{3}{4}$	8	8
Ric. Fricket, m.	$1\frac{3}{4}$	7	12
Jo. Goodale, m.	$1\frac{1}{2}$	8	13
Jo. Cooke, m.	$3\frac{1}{4}$	8	$8\frac{1}{2}$
Ed. Lancham, m.	$2\frac{1}{4}$	7	$9\frac{1}{2}$
Robt. Powell, m.	$3\frac{1}{2}$	5	$13\frac{1}{4}$
Thos. Bartlett, m.	$5\frac{3}{4}$	13 [1]	...

A complete list of copyholders. They all have stinted common for sheep in the common of Wellow.

HUMBERSTON, LINCOLNSHIRE

Land Rev., M. B. 256, ff. 272–285. 5 Jas. I

Liberi Tenentes [2]	Enclosed	Arable in the Open Common Fields		Common Meadow
		East field	West field	
Vincent Sheffeld, m.	3	30	30	28
Wm. Nutsey, m.	9	20	20	$22\frac{1}{2}$
Mich. Spencer, m.	$2\frac{3}{4}$	16	16	11
Rich. Allenson, m.	4	7	6	$11\frac{1}{2}$
Thos. Hawnby, m.	$\frac{3}{4}$	9	$7\frac{1}{2}$	2
Wm. Wentworth, m.	$\frac{1}{2}$	14	14	8
Saml. Waterhouse, m.	1	30	30	5
Wm. Webster, m., 3 cott.	$10\frac{1}{2}$	44	44	17
Eliz. Dickinson, m.	$\frac{1}{2}$	14	16	$19\frac{1}{2}$
Wm. Wraye.	1	10	10	12

Some of the tenants have stinted common of pasture in "le Southe marshe."

[1] "In communibus campis."   [2] There are no copyholders.

ALVINGHAM, LINCOLNSHIRE

Land Rev., M. B. 265, ff. 1–13. 1608

Lessees[1]	Enclosed Md. and Past.	Arable in the Open Common Fields		Common Meadow
		East field	West field	
Wm. Hennage, Esq., m., cott.	30	35	35	..[2]
Ric. Horsard, m.	4½	30	30	15
Jo. Yarburghe, cott.	11	5	5	..
Ric. Mackerell, m.	3	16	16	6
Abm. Blainchard, m.	4	20	20	8
Wm. Horford, m.	12	26	30	..
Wm. Horford, 2 cott.	1½	3	3	..

BOWER HENTON, A DECENNA OF MARTOCK, SOMERSET

Land Rev., M. B. 203, ff. 306–315. 1-2 Philip and Mary

Custumarii	Enclosed		Arable in the Open Common Fields			Common Meadow
	Arab.	Past.	Campus Australis	Campus Orientalis	Campus Occidentalis	
Hugo Geale, m.	¾	7½[3]	6	6	6	3⅛
John Prest, m.	2½	13	12	12½	10½	5½
John Gawller, m.	¼	19½[4]	12	12	11	6¼
Margareta Borow, m.	3⅛	11	8½	7	7¼	3⅝
Alicia Adams, m.	2	6[3]	11¼	9	7¼	4¼
Johes. Gele, m.	½	9[5]	6	5¾	5¾	4¾
Alicia Punfolde, m.	¾	16½[6]	10½	9	8	4¾
Willielmus Genes, m.	½	7½	6½	8	7	4
Hugo Poole, m.	1½	8	9	8¼	8	4
Jo. Symond, m.	1	16¼[7]	5½	5½	5¼	4⅝

There are nineteen similar holdings. Tenants have unstinted common of pasture in Whetmore.

[1] There are no copyholders.
[2] "Tythe haye in 80 acres."
[3] "De novo incluse."
[4] 2½ "de novo incluse."
[5] 3 "in le estfild."
[6] 4½ "de novo incluse."
[7] 12 "in le Southefilde."

HINTON ST. MARY, DORSET

Land Rev., M. B. 241, ff. 84 sq. [Eliz.]

Copyholders	Enclosed Past.	Enclosed Md. or Past.	Arable in the Open Common Fields North field	Arable in the Open Common Fields South field	Arable in the Open Common Fields West field	Common Meadow
Geo. Yonge, m.	4	16	15	7½	10½	5½
Ric. and Wm. Shorte, 2 m., 1 cott.	..	38	14½	12	13	7
Widow Stacye, cott.	..	..	2	4¾	2	1
Thos. Lambert, cott.	4¼	..	..	2	4	1
Margaret Webb, 2 m.	6	13	11	11	11	3½
Alice and Rich. Cowpe, cott.	4	1	..	..	..	3¼
Isabell Gancer, cott.	..	11	..	..	..	..
Elene More, m.	3	11	4½	5	2	2
Robt. Stacye, m.	..	7½	3½	5	4½	1
Wm. Castleman and Jo. Gardyner, m.	11	7	10	6	6	7
Ric. and Thos. Castleman, m.	..	16½	5	11	8	2½

There are fifteen other copyholds. Each copyholder has "communem in the fylds for his cattell and shepe."

ASHTON KEYNES, WILTSHIRE

Exch. Aug. Of., M. B. 422, ff. 48–79. 1 Jas. I

Custumarii	Enclosed Arab.	Enclosed Past.	Arable in the Open Common Fields East field	Arable in the Open Common Fields North field	Arable in the Open Common Fields West-ham	Common Meadow
Ant. Ferres, m., 1 virg.	1¼	18	12	11	16	..
Johanna Archard, m., 1 virg.	1	12	3	10	9	⅝
Jo. Cove, m., 1 virg.	½	17	8	8	7	7
M. Chapperlyn, m., 1 virg.	4½	5	15	11	8	2½
Wm. Clyfford, m., 1 virg.	3	6	8	10½	8	13½
Jo. Rowley, m., ¼ virg.	½	..	..	..	5	3
Ric. Sawyer, m., ½ virg.	½	2½	6	3¾	4½	3
Joanna Syninge, m., ½ virg.	1	6	..	6	10	1
Geo. Androwes, m., ½ virg.	2	..	9¾	6½	7	1¾
Benetta George, m., ¼ virg.	1	2	7½	3¼	4¼	1
Ellena Hardinge, m., ½ virg.	3	7	6	7	10	1¾
Jo. Hardinge, m., ½ virg.	2	6	7	7	10	2

There are several other copyholders and several freeholders. The copyholders have unstinted common of pasture in the common fields, the common meadows, and the forest of Braydon.

APPENDIX I

Ansty, Hampshire
Exch. Aug. Of., M. B. 56, ff. 67–76. 10 Hen. VI

Tenants	In Netherstrete	Arable in the Open Common Fields		
		South field	Middel field	East field
Edwatte Rithe, 10 acres pasture (4)[1]	5	3 (4)	$4\frac{1}{2}$ (7)	$3\frac{1}{4}$ (6)
Jo. Ansty, m.	..	$\frac{3}{4}$ (2)	1 (2)	$\frac{1}{2}$ (1)
Thos. Cochman, m.	..	1 (3)	1 (2)	2 (3)
Robt. Kyng, 2 m., toft, croft.	$\frac{1}{4}$	$2\frac{1}{2}$ (3)	$1\frac{3}{4}$ (4)	4 (8)
Phil. Gredare, m.	1 (2)	$4\frac{3}{4}$ (7)	4 (8)	$4\frac{1}{4}$ (4)
Steph. Dyere, m.	3	$4\frac{1}{4}$ (7)	$3\frac{1}{2}$ (5)	2 (4)
Jo. Tortyngton, toft.	$1\frac{1}{4}$	$19\frac{1}{2}$ (7)	8 (11)	$11\frac{3}{4}$ (14)
Wm. Sawyer, m., croft.	..	$3\frac{1}{2}$ (5)	3 (7)	4 (9)
Wm. Asselot, toft.	$\frac{1}{4}$	1 (1)	$1\frac{3}{4}$ (4)	$1\frac{3}{4}$ (5)
Wm. Wollane	4 (3)	$5\frac{1}{2}$ (5)	....	......
Jo. Godard, toft.	$\frac{1}{2}$	1 (1)	$1\frac{1}{2}$ (2)	1 (1)
Jo. Clerk, m., toft., garden	$1\frac{1}{2}$	$7\frac{3}{4}$ (9)	$9\frac{3}{4}$ (11)	$12\frac{1}{2}$ (11)
Jo. Ken, 3 m., 4 gardens, toft, croft, together containing $7\frac{3}{4}$ acres	$5\frac{1}{4}$ (5)	$12\frac{1}{2}$ (18)	18 (27)	$18\frac{1}{4}$ (31)

Alfriston, Sussex
Exch. Aug. Of., M. B. 56, ff. 257b–266. 11 Hen. VI

Copyholders	Arable in the Open Common Fields			Common Meadow
	North leyne	Middil leyne	South leyne	
John Syger, cott., $\frac{1}{2}$ wista	$3\frac{3}{4}$ (6)[1]	$2\frac{1}{2}$ (6)	2 (4)	$\frac{1}{2}$
Rich. Man, tent., $\frac{1}{2}$ wista	2 (3)	3 (4)	$2\frac{1}{2}$ (4)	$\frac{1}{2}$
John Bydon, tent., 1 wista	$6\frac{3}{4}$ (9)	5 (6)	$6\frac{1}{2}$ (7)	1
Simon Benet, tent., 1 wista	$5\frac{1}{2}$ (7)	$5\frac{3}{4}$ (7)	$5\frac{3}{4}$ (8)	1
Thos. Smyth, tent., $\frac{1}{2}$ wista	$2\frac{3}{4}$ (5)	$2\frac{1}{2}$ (5)	3 (5)	$\frac{1}{2}$
Philip Younge, tent., 1 wista	$6\frac{1}{4}$ (8)	$5\frac{1}{2}$ (6)	$6\frac{3}{4}$ (10)	$1\frac{1}{4}$
Demesne at farm				
Rich. Man, $\frac{1}{2}$ wista	$3\frac{1}{4}$ (7)	$2\frac{1}{4}$ (6)	$2\frac{3}{4}$ (4)	..
Rich. Chukke, $\frac{1}{2}$ wista	$3\frac{1}{4}$ (7)	$2\frac{1}{4}$ (6)	$2\frac{3}{4}$ (7)	..

This is a complete list of the wistae, but there were several cottagers.

[1] The figures in parentheses indicate the number of parcels.

SALFORD, BEDFORDSHIRE. All Souls Typus Collegii, Map of 1595

Tenants of the College Grounds	Enclosed	Arable in the Open Common Fields: Brook field	Middle field	Wood field	"Meadow in the Fields"	"Pasture and Lea Ground"
Henry House........	17¼	21 (31)[1]	15½ (28)	24½ (33)	4½	6¾
Widow Crowley.....	5	11¾ (17)	9 (11)	11 (19)	5¼	12¼
Mr. Francklinge.....	10½	19½ (27)	9½ (18)	18¾ (22)	4½	16
Widow Perse	4	30¼ (47)	22¼ (31)	33¾ (39)	9¼	24¾
Ric. Odell..........	5	10½ (24)	8½ (18)	11¼ (18)	4½	13
Robt. Cowper.......	2½	11½ (23)	6¼ (13)	12½ (22)	4½	12½
Robt. Woodwarde...	11¼	10¼ (14)	9¼ (13)	9 (11)	4½	4
The Miller..........	...	....	....	....	3¼	...
Robt. Freeman......	...	1 (2)	....	....	...	...
Jo. Crouche.........	...	½ (1)	1¾ (3)	....	...	...
Wm. Briar..........	2¼	1½ (4)	2¼ (3)	1 (2)	¾	4¾
Martha Langford ...	272¼[2]	....	....	....	24	...
Freeholders						
The Vicar..........	4¾	6¼ (17)	5¼ (12)	7¾ (12)	5¼	1¼
Thos. Pedder	4	1½ (3)	1½ (3)	2 (3)	3½[3]	...
Edw. Butterfield	¼	4½ (10)	½ (1)	2 (5)	1[3]	...
Widow Letter	...	2 (4)	1¼ (4)	½ (2)	...	...

A complete list.

WELFORD, NORTHAMPTONSHIRE (two manors)

Bodl., Gough MS., Northants. 2. 1602 (XVIII cen. copy)

	Enclosed	Arable in the Open Common Fields: Hemplow field	Middle field	Abbey field
Manor of Wm. Saunders				
Wm. Saunders, Gent., demesne...	4 large closes	77¾	83½	79½
"Tenants at Will"				
Mary Symes, m., 2½ virg.	½	13¼	11¾	16
Randall Wilkinson, m., 2 virg....	¼	10¾	13¾	13
Ric. Willis, m., 1½ virg..........	1	9	9½	9
Thos. Brett, m., 1¾ virg.........	¼	10¾	10¾	10½
Robt. Eyle, m., 1 virg...........	⅛	5	5¼	6
Manor of the Queen, formerly of Sulby Monastery				
"The Queen's Patentees"				
Roger Brewster, m., 3½ virg......	¾	16¼	24¼	24¾
Katherine Watts, m., 1 virg......	⅛	4¼	5	5¾
Francis Vanse, m., 1¼ virg.	⅛	8	8¾	6¾
Ed. Horton, m., 1 virg..........	¼	8½	7¾	8½
"Ancient Freeholders"				
Wm. Sturgis, m., 1 virg..........	..	5¾	5¼	4¼
Robt. Moore, 2 m., 1½ virg.......	⅛	8¼	9	8¾
Jo. Cox, m., 1 virg..............	⅛	4¾	4½	7¾
Theron Symes, m., ½ virg........	..	1¾	3¼	4½
Thos. Noble, m., ¼ virg..........	..	2	1	¾

There are other tenants of each class. The strips of meadow in the fields have been omitted in all cases.

[1] The figures in parentheses indicate the number of parcels.

[2] Of this, 160¼ acres are arable and lie mainly in Middle field, perhaps unenclosed.

[3] "Meadow and pasture in the fields."

LUTTERWORTH, LEICESTERSHIRE. Land Rev., M. B. 255, ff. 117–172. 5 Jas. I

Tenants "per Letteras Patentes"	Enclosed	Arable in the Open Common Fields: Strete field	Cowhey-don field	Thorne-borough field	Meadow in: Strete field	Cowhey-don field	Thorne-borough field	"Leis" in: Strete field	Cowhey-don field	Thorne-borough field	In bruerio vocato Cowpasture
Wm. Hayward,[1] m., 3 virg.	¾	12	12	12	3	2½	3½	1	1	1	1¼
Edw. Neale, m., 3 virg.	3¼	12	14	12	3	2½	2½	1½	3	1½	4
Joh. Neale, m., 3 virg.	2½	12½	12	12½	3	2½	3	2¾	2¾	2	4
Wm. Heyward, m., 1 virg.	1	4	4¼	4	1	¾	1¼	⅜	½	½	1
Franc. Tayler, m., 1 virg.	1¼	6	6½	5	2 ("in campis")		..	¼	½	¾	..
Thos. Wood, m., 3 virg.	1¾	12¼	13¼	12½	3	2½	3	1¼	1½	½	2
Wm. Cooke, m., 2 virg.	⅜	8	8	8	1½	2	1¾	½	1	½	1

There are many similar holdings.

ROLLESTON, STAFFORDSHIRE. D. of Lanc., M. B. 109, ff. 57–151. 1 Eliz.

Liberi Tenentes	Enclosed	Arable in the Open Common Fields: Doddes-low field	More field	Crathorne field	Falling-pit field	Dove field	Nether field	Miscel.	Common Meadow
Joh. Shepard, m., garden	½	4½ and More		6 and Fallingpit		7 and Nether		1	1¾
Ric. Watson, m., gard.	½	½	5¾	1	4¾	3¼	½	..	6½
Nich. Caldwall, m., 1 virg.	½	..	3½	3½	3	3	..	..	1¾
Law. Watson, m., gard.	½	8	..	..	..	8½	..	..	2⅛
Thos. Statham, m., gard.	½	3	..	2½	..	2½	..	..	½
Joh. Cortell, m.	½	..	5	..	6	5	..	..	2
Humf. Carter, m., gard.	½	6	..	2½	..	2	2	..	1¼
Humf. Carter, m., gard.	½	½	3½	..	6	4	..	..	2
Robt. Bonde, m.	½	1	4¾	½	4½	3¼	½	..	2
Thos. Bonde, m.	½	5½	..	6	1	4½ and Nether		..	2
Wm. Golde, m.	½	5	..	4	..	4¾	..	..	2½
Agnes Statham, m.	¾	2	1½	..	4½	1	..	..	½
Margareta Prestbury, m.	¾	5¼	¾	4	1	1½	..	1½	3½
Thos. Raye, m.	½	¾	4¼	½	3¾	3½	½	¼	2

There are many similar holdings.

[1] Wm. Hayward has "communia pasture pro ix averiis et v^{xx}v ovibus in campis," the usual stint for a holding of three virgates.

446 APPENDIX I

Elloughton, Yorkshire

Land Rev., M. B. 229, ff. 74–86. 6 Jas. I

Copyholders	Enclosed	Arable in the Open Common Fields			Common Meadow[1]
		Southeast field	Middle field	Milne field	
Franc. Scarfe, m., 3 cott...	..	10	8	10	8
Wm. Carhill, m., cott.	$\frac{1}{2}$	15	12	15	12
Jo. Simpson, m...........	6	10	8	10	8
Jane Bacon, m...........	1	$7\frac{1}{2}$	6	$7\frac{1}{2}$	6
Peter Bower, m...........	..	10	8	10	8
Hamond Kelde, m., 4 oxgangs.................	..	10	8	10	8
Thos. Simpson, m., 2 oxgs..	..	5	4	5	4
Rich. Bentlye, 4 cott., 2 oxgs.	..	5	4	5	4
Robt. Carlille, $1\frac{1}{2}$ oxgs. ...	..	$3\frac{1}{4}$	3	$3\frac{1}{4}$	3
Wm. Kirke, cott., $\frac{1}{2}$ oxg. ..	..	$1\frac{1}{4}$	1	$1\frac{1}{4}$	1
Francis Thorley, 1 oxg. ...	..	$2\frac{1}{2}$	2	$2\frac{1}{2}$	2

There are many other holdings, all smaller than $1\frac{1}{2}$ oxgangs. Each oxgang has common " for 1 draught, 2 kine, and 1 yonge beast."

Ingleton, Durham

Land Rev., M. B. 192, ff. 16b–18. 5 Jas. I

Tenants "ex Litteris Patentibus"	Enclosed		Arable in the Open Common Fields			Common Meadow
	Arab.	Md.	North or Rigg field	East field	South field	
Robt. Shawe, tent.........	$1\frac{1}{4}$	5	7	11	8	3
Geo. Marley, tent..........	$\frac{3}{4}$	$1\frac{3}{4}$	4	6	$4\frac{1}{4}$	2
Barth. Horne, tent.........	$\frac{3}{8}$	$1\frac{3}{8}$	$3\frac{1}{2}$	6	$5\frac{1}{2}$	$1\frac{1}{4}$
Pet. Marley, tent..........	$\frac{1}{2}$	$2\frac{1}{2}$	11	$7\frac{3}{4}$	$7\frac{1}{2}$	$2\frac{3}{4}$
Geo. Marley, tent..........	1	..	6	$5\frac{1}{2}$	$5\frac{1}{4}$	$2\frac{1}{4}$
Milo Waide, tent..........	$\frac{3}{4}$	3	6	6	5	1
Pet. Dixon, tent...........	$\frac{3}{4}$	$3\frac{1}{2}$	6	$5\frac{1}{2}$	3	$2\frac{1}{4}$
Jo. Waide, tent............	$\frac{3}{4}$	$3\frac{1}{2}$	6	$5\frac{1}{2}$	3	$2\frac{1}{4}$
Wm. Simpson, $\frac{1}{2}$ tent.......	$\frac{3}{8}$	2	$2\frac{1}{2}$	5	5	1
Robt. Paverell, $\frac{1}{2}$ tent.	$\frac{3}{8}$	2	$2\frac{1}{2}$	5	5	1
Edw. Middleton, tent.	$3\frac{1}{4}$	..	$4\frac{1}{2}$	$5\frac{1}{2}$	$4\frac{1}{2}$	$1\frac{3}{4}$

A complete list of patentees. There are freeholders, but no copyholders. The patentees have stinted pasture in "le town pasture" (also once called "le towne feilds") and in "le Faughs."

---

[1] These acres are "on the salt marsh."

APPENDIX I

STOKE PRIOR, HEREFORDSHIRE. Land Rev., M. B. 217, ff. 349-388. 6 Jas. I

| Custumarii[1] | Arab. | Enclosed Md. | Past. | Arable in the Open Common Fields ||||||| Common Meadow |
				Blakardyn	Elfords field	Church field	Grettwall field	Shuttocks field	Churchwall field	Aspe field	
Thos. Kynton, m., Noven's Lands and Smythe's Lands	½	¾	1¼	18	16	8	8	2	..	..	1
John Child, m.	1	1	7½	20	20	20	..	..	..	..	1
Egidius Caldewell, m.	1	..	1	16	18	8	..	..	8	..	2½
Phil. Caldwall, m.	1½	1½	6	16	18	15	..	..	..	..	2
John Davyes, m.	¼	3¼	6	14	10	3	..	3	..	11	1

RISBURY, A MEMBER OF STOKE PRIOR

| Custumarii | Arab. | Enclosed Md. | Past. | Arable in the Open Common Fields ||||| Common Meadow |
				Muston field	Mere field	Inn field	Ornall field		
Wm. Stansbye, m.	2	2½	5	16	16	16	..	..
Jo. Stannesby, m.	½	2	2¾	4	4	4	..	..
Joh. Stann, m.	2	7	16½	30	30	30	..	..
Liber tenens								
John Evans, m.	1	6	15½	24	24	5	15	2½

HENOR, PART OF THE MANOR OF STOKE PRIOR

| Liber tenens | Arab. | Enclosed Md. | Past. | Arable in the Open Common Fields ||| |
				Fairemile field	Great field	Hallibrook field
Wm. Burgin, m.	1	10½	23¼	30	30	28
2 cott. and mill	2	1	4	..	..	..

[1] A complete list of copyholds, except two small ones without messuages. There are several large freeholds with areas similarly divided among the fields. The Risbury list is complete.

STOCKTON, HEREFORDSHIRE

Land Rev., M. B. 217, ff. 296–348. 6 Jas. I

Custumarii [1]	Enclosed			Arable in the Open Common Fields								Common Meadow
	Arab.	Md.	Past.	Rowleys field	Church field	Meare field	Moore field	Bache field	Rade field	Grawntons field	Kyrmeltons field	Polliats Croft	
Ric. Carpenter, m.	½	..	..	4	4	..	..	..	..	..	4	..	..
Ric. Wanckleton, m.	gard.	5	1¾	15	..	10	3	1	13	1	..	..	3
Walter Colman, m.	¾	4	3¾	8	1	1	6	8	12	2	6	..	1
Jo. Hale, m.	gard.	..	3½	16	..	8	8	..	14	..	..	..	2½
Wm. Powle, m.	gard.	4½	8	20	..	12 and Moore		2	13	..	..	2	1
Walt. Bilwyn, m.	½	..	..	3	..	..	3	..	3	..	..	..	..
Humf. Bilwyn, 2 m.	2	..	5	16	..	3	16	..	16	..	..	..	2
Wm. Yeomans, m.	gard.	2	5¾	18	..	..	2	16	18	..	..	..	1¾
Wm. Bach, 2 m.	1	..	4¼	20	..	..	20	..	20	..	..	..	2
Joh. Musgrove, m.[2]	1	1½	4	10	..	5	..	10	10	1	..	..	1

HAMNASHE, PART OF THE MANOR OF STOCKTON (f. 321)

Custumarii [3]	Enclosed			Arable in the Open Common Fields			
	Arab.	Md.	Past.	Hamnash field	Hamnash Over field	Wales et Litel field	Campus adiacens Hallibrooke field
Jo. Cornwall, m.	2	2	9½	12	10	8	..
Jo. Bideawhile, m.	gard.	3	4	10	..	10	10
Joh. Goodier, m.	3	1	9½	10	13	12	..

[1] This list includes all the customary tenants except Anne Hardwick, who has a messuage, 8¾ acres of meadow and pasture, and 35 acres of arable assigned without division to six fields.
[2] This holding is separated from the others. For it alone there is declared to have been " communia pastura in communibus campis predictis pro omnibus suis ovibus et averiis (f. 330)."
[3] The list of customary tenants is complete.

APPENDIX I

KIMBOLTON, PART OF THE MANOR OF STOCKTON (ff. 318–321)

Custumarii[1]	Enclosed			Arable in the Open Common Fields					
	Arab.	Md.	Past.	Church field	Midle field	Criniden field	Raide field	Hardwick field	Hopmonway field
Wm. Jeffries, m.	gard.	2	15½	5	4	6	20	14	..
Jo. Waucklen, m.	gard.	5	9¼	10	7	3	20	..	20
Hugo Waucken, m.	gard.	11½	12½	2	10	10	20	..	26
Jo. Morris, m.	gard.	7	15	18	13	6	40	..	40
Thos. Goodeyere, gent.	..	..	1	..	..	2	4	..	3

MAWLEY AND PRYSLEY, PART OF THE MANOR OF CLEOBURY, SHROPSHIRE

Land Rev., M. B. 185, f. 88. [21 Eliz.]

Custumarii	Enclosed			Arable in the Open Common Fields		
	Arab.	Md.	Past.	Cros field	Longecrosse-pit field	Neather field
Wm. Fennor, m., 1 virg.	4½	2	16½	7	8	4
Edw. a Wyer, 4m., 2 virg.	6	11½	53½	10	12	10
Johanna Wyer, m.	5	2	17	4	5	2
Eliz. a Wyer, m.	15	8	64½	8	10	12
Joh. a Wyer, cott.	..	..	1	48 (past. and arab.)		

The list is complete, save for one close of pasture held by a gentleman at a rent of 2 s.

[1] The list of customary tenants is complete.

APPENDIX II

EVIDENCE, LARGELY EARLY, BEARING UPON THE EXTENT OF THE TWO- AND THREE-FIELD SYSTEM

BEDFORDSHIRE. TWO-FIELD TOWNSHIPS

Township	Description
Chalgrave	Grant of 7¼ acres in 6 parcels *in uno campo* and 6½ acres in 5 parcels *in alio campo*. Grant of 3 acres in 4 parcels *in campo del West* and 3 acres in 4 parcels *in campo del Est*.[1]
Dean	Grant of a messuage, a croft, and 4 acres of arable, of which "due acre iacent in Wdefeld et due in campis versus Scelton."[2]
Flitwick	Grant of "unam acram in campo de Flittewic cuius dimidia acra iacet in Rugweifurlong in campo del Est et altera iacet in campo de West."[3]
Flitwick	Survey. Each tenant's arable is divided almost equally between East field and West field.[4]
Hinwick	Grant of "duas perticatas terre, viz., unam perticatam terre in campo orientali super Watterlonde et unam perticatam terre in campo occidentali super Raveneswelle."[5]
Houghton Regis	Grant of a manse, a croft of 2 acres, and a half-virgate of demesne comprising 8 acres in 3 places in one field (*in eodem campo*) and 8 acres in 2 places *in campo del Nord*.[6]
Southill	Grant of 11½ acres from one-third of 1¼ virgates, scil., "in Northfelde tres acras versus aquilonem et in Suthfelde quinque acras versus austrum et in crofta que fuit Matillis La Dele tres acras et dimidiam."[7]
Tebworth	Grant of 1¼ acres in 2 parcels *in campo occidentali* and ¾ acre in 2 parcels *in campo orientali*. Grant of 4 acres in 3 parcels and *in alio campo* 6 acres in 5 parcels.[8]
Toddington (" cum Hare est una villa:" *Feudal Aids*, i. 21)	Plea *inter alia* " de xv acris in uno campo et de xv acris in alio campo."[9] Detailed terrier of Dunstable lands in Hare: *in campo de North* 88 acres in 123 parcels *in campo de Suth* 88¼ acres similarly subdivided.[10]

[1] Harl. MS. 1885, ff. 37, 40. [Early XIV cen.]
[2] Ped. Fin., 1-9-13. 3 Hen. III.
[3] Harl. MS. 1885, f. 52b. [XIII cen.]
[4] Exch. Aug. Of., M. B. 358, f. 40 sq. 6 Jas. I.
[5] Ped. Fin., 1-14-39. 12 Hen. III.
[6] Harl. MS. 1885, f. 35. [Early XIII cen.]
[7] Ped. Fin., 1-8-10. 15 John.
[8] Harl. MS. 1885, f. 53. [XIII cen.]
[9] Ped. Fin., 1-2-30. 9 Rich. I.
[10] Harl. MS. 1885, ff. 9-10. [Late XIII cen.]

APPENDIX II 451

Township	Description
"Wadelawe" [Odell?]	Grant of 5 acres in 2 parcels *in campo occidentali* and 5 acres in 2 parcels *in campo orientali*.[1]
Wrestlingworth	Grant of 1 acre *in campo qui iacet versus Sutton* and ½ acre *in campo qui iacet versus Tadelawe*.[2]
Eyworth	Extent of the demesne arable, of which "medietas potest seminari per annum ... et alia medietas nihil valet quia iacet ad warectam in communi."[3]
Stagsden	Extent of the acres of demesne arable "iacentes in communi, quarum medietas que potest seminari per annum valet ... et alia medietas nihil valet quia iacet in communi."[4]
Sundon	Extent of the demesne arable, of which "medietas quolibet anno potest seminari....[Est] quedam pastura que quolibet altero anno est separalis in le Westfeld...."[5]
Warden	The acres of demesne arable "iacent in communi...; medietas dicte terre seminanda valet per annum ... et alia medietas nihil valet quia iacet in communi campo warectato...."[6]
Wrestlingworth	Extent of the demesne arable. "Medietas dicti terre potest seminari per annum ... et alia medietas iacet warecta quolibet anno et nihil valet tunc quia iacet in communi."[7]

BEDFORDSHIRE. THREE-FIELD TOWNSHIPS

Campton	Incomplete survey. The arable of a leasehold comprises 30 acres in High field, 30 in Chickson field, and 30 in Benhill field.[8]
Houghton Regis	Grant *inter alia*, and not consecutively, of ½ acre and ½ acre *in campo del Suth* 1¾ acres in 4 parcels *in campo del North* 2½ acres in 3 parcels *in campo del West*.[9]
[Northill]	Terriers of the lands of the "Colledg of Norrell" showing them always divided among the same three fields, e. g., in Padworth field 3½ acres in 3 parcels in Bamworth field 4 acres in 8 parcels in Ladywood field 4¼ acres in 5 parcels.[10]
Salford	Map of 1595 and several terriers. A terrier of lands in the occupation of Thos. Whyler describes

[1] Harl. MS. 1885, f. 63*b*. [Early XIII cen.]
[2] Ped. Fin., 1-2-21. 9 Rich. I.
[3] C. Inq. p. Mort., Edw. III, F. 37 (22). 8 Edw. III.
[4] Ibid., F. 38 (14). 8 Edw. III.
[5] Ibid., F. 56 (1). 12 Edw. III.
[6] Ibid., F. 42 (22). 9 Edw. III.
[7] Ibid., F. 65 (7). 15 Edw. III.
[8] Exch. Aug. Of., M. B. 396, ff. 34-53. 3 Jas. I.
[9] Harl. MS. 1885, f. 54*b*. [Early XIII cen.]
[10] Rents. & Survs., Portf. 23/61. 30 Eliz. and 1612.

APPENDIX II

Township	Description
Salford (*continued*)	11¼ acres in 11 parcels in the Brooke field 14¾ acres in 11 parcels in the Myddell field 14¾ acres in 16 parcels in the Wood (also New, or Upper) field.[1]
Souldrop	Grant of 3 acres from a virgate "in campis de Suthrop, scil., in Northfeld, i acram in Lusemere et in Westfeld unam acram in Sortebrache et in Suthfeld i acram in Rawedeheg."[2]
Tilsworth	The enclosure award allots 911½ acres, which lie in Lower field, Middle field, and Upper field.[3]
Wilden	Grant from 1½ hides, — "in Suthfeld decem acras et in Westfeld novem acras et in Estfeld septem acras."[4]
Cardington	Extent of the acres of demesne arable "iacentes in communi unde due partes possunt seminari per annum."[5]
Sudbury	Extent of the acres of arable demesne "iacentes in communi unde due partes possunt seminari per annum."[6]
Sutton	Extent of the acres of arable demesne "iacentes in communi unde due partes possunt seminari per annum. ... Et tertia pars iacet ad warectam."[7]
Wootton	Extent of the acres of demesne arable "que iacent in communi unde due partes possunt seminari per annum."[8]

BERKSHIRE. TWO-FIELD TOWNSHIPS

Township	Description
Ashbury	Survey. Tenants' holdings always lie equally divided between East field and West field.[9]
Bassildon	Injured survey of one-third of the manor. There are the same number of furlongs in East field and West field.[10]
Bockhampton	Grant of a virgate "que sic iacet dispersa per acras in campo," viz., *in campo aquilonari* 18¼ acres in 14 parcels *in campo australi* 19 acres in 8 parcels.[11]
Chievely	Grant of two half-acres *in campo occidentali* and two half-acres *in campo orientali.*[12]

[1] All Souls MSS., Terriers 3a, 3c, and Typus Collegii, i, map. 23. [Late Eliz.]
[2] Ped. Fin., 1-4-10. 4 John.
[3] C. P. Recov. Ro., 8 Geo. III, Trin. 1768.
[4] Ped. Fin., 1-7-19. 9 John.
[5] C. Inq. p. Mort., Edw. III, F. 52 (5). 11 Edw. III.
[6] Ibid., F. 38 (26). 8 Edw. III.
[7] Ibid., F. 44 (6). 9 Edw. III.
[8] Ibid., F. 39 (16). 8 Edw. III.
[9] Harl. MS. 3961, ff. 117-33. 10 Hen. VIII.
[10] Rents. & Survs., Portf. 5/16. 7 Hen. IV.
[11] *Cartl. St. Frideswide* (ed. S. R. Wigram), ii. 315. [Early XIII cen.]
[12] Ped. Fin., 7-13-3. 23 Hen. III.

APPENDIX II 453

Township	Description
Clapcote	Grant of a tenement, "... scil., vi acras terre arrabilis in uno campo ... et vi in alio cum insula prati ad eandem pertinente."[1]
Coleshill	Grant of " sex acras terre et i acram prati ... scil., tres acras in campo australi et tres acras in campo boriali."[2]
Farnborough	Enclosure award, affecting (apart from 45 acres of down land) 426½ acres in the North field and 478 acres in the South field.[3]
Knighton	Grant of one-half hide of demesne, viz., *in campo orientali* 28 acres in 3 places *in campo quoque occidentali* 32 acres in 4 places and 6 acres of meadow.[4]
Milton	The enclosure map shows two large open fields called North field and South field, lying to the north and south of the village. The award encloses 129 acres in the former, 327 in the latter.[5]
Brook, West	An extent in which 76 acres of demesne arable lie in East field and West field and " Ovenham " field (Ownham is an adjacent hamlet), and " iacent in communi tempore aperto omnibus tenentibus domini."[6]
Cookham	Extent of the demesne arable, of which " medietas potest seminari per annum ... et si non seminatur nihil valet, quia iacet in communi. ... "[7]
Shaw	Extent. " Sunt in dominico due carucate terre ... quarum medietas seminari potest per annum ... et alia medietas ... iacet ad warectam et in campo communi unde pastura est communis."[8]
Upton	Extent of the demesne arable, of which " medietas seminari potest per annum. ... De terra iacente ad warectam nihil inde percipi potest quia in campo communi et pastura communis."[9]
Wittenham, West	Extent of the demesne arable, of which " medietas per annum seminari potest. ... Et alia medietas iacet ad warectam in communi campo. Ita quod nulla inde pastura vendi potest."[10]

[1] *Cartl. St. Frideswide* (ed. Wigram), ii. 365. [1235–48.]
[2] Ped. Fin., 7–2–8. 1 John.
[3] C. P. Rec. Ro., 18 Geo. III, Hil. 1777.
[4] *Cartl. St. Frideswide* (ed. Wigram), ii. 302. [c. 1150–60.]
[5] C. P. Recov. Ro., 50 Geo. III, Trin. 1810.
[6] Rents. & Survs., Ro. 46. 14 Edw. III.
[7] C. Inq. p. Mort., quoted in *Bibl. Topog. Brit.*, iv. 138. 25 Edw. III.
[8] C. Inq. p. Mort., Edw. III, F. 67 (4). 16 Edw. III.
[9] Ibid., F 52 (7). 11 Edw. III.
[10] Ibid., F. 51 (3). 11 Edw. III.

APPENDIX II

BERKSHIRE. THREE-FIELD TOWNSHIPS

Township	Description
"Ballattesfeld"	A short and partly illegible terrier. In North field are 4 entries with a total of 20¼ acres; in West field are 5 entries, in East field 2 entries, the totals being illegible.[1]
Great Coxwell	A terrier of what is probably the demesne, showing in the first field (unnamed) 105 acres in 13 places *in campo boreali* 121 acres in 18 places *in campo orientali* 172 acres in 23 places besides 53 acres consolidated and 22 acres in crofts.[2]
"Hullefeld"	A terrier, locating in North field 5¾ acres in 8 places in West field 6¼ acres in 3 places in East field 3¼ acres in 2 places.[3]
Stanford in the Vale	A terrier of two yard-lands, which comprise a messuage, an orchard, two closes, "half a hide of meade in the Lott meade," common of pasture, and in Nye or West field 10¾ acres in 21 parcels in North field 11¾ acres in 24 parcels in East field 9½ acres in 22 parcels.[4]

BUCKINGHAMSHIRE. TWO-FIELD TOWNSHIPS

Bradwell	Grant of " dimidiam virgatam terre ... scil., xiii acras et unam rodam in uno campo et xi acras in alio campo."[5]
Claydon	Grant of 9½ acres of arable and meadow, scil., *in campo orientali* 6 acres in 7 parcels *et in alio campo, scil., in occidentali* 3½ acres in 5 parcels.[6]
Claydon, Steeple	Terrier of " two yardlands of glebe lands contayning in number Three score and one ridges or lands arable wherof Thirtie and one are lying and being in the Mill field. [11 acres, 9 lands, and 3 'mowing Hades '] and other Thirtie in the Wood field [14 acres, 2 lands, and 1 'mowing hade in Puddle ']".[7]
Drayton Parslow	Agreement that there shall be pasture for a certain number of cattle " quando campus de Draiton qui est versus austrum iacebit ad warectas," and that a certain " cultura " shall not be plowed and sown " nisi quando homines de Draiton arabunt et

[1] Rents. & Survs., Portf. 1/1. 14 [Edw. III.]
[2] Cott. MS., Nero A XII, f. 132. [Early XV cen.]
[3] Rents. & Survs., Portf. 1/1. 14 [Edw. III.]
[4] Add. MS. 36903, f. 11. 27 Eliz.
[5] Cott. MS., Tib. E V, f. 214b. [Transcript in XIV cen. cartl.]
[6] Rents. & Survs., Ro. 75. 9 Hen. VI.
[7] Ibid., Portf. 22/99. 1639.

APPENDIX II 455

Township	Description
Drayton, Parslow (*continued*)	seminabunt predictum campum de Draiton scil. quolibet secundo anno."[1]
Hadenham	Grant of 8 acres, " quarum quatuor iacent in campo versus occidentem scil. due in Cotland et due in hutland et alie quatuor in campo versus orientem scil. due in Cotland et due in hutland."[2]
" Helpestrop "	Grant of 4 acres of arable, viz., *in campo versus North* 4½ buttes and 1 acre *in campo versus Suth* 2 acres in 4 parcels.[3]
Shalstone	Grant of a half-virgate, viz., 4½ acres in the East field 4½ acres in the West field.[4]
Stewkley	Grant of 160 acres of demesne arable, of which 80 are *in campo de Suhelt* in 11 "culture," and 80 are *in campo del Est* in 13 "culture."[5]
Stoke	Exchange of 4 acres of meadow for " 6 acras terre in campo de Nord iuxta croftam ... et 6 acras terre in campo de Sud " in 3 parcels.[6]
Aston Clinton	Extent of the demesne, of which "medietas seminari potest per annum... Et si [acre] non seminantur nichil valent eo quod pastura inde est communis omnibus tenentibus predicti manerii."[7]

BUCKINGHAMSHIRE. THREE-FIELD TOWNSHIPS

Adstock	The enclosure award and map show three large open fields, Breach, Newell, and Haskell.[8]
Bierton	Survey in which the open-field acres of the customary holdings are equally divided among Hoods field, Middle field, and Fenne field.[9]
Borstall	Confirmation of a grant of the tithes " de tribus campis de Borstall... vocatis Frithfeld, Cowhousefeld et Arnegrovefeld."[10]
Claydon St. Botolph	Terrier describing arable, viz., in the Wode field 10½ acres in 15 parcels in the East field 8½ acres in 11 parcels in the North field 7½ acres in 15 parcels.[11]
Hulcot	Terrier of lands held by a lessee of All Souls College, viz., 30 parcels in Esseldon field containing 10 acres, 4 butts, 8 leys

[1] Ped. Fin., 14-13-4. 6 Hen. III.
[2] Ped. Fin., 14-1-6. 7 Rich. I.
[3] Harl. MS. 1885, f. 53. [Copy of *c.* 1300.]
[4] R. Ussher, *One Hundred and Sixty-two Deeds* (privately printed), p. 7. [*c.* 1250.]
[5] Pipe Roll. Soc., 1894, *Publ.*, xvii, no. 138. 7 Rich. I.
[6] Ped. Fin., 14-3-14. 10 Rich. I.
[7] C. Inq. p. Mort., Edw. III, F. 64 (10). 15 Edw. III.
[8] C. P. Recov. Ro., 39 Geo. III, Trin. 1797.
[9] Land Rev., M. B. 200, ff. 101-120. 6 Jas. I.
[10] White Kennett, *Parochial Antiquities*, ii. 381. [28 Hen. VI.]
[11] Rents. & Survs., Portf. 5/20. [Hen. VIII.]

Township	Description
Hulcot (*continued*)	17 parcels in North field containing 6¼ acres, 2 londs, 1 ley 18 parcels in Moore field containing 6 acres, 1 butt, 11 leys.[1]
Ilmere	Extent of the demesne arable which was probably in open field, viz., *in prima seisona* 35½ acres in 6 places *in secunda seisona* 63 acres in 9 places *in tertia seisona* 47¾ acres in 3 places.[2]
Maids Norton	Terrier of the farm of the College, viz., in Hollowaye field 69 lands, 63 butts, 7¼ acres, 3 leys, in 59 parcels in the Meade field 51 lands, 41 butts, 23½ acres, 13 leys, in 71 parcels in the Chattell field 60 lands, 55 butts, 12½ acres, 12 leys, in 67 parcels. The contemporary map shows the same three fields and a small one called Rodwell field.[3]
Olney	Areas affected by the enclosure award lay in six large fields, called Limekiln, Hyde, West, Middle, Woodlands, and Beneath the Town.[4]
Padbury	A map showing three large fields, East, West, and Hedge.[5]
Stewkley	Grant of 18½ acres of arable, scil., "sex acr[as] ex north' parte campi de terra coci ... et sex acras ex est parte campi iuxta le bonde mon land ... et sex acras ex suth parti campi super Lanedistubbing' et dimidiam acram ad mansum."[6]
Wingrave	Terrier of the copyhold of Thomas Broke, the parcels being nearly always half-acres, viz., in South field in 17 furlongs, 25 acres of arable, 8½ acres of "layes" in Rowsam field in 6 furlongs, 2½ acres of arable, 2 acres of "layes" in North field in 23 furlongs, 29 acres of arable, 12¼ acres of "layes" in East field in 18 furlongs, 24 acres of arable, 2 acres of "layes" meadow ground in 15 furlongs [no areas given].[7]
Adstock	Extent of the demesne arable which lies "in communi unde due partes possunt seminari ... et tertia pars nichil valet quia iacebit ad warectam."[8]

[1] All Souls MSS., Terrier 2. 1575.
[2] Rents. & Survs., Ro. 79. 11 Edw. III.
[3] All Souls MSS., Terrier 9, and Typus Collegii, i, map 2. 1592.
[4] C. P. Recov. Ro., 8 Geo. III, Trin. 1768.
[5] All Souls Typus Collegii, i, map 1. 1591.
[6] Harl. MS. 3640, f. 52. [XIV cen. cartl.]
[7] Rents. & Survs., Portf. 5/76. [Hen. VIII.]
[8] C. Inq. p. Mort., Edw. III, F. 47 (8). 10 Edw. III.

Township	Description
Aylesbury	Extent of the acres of demesne arable " iacentes in communi unde due partes possunt seminari per annum...." [1]
Buckland	Extent of the acres of demesne arable " iacentes in communi unde due partes possunt seminari per annum...." [2]
Caldecot	Extent of 60 acres of demesne arable, " de quibus seminabantur hoc anno ... tam semine hiemali quam quadragesimali xl acre et residuum iacet ad warectam et in communi." [3]
Soulbury	Extent of the acres of demesne arable, "unde due partes possunt seminari per annum ... et residuum nichil valet quia iacet in communi." [4]
Wolverton	Extent. " Sunt in dominico cc acre terre arabilis que valent per annum lxvi s. viii d. et non plus quia tertia pars dicte terre est warecta et iacet in communi et est nullius valoris...." [5]

CAMBRIDGESHIRE. TWO-FIELD TOWNSHIPS

Abington	Grant (defaced) of about 4 acres, viz., *in campo* ... 8 rods in 8 parcels *in campo orientali* 9 rods in 7 parcels.[6]
Boxworth	Transfer of 10 acres " in campis de Bokesworth," viz., *in campo boriali* 30 selions *in campo australi* 12 selions.[7]
Tadlow	Grant (defaced) of 22 acres, viz., *in campo orientali* about 11 acres in 11 parcels *in campo occidentali* about 11 acres in 13 parcels.[8]
Boxworth	Extent. " Sunt in dominico clx acre terre arabilis de quibus possunt seminari per annum iiii^{xx}. ... Et alie iiii^{xx} acre nichil valent quia iacent in communi campo." [9]
Litlington	Extent. " Sunt in dominico cliii acre terre arabilis de quibus possunt seminari per annum lxxvi acre et dimidia. Et alie lxxvi acre et dimidia que non seminantur nichil valent per annum quia iacent in communi." [10]

[1] C. Inq. p. Mort., Edw. III, F. 55 (18). 12 Edw. III.
[2] Ibid., F. 37 (22). [8] Edw. III.
[3] Ibid., F. 61 (14). 14 Edw. III.
[4] Ibid., F. 37 (22). [8] Edw. III.
[5] Ibid., F. 64 (23). 15 Edw. III.
[6] Ped. Fin., 23-4-47. 4 John.
[7] Harl. MS. 3697. [XV cen. copy.]
[8] Ped. Fin., 23-9-23. 3 Hen. III.
[9] C. Inq. p. Mort., Edw. III, F. 51 (11). 11 Edw. III.
[10] Ibid.

CAMBRIDGESHIRE. THREE-FIELD TOWNSHIPS

Township	Description
Barnwell	Terrier of several holdings, all of which divide their arable acres among North field, Stony field, and White field. Sometimes the division is nearly equal (7, 7, 10; 9, 9, 9), sometimes less so (9½, 7¼, 4¾).[1]
"Beche"	Grant of "ix acras terre, iii in quolibet campo."[2]
Chesterton	Indenture regarding 7 acres, "quarum tres dimidie acre iacent in campo versus Middleton... et tres acre in Middelfeld ... et in tertio campo in foxholl' dimidia acra. ..."[3]
Chippenham	Grant of a croft and 15 acres of arable, viz., in "Norcampo" 5 acres in 4 parcels in "Soutcampo" 3¼ acres in 3 parcels in West field 6¾ acres in 6 parcels.[4]
Cottenham	Grant of 10½ acres, scil., "in campo qui vocatur Altebur... et in campo qui vocatur Foxholefeld iii acre et in campo qui vocatur Lawefeld...."[5]
Cottenham	Grant of three rods, one in each of the above named fields.[6]
Downham	Extent. The demesne arable comprises "in campo qui vocatur aggrave octoviginti acre in campo qui vocatur Westfeld septemviginti et decem acre in campo qui vocatur Estfeld sexviginti et quatuordecim acre et dimidia."[7]
Foxton	Terrier of the lands of St. Michael's College, Cambridge: in the North field 17¾ acres in 24 parcels in "Chawdwel" field 11¼ acres in 16 parcels in a field called Down 18 acres in 34 parcels.[8]
Gamlingay	Map and accompanying schedule showing that demesne and tenants' holdings were divided pretty evenly among three fields. The demesne arable of the manor of "Avenelles" comprised 118½ acres in the East field, 106 acres in the Middle field, 12 acres in the Sandes field, and 88 acres in the South field.[9]
Harlton	Terrier of the lands of St. Michael's College, Cambridge: in the North field 11¾ acres in 22 parcels in the West field 11⅝ acres in 27 parcels in the "hy" field 9 acres in 18 parcels.[10]

[1] Rents. & Survs., Portf. 5/78. [Rich. II.]
[2] Cott. MS., Titus A I, f. 51b. [Copy in cartl. temp. Edw. I.]
[3] Merton Col. MSS., Charter 1546. 42 Hen. III.
[4] Harl. MS. 3697, f. 155. [Copy in cartl. of 138⅞.]
[5] Ped. Fin., 23-5-27. 4 John.
[6] Ibid., 23-12-39. 12 Hen. III.
[7] Cott. MS., Claud. C XI, f. 34. 1278.
[8] Rents. & Survs., Ro. 7. [XVI cen.]
[9] Merton Col. MSS., map and schedule. 1601.
[10] Rents. & Survs., Ro. 7. [XVI cen.]

APPENDIX II

Township	Description
Haslingfield	Terrier of the lands of St. Michael's College, Cambridge: in the field called " Dawland " 8¾ acres in 25 parcels in the field called " Rowlay " 11¼ acres in 25 parcels in the field called " Downefelde " 9¼ acres in 19 parcels.[1]
Hinxton	Terrier of the lands of St. Michael's College, Cambridge: in the West field 8⅝ acres in " ly Medylfeld " 9¼ acres in " ly Chyrchfeld " 8¾ acres.[2]
Litlington	Survey. The demesne arable comprises 41 acres in Westwoode field, 31 in Grenedon field, and 35 in Hyndon field, together with 125 acres not in the fields.[3]
Littleport	Extent. The demesne arable comprises " in campo qui vocatur Westfeld quinque viginti acre et quinque rode terre in Suthfeld quinque viginti acre in Estfeld quater viginti acre et tres rode." [4]
" Lyndon "	Extent. The demesne arable comprises " in campo qui vocatur Northay ducente et quadraginta acre in campo qui vocatur Middelfeld ducente et quater-viginti acre in campo qui vocatur Sephey ducente et quater-viginti et novem acre in campo qui vocatur Snota viginti quinque acre." [5]
Shudy Camps	Grant of half a woodland and " tres acras terre arabilis quarum tres rode et dimidia iacent in Cherchefeld . . . et in alio campo una acra . . . et in tertio campo una acra et dimidia roda." [6]
Swaffham Prior	Terrier of two holdings with acres allotted to " Dychefelde," " Middelfeld," and " Begdalfyld," as follows: 3¼ in 12 parcels, 2¾ in 7 parcels, 1¾ in 7 parcels 1⅞ in 5 parcels, 1½ in 3 parcels, 1¾ in 3 parcels.[7]
Thriplow	Extent. The demesne arable comprises " in campo qui vocatur Kirkefeld quinque viginti et octo acre et una roda in Hethfeld septem viginti et tresdecim acre in Westfeld quinque viginti et undecim acre." [8]
Wilburton	Extent. The demesne arable comprises " in campo vocato Estfeld quater viginti et sexdecim acre

[1] Rents. & Survs., Ro. 7. [XVI cen.]
[2] Ibid.
[3] Rents. & Survs., Portf. 6/18. [29] Hen. VIII.
[4] Cott. MS., Claud. C XI, f. 38b. 1278.
[5] Ibid., f. 53. 1278.
[6] Ped. Fin., 23–9–22. 3 Hen. III.
[7] Rents. & Survs., Portf. 22/43. 1566.
[8] Cott. MS., Claud. C XI, f. 132. 1278.

Township	Description
Wilburton (*continued*)	in campo vocato Suthfeld sexaginta et duodecim acre in campo vocato Nortfeld centum et octo acre."[1]
Willingham	Extent. The arable demesne comprises "in campo qui vocatur Westfeld quater viginti et due acre in campo qui vocatur Middelfeld quater viginti et undecim acre in campo qui vocatur Belasis quinque viginti et sex acre."[2]
Willingham	Terriers of small holdings which have parcels in Cadwin field, Belsaies field, and West field. The terrier of a half-yardland in an eighteenth-century hand allots acres equally to these three fields.[3]
Fen Ditton	Extent of the demesne arable, "de quibus tertia pars iacet quolibet anno ad warectam. Et de residuo quelibet acra valet per annum vi d. quum seminantur et quum non seminantur tunc nihil valent quia tunc iacent in communi per totum annum."[4]
Gransden	Extent of the demesne arable, "unde due partes earundem seminantur per annum ... et tertia pars earundem nichil valet quia iacet in communi ad warectam."[5]
Madingley	Extent of the demesne arable, "quarum due partes possunt seminari quolibet anno et ... tertia pars nichil valet quia iacet quolibet anno ad warectam et in communi.[6]
Whaddon	Extent. "Sunt ibidem iiii^{xx} acre terre arabilis et inde seminabantur hoc anno lx acre terre et residuum iacet in communi."[7]

DERBYSHIRE. THREE-FIELD TOWNSHIPS

Glapwell	Grant *inter alia* of "duas culturas ... in campo qui vocatur Northfeld et unam culturam in campo qui vocatur Suthfeld ... et duas acras et dimidiam in Estfeld. ..."[8]
Osmaston	Specification of the common of pasture which the abbot and convent of Derley may have "quum campi de Osmudestone et de lutchurch iacent in warecto versus Derewente ... in anno subsequente ... quum campi de Osmundestone et de Lutchurch iacent in warecto versus Normanton et Codintone

[1] Cott. MS., Claud C XI, f. 49. 1278.
[2] Ibid., f. 111. 1278.
[3] Add. MS. 14049. [XV cen.]
[4] Add. MS. 6165, f. 134*b*. 30 Edw. III.
[5] C. Inq. p. Mort, Edw. III, F. 43 (10). 9 Edw. III.
[6] Ibid., F. 46 (33). 10 Edw. III.
[7] Ibid., F. 64 (20). 15 Edw. III.
[8] Cott. MS., Titus C XII, f. 120. [Late XIII cen. copy.]

Township	Description
Osmaston (*continued*)	et in tertio anno ... quum campus de Lutchurch iacet in warecto versus Derb'."[1]
Shirebroke	Transfer of a toft and a bovate " prout iacet in tribus stadiis camporum ... cuius bovate
unum stadium abuttat super parcum de pleseley in campo australi
et alterum stadium abuttat super akkyr hedge in uno fine et villam predictam in campo qui vocatur Tonnefeld
et tertium stadium iacet in campo occidentali ...
et ista bovata extendit se ab austro in boriam ubique exceptis duobus buttis qui iacent in campo boriali."[2]

DORSET. TWO-FIELD TOWNSHIPS

Afflington	Extent (defaced). The demesne arable comprises
in campo horientali 66 acres and
in campo occidentali [82 acres].[3]
Blandford	Grant of " vii acre terre in uno campo et vii in alio campo," and of " quatuor acre terre in uno campo et iiii in alio."[4]
Crichel	Grant of 8 acres, scil.,
4 acres in one field in 4 parcels and
4 acres in the other field in 3 parcels.[5]
Eastington	Terrier of the lands of Christchurch Priory, Hants.
In the North field are 105¼ acres in 11 " culture "
in the South field, 108 acres in 21 " culture."[6]
Gufsage	Grant of a virgate from the demesne, scil.,
" decem acras in uno campo
et decem acras in alio campo
et unam acram prati ... et unam acram ad faciendum curtillagium."[7]
Ilsington	Grant of a messuage and " tres acre in uno campo et duas acras in alio,
in campo orientali ... [5 parcels]
in campo occidentali ... [2 parcels]."[8]
Nyland	Survey showing six holdings, the acres of which are almost equally divided between North field and South field, i.e., 12 *vs.* 12, 9 *vs.* 8, 10 *vs.* 7, 9½ *vs.* 7½, 22 *vs.* 20, 14 *vs.* 14.[9]
Piddle	Grant of one-third of three hides, viz., " a third part of the West field towards the south, with the yards and crofts which are in the same field towards the

[1] Cott. MS., Titus C XII., f. 62b. [1247-1353.]
[2] Exch. K. R., M. B. 23, f. 79.   2 Hen. V.
[3] Rents. & Survs., Portf. 7/9. [Edw. III.]
[4] Cott. MS., Otho B XIV, f. 44. [XV cen. copy.]
[5] E. A. and G. S. Fry, *Dorset Fines* (1896), p. 10. 4 John.
[6] Cott. MS., Tib. D VI, f. 235. 34 Edw. [I].
[7] Ibid., f. 119. [XV cen. copy.]
[8] Ibid., f. 163. [XV cen. copy.]
[9] Land Rev., M. B. 214, ff. 91-92. 6 Jas. I.

Township	Description
Piddle (*continued*)	south of the little garden and a third part of the East field towards the south. ...." [1]
Piddletown	Grant of " centum acras colendas, viz., quadraginta in uno campo et quadraginta in alio campo de dominico meo." [2]
Tatton	Grant of two and one-half virgates, viz., *in campo occidentali* 30 acres in 21 parcels *in campo orientali* 32 acres similarly subdivided.[3]
Westhill	Grant of four acres, scil., " duas acras terre in uno campo et duas acras in alio campo." [4]
Wimborne, Upper	Grant of a messuage and a half-virgate, scil., " quinque acras in uno campo et v in alio. " [5]

DORSET. THREE-FIELD TOWNSHIPS

Abbotsbury Graston Hilton Portisham Tolpiddle Witherston " Wotton "	Extents of the demesne lands of these manors. Relative to the demesne in each instance occurs the phrase, " due partes ... possunt seminari per annum ... et tertia pars iacet ad warectam et in communi et ideo nullius valoris." [6]
Affpiddle Bloxworth Cerne Hawkchurch Mintern Symondsbury Winfrith " Wydesford "	Extents of the lands of Cerne abbey. Relative to the demesne in each instance occurs the phrase quoted above.[7]

DURHAM. THREE-FIELD TOWNSHIPS

Township	Description
Bolam	Terrier of seven holdings. Each has an almost equal amount of arable and meadow in East field, West field, and North field.[8]
Gainford	Terrier of many holdings. Each is predominantly arable and is evenly divided between West field, Middle field, and East field.[9]
Langton	Terrier of several holdings. Each is largely arable and is divided evenly between West field, South field, and North field.[10]

[1] Fry, *Dorset Fines*, p. 17. 13 John.
[2] Cott. MS., Tib. D VI, f. 140. [XV cen. copy.]
[3] Ibid., f. 169. [XV cen. copy.]
[4] Ibid., f. 122. [XV cen. copy.]
[5] Ibid., f. 116. [XV cen. copy.]
[6] Rents. & Survs., Portf. 7/10. 17 Edw. III.
[7] Add. MS. 6165, ff. 33–35. 30 Edw. III.
[8] Land Rev., M. B. 192, f. 31 sq. 5 Jas. I.
[9] Land Rev., M. B. 193, ff. 15–19. 5 Jas. I.
[10] Ibid., ff. 22–24. 5 Jas. I.

APPENDIX II

Township	Description
Long Newton	Terrier of many holdings. Each is largely arable and is divided evenly between North field, South field, and West field.[1]
Raby	Terrier of several leaseholds. Each is largely arable and is evenly divided among three fields, two of which are subdivided, viz., West field and Crundledyke field, Chapell field, Crabtree field and High field.[2]
Shotton	Terrier of seven holdings. Each is largely arable and the acres are almost exactly divided between West field, East field, and Middle field.[3]
Summerhouse	Terrier of a few large holdings. The arable of each is divided between Low field, South field, and North field.[4]
Wackerfield	Terrier of several holdings. They are largely arable and their arable is evenly divided between West field, High field, and East field.[5]
Wellington	Terrier of several holdings. A few retain some open-field arable divided between Hither Crownel field, Far Crownel field, and le Langlands.[6]
Whitworth	Terrier of several holdings. They still have considerable arable, which is pretty evenly divided between West field, Broome field, and Farnehill field.[7]
Whorlton	Terrier of several holdings, rather more than one-half of each being equally divided between West field, Bennow field, and Lowe field.[8]
Wick, West	A long terrier. Arable predominates in each holding, and is often evenly divided between Middle field, Low field, and High field.[9]
Wolviston	Transfer of 6 acres of arable in the fields, viz., *in campo boriali* 5 roods and 2 selions in 2 places *in campo australi* 9 selions in 2 places *in campo occidentali* 7 selions in 1 place.[10]

GLOUCESTERSHIRE. TWO-FIELD TOWNSHIPS

Ablington	A short survey. In three instances the arable is divided evenly between East field and West field, e.g., 210 acres *vs.* 200 acres.[11]
Alderton	Transfer of 4 acres of arable *in uno campo* in 5 parcels and of 4 *in alio* in 4 parcels.[12]

[1] Land Rev., M. B. 193, f. 29 sq. 5 Jas I.
[2] Land Rev., M. B. 192, f. 5 sq. 5 Jas. I.
[3] Ibid., f. 29b. 6 Jas. I.
[4] Ibid., f. 22b. 5 Jas. I.
[5] Ibid., ff. 18b, 19. 5 Jas. I.
[6] Ibid., f. 50b. 5 Jas. I.
[7] Ibid., f. 64. 5 Jas. I.
[8] Land Rev., M.B. 193, f. 8b. 5 Jas. I.
[9] Ibid., f. 5b. 5 Jas. I.
[10] *Feodarium Prioratus Dunelmensis* (Surtees Soc., 1871), p. 32, n. 2. 1325.
[11] Rents. & Survs., Portf. 2/46. 1 Edw. VI.
[12] *Cart. St. Pet. Glouc.* (Rolls Series), i. 167. [Mid. XIII cen.]

APPENDIX II

Township	Description
"Eston" [Cold Ashton]	Grant of "servitium ... de octo acris terre in uno campo et de octo acris in alio campo."[1]
Aston Sub-edge	The abbot has the chapel and "in uno campo xv acras terre et in alio xiiii."[2]
Badminton	Transfer of 14¼ acres of arable, viz., 8 acres *in campo orientali* and 6¼ acres *in campo occidentali*. The numerous parcels are described, and other holdings are similarly divided between the fields.[3]
Bagendon	Transfer of a messuage, a close, and 2 acres of arable "in utroque campo."[4]
Bisley	Long survey. Most customary holdings are evenly divided between Battlescombe field and Stankcombe (sometimes Stankham) field.[5]
Cherrington	Rental. "Una virgata terre iacet in campis, viz., "in campo vocato le Northfeld xxx acre terre arabilis et in campo vocato le Southfeld xxx acre terre arabilis."[6]
Cowley	Transfer of a messuage and 14 acres of arable, viz., "septem acras terre in campo meridionali et alias septem acras in campo occidentali."[7]
Dorsington	Transfer of a messuage, a croft, and a virgate of land, the latter comprising 13 acres *in campo orientali* in 15 parcels and 13 acres *in campo occidentali* in 13 parcels.[8]
Duntisborne Abbots	Extent. "Robert Abovetun tenet unam virgatam terre continentem quadraginta quatuor acras in utroque campo."[9]
East Leach and Fyfield	Transfer of a messuage in East Leach, and of "viii acras in uno campo de Fishyde [Fyfield], scil., in campo versus North [in 16 parcels] et totidem in alio, scil., in campo versus Suth [in 17 parcels]". Transfer of 13 acres of arable in East Leach, divided almost evenly between South field and North field.[10]
Hampen	Transfer of 6 acres of arable in 6 parcels and *in alio campo* of 6 acres in 2 parcels. The same land is again described as "sex acras in uno campo et sex acras in alio campo."[11]

[1] *Reg. Monast. de Winchelcumba* (ed. D. Royce), i. 233. [XIII cen.]
[2] *Eynsham Cartl.* (ed. H. E. Salter), i. 137. [1180–84.]
[3] Exch. Aug. Of., M. B. 61, f. 5. [1235–49.]
[4] *Glouc. Inq. p. Mort.*, British Record Soc., Index Library, ix. 31. 2 Chas. I.
[5] Exch. Aug. Of., M. B. 394, ff. 78–126. 6 Jas. I.
[6] Rents. & Survs., Portf. 7/70, f. 13. [Hen. VIII.]
[7] Exch. Aug. Of., M. B. 61, f. 23. [1249–62.]
[8] Harl. MS. 4028. 1 Edw. II.
[9] *Cartl. St. Pet. Glouc.* (Rolls Series) iii. 194. [1266–67.]
[10] Ibid., i. 271, 274. [1212–43, 1263–84.]
[11] *Reg. Monast. de Winchelcumba* (ed. Royce), i. 151–54. [c. 1200.]

Township	Description
Hawkesbury	Transfer of a messuage and a croft, " cum duabus acris in uno campo et duabus in alio."[1]
Hitcot	The abbot has " in utroque campo duas acras."[2]
Mikleton	" Et habet vicarius in utroque campo unam hidam."[3]
Norton	Extent of the demesne, which comprises " ccvii acre terre arabilis in utroque campo."[4]
Sevenhampton	Extent of the demesne arable, which comprises 419½ acres *in campo Haustrali* " in diversis culturis " and 390 acres *in campo orientali*.[5]
Sherborne	A series of charters showing holdings equally divided between North field and South field (cf. especially p. 268).[6]
Shipton	Grant of 30 acres from a virgate, scil., *in campo qui vocatur Estfeld* 15 acres in 12 parcels *in alio campo qui vocatur Westfeld* 15 acres in 11 parcels.[7]
Stratton	Extent. The demesne arable lies in two fields, viz., " in campo orientali et boriali in diversis culturis ...xlviii acre.... Et in eisdem campis l acre ... et in eisdem campis l acre. ... "[8]
Tresham	Grant of 2½ acres, viz., *in campo australi* 1½ acres in 2 parcels *in campo boriali* 1 acre in 1 parcel.[9]
Whittington	Grant of common of pasture " super totam terram ... [except] in uno campo culturam de Cumbe et in alio campo culturam de Wicham."[10]
Yanworth	Transfer of 5 acres *in uno campo* in 2 parcels and 5 acres *in alio campo* in 5 parcels. Transfer of 4 acres *in campo del Suth* and 4 acres *in campo de North*.[11]
Fairford	Extent. " Sunt in dominico cccciiii^{xx} acre terre arabilis quarum ccxl acre seminate fuerunt ante [20 July]. ... Et ccxl non possunt extendi quia iacent ad warectam et in communi."[12]

GLOUCESTERSHIRE. THREE-FIELD TOWNSHIPS

Minchin Hampton	Extent of the demesne arable, " de quibus due partes possunt seminari per annum ... et tertia pars nihil valet quia iacet warecta et in communi."[13]

[1] Exch. Aug. Of., M. B. 61, f. 7. [1234–49.]
[2] *Eynsham Cartl.* (ed. Salter), i. 137. [1180–84.]
[3] Ibid.
[4] C. Inq. p. Mort., Edw. III, F. 2 (15). 1 Edw. III.
[5] Rents. & Survs., Portf. 16/66. 22 Edw. I.
[6] *Reg. Monast. de Winchelcumba* (ed. Royce), ii. 234–68. [XIII-XIV cen.]
[7] Ped. Fin., 73-12-207. 20 Hen. III.
[8] Rents. & Survs., Portf. 16/66. 22 Edw. I.
[9] Exch. Aug. Of., M. B. 61, f.7b. [1249–62.]
[10] Ped. Fin., 73-3-62. 10 John.
[11] *Reg. Monast. de Winchelcumba* (ed. Royce), ii. 320, 371. [XIII cen.]
[12] C. Inq. p. Mort., Edw. III, F. 51 (2). 11 Edw. III.
[13] Ibid., F. 39 (6). 8 Edw. III.

Township	Description
Oldland	Extent of the demesne arable, of which " tertia pars iacet quolibet anno ad warectam et in communi."[1]
Oxenton	Extent of the demesne arable, " unde due partes seminantur et tertia pars iacet ad warectam et frisca hoc anno."[2]
Shipton Moyne	Extent of the demesne arable which lies "in communi, unde due partes seminabantur ante [26 April] et tertia pars iacet ad warectam et in communi."[3]
Sudbury	Extent. " Sunt in dominico cccclx acre terre arabilis quarum cccvii acre seminate fuerunt ante [20 July]. ... Et cliii acre non possunt extendi quia iacent ad warectam et in communi."[4]
Stoke Orchard	Extent. Phraseology as in Sudbury extent. In demesne are 100 acres, of which 60 are sown.[5]
Tewkesbury	Extent. Phraseology as in Sudbury extent. In demesne are 400 acres, of which 270 are sown.[6]
Tockington	Extent. " Sunt in dominico cxxii acre terre arabilis. ... De quibus seminabantur hoc anno ante [10 July] iiii^{xx} acre semine yemali et quadragesimali et residuum iacet ad warectam et in communi."[7]

HAMPSHIRE AND THE ISLE OF WIGHT. TWO-FIELD TOWNSHIPS

Township	Description
Barton Stacy	The enclosure award allots 1807 acres lying in two great fields of similar size, named East and West.[8]
Bullington	Grant of 8½ acres in 12 parcels *in campo australi* and 9½ acres in 10 parcels *in campo boriali*. A virgate comprises 10 acres *in campo aquilonari* and 12 acres *in campo australi*.[9]
Forton	Exchange of a mill, a meadow, and four acres of arable, of which " due sunt in campo de Northt et due sunt in campo de Soudh."[10]
Hinton	Lands of the priory of Twineham comprise a croft of 5¾ acres and 172¾ acres in 14 " culture " in North field and 144¾ acres in 9 " culture " in South field.[11]
Middleton	Grant of " duo curtilagia ... et octo acre terre unde tres acre iacent in campo boriali [in 2 parcels] et quinque acre iacent in campo australi [in 5 parcels]."[12]

[1] C. Inq. p. Mort., Edw. III, F. 61 (13). 14 Edw. III.
[2] Ibid., F. 56 (1). 12 Edw. III.
[3] Ibid., F. 62 (6). 14 Edw. III.
[4] Ibid., F. 51 (12). 11 Edw. III.
[5] Ibid.
[6] Ibid.
[7] Ibid., F. 51 (8). 11 Edw. III.
[8] Chancery Close Ro. 1757.
[9] Egerton MS. 2104, ff. 59b, 160. [XIII cen. copy.]
[10] Ibid., f. 34b. 1234.
[11] Cott. MS., Tib. D VI, vol. ii. f. 71. 34 Edw. I.
[12] Egerton MS. 2104, f. 139. 11 Edw. II.

Township	Description
Week	Grant of " duas acras et dimidiam in utroque campo et unam virgatam prati
et duas acras quas Martinus tenuit scil. in utroque campo unam
et duas acras quas Sewynas tenuit scil. in utroque campo unam." [1]
Wherwell	Grant of 4 acres " in campo de Wherwell quarum una iacet in campo orientali in cultura que vocatur dodenam
et in campo occidentali tres acre [in 3 parcels]." [2]
Newtown, Isle of Wight	Extent of demesne arable, viz.,
" in cultura in quodam campo quod vocatur le Suthfeld continentur xxxii acre i roda
et in cultura que vocatur le Nortfeld continentur xxx acre et dimidia terre." [3]
Somerford, Isle of Wight	Lands of the prior of Twynham comprise
a croft of 6½ acres and
in 6 " culture " in West field 30 acres and
in 9 " culture " in East field 98½ acres. [4]

HAMPSHIRE. THREE-FIELD TOWNSHIPS

Andover Charlton Enham	A long series of small grants of land in each of these townships. Three fields appear in each, viz., at Andover, East, South, and West fields at Charlton, North, South, and West fields at Enham, North, South, and East fields. [5]
Bradley	Grant of 2½ acres *in campo qui vocatur Westfeld* and 2½ acres *in campo qui vocatur Northfeld* and 2½ acres *in campo qui vocatur Estfeld*. [6]
Drayton in the parish of Barton Stacy	Lands from which tithes are due are enumerated at length. Apart from a few crofts they lie in ' le Estfeld," " le Sowthfelde," and " Westfeld." [7]
Faccombe	Terrier of the parsonage glebe, which is " accounted one yard land." Besides crofts of 2½ acres, it comprises
in Middle field 13½ acres in 11 parcels
in South field 7½ acres in 6 parcels
in North field 10 acres in 13 parcels. [8]
Oakley, [Church]	Compotus rolls. In 1338 the *campus orientalis* and *campus australis* are sown, in 1398 the West field. [9]

[1] Egerton MS. 2104, ff. 38b, 210b. 33 Hen. III.
[2] Ibid., f. 155. [XIII cen. copy.]
[3] Rents. & Survs., Ro. 579. 28 Edw. I.
[4] Cott. MS., Tib. D VI, vol. ii. f. 71b. 34 Edw. I.
[5] Magd. Coll., Brackley Deeds. [Late XIII cen.]
[6] Ped. Fin., 203-8-55. 33 Hen. III.
[7] Egerton MS. 2104, f. 204. [Edw. III.]
[8] Rents. & Survs., Portf. 1/31. 16 Jas. I.
[9] G. W. Kitchin, *Manor of Manydown* (Hampshire Rec. Soc., 1895), pp. 151, 160. 1338, 1398.

Township	Description
"Est Acle" [Oakley?]	Grant of 16 acres of arable, "unde sex acre terre iacent in cultura que vocatur Sudfeld ... septem acre terre iacent in campo qui vocatur Nordfeld ... [in 2 parcels] et tres acre terre iacent in campo qui vocatur Estfeld."[1]
Wroxhall, Isle of Wight	Extent of the demesne arable, which comprises *in campo australi* 44 acres *in campo orientali* 82¾ acres *in campo boriali* 81¾ acres.[2]
Hinton	Extent. "Sunt in dominico ccxxx acre terre arabilis de quibus clx acre possunt seminari per annum ... et lxx ... non possunt extendi quia iacent ad warectam et in communi per totum annum."[3]
Lasham	Extent. "Sunt ibidem iiii^{xx}x acre terre arabilis de quibus lx acre ... possunt seminari per annum ... et xxx acre non possunt extendi quia iacent ad warectam et in communi per totum annum."[4]
Nutley	Extent. "Sunt in dominico cciiii acre terre arabilis de quibus cxxxvi acre possunt seminari per annum. ... Et lxvii acre ... non possunt extendi quia iacent ad warectam et in communi per totum annum."[5]
Tystede	Extent. "Sunt ibidem ciiii^{xx}iiii acre de quibus cxxiii acre possunt seminari per annum. ... Et lxi acre ... non possunt extendi quia iacent ad warectam et in communi per totum annum."[6]

HEREFORDSHIRE. THREE-FIELD TOWNSHIPS

Asperton and Stretton Grandison	Extent. The three carucates of demesne arable are worth only a certain amount, "quia tertia pars iacet quolibet anno ad warectam et in communi."[7]
Bickerton	Extent. A carucate of demesne arable is worth only 20 s., "quia tertia pars iacet ad warectam et in communi."[8]
Castle Richard	Extent of two carucates of demesne arable. "Et tertia pars earundem iacet quolibet anno ad warectam et in communi."[9]

[1] Ped. Fin., 203-8-9. 31 Hen. III.
[2] Rents & Survs., Ro. 579. 28 Edw. I.
[3] C. Inq. p. Mort., Edw. III, F. 37 (20). 8 Edw. III.
[4] Ibid., F. 46 (14). 10 Edw. III.
[5] Ibid., F. 41 (10). 8 Edw. III.
[6] Ibid.
[7] C. Inq. p. Mort., Edw. III, F. 43 (4). 9 Edw. III.
[8] Ibid., F. 44 (5). 9 Edw. III.
[9] Ibid., F. 62 (7). 14 Edw. III.

Township	Description
Eyton	Extent. Two carucates of demesne arable are worth only a certain amount, "quia tertia pars iacet quolibet anno ad warectam et in communi."[1]

HUNTINGDONSHIRE. TWO-FIELD TOWNSHIPS

Gransden	Grant of three acres, of which 1½ lie *in campo qui vocatur Estfeld* and 1½ lie *in campo qui vocatur Westfeld*.[2]
Hemingford	Extent of a messuage, garden, 10 acres of meadow, and 100 acres of arable " in communi, de quibus possunt seminari per annum lx acre. . . . "[3]
Toseland	Extent of a messuage, 8 acres of meadow, 4 of pasture, 6 of woodland, and 200 of arable " in communi, de quibus possunt seminari per annum c acre. . . . "[4]
Yelling	Extent of a messuage and 120 acres of arable " in communi, de quibus possunt seminari lx per annum."[5]

HUNTINGDONSHIRE. THREE-FIELD TOWNSHIPS

Coppingford	Transfer of a messuage and 6¾ acres, of which " due acre et una roda iacent in campo versus Bokeswrth due acre et una roda iacent in campo versus Hamerton et due acre et una roda iacent in campo versus Aylbriktesl'."[6]
Folksworth	A survey arranged by fields and furlongs. There are three rather large fields and one small one, viz., West field with 12 furlongs, Bulmere field with 7 furlongs, Middle field with 18 furlongs, and the " fylde to Styltonwarde " with 4 small furlongs.[7]
Hammerton	Transfer of 8 acres (part of a virgate), of which " tres acre iacent in campo versus Copmanneford tres acre iacent in campo versus Salenegrove et due acre iacent in campo versus Wynewyk."[8]
Keyston	Survey. The arable of the holdings is divided among three fields, e.g., " terra arabilis et leylonde in Mill field per estimationem xvii acre terra arabilis et leylonde in Gotteredge field per estimationem xv acre terra arabilis et leylonde in Morton field per estimationem xviii acre."[9]

[1] C. Inq. p. Mort., Edw. III, F. 43 (4). 9 Edw. III.
[2] Ped. Fin., 92-9-163. 32 Hen. III.
[3] C. Inq. p. Mort., Edw. II, F. 82 (9). 17 Edw. II.
[4] Ibid.
[5] Ibid.
[6] Ped. Fin., 92-9-175. 32 Hen. III.
[7] Add. MS. 29611. 4 Edw. VI.
[8] Ped. Fin., 92-10-188. 32 Hen. III.
[9] Land Rev., M. B. 216, ff. 71-104. 3 Jas. I.

470   APPENDIX II

Township	Description
Overton	Transfer of 12 acres which are part of a virgate, viz., in various furlongs 4 acres in 4 parcels in Middel field 4 acres in 5 parcels *in campo australi* 4 acres in 5 parcels.[1]
Weston	Transfer of 20 acres, of which "septem iacent in campo qui vocatur Estfeld septem iacent in campo qui vocatur Westfeld quatuor iacent in campo qui vocatur Brokfeld et due iacent in campo qui vocatur Badelisford."[2]
St. Neots	Extent. The prior " tenet in dominico suo ibidem vii^{c}xx acre terre arabilis quarum ... ii^{c}xl acre ... iacent ad warectam et in communi et nihil valent per annum."[3]

LEICESTERSHIRE. TWO-FIELD TOWNSHIPS

Cotes Daval	Grant of 10 acres of demesne arable *in uno campo* and 10 acres *in alio campo*.[4]
Croxton	General description of the arable of the township with reference to *campus orientalis* and *campus occidentalis*.[5]
Gilmorton	The enclosure award allots 784 acres in Usser field, 548 in Mill field, and 632 in Ridgway field.[6]
Newton Harcourt	Grant of " quinque acras de dominico, scil., duas acras terre et dimidiam ex una parte predicte ville et duas acras terre et dimidiam ex altera parte predicte ville in illa parte, scilicet, versus occidentem [1½, ½, ½ acres] ex altera parte versus orientem [1½, 1 acres]."[7]
Owston	Transfer of the half of a virgate, scil., " in campis ex parte aquilonali ville versus solem et ex parte australi remotius a sole."[8]
Segrave	Transfer of " una acra ex una parte campi et alia acra ex altera parte."[9]
Slawton and Othorp	Terrier showing, in many parcels, *in campo occidentali* 48½ acres *in campo orientali* 38¾ acres *in campo de Outhorp'* 14¾ acres.[10]

[1] Ped. Fin., 92-2-26. 10 John.
[2] Ibib., 92-10-181. 32 Hen. III.
[3] Add. MS. 6164, f. 420. 44 Edw. III.
[4] Cott. MS., Vitel. A I, f. 106. [XV cen. copy.]
[5] John Nichols, *Leicestershire*, ii, App., p. 81, Memd. 1258.
[6] C. P. Recov. Ro., 19 Geo. III, Mich. 1778.
[7] Cott. MS., Nero C XII, f. 92. [XIV cen. copy.]
[8] Ped. Fin., 121-10-78. 6 Hen. III.
[9] Nichols, *Leicestershire*, ii, App., p. 111, from Segrave Cartl., Harl. MS. 4748. [1201-41.]
[10] Cott. MS., Claud. C V. [Late XIV cen.]

APPENDIX II

Township	Description
Sysonby	Transfer of " tres virgatas terre, scil., ... xxv acras in uno campo et in altero totidem." [1]
Sysonby	Transfer of a half-carucate, scil., " xx [acras] in uno campo et xx in alio campo et x acras ubi est situs loci domorum suarum." [2]
Thurnby	Transfer of 45 acres of arable, viz., *in uno campo* 19 acres in 15 parcels *in alio campo* 26 acres in 13 parcels.[3]
Twyford	Grant of " unam partem tofti mei et ... duas seliones in campo occidentali et unam rodam in campo orientali." [4]

LEICESTERSHIRE. THREE-FIELD TOWNSHIPS

Arnesby	Terrier of two yard-lands, viz., in the Brooke field 11 acres of arable, 7 acres of layes, meadow, pastures, and feeding in the East field 12 acres of arable, 7½ acres of layes, meadow, pastures, and feeding in the North field 11 acres of arable, 8 acres of layes, meadow, pastures, and feeding.[5]
Barsby and S. Croxton	Areas allotted by the enclosure award lie in three large fields called Nether, Middle, and Upper.[6]
Beeby	Transfer of " medietatem unius bovate, scil., unam acram et tres rodas in campis versus australem partem et tres dimidias acras in Halwefeld' ... et unam acram et tres rodas in Linkefeld...." [7]
Bowden, Great	Areas allotted by the enclosure award lie in East, West, and South fields.[8]
Burton Lazars	Transfer of 20 acres of arable, of which 6¾ are fallow.[9]
Burton Lazars	" Registrum omnium terrarum arabilium ... prioris de lewes," viz., in the West field about 10⅜ acres in 11 parcels in the Nether field about 23 acres in 24 parcels in the Manmylne field about 28¼ acres in 20 parcels in the Over field about 29 acres in 28 parcels.[10]

---

[1] Nichols, *Leicestershire*, ii, App., p. 137, from Cartl. Gerendon Abbey, Lansd. MS. 415, f. 1b. [XIII cen. copy.]

[2] Ibid., App., p. 138, from Lansd. MS. 415, f. 12. [XIII cen. copy.]

[3] Cott. MS., Calig. A XII, f. 143. [XIII cen. copy.]

[4] Cott. MS., Nero C XII, f. 85b. [Copy of 1404.]

[5] Rents. & Survs., Portf. ½, memb. 4. [Chas. I.]

[6] C. P. Recov. Ro., 39 Geo. III, Trin. 1798.

[7] Ped. Fin., 121–6–126. 10 John.

[8] C. P. Recov. Ro., 17 Geo. III, Trin. 1777.

[9] Anc. Deeds, A 1437. [Early XIV cen.]

[10] Add. MS. 8930, ff. 41–46. 1493.

APPENDIX II

Township	Description
Clawson	Survey showing the arable of the tenants' holdings " in tribus campis ibidem vocatis le peazefeld, the wheatfeld, and the fallow feld." [1]
Etton	Grant of a part of the sixth of a hide, scil., " iiii acras et dimidiam inter Etton et Elinton in Estfeld et iiii acras et unam rodam in Sudhfeld et ii acras et dimidiam in Aldefeld et v acras et dimidiam in Westfeld et ii acras prati in Lenstang." [2]
Freeby	Terrier of two virgates, viz., in Sowthefyld 16 acres in 25 parcels in Northfyld 15½ acres in 19 parcels, one of which contains 32 selions in Westfyld and Estfyld 14¼ acres in 29 parcels. [3]
Kirkby Bellars	Grant of a toft and 8¾ acres of arable, viz., *in campo occidentali* 2⅛ acres in 7 parcels *in medio campo* 4⅜ acres in 11 parcels *in campo orientali* 2¼ acres in 7 parcels. [4]
Melton Mowbray	Terrier of three large holdings evenly divided between " Fallowfyld [or] Sowthefeld, Peysefyld [or] Westfyld, Whetefyld [or] Northefyld." [5]
Norton	View of the lands of the abbot of Owston, viz., *in campo boriali* 69⅜ acres in 187 selions lying in 14 furlongs *in campo occidentali* 70¼ acres in 268 selions lying in 23 furlongs *in campo australi* 51⅛ acres in 246 selions lying in 35 furlongs. [6]
Rollestone	" Territor' Prioris domus Cartusiensis Londonie," whose lands comprised *in campo orientali* 173 acres in 43 parcels *in campo boriali* 169½ acres in 63 parcels *in campo occidentali* 269¼ acres in 67 parcels. [7]
Rotherby	Terrier of the demesne lands of the priory of Chalcombe, which comprised in East field 68½ acres in Middle field 62 acres in " le heyfeld " 68 acres. [8]
Stonesby	The enclosure award and plan show three large fields, Mill field, Waltham Gate field, and Gasthorp Gate field. [9]

[1] Land Rev., M. B. 220, ff. 66–76. 7 Jas. I.
[2] Ped. Fin., 121-3-57. 2 John.
[3] Add. MS. 8930, ff. 8–11. 24 Hen. VII.
[4] Cott. MS., Nero C XII, f. 59b. [Copy of 1484.]
[5] Add. MS. 8930, f. 12 sq. 24 Hen. VII.
[6] Rents. & Survs., Ro. 386. 1360.
[7] Ibid, Ro. 388. 13 Hen. VII.
[8] Exch. Aug. Of., M. B. 378, f. 17b. [Edw. III.]
[9] C. P. Recov. Ro., 21 Geo. III, Easter. 1781.

APPENDIX II 473

Township	Description
Stoughton	"Memorandum quod sunt in dominico ccxiii acre per maius centum ... viz.,
	in campo boreali versus Thurnby v^{xx} acre viz., iiii^{xx} infra fossam et xx extra fossam
	in campo orientali vocato Longwong [60 acres including a close of 16 acres]
	in campo australi [93 acres]."[1]
Theddingworth	Terrier of " ... Addygtons Land lately dyssessed," which comprised
	in Santles field 27 acres of arable in 43 parcels and 18 acres of meadow in 25 parcels
	in Nonhylls field 22 acres of arable in 40 parcels and 11½ acres of meadow in 20 parcels
	in Gostyll field 29 acres of arable in 44 parcels and 19 acres of meadow in 33 parcels.[2]
Twyford	The allotments of the enclosure award lie in Nether, Spinney, and Mill fields.[3]
Walton and Kimcote	Terrier of the lands of Gabriell Pulteney, Esq., which comprised 5½ yard-lands "in very greate measure," viz.,
	in the North field 38 acres of arable and 6 acres of meadow
	in the Middle field 36½ acres of arable and 5¾ acres of meadow
	in the South field 35¾ acres of arable and 8 acres of meadow.[4]

LINCOLNSHIRE. TWO-FIELD TOWNSHIPS

Aylesby	The township contained 29 bovates, "ita quod quelibet bovata contineat in se sexdecim acras terre, scil., octo acras ex una parte ville et totidem ex altera."[5]
Barnetby upon the Wolds	The enclosure award divides some 2000 acres almost equally between the North field and the South field.[6]
Barton [upon Humber]	Grant of a half-bovate, containing
	in campo occidentali 5 acres in 2 parcels
	in campo orientali 5 acres in 4 parcels.[7]
Benniworth	Terrier of " quindecies xx^{ti} acras terre arabilis ex una parte ipsius ville et quatuordecies xx^{ti} et xv acras et unam perticatam et quinque fal[las] terre arabilis ex altera parte ville." There is no field rubric

[1] Nichols, *Leicestershire*, i, App., p. 96, from Rent. Monast. S. Marie de Pratis Leycestrie. [7 Hen. IV.]
[2] Rents. & Survs. Portf. 10/22. [Eliz.]
[3] C. P. Recov. Ro., 39 Geo. III, Trin. 1796.
[4] Rents. & Survs., Ro. 909. 34 Eliz.
[5] *Cartl. Prior. de Gyseburne* (Surtees Soc., 1891), ii. 315. [1218-21.]
[6] C. P. Recov. Ro., 8 Geo. III, Trin. 1766.
[7] Cott. MS., Vesp. E XX, f. 155.

Township	Description
Benniworth (*continued*)	at the beginning of the enumeration, but halfway through occurs " in orientali campo."[1]
Branston	Grant of $3\frac{7}{8}$ acres in 2 parcels *in campo orientali* and $2\frac{1}{8}$ acres in 2 parcels *in campo occidentali*.[2]
Burwell	" Robt. Taler holdeth in *both* the feildes of Burwell ..." [amount torn].[3]
Claxby	Survey of the manor. Tenants' holdings are divided between North field (not always named) and South field.[4]
Claxby	Grant of one-half of a bovate, viz., 4 acres on the south of the vill and 4 on the north.[5]
Coates	Grant of " quinque acras terre ex una parte ville ... et quinque ex altera."
	Grant of " unam bovatam terre de dominico meo in campo de Chottes, scil., x acras terre ex una parte ville et x ex altera."[6]
Cockerington, South	Schedule of attainted lands. " ... Oonclosse ... and also he holdyth iii acres of arrable lande lieng in ye Estfeld and oon acre in ye Westfeld."[7]
Croxton	The enclosure award and plan show two large and approximately equal fields, East and West.[8]
Grimblethorp	Grant of " decem acras terre arabilis ... scil., v in orientali campo [in 3 parcels] et v in occidentali campo [in 6 parcels]."[9]
Grimsby, Great	Survey. " Et decem acre seminate, id est, decem acre ex una parte ville et decem ex altera parte faciunt unam bovatam. ... "[10]
Haburgh	The enclosure award allots 2500 acres in North field, South field, and the Marsh.[11]
Legbourne	Survey of the manor. Often one-third of a holding is enclosed pasture, but the remainder is divided between East field and West field.[12]
Limber, Great	Assignment of dower. The arable transferred comprises *in campo orientali* 5 acres in 9 parcels *in campo occidentali* 5 acres in 8 parcels.[13]
Norton	Grant of 8 acres, of which 4 lie *ad umbram* and 4 *ad solem*.[14]

[1] Cott. MS., Vesp. E XVIII, f. 27. [XIII cen. copy.]
[2] Ibid., f. 60b. [XIII cen. copy.]
[3] Rents. & Survs., Portf. 10/57, f. 4b. 6 Jas. I.
[4] Exch. K. R., M. B. 43, ff. 1–36. 39 Eliz.
[5] W. O. Massingberd and W. Boyd, *Final Concords* (London, 1896), i. 27. 4 John.
[6] Harl. MS. 3640, f. 98. [Late XIV cen. copy.]
[7] Rents. & Survs., Portf. 10/38. 38 Hen. VIII.
[8] C. P. Recov. Ro., 52 Geo. III, Trin. 1809.
[9] Cott. MS., Vesp. E XVIII, f. 96. [XIII cen. copy.]
[10] Rents. & Survs., Ro. 406. 7 Hen. VII.
[11] C. P. Recov. Ro., 1 Geo. IV, Trin. 1820.
[12] Rents. & Survs., Portf. 10/57. 6 Jas. I.
[13] C. Inq. p. Mort., Edw. III, F. 37 (12). 7 Edw. III.
[14] Massingberd and Boyd, *Final Concords*, i. 284. 18 Hen. III.

APPENDIX II 475

Township	Description
Stainton	Grant of three selions, of which one lies *in aquilonali campo* and two lie *in australi campo*.[1]
Stubton	From a bovate are granted 4 acres of land in the North field and 3¾ acres in the South field.[2]
Thurlby	The demesne arable lies *in campo qui dicitur Westfeld* in 6 " culture " and *in campo qui dicitur Estfeld* in 7 " pecie."[3]
Ulceby	Grant of a half-bovate, scil., " v acras ex una parte ville et v acras ex alia parte." Two bovates lie *in Est campo* in 6 parcels and *in West campo* in 6 parcels.[4]
Winceby	Grant of a half-bovate which lies " in boriali parte ville ... in tribus locis et in meridionali parte in quatuor locis."[5]
Baston	Extent. " De predictis iii^{xx}x acris terre arabilis xxx acre possunt seminari per annum.... Et residuum iacebit ad warectam et tunc nichil valet quia in communi."[6]
Grimsby, [Great] Cabourne Swallow Tetney Thorganby Weelsby	Extents. At Grimsby " sunt ... de terris dominicis iiii^{xx}iiii acre terre arabilis de quibus xlii acre possunt seminari quolibet anno et alie xlii acre iacent quolibet anno ad warectam et in communi." The demesne lands of the other manors of the abbot of Grimsby are similarly described.[7]
Harrington	Extent. " Sunt in dominico iiii^{xx} acre terre arabilis ... et xl acre possunt seminari per annum ... et residuum iacet ad warectam et tunc nichil valet quia in communi."[8]
Leadenham	Extent. " Sunt in dominico xii bovate terre unde vi bovate possunt seminari per annum.... Et residue vi bovate terre iacebunt ad warectam et tunc nichil valent quia in communi."[9]
Leasingham	Extent. " Sunt in dominico clx acre terre arabilis de quibus possunt seminari per annum iiii^{xx} ... et totum residuum nichil valet quia iacet warectum et in communi."[10]

[1] Cott. MS., Vesp. E XVIII. [XIII cen. copy.]
[2] Massingberd and Boyd, *Final Concords*, i. 244. 15 Hen. III.
[3] Cott. MS., Nero C VII, f. 149. [1 Hen. IV.]
[4] Cott. MS., Vesp., E XVIII, ff. 15, 12. [1280–1318, and late XIII cen. copy.]
[5] Cott. MS., Vesp. E XX, f. 116. [XIV cen. copy.]
[6] C. Inq. p. Mort., Edw. III, F. 38 (28). 8 Edw. III.
[7] Add. MS. 6165, ff. 63–65. 5 Hen. IV.
[8] C. Inq. p. Mort., Edw. III, F. 39 (3). 8 Edw. III.
[9] Ibid., F. 40 (8). 8 Edw. III.
[10] Ibid., F. 51 (5). 11 Edw. III.

Township	Description
Morton	Extent. " Sunt in dominico clx acre terre arabilis unde medietas potest seminari per annum ... et residuum nihil valet per annum quia in communi campo."[1]
Ringstone	Extent. Phraseology and areas are like those of the extent of Leasingham.[2]
Searby	Extent. " Sunt in dominico cx acre terre quorum lv acre quolibet anno seminabiles ... et lv iacent ad warectam et in communi et ideo nullius valoris."[3]

LINCOLNSHIRE. THREE-FIELD TOWNSHIPS

Burton iuxta Lincoln	Extent. " Quelebet bovata continet xii acras terre unde possunt seminari quolibet anno ... viii acre ... et residuum ... quod iacet ad warectam nichil valet pro eo quod iacet in communi."[4]
Canwick	Extent. " Sunt in dominico vi^{xx}viii acre terre arabilis de quibus due partes possunt seminari per annum et residuum nichil valet quia iacet ad warectam et in communi."[5]
Doddington	Extent. " Sunt in dominico vi^{xx} acre terre arabilis quarum iiii^{xx} possunt seminari per annum ... et residuum iacet ad warectam et in communi."[6]
Gedney and Gouxhill	Extent. " Sunt in dominico vi^{xx} acre terre arabilis quarum iiii^{xx} possunt seminari per annum et valet acra quum seminatur xii d. ... et quum non seminatur iacet warecta et in communi."[7]
Kexby	Extent of the demesne, which comprises " in campo australi xliii acre terre arabilis et dimidia in campo occidentali xxxii acre terre arabilis in campo boriali xxxvii acre et tres rode terre arabilis."[8]
Scotter	Extent of the demesne, which comprises 20 parcels lying *in campo qui dicitur Midilfeld* 26 parcels lying *in campo boriali* 21 parcels lying *in campo occidentali*.[9]
Stowe	The enclosure award makes allotments in three large fields, Skelton, Normanby, and West.[10]
Upton	Extent of the demesne, which comprises " in campo orientali xxiiii acre et dimidia in campo boriali xvi acre in campo occidentali xxi acre."[11]

[1] C. Inq. p. Mort., Edw. III, F. 37 (22). 8 Edw. III.
[2] Ibid., F. 51 (5). 11 Edw. III.
[3] Ibid., F. 59 (8). 13 Edw. III.
[4] Ibid., F. 68 (14). 17 Edw. III.
[5] Ibid., F. 54 (11). 12 Edw. III.
[6] Ibid., F. 52 (5). 11 Edw. III.
[7] Ibid., F. 54 (10). 12 Edw. III.
[8] Rents. & Survs., Ro. 409. 18 Edw. I.
[9] Cott. MS., Nero C VII, f. 210. [1 Hen. IV.]
[10] C. P. Recov. Ro., 49 Geo. III, Hil. 1808.
[11] Rents. & Survs., Ro. 409. [18 Edw. I.]

APPENDIX II 477

NORTHAMPTONSHIRE. TWO-FIELD TOWNSHIPS

Township	Description
Adstone	Grant of 17¼ acres of arable, viz., 7¾ acres *in campo occidentali* in 19 parcels and 9½ acres *in campo horientali* in 26 parcels.[1]
Althorp	Grant of a messuage and 4 acres of arable, viz., due *in una parte campi* in 3 parcels et due *in alia parte campi* in 3 parcels.[2]
Astcote Pattishall " Edeweneskote "	Transfer of " quatuor viginti acras terre in campis de Pateshulle, Acheskote et Edeweneskote, viz., quadraginta acre de warecta et quadraginta acre terre seminate."[3]
Bodington	Grant of 4 acres of arable, viz., 2 in the eastern field and 2 in the western field.[4]
Brackley	Grant of 8 acres in the fields, viz., in the South field 4 acres in 7 parcels in the North field 4 acres in 7 parcels.[5]
Brackley	Grant of 20 acres in the fields, viz., in the North field 10 acres in half-acre parcels in the South field 6½ acres in half-acre parcels in the Baulond 2 acres and in a croft 1 acre.[6]
Canons Ashby	" Tarrer of Thomas Gaywood's londs yn the Estfelld & the Westfelld," viz., in the East field 2 acres and 90 " yards " in 9 parcels in the West field 20 acres and 42 "yards" in 8 parcels.[7]
Chalcombe	A long terrier of the lands of Chalcombe Priory. The arable is divided into two series of parcels, one comprising 76¾ acres, the other 77⅛. Owing to an injury to the manuscript, the names of the two fields are missing in connection with the arable, but are given in connection with the meadow as East and West.[8]
Culworth	Grant of 62 acres, viz., a croft of 4 acres, and *in campo boriali* 28 acres in 72 parcels *in campo australi* 30 acres similarly subdivided.[9]
Drayton	Grant of 4½ acres in the fields, viz., *in campo de Norht* 2½ acres in 5 parcels *et in campo de Su'* 2¼ acres in 5 parcels.[10]

[1] Anc. Deeds, B 239. [Early XIV cen.]
[2] Cott. MS., Tib. E V, f. 49b. [XIV cen. copy.]
[3] Harl. Char. 57 F 24. 5 Edw. I.
[4] Magd. Coll., Brackley Deeds, C 97. [1190–1200.]
[5] Ibid., 4. [c. 1260.]
[6] Ibid., 63. [c. 1260.]
[7] Rents. & Survs., Portf. 13/20. [Hen. VIII.]
[8] Exch. Aug. Of., M. B. 378. [Mid. XIV cen.]
[9] Add. Char. 44285. 24 Edw. I.
[10] Cott. MS., Claud. D XII, f. 40. [XIII cen. copy.]

APPENDIX II

Township	Description
Evenley	Grant of 2 acres in the fields, viz., in the Eastern field 1½ acres and in the Western field ½ acre. There are several similar charters.[1]
Evenley	Transfer of 10 acres, viz., 5 acres of arable in the East field and 5 acres of arable in the West field.[2]
Farthinghoe	Grant of all the land which Hallough held, viz., a messuage, croft, and 6 acres in one field and 6 in another.[3]
Fawsley	Grant of " due acre terre de eorum dominico ... scil., in campo aquilonari unam acram in campo australi unam acram."[4]
Flower	Grant from the demesne of " decem acre terre in campo orientali et decem acre terre in campo occidentali et quinque in campo orientali et quinque acre in campo occidentali."[5]
Gayton	Grant of 16¼ acres of arable, viz., *in campo occidentali* 7¾ acres in 12 parcels (the charter is injured and the remaining 8½ acres do not appear) " cum duabus rodis prati quarum una iacet in campo occidentali et altera in campo orientali."[6]
Haddon, West	Terrier of " totam terram subscriptam in campis, viz., in campo australi [several parcels] in campo boriali [several parcels]."[7]
Harleston	Grant of 12 acres, viz., 6 acres *in campo australi* and 6 *in campo boriali*. Grant of 7½ acres, viz., *in campo australi* 3½ acres in 7 parcels *in campo boriali* 4 acres in 8 parcels.[8]
Harpole	Grant of a toft and [field name omitted] 2½ acres in 3 parcels and *in alio campo* a head-acre and 3 half-acres.[9]
Harrowdon	A long but probably incomplete terrier locating parcels in " Estfeld " and in " Westfeld."[10]

[1] Magd. Coll., Brackley Deeds, B 178. [c. 1220–30.]
[2] Bodl., Rawl. B 408, f. 45b. [XV cen copy, in English, of an old charter.]
[3] Magd. Coll., Brackley Deeds. B 246. [c. 1200.]
[4] Cott. MS.. Claud. D XII. f. 104. [1231–64.]
[5] Ped. Fin., 171-7-86. 4 John.
[6] Cott. MS., Tib. E V, ff. 130. 130b. [XIV cen. copy.]
[7] Cott. MS., Claud. D XII, f. 120b. [XV cen. copy.]
[8] Cott. MS., Nero C XII, ff. 174, 176. [Late XIII cen. copy.]
[9] Cott. MS., Tib. E V, f. 29b. [XIV cen. copy.]
[10] Cott. MS., Vesp. E XVII, f. 320. [Late XV cen.]

APPENDIX II 479

Township	Description
Heyford, Nether	Grant of 3 acres, scil.,
	in occidentali campo 2 acres in 4 parcels
	in orientali campo 1 acre in 2 parcels.[1]
Holdenby	Grant of 36 acres of arable " de libero meo dominico, scil.,
	decem et octo acre in campo orientali
	et decem et octo acre in campo occidentali."[2]
Hothorp	Grant for a chapel " de singulis virgatis, unam acram ex una parte ville et aliam ex altera."[3]
Kislingbury	Grant of 2 acres, scil.,
	in campo occidentali 1 acre in 2 parcels
	in campo orientali 1 acre in 2 parcels.[4]
Preston	Grant of 95 acres of demesne arable and 9½ acres of demesne meadow, scil.,
	in campo versus aquilonem 44 acres of arable and 6¾ acres of meadow
	in campo versus australem 51 acres of arable and 2¾ acres of meadow.[5]
Radstone	Grant of 2 acres, viz.,
	in the South field 1 acre in 2 parcels
	in the North field 1 acre in 2 parcels.[6]
Rushton	Grant of a mill, " cum duabus acris terre, una in uno campo et alia in alio."[7]
Southorp	Transfer of a half-bovate comprising
	in campo boriali 8 selions in 4 places and
	in campo australi 6 selions in 4 places.[8]
Staverton	Many charters detailing land in *campus borialis* and *campus australis*, e.g.,
	in campo boriali 5½ acres in 7 parcels
	in campo australi 5 acres in 9 parcels (f. 73).[9]
Sulgrave	Transfer of " unam virgatam ... domum et cotagium, scil.,
	unam acram terre cum dimidia acra terre in uno campo
	et tantum in alio campo
	et item sex acras terre de inlonde in uno campo et unam acram prati
	et sex acras in alio campo et unam acram prati."[10]
Thrup	Terrier of the demesne of the prior of Chalcumbe, which comprised

[1] Cott. MS., Tib. E V, f. 36b. [XIV cen. copy.]
[2] Add. Char. 21897. [XIII cen.]
[3] Add. Char. 22012. [Early XIII cen.]
[4] Cott. MS., Tib. E V, f. 30b. [XIV cen. copy.]
[5] Ped. Fin., 171-3-62; printed, Pipe Roll Soc., Publ., 1900, xxiv, no. 225. 10 Rich. I.
[6] Magd. Coll., Brackley Deeds, B 148. [c. 1220-30.]
[7] Cott. MS., Tib. E V, f. 84. [XIV cen. copy.]
[8] Exch. Treas. Recpt., M. B. 71, f. 52. 1303.
[9] Cott. MS., Claud. D XII, ff. 73-82. [XV cen. copies.]
[10] Cott. MS., Vesp. E XVII, f. 163b. [c. 1260.]

Township	Description
Thrup (*continued*)	*in campo boriali* 49½ acres in 68 parcels *in campo australi* 42¼ acres in 61 parcels.[1]
Thurlaston	Transfer of 4 acres of arable, viz., 2 acres in 4 parcels *ex una parte campi* 2 acres in 4 parcels *ex alia parte campi*.[2]
Weedon	A series of leases assigning to tenants parcels in North field and South field. The longest (no. 40) describes in North field 82 parcels consisting of 44 lands, 23 butts, 29 " yards," and 20 leys in South field 85 parcels consisting of 37 lands, 18 butts, 50 " yards," and 26 leys.[3]
Welton	Grant of 1½ acres of arable, viz., *in campo occidentali* ½ acre in 1 parcel *in campo orientali* 1 acre in 2 parcels.[4]
Welton	Transfer of 6 acres of arable, scil., " iii in uno campo et iii in alio."[5]
Woodford	Transfer of 24 acres of arable in the fields, viz., " xii acras in uno campo et xii acras in altero campo."[6]
Pattishall	Extent of the demesne arable, " unde medietas potest seminari per annum. . . . Et altera medietas nichil valet per annum quia iacet ad warectam et tunc est communis."[7]
Stowe	Extent. " Item sunt ibidem in dominico ccxl acre terre arabilis de quibus possunt seminari per annum cxx et . . . cxx acre terre iacent ad warectam que nihil valent per annum quia dicta warecta est communis omnibus tenentibus ibidem."[8]
Warden, [Chipping]	Extent of the demesne arable, "de quibus medietas potest seminari per annum et . . . alia medietas que iacet ad warectam nichil valet per annum quia in communi campo."[9]
Weston iuxta Weedon	Extent. " Predicta virgata continet xxxii acras terre quarum medietas potest seminari per annum. . . . Et alia medietas quolibet anno ad warectam cuius proficuus nichil valet quia iacet in communi campo."[10]

[1] Exch. Aug. Of., M. B. 378, f. 29. [Mid. XIV cen.]
[2] Cott. MS., Calig. A XIII, f. 117b. [Late XIV cen. copy.]
[3] All Souls MSS., Terriers 37, 38, 40. 1586-87.
[4] Anc. Deeds, A 3065. 5 Edw. III.
[5] Cott. MS., Claud. D XII, f. 82b. [XV cen. copy.]
[6] Cott. MS., Vesp. E XVII, f. 128b. [XV cen. copy.]
[7] C. Inq. p. Mort., Edw. III, F. 47 (5). 10 Edw. III.
[8] Ibid., F. 1 (12). 1 Edw. III.
[9] Ibid., F. 40 (6). 8 Edw. III.
[10] Ibid., F. 68 (22). 17 Edw. III.

APPENDIX II

NORTHAMPTONSHIRE. THREE-FIELD TOWNSHIPS

Township	Description
Adstone, or Canons Ashby	Terrier of land belonging to the prior of Ashby, viz., [field rubric omitted] 6¾ acres in 12 parcels *in campo borientali* 6¾ acres in 12 parcels *in campo occidentali* 8 acres in 15 parcels.[1]
Barnack and Bainton	Retention of xv acres from one carucate, " unde v acre iacent in campo inter Bernake et Pilesgate et alie v acre iacent in campo inter viam de Stanford et Wel... e [Walcot ?] et iiii acre versus occidentalem partem ville de Bernake et i acra versus aquilonem." [2]
Barnack	A series of charters incidentally mentioning " campus australis," " campus orientalis," " campus occidentalis," and " Aldefeld." No holding is of any size and none is divided among these fields.[3]
Blakesley	Terrier of a virgate which comprises *in campo orientali* 6¾ acres in 11 parcels *in campo borientali* 7½ acres in 13 parcels *in campo occidentali* 6¾ acres in 15 parcels.[4]
Bozeat	Terrier of All Souls lands in the tenure of Wm. Bettell, showing, along with 2¼ acres of meadow, in the Wood field 20 lands in 14 parcels in the Diche field 19 lands in 17 parcels in Sandwell field 1 acre and 24 lands in 17 parcels.[5]
Braybrook	Grant of " triginta acras terre de domenico ... scil., undecim acras in campo versus Oxendun' [Oxendon Magna] undecim acras in campo versus Bugedon' [Bowden] undecim acras in campo versus Deresburc [Desborough]." [6]
Braybrook	Two charters locating small parcels "in campo orientali, in campo occidentali, et in aquilonali parte campi." [7]
Braybrook	A terrier of the glebe, which comprises in Loteland field 29 parcels of arable and 12 leys in Hedickes field 29 parcels of arable and 17 leys in Arnsborrow and Black fields 27 parcels of arable and 4 leys.[8]
Brixworth	Rental of ten virgates of land, each of which is described in detail. One comprises ½ acre of meadow and in Schotenwelle field 6 acres in 11 parcels in Demmyswelle field 6 acres in 11 parcels in Whaddon field 7¾ acres in 11 parcels.[9]

[1] Rents. & Survs., Portf. 13/19. [Hen. VIII.]
[2] Ped. Fin., 171-2-24; printed, Pipe Roll Soc., Publ., 1898, xxiii, no. 117. 9 Rich. I.
[3] Cott. MS., Faust. B III, ff. 58b, 63b, 65b, 75b. 4 Edw. II and 2 Edw. III.
[4] Rents. & Survs., Portf. 13/15. [Hen. VII.]
[5] All Souls MSS., Terrier 35. 1580.
[6] Ped. Fin., 171-12-210. 9 John.
[7] Cott. MS., Calig. A XII, ff. 105, 106. [XIII cen. copy.]
[8] Stowe MS. 795, f. 219. 1631.
[9] Cott. MS., Vesp. E XVII, ff. 183-194. 1 Hen. VI.

Township	Description
Clopton	Grant of 42¼ acres " de predicta virgata terre, scil., xvi acre et una roda inter le Frid et exitum de Clapetom versus occidentem xiii acre et una roda inter le predictum exitum et feodum Sancti Gutlaci xiii acre et una roda inter le predictum feodum Sancti Gutlaci et Swinehaw."[1]
Culworth	Grant of two acres, viz., *in campo australi* ½, ¼ *in campo occidentali* ¼ *in campo boriali* ½, ¼, ¼.[2]
Culworth	Terrier of a yard-land, which contains in the South field 12 " yerdes " and ½ acre in 13 parcels in the West field 19 " yerdes " and ½ acre in the North field 34 "yerdes" and ½ acre in 30 parcels.[3]
Desborough	Grant of common of pasture " per totum campum de Deseborug tam in boscis quam in campis tribus."[4]
Draughton	Grant of a cottage and " sex acre terre arabilis de terris dominicalibus ... quarum due acre simul iacent in campo australi septem rode simul iacent in campo boriali sex rode simul iacent in campo orientali et quinque dimidie rode simul iacent super eandem culturam."[5]
Drayton	A survey assigning to West field 529 acres to North field 573 acres and to East field 414 acres. The copyholds and leaseholds are divided among the same three fields.[6]
Evenley	A terrier of two yard-lands, which comprise in the West field 17 acres in 15 parcels in the South field 13 acres in 13 parcels in the East field 17 acres in 16 parcels.[7]
Hardingstone	Grant of 3 acres of arable, viz., *in campo australi* 1 acre in 2 parcels *in campo medio* 1 acre in 2 parcels *in campo* [septen] *trionali* 1 acre in 2 parcels.[8]
Holdenby	Transfer of several parcels of arable located in the West field, called Langeland field, in the Wood field, and in Cargatt field.[9]

[1] Ped. Fin., 171-7-80. 4 John.
[2] Add. Char. 44291. 7 Edw. III.
[3] Add. Char. 44354. [Early XVI cen.]
[4] Cott. MS., Otho B XIV, f. 169. [Late XIV cen. copy.]
[5] Add. Char. 81829. 6 Hen. V.
[6] D. of Lanc., M. B. 113, ff. 34 sq. 13 Eliz.
[7] Rents. & Survs., Portf. 13/25. [Hen. VIII or later.]
[8] Cott. MS., Tib. E V, ff. 115, 115b. [XIV cen. copy.]
[9] Add. Char. 21991. 32 Hen. VIII.

APPENDIX II 483

Township	Description
Holcot	The numerous parcels of a half-virgate are divided among East field, Middle field, and West field.[1]
Isham	Terrier of a virgate, which comprises *in campo australi* 9 acres in 1 parcel *in magno campo boriali* 8¼ acres in 4 parcels *in campo occidentali* 5¾ acres in 3 parcels.[2]
Kislingbury	Transfer of 10 acres of arable, viz., *in campo orientali* 3¾ acres in 11 parcels *in campo australi* 4¾ acres in 13 parcels *in campo occidentali* 1¼ acres in 4 parcels.[3]
Kislingbury	Long terrier of the demesne arable, which lies in many parcels in East field, South field, and West field.[4]
Maidwell	Grant of two cottages and 4½ acres lying " in campo boreali, in campo occidentali, et in campo australi."[5]
Market Harborough and Great Bowden	Transfer of 20 acres in the fields of Great Bowden almost equally divided among South field, East field, and North field. Other charters divide their acres similarly (pp. 178, 196).[6]
Moulton	A long terrier with apparently even division of parcels among North field, South field, and West field.[7]
Northampton	Terrier of a carucate lying in the fields of Northampton, and comprising *in campo orientali* 32 acres *in le Middelfeld* 17¼ acres *in campo boriali* 31¼ acres.[8]
Oundle	Extent of the demesne arable, which consists of 65½ acres in 5 "culture" *in campo qui vocatur Inhamfeld* 94¾ acres in 9 "pecie" *in campo qui vocatur Howefeld* 96¼ acres in 10 "pecie" *in campo qui vocatur Holmfeld*.[9]
Paulers Pury	Richard Stuemyn holds a messuage, two cottages, one-half acre, and " xxx acras terre errabilis et ii acras prati iacentes in communibus campis predictis vocatis parkefeild, Totehilfeild, et Myddelfeild."[10]
Pilsgate	Terriers of several holdings, with considerable variation in the distribution of acres among fields. One terrier (f. 145) has about the same number of acres in Nether field, West field, and South field.[11]

[1] Cott. MS., Vesp. E XVII, f. 52. 10 Edw. II.
[2] Ibid., f. 315b. [Mid. XV cen.]
[3] Add. Char. 22078. 14 Edw. III.
[4] Cott. MS., Vesp. E XVII, f. 300. [Mid. XV cen.]
[5] Add. Char. 22269. 6 Hen. VI.
[6] J. E. Stocks and W. B. Bragg, *Market Harborough Parish Records* (London, 1890), p. 161. 1343.
[7] Cott. MS., Vesp. E XVII, ff. 304–309. 9 Hen. VI.
[8] Ibid., f. 299b. [Mid. XV cen.]
[9] Cott. MS., Nero C VII, f. 154b. 1 Hen. IV.
[10] Exch. Aug. Of., M. B. 419, f. 3. 32 Hen. VIII.
[11] Cott. MS., Faust. B III, ff. 145–152. [Mid. XIV cen.]

Township	Description
Walcot	Extent of the demesne arable, which consists of 19 acres in 7 places *in campo occidentali* 15 acres in 4 places *in campo australi* 53 acres in 10 places *in campo orientali* (the last including 29 acres in *Northdykcroft*).[1]
Wellingborough	A terrier, which locates many rood parcels as follows: in East field 33 roods and 10 " todel " in Up field 21 roods and 11 " todel " in West field 32 roods and 4½ " todel."[2]
Braybrook	Extent of the demesne arable, "quarum due partes possunt seminari per annum ... et tertia pars ... que iacet ad warectam nichil valet quia in communi campo."[3]
Clapton	Extent of the demesne arable. " Et tertia pars eiusdem terre quolibet anno iacet ad warectam et est nullius valoris quia in communi campo."[4]
Corby	Extent. "Sunt in dominico ix^{xx} acre terre arabilis unde cxx possunt seminari per annum ... et residuum iacet ad warectam et tunc nichil valet quia in communi."[5]
Dallington	Extent of the demesne arable. " Et tertia pars eiusdem terre quolibet anno iacet ad warectam et est nullius valoris quia in communi campo."[6]
Deanshanger	Extent of the demesne arable, " de quibus due partes possunt seminari per annum.... Et tertia pars que iacet ad warectam nichil valet per annum quia semper tempore warecta [terra] est communis omnibus tenentibus ibidem."[7]
Harpole	Extent of the demesne arable, of which " tertia pars... quando iacet ad warectam nichil valet quia tunc est communis."[8]
Higham Ferrers	Extent of the demesne arable, of which two-thirds may be sown yearly. " Et dicunt quod residuum de terra predicta iacens ad warectam nihil valet per annum quia tempore warecta [terra] est communis omnibus tenentibus ibidem."[9]
Rushden	Extent of the demesne arable, of which two-thirds may be sown yearly. " Et warecte terre residue nihil valent quia semper tempore warecta [haec terra] est communis omnibus tenentibus ibidem."[10]

[1] Cott. MS., Nero C VII, f. 164b. 1 Hen. IV.
[2] Cott. MS., Vesp. E XVII, f. 309b. 10 Hen. VI.
[3] C. Inq. p. Mort., Edw. III, F. 40 (6). 8 Edw. III.
[4] Cott. MS., Cleop. C II, f. 123. 20 Edw. III.
[5] C. Inq. p. Mort., Edw. III, F. 44 (6). 9 Edw. III.
[6] Cott. MS., Cleop. C II, f. 123. 20 Edw. III.
[7] C. Inq. p. Mort., Edw. III, F. 47 (8). 10 Edw. III.
[8] Ibid., F. 53 (18). 12 Edw. III.
[9] Ibid., F. 6. 1 Edw. III.
[10] Ibid.

Township	Description
Wodenhoe	Extent of the demesne arable, "quarum due partes possunt seminari per annum ... et tertia pars ... iacebit ad warectam et tunc nichil valet per annum quia in communi campo."[1]

NOTTINGHAMSHIRE. THREE-FIELD TOWNSHIPS

Cotgrave	Two bovates are withheld from a grant of one and one-fourth carucates, scil., "una bovata terre quam Warinus de Boville tenuit et que iacet in campo aquilonari et orientali et una bovata terre quam Walterus Capel tenuit et que iacet in campo occidentali."[2]
Hucknall Torkard	Transfer of toft, croft, and 2¾ acres of arable, viz., 1 acre in the West field ½ acre in the East field ¼ acre *ad capud orientale ville* 1 acre in the South field in 2 parcels.[3]
Keyworth	Long terrier of a holding, specifying many selions but in part illegible. The parcels seem to be pretty evenly divided among "Brokfeld, campus de Senygow, and le Wooldefelt."[4]
Kirton	Grant of 3½ acres, of which "una acra iacet in Nortfeld ... et altera in Holims et una acra in Suthfeld ... et dimidia acra in medio campo." Each of these three fields is mentioned singly in three following charters.[5]
Knighton	Grant of two bovates, scil., *in campo occidentali* 5¼ acres in 11 parcels *in campo aquilonari* 7¾ acres in 9 parcels *in Middelfeld* 6 acres in 9 parcels.[6]
Walkeringham	Terriers of the holdings of eleven tenants, the acres being often divided equally among three fields, West, North, and East.[7]
Wandesley	Extent. "Sunt in dominico xx acre terre ... de quibus tertia pars iacet quolibet anno in warecto et pastura inde nihil valet quia iacet in communi."[8]

[1] C. Inq. p. Mort., Edw. III, F. 43 (10). 9 Edw. III.
[2] Ped. Fin., 182–5–142. 16 Hen. III.
[3] Exch. K. R., M. B. 23, f. 66. 14 Hen. IV.
[4] Cott. MS., Titus C XII, f. 147. [XIV cen. copy.]
[5] Harl. MS. 1063, f. 185*b*. [XVII cen. copy of an early charter.]
[6] Ped. Fin., 182–4–84. 10 Hen. III.
[7] Rents. & Survs., Portf. 13/87. 1608.
[8] C. Inq. p. Mort., Edw. III, 50 (24). 11 Edw. III.

OXFORDSHIRE. TWO-FIELD TOWNSHIPS

Township	Description
Barford	Terrier of a messuage and one virgate of arable in the fields, viz., *in campo occidentali* 10¼ acres in 22 parcels *in campo orientali* 9½ acres and 3 " forere " in 24 parcels.[1]
Bensington	Grant to the Templars of a messuage and 10¼ acres *in campo australi* in 13 furlongs and 11¼ acres *in campo aquilonari* in 13 furlongs.[2]
Bicester [King's End]	Extent, with certain holdings described in detail. One half-virgate (e.g., p. 573) contains in South field 10½ acres in 21 parcels in North field 11¼ acres in 23 parcels.[3]
Bletchingdon	Grant of 32¼ acres in many parcels, viz., " in one felde " 14 acres and " in anothyr felde " 17¼ acres.[4]
Burford	Terrier of the glebe, which comprises " in le Estfyld " 47½ acres in 51 parcels " in le Westfylde " 48 acres in 52 parcels.[5]
Chadlington	Transfer of four holdings, the acres of which are equally divided between East field and West field, viz., 2 *vs.* 1¼, 27 *vs.* 27, 15 *vs.* 14, 6 *vs.* 5½.[6]
Chadlington	Grant of " duas acras in uno campo et duas in alio." [7]
Chipping Norton	Valor. " Sunt ibidem ii carucate terre continentes per estimationem clx acre terre arabilis quarum iiii^{xx} acre iacent quolibet anno ad warectam... et iiii^{xx} acre quolibet anno seminande ad utramque sementem." [8]
Churchill	" Habebit eciam dictus Vicarius v acras terre arabilis in uno campo et v in alio, cum prato ad easdem pro rata pertinente." [9]
Cleveley	Transfer of a messuage, 3 acres enclosed, and *in campo orientali* 17¾ acres in 17 parcels *in campo occidentali* 12 acres in 16 parcels.[10]
Cornwell	Grant of 4 acres "in Cumbe near Heanhulle in one field and in another field 4 acres towards the way to Kaigham." [11]

[1] Exch. Aug. Of., M. B. 378. [Mid. XIV cen.]
[2] Bodl., Wood Donat. 10, f. 99. [Copy of Edw. I.]
[3] Kennett, *Parochial Antiquities*, i. 565–578. 19 Edw. II.
[4] Bodl., Rawl. B 408, f. 8*b*. [XV cen. transcript of a deed of Edw. I.]
[5] Bodl., Oxfordshire Archdeaconry Papers. 1576.
[6] *Eynsham Cartl.* (ed. Salter), i. 247. [1264–68.]
[7] Ibid. 107. [*c.* 1173.]
[8] Rents. & Survs., Ro. 33. [Hen. VI.]
[9] *Cartl. St. Frideswide* (ed. Wigram), ii. 286. 1340.
[10] *Reg. Monast. de Winchelcumba* (ed. Royce), ii. 172. [*c.* 1280.]
[11] W. H. Turner and H. O. Coxe, *Calendar of Charters and Rolls in the Bodleian Library* (Oxford, 1878), p. 324. [*c.* 1220.]

APPENDIX II 487

Township	Description
Enstone	Grant of 6 acres in the field, viz., *in campo del Suht* 3 acres in 6 parcels *in campo del Norht* 3 acres in 5 parcels.[1]
Faringdon, Little	Description of the arable demesne, of which 230¼ acres are *in campo boriali* and 156½ acres *in campo australi*.[2]
Fulwell	Grant of one acre of arable *in campo orientali* and one *in campo occidentali*.[3]
Hampton Gay	Grant of " 1 halfe hyde of londe in Gaihampton conteynynge xxv acres of land in on feelde and also many in an othyr feelde."[4]
Hanwell	Transfer of 4 acres, of which " due acre iacent coniunctim in uno campo ... et alie due acre iacent coniunctim in alio campo."[5]
Hensington	Grant of a messuage and 4 acres of arable, viz., 2 acres in the North field and 2 in the South field.[6]
Hook Norton	Grant of a messuage, ½ acre of meadow, and 2 acres in the East field in 4 parcels and 2 acres in another field.[7]
Kirtlington	Grant of part of a half-virgate, viz., 1 acre of meadow and *in uno campo* 5¾ acres in 7 parcels and *in alio campo* 5¼ acres in 6 parcels.[8]
Lidstone	Grant of 4 acres, scil., " in uno campo duas acras et in alio campo duas acras."[9]
Middleton	Grant of 4 acres, of which 2 lie in the East field and 2 in the West field.[10]
Milcombe	Grant of 16 acres of arable in the fields, viz., 7 acres in the South field in 7 parcels 1 acre in the West field in 1 parcel 8 acres in the North field in 8 parcels.[11]
Milton-under-Wychwood	Grant of 5 acres *in uno campo* in 5 parcels and 5 acres *in alio campo* in 5 parcels.[12]
Newington " Juel "	" John Busseby tenet i acram terre ... in utroque campo.... Hugo Gilbert tenet i messuagium et unam acram terre in uno campo et duas terre in alio."[13]

[1] *Reg. Monast. de Winchelcumba* (ed. Royce), ii. 175. [c. 1240.]
[2] Cott. MS., Nero A XII, f. 100b. [XV cen. copy.]
[3] Kennett, *Parochial Antiquities*, i. 306. [20 Hen. III.]
[4] Bodl., Rawl. B 408, f. 9b. [XV cen. copy of a deed of Edw. I.]
[5] Add. Char. 22016. 16 Rich. II.
[6] Edward Marshall, *Early History of Woodstock Manor* (Oxford, 1873), p. 408. [XIII cen.]
[7] Turner and Coxe, *Calendar of Charters*, etc., p. 329. [c. 1200–10.]
[8] *Cartl. St. Frideswide* (ed. Wigram), ii. 215. [c. 1210–28.]
[9] *Reg. Monast. de Winchelcumba* (ed. Royce), ii. 182. [c. 1235.]
[10] Bodl., Rawl. B 408, f. 91. [Edw. I.]
[11] Ibid., f. 88b. [c. 1216–30.]
[12] *Eynsham Cartl.* (ed. Salter), p. 117. [1160–80.]
[13] *Hundred Rolls*, ii. 840a–b. 7 Edw. I.

Township	Description
Newington, South	Grant of one virgate from the demesne, viz., *in campo aquilonali* 10 acres in 12 parcels *in campo australi* 11 acres in 11 parcels.[1]
Newnham Murren	One holding comprises " terra arabilis in communibus campis ibidem, viz., South field per estimationem lxv acre North field per estimationem lxxxvi acre."[2]
Piddington	Grant *inter alia* of tithes from 2 acres of demesne meadow, " scil., quando occidentalis campus seminatur duas primas acras de prato quod dicitur Westmede, quando vero orientalis campus seminatur latitudinem duarum acrarum in prato quod dicitur Langdale."[3]
Rollright	Dower of Johanna, wife of Adam le Despencer, comprises ⅓ of a messuage, a pond, and *in campo boriali* 80 acres of arable and *in campo australi* 60 acres of arable.[4]
Rousham	Grant of a messuage and 9 acres of demesne in one field and 9 acres in another.[5]
Sandford	Grant of one acre in the North field and one acre in the South field.[6]
Shipton	Grant to the Templars of a croft and 2 acres *in campo versus orientem* in 2 parcels and 2 acres *in alio campo* in 3 parcels.[7]
Sibford	Grant to Templars of 10 acres of demesne arable, viz., 5 acres *in uno campo* and 5 acres *in alio*.[8]
Stratton Audley	Grant of 40 acres of demesne, scil., *in uno campo* 20 acres in 5 furlongs and *in altero campo* 20 acres.[9]
Tew, Great	" Adam Prat tenet 1 cottagium et ii acras terre in utroque campo." [10]
Wilcot	Terrier of two virgates held by the Templars: in East field 24 acres in 10 places in West field 24 acres in 13 places.[11]

[1] Harl. MS. 4028. [XVIII cen. copy of an early charter.]
[2] Exch. Aug. Of., M. B. 388, f. 67. 6 Jas. I.
[3] Kennett, *Parochial Antiquities*, i. 103. 6–7 Hen. I.
[4] C. Inq. p. Mort., Edw. I, F. 134(3) 34 Edw. I.
[5] Turner and Coxe, *Calendar of Charters*, etc., p. 362. [c. 1200.]
[6] Ibid., 363. [4] 4 Edw. II.
[7] Bodl., Wood Donat. 10, f. 76. [Edw. I.]
[8] Ibid., f. 94. [Edw. I.]
[9] Kennett, *Parochial Antiquities*, i. 188. [29 Hen. II.]
[10] *Hundred Rolls*, ii. 846a. 7 Edw. I.
[11] Bodl., Wood Donat. 10, f. 38b. [Edw. I.]

APPENDIX II 489

OXFORDSHIRE. GLEBE TERRIERS OF THE SEVENTEENTH CENTURY
(*Bodl., Oxfordshire Archdeaconry Papers*)

Township	Description of the Glebe
Alkerton	"Lands and Layes on the South side 21, on the North side 19."[1]
Alvescot	In a field called Barrell 4½ acres in 5 parcels in a field called Woo-Lands 1 acre in 2 parcels. In the Upper Fields are 17 acres of ground, viz., 9½ acres in the East field in 18 parcels and 7½ acres in the West field in 14 parcels.[2]
Ardley	Two yard-lands, viz., in West field 22 acres in 9 parcels in East field 26 acres, 8 butts, in 10 parcels.[3]
Astall	In the East field 3½ acres in 3 parcels in the West field 4 acres in 4 parcels.[4]
Aston, Steeple	In the lower field 12 acres in 5 parcels in the upper field 9 acres in 4 parcels.[5]
Brize Norton	In the grove field 36¼ acres, 5 lands, in 34 parcels in Ringborough field 47 acres, 4 lands, in 35 parcels.[6]
Broughton Poggs	Two yard-lands, viz., in the East field 31¾ acres in 31 parcels in the West field 37 acres in 30 parcels, with 2 crofts containing 4 acres.[7]
Cottisford	"The glebe land in Cottesford fielde lies for 4 yard lands divided into two fields: whereof one field, which lieth Eastward, containeth 39 acres, and the other field, which lieth Westward, containeth 21 acres."[8]
Duns Tew	Two yard-lands described in two (not very legible) columns, containing many parcels and headed "on the North side" and "on the South side."[9]
Fulbrook	In the West field 33 acres in 25 parcels in the East field 30½ acres in 20 parcels.[10]
Glimpton	On the north side of the town in the common fields 27½ acres, 7 lands, 1 butt, in 25 parcels on the south side of Glimpton in the common fields 14 acres, 3 lands, in 9 parcels on the upper side of Woodstocke way [also on the south] 16 acres, 9 lands, in 13 parcels. There are closes, viz., 5 acres of arable, 1 acre of meadow, 4 acres of pasture, 19 acres of heath.[11]
Kencott	In the West field 35½ acres in 21 parcels, and in the Hitching 7¼ acres, and in the "new broake ground" 1½ acres to be sown with this field

[1] 1647.
[2] 1685.
[3] [Before 1679.]
[4] 1634.
[5] 1684.
[6] 1685.
[7] 1685.
[8] Late XVII cen. Printed by J. C. Blomfield, *History of the Present Deanery of Bicester* (8 pts., London, 1882–94). iii. 33.
[9] 1634.   [10] 1685.   [11] 1685.

Township	Description of the Glebe
Kencott (*continued*)	in the East field 32 acres in 24 parcels, and in the Hitching 9¾ acres to be sown with this field.[1]
Middleton Stony	In the South fields 32¼ acres and 14 lands in 31 parcels in the North field 29 acres and 14 lands in 30 parcels.[2]
Shutford, West	In the South field 46 acres in 18 parcels in the North field 46 acres in 27 parcels, with one other parcel of 20 acres.[3]
Stoke, South	In the little North field 3½ acres in 5 parcels, in the great North field 13¼ acres in 28 parcels in the great South field 11¼ acres in 23 parcels. The total constitutes two yard-lands.[4]
Tackley	In the South field 21 acres in 11 parcels in the North field 20 acres in 11 parcels. There are closes of arable containing 6⅞ acres.[5]
Westwell	In the East field 44 acres in 22 parcels (one of them containing 9 acres) in the West field 37 acres in 22 parcels.[6]

OXFORDSHIRE. THREE-FIELD TOWNSHIPS

Township	Description
Aston and Cote	" There are ... several Leyes of greensward lying in the common fields, two years mowed and the other fed [the fields being Holliwell field, Windmill field, and Kingsway field.] "[7]
Stoke, South	Extent of the demesne arable in three fields, " unde duo seminantur annuatim, tertius vero iacet warectus," viz., *in campo australi* 126 acres *in campo medio* 73¼ acres *in campo aquilonari* 64½ acres.[8]
Stoke Talmage	Grant to the Templars of one-half hide of demesne arable, viz., *in campo aquilonari* 15 acres in 3 places *in campo australi* 15 acres in 5 places *in campo occidentali qui vocatur chelfelde* 15 acres in 14 places.[9]
Tew, Great	Grant of 7 acres, viz., 1¾ acres in the North field in 4 parcels 1¾ acres in the West field in 4 parcels 3 acres in the South field in 8 parcels ½ acre in the mede in the East field.[10]
Thomley	Grant of 5 acres, 1 butt, and ¼ of a meadow. The arable comprises

[1] 1634. [3] 1634. [5] 1634. [6] *Eynsham Cartl.* (ed. Salter), ii. 118–128.
[2] 1679. [4] 1685. [6] 1601. 1366.
[7] J. A. Giles, *History of Bampton*, Supplement, pp. 3, 7. 1657. [8] Bodl., Wood Donat. 10, f. 53. [Edw. I.]
[10] Bodl., Rawl. B 408, f. 165. [XV cen. copy.]

APPENDIX II 491

Township	Description
Thomley (*continued*)	in the furlongs Bremor Gorstilond and Harsaeforlong 2 acres in 3 parcels *in alio campo* 2½ acres in 4 parcels *in alio campo* toward Wormehale ½ acre and in Shortlond 1 butt.[1]
Bampton	Extent. " Sunt ibidem tres carucate terre continentes in se cxcvi acras unde due partes possunt quolibet anno seminari ... et tertia pars nihil valet quia iacet ad warectam et in communi." [2]
Finmere	Extent. " In dominico sunt cc acre terre arrabilis unde due partes possunt seminari." [3]
Kidlington	Extent. " Sunt duo carucate terre in dominico que continent cxlvi acre terre quarum due partes omnimodis annis seminari possunt et tertia pars iacet ad warectam et in campo communi ita quod nihil inde percipi potest." [4]
Marston	" Est quedam terra in campo de Merston ultra dominicum que ... semper in tertio anno nichil reddit quia iacet warecta." [5]

OXFORDSHIRE. GLEBE TERRIERS OF THE SEVENTEENTH CENTURY
(*Bodl., Oxfordshire Archdeaconry Papers*)

Township	Description of the Glebe
Bladon	Two yard-lands, viz., in Church field 6 acres and 10 lands in 16 parcels in the Down field 21 lands in 19 parcels in Burley field 27 lands, 2 acres, 2 butts, in 25 parcels.[6]
Bourton, Black	In West Brook field next Alvescott 2½ acres in 5 parcels in the Downe field next Bourton 3½ acres in 6 parcels in the Downe field next Norton 5¼ acres in 8 parcels.[7]
Britwell Salome	In the West field 18 acres in 10 parcels in the Hill field 6 acres in 6 parcels in the East field 6 acres in 6 parcels.[8]
Chinnor	In West Pond field 4 acres in 3 parcels in Great Winnoll field 1 acre, 1 land, in 2 parcels in Chinnor field 2 lands in 2 parcels.[9]
Culham	In Ham field 4¾ acres in 10 parcels in Middle field 6¼ acres in 13 parcels in Cositer field 6 acres in 11 parcels.[10]
Cuxham	In the South field 10 acres in 13 parcels (" two of these acres are now tilled with the West field ")

[1] *Cartl. St. Frideswide* (ed. Wigram), ii. 158. [*c.* 1210–20.]
[2] Giles, *History of Bampton*, p. 138. 36 Edw. III.
[3] C. Inq. p. Mort., Edw. III, F. 56 (1). 12 Edw. III.
[4] Ibid., F. 51 (3). 11 Edw. III.
[5] *Hundred Rolls*, ii. 711b. 7 Edw. I.
[6] [*c.* 1685.] [7] 1634. [8] 1635. [9] 1635. [10] 1685.

Township	Description of the Glebe
Cuxham (*continued*)	in the West field 9 acres in 7 parcels in the North field 9½ acres in 7 parcels.[1]
Finmere	In South field 22 lands, 3 butts, 3 leys, in 27 parcels in Mill field 25 lands, 4 butts, 6 leys, in 35 parcels in the field next Fullwell 20 lands, 3 butts, 3 leys, 2 yards, 2 " foreshorters," in 30 parcels.[2]
Handborough	In the Church field 12 acres in 12 parcels in the Mill field 15 acres in 24 parcels in the South field 12 acres in 15 parcels. Also three closes of 7 acres and 10 acres of meadow.[3]
Hardwick	In the Heeth field 14 lands, 4 " heads," in 3 parcels in the Mill field 20 lands in 7 parcels in Tinkers field 13 lands in 4 parcels.[4]
Heath	[In the first field] 9 lands, 1 acre, 1 butt, in 6 parcels in the second field 7 lands, 1 acre, 1 butt, 1 yard, in 10 parcels in the third field 11 lands, 3 acres, 1 butt, 2 yards, in 12 parcels.[5]
Lewknor	In the field next to Aston 5 acres, 4 lands, 10 yards, 2 butts, in 16 parcels in the field next Shuerborn 5 acres, 12 lands, 1 yard, in 14 parcels in the middle field 8 acres, 5 lands, 5 yards, in 10 parcels.[6]
Newton Purcell	In the field toward Finmere (North field), 5 acres, 1 yard, in 6 parcels in the field butting upon broadmeadow (South field) 7 acres, 1 yard, 2 lands, 2 butts, in 11 parcels in the field adjoining Willoston lordship (West field) 7 acres, 4 yards, 2 lands, in 10 parcels.[7]
Waterstock	In Conygere field (South field) 5½ acres in 8 parcels in Gravewaye field (East field) 3½ acres in 5 parcels in Ham field (North field) 2½ acres in 4 parcels.[8]
Wendlebury	In the field toward Bisseter 17 lands, 3 butts, in 19 parcels in the field toward Charleton 12 lands in 12 parcels in the field toward Weston 15 lands, 3 yards, 1 ley, in 13 parcels.[9]
Weston, South	In Cope field 8 acres in 10 parcels in Stonie field 6¾ acres in 7 parcels, and 2 butts of grass ground in Moshill field 17¼ acres in 15 parcels, and 1 acre of grass ground.[10]
Whitchurch	In the Parke field 40½ acres in 23 parcels in the West field 3¼ acres in 3 parcels

[1] 1634. [3] [Early XVII cen.] [5] [c. 1601.] [7] 1634. [9] 1679.
[2] 1634. [4] 1601. [6] 1704. [8] 1601. [10] 1635.

APPENDIX II 493

Township	Description of the Glebe
Whitchurch (*continued*)	in the East field 33½ acres in 20 parcels
	in the Moore End field 16 acres in 13 parcels
	in Bozden field 10¼ acres in 14 parcels.
	These fields were probably grouped as three.[1]
Wooton	In the North field 25 acres in 8 parcels and probably 11¾ acres of meadow
	at Greene Hitch 23 acres in 13 parcels
	in the West field 19 acres in 17 parcels with 1 ley.[2]

OXFORDSHIRE. TOWNSHIPS DIVIDED INTO "QUARTERS," USUALLY FOUR
(*Bodl., Oxfordshire Archdeaconry Papers*)

Adderbury, West	In the Flags quarter 3 acres, 1 land, in 4 parcels
	in Poinfurlong quarter 4 acres, 3 yards, in 4 parcels
	in Berryll quarter 2 acres in 1 parcel
	in Langland quarter 6 acres in 4 parcels.[3]
Hanwell	In Debtcombe quarter 7 lands, 2 butts, 2 leas, 3 pikes, 1 yard, in 14 parcels
	in the Westfield quarter 6 lands, 2 butts, 5 leas, 1 headacre, ½ piece greensward, in 8 parcels
	in Lotrum quarter 8 lands, 4 butts, 3 yards, 1 headacre, 1 "hade," in 16 parcels
	in Pit-Acre quarter 7 lands, 2 butts, 3 leas, 1 "hade," in 11 parcels.[4]
Heyford, Upper	In the field adjoining Nether Heyford and Calcott (Dean field in 1685) 21 ridges
	in another field called Stanhill 20 ridges
	in Lobdane-tree field (Elmn field in 1685) 22 ridges
	in the field between the ways called Somerton way and above Oxford way 17 ridges
	in the lower field 11 ridges.[5]
Kingham	In the Ryeworth quarter 4 acres, 52 ridges (1 ridge usually equals ½ acre), in 9 parcels
	in Wythcombe quarter 4 acres, 27 ridges, 24 butts, in 11 parcels
	in Broadmore quarter 2 acres, 59 ridges, 11 butts, in 8 parcels
	in Brookside or Adbridge quarter 4 acres, 7 ridges, 10 butts, in 6 parcels
	in Townehill quarter "every yeares land," 2 acres, 6 ridges, and Rye close, in 4 parcels.[6]
Somerton	In the wheat field lying toward Frittwell moor, 3 acres, 8 lands, 8 butts, in 10 parcels
	in the second field butting on the south side Ardley way, 7 acres, 7 lands, 4 butts, in 13 parcels

[1] 1635. [3] 1722. [5] 1679. Printed by Blomfield, *Bicester*, vi [pt. ii],59–60.
[2] 1601. [4] 1680. [6] 1685.

Township	Description of the Glebe
Somerton (*continued*)	in the third field adjoining the way leading to Bister, 7 acres, 16 lands, 3 butts, in 12 parcels
	in the fourth field lying on the south side Bister way, 5 acres, 10 lands, 4 butts, in 12 parcels.[1]
Standlake	In Standlake Little field 6 acres in 8 parcels
	in the Richland field 8 acres in 16 parcels
	in the North field 9¾ acres in 19 parcels
	in the South field 6 acres in 12 parcels.[2]
Tadmarton	In Blackland quarter 1 acre, 9 lands, 1 lea, in 11 parcels
	in Fulling mill quarter 3¼ acres, 7 lands, 4 leas, 3 butts, in 17 parcels
	in Leabrouch quarter 10 acres, 2 lands, 2 leas, in 14 parcels
	in Rattnill quarter 4¼ acres, 6 lands, 1 lea, 2 butts, in 14 parcels, and 9 "lottes" of furzes in the heath.[3]
Wigginton	In Milcome quarter 2 acres, 4 lands, in 6 parcels
	in Petsbush quarter 14 lands in 14 parcels
	in Midnill quarter 5 lands, 3 yards, 1 butt, in 8 parcels, and 13 leys or other parcels of grass ground
	in South quarter 7 lands, 2 yards in 7 parcels.[4]

RUTLAND. THREE-FIELD TOWNSHIPS

Township	Description
Essendine	Extent. " Sunt ibidem cl acre terre arabilis in dominico unde possunt seminari per annum c acre. . . . Et residuum iacebit ad warectam et tunc nihil valet quia in communi."[5]
Tinwell	Extent of the demesne arable, which comprises
	in " campus qui dicitur Estfeld [5] pecie "
	in " Middilfeld [8] pecie "
	in " campus qui dicitur Westfeld [13] pecie "
	in " campus qui dicitur Inglethorpfeld [2] pecie."[6]
Whissendine	Extent of the demesne arable, " de quibus due partes possunt seminari per annum ... et warecta inde nihil valet per annum quia tunc est communis omnibus tenentibus."[7]

SOMERSET. TWO-FIELD TOWNSHIPS

Baltonsborough	Extent of the demesne, of which 131 acres lie *in campo occidentali* and 88 acres lie *in campo orientali*.[8]

[1] 1634. [2] 1685. [3] 1676. [4] 1685. [5] C. Inq. p. Mort., Edw. III, F. 37 (22). [6] Cott. MS., Nero C VII, f. 150. 1 Hen. IV. [7] C. Inq. p. Mort., Edw. III, F. 46 (31). 10 Edw. III. [8] *Rent. et Cust. Monast. Glastoniae*, Somerset Rec. Soc., [*Publ.*], 1891, v. 195. [1252–61.]

APPENDIX II

Township	Description
Barwick	Grant of 7 acres from one-third of a virgate, viz., 3½ acres *in campo versus orientem* and 3½ acres *in campo versus occidentem*.[1]
Bath Easton	Grant of a grove and 16¾ acres of arable, viz., 8½ acres in the East field in 4 parcels 7¼ acres in the West field in 9 parcels.[2]
" Berewe "	Grant. " De illa dimidia virgata terre que remanet ... iacent decem acre in Sudfeld et octo acre in Nordfeld."[3]
Bratton St. Maur	Grant of a messuage and 60½ acres of land in the West field and 59 acres of land in the East field.[4]
Camel " Rumare "	Grant of two messuages, two crofts, 2½ acres of meadow, and 27 acres in 20 parcels in the North field and 25 acres in 19 parcels in the South field.[5]
Cameley	Confirmation of a grant of a messuage, 2 acres of meadow, and 6 acres of arable in one field and 6 acres in the other field.[6]
Charlton Musgrove	Grant of 3 acres of meadow and 18 acres of arable, viz., 10 acres in 10 parcels in the North field and 8 acres in 7 parcels in the South field.[7]
Coleford	Grant *inter alia* of two mills " cum omnibus pertinentibus suis, scil., " cum sex acris terre iacentibus in campo orientali et sex acris terre iacentibus in campo occidentali."[8]
Compton	Grant of a messuage, a grove, a meadow, and " quinque acras terre in campo qui dicitur Suthfeld [in 4 parcels] et quinque acras terre in campo qui dicitur Estfeld [in 4 parcels]."[9]
Curry Rivell	Grant of a rent of 3s. paid for 9 acres of arable, viz., in the West field 5 acres and in the East field 4 acres.[10]
English Combe	Grant of 4 acres of arable, whereof 2 acres are in the East field and 2 acres are in the West field.[11]

[1] Ped. Fin., 196-3-23. 3 Hen. III.
[2] *Chartl. Bath Priory*, Somerset Rec. Soc., [Publ.], 1893, vii. 82. [1295-1300.]
[3] Ped. Fin., 196-2-63. 5 John.
[4] *Cartl. Bruton Priory*, Somerset Rec. Soc., [Publ.], 1894, viii. 22. [Before 1232.]
[5] *Cartl. Muchelney Abbey*, ibid., 1899, xiv. 68. [1240.]
[6] *Cartl. Buckland Priory*, ibid., 1909, xxv. 56. 1201.
[7] *Cartl. Bruton Priory*, ibid., 1894, viii. 44. [1256-67.]
[8] Ped. Fin., 196-4-89. 18 Hen. III.
[9] Ibid., 196-3-71. 10 Hen. III.
[10] *Cartl. Muchelney Abbey*, Somerset Rec. Soc., [Publ.], xiv. 66. [Early XIII cen.]
[11] *Chartl. Bath Priory*, ibid., vii. 165. 15 Edw. III.

Township	Description
Lovington	Grant with the advowson of the church of one hide of land, viz., four score acres in one field and four score in the other.[1]
Lyncombe	Grant of 2 acres in the lower field and 2 acres in the upper field.[2]
Mells	Extent of the demesne, of which 219 acres lay *in campo orientali* and 209½ acres lay *in campo occidentali*.[3]
Somerton	Grant of 8 acres of arable, viz., in the East field 4 acres in 3 parcels and in the West field 4 acres in 1 parcel.[4]
Swell	Grant of one-fourth of a virgate, viz., 7½ acres in Babforlang and Crouforlang and 7½ acres in the other field in 5 parcels.[5]
Weston	Description of a virgate of land, of which 15 acres are in the South field in 7 parcels and 15 acres in the North field in 8 parcels.[6]
Wookey Hole	Grant of houses, meadow, and 14 acres of arable land in one field and 13 acres in the other.[7]

SOMERSET. THREE-FIELD TOWNSHIPS

Brompton	Extent of the demesne arable, which comprises *in campo occidentali* 49 acres sown with wheat *in campo boriali* 60 acres sown with spring corn *in campo orientali* 39 acres lying fallow.[8]
Little Marston	Transfer of a half-virgate, which comprises 18½ acres, viz., in the East field 5½ acres in the East [*sic*] field 7 acres in the South field 6 acres.[9]
"Huredcote" Ilton Lyng Sutton	Extents of the manors of Athelney abbey. Relative to the demesne arable of each, "due partes possunt seminari per annum... et tertia pars iacet in communi et ad warectam et ideo nullius valoris."[10]

[1] *Cal. MSS. of Dean and Chapter of Wells* (Hist. MSS. Com., 1907), i. 41. n. d.
[2] *Chartl. Bath Priory*, Somerset Rec. Soc., [*Publ.*], vii. 77. [1261–91.]
[3] *Rent. et Cust. Monast. Glastoniae*, ibid., v. 219. [1252–61.]
[4] *Cartl. Muchelney Abbey*, ibid., xiv. 66. [Before 1282.]
[5] *Cartl. Bruton Priory*, ibid., viii. 40. [1206–23.]
[6] *Chartl. Bath Priory*, ibid., vii. 77. [Early XIV cen.]
[7] *Cal. MSS. of Dean and Chapter of Wells*, i. 107. 1294.
[8] Rents. & Survs., Ro. 564. 16 Edw. III.
[9] *Cartl. Muchelney Abbey*, Somerset Rec. Soc., [*Publ.*], xiv. 71. 1241.
[10] Add. MS. 6065, ff. 9, 10. 23 Edw. I.

APPENDIX II

STAFFORDSHIRE. THREE-FIELD TOWNSHIPS

Township	Description
Barton	An extensive survey arranged by fields and devoting to Wheat field 17 folios, to Walkers field 26, to Rowghemedowe field 4, and to Arloo field 24.[1]
Bonehill	Two terriers describe parcels in three common fields. John Warde, for example, has 22 selions in Marlepit field, 6 in Uppal field, and 5 in Morowe field.[2]
Bromley, Abbots	The enclosure award allots, in addition to 818 acres of waste, 39½ acres in Ley field, 40 in Hollywell field, and 17 in Mickledale field. Each of the three fields is surrounded by enclosures.[3]
Clayton Griffin	Extent describing " in campo de Middulhulfeld et le Heygrene quatuordecim acras terre regales... in campo inter villam Novi Castri et villam de Clayton undecim acras terre... in campo de Fullwallefeld et assarto quondam domini de Clayton quatuordecim acras terre." [4]
Hilderston	Transfer of a messuage and a half-virgate, scil., " tres acras et dimidiam que iacent in campo de Holewei... et tres acras et dimidiam in campo de Blakedone... et tres acras in campo de Whecolnelfeld...." [5]
Leek	Transfer of a messuage and 12 acres of demesne arable: *in campo occidentali* 21 selions in 3 parcels *in campo aquilonali* 18 selions in 3 parcels *in campo orientali* 15 selions in 3 parcels.[6]
Trysull and Siesdon	The enclosure award and map show three non-adjacent fields, each containing many parcels. The total area re-allotted comprises 52½ acres in Whitney field, 28 acres in Newlunt field, and 20½ acres in Budbrooke field.[7]
Tutbury	A survey showing three fields, Castelhay, Mill, and Middle, but they are not large and their tenants are few.[8]
" Berkeswych " Brewood Heywood Draycott, Derbyshire	Extents of the manors of the bishop of Coventry and Lichfield. Relative to the arable of each, " de tertia parte nihil potuit levari quia iacebat frisca et inculta et in communibus campis ad warectam."[9]

[1] D. of Lanc., M. B. 110. [1 Eliz.]
[2] Land Rev., M. B. 185, f. 66. 21 Eliz.
[3] C. P. Recov. Ro., 39 Geo. III, Trin. 1799.
[4] *Chartl. Priory of Trentham*, Wm. Salt Archaeol. Soc., *Colls.*, 1890, xi. 328. 34 Edw. I.
[5] Ped. Fin., 208-4-47. 21 Hen. III.
[6] Cott. MS., Vesp. E XXVI, f. 46b. [XIV cen. copy.]
[7] C. P. Recov. Ro., 20 Geo. III, Easter. 1780.
[8] D. of Lanc., M. B. 109, ff. 35-42b. 1 Eliz.
[9] Add. MS. 6165, ff. 97-104. 31 Edw. III.

Township	Description
Bradley	Extent of the demesne arable, of which "tertia pars iacet quolibet anno ad warectam et in communi."[1]
Essington	Two and one-half virgates are valued at only half a mark because the soil is sandy, and "tertia pars eiusdem terre quolibet anno iacet ad warectam et in communi."[2]
Madeley	Extent. "Sunt ibidem ciiii^{xx} acre terre arabilis de quibus cxx seminantur quolibet anno ... et lx acre nihil valent per annum quia iacent quolibet anno ad warectam et in communi."[3]
Mere End	Extent of a carucate of arable, of which "tertia pars iacet quolibet anno ad warectam et in communi."[4]
Norbury	Extent of a carucate of arable, of which "tertia pars iacet quolibet anno ad warectam et in communi."[5]
Swinnerton	Extent of three carucates of arable, of which "tertia pars iacet quolibet anno ad warectam et in communi."[6]

SUSSEX. TWO-FIELD TOWNSHIPS

Amberley	Extent of 176 acres of demesne arable, of which 53 are sown with wheat, 24 with barley, 10 with beans, peas, and vetches, 5 with oats. "Et valet acra per annum si debet dimitti ad firmam iiii d. et non plus quia iacent in communi ...."[7]
Broadwater	The enclosure plan shows, apart from many old enclosures, two open common fields called North and South.[8]

SUSSEX. THREE-FIELD TOWNSHIPS

Alciston	Terrier of the manor showing the parcels of the holdings divided between Westleyne, Middleleyne, and Eastleyne.[9]
Angmering	The enclosure plan shows three compact open arable fields named West, Middle, and East.[10]
Atherington	A reproduction of the plan of 1606 shows one-third of the manor still in three open fields named West, Mead, and Mill.[11]
Eartham	The enclosure award and plan show six large open arable fields with areas as follows: North field 80

[1] C. Inq. p. Mort., Edw. III, F. 51 (7). 11 Edw. III.
[2] Ibid., F. 61 (18). 14 Edw. III.
[3] Ibid., F. 51 (7). 11 Edw. III.
[4] Ibid., F. 54 (9). 12 Edw. III.
[5] Ibid., F. 66 (25). 16 Edw. III.
[6] Ibid., F. 54 (8). 12 Edw. III.
[7] Add. MS. 6165, f. 109b. 11 Rich. II.
[8] C. P. Recov. Ro., 51 Geo. III, Hil. 1810.
[9] Exch. Aug. Of., M. B. 56, ff. 246-248. 11 Hen. VI.
[10] C. P. Recov. Ro., 52 Geo. III, Trin. 1809.
[11] Sussex Archaeol. Colls., 1901, xliv. 147. 1606.

APPENDIX II

Township	Description
Eartham (*continued*)	acres, Middle field 93, Church field 92, Hodge Lee field 82, Mill field 102, Boar's Hill field 46.[1]
Nutbourne	The enclosure plan shows three common fields called Weston, Mill Pond, and Hat Coppice.[2]
Ovingdean	Terrier of the lands of the prior of Lewes, who has this year arable to be sown " in qualibet leyna, viz., in le Northleyne cum frumento ... xxv acras in le Sowthleyne cum ordeo ... xxxi acras ... be Estoune terra frista ... xxix acras ... " and in four other places 8 acres.[3]
Prinsted	A complete and detailed terrier of the township. The open-field arable of each tenant is divided equally among three groups of fields, viz.: Eastope, the Steane, and Langlands (" this Laine"); Moggland fields; Southerdeane and Plumer (" this Laine ").[4]
Worthing	The enclosure plan shows three large open-field areas, East field, Home and Middle fields, and West field.[5]
Heighton	Extent of 60 acres of demesne arable, "de quibus seminabantur hoc anno ... xl acre et residuum iacet ad warectam et in communi."[6]
Rackham	Extent of 160 acres of demesne arable, of which 50 are sown with wheat, 20 with barley, 16 with oats, 20 with peas and vetches. " Et valet acra per annum iiii *d*. et non plus quia iacent in communi."[7]

WARWICKSHIRE. TWO-FIELD TOWNSHIPS

Chapel Ascote	Grant of 6 acres in the field, of which " tres sunt in parte [defaced] ... campi et tres alie in illa parte campi qui est inter Astanescote et hodenhulle...."[8]
Church Over	Transfer of demesne arable " in uno campo... et in alio campo, scil., in aquilonali parte [52 acres in 10 parcels] in australi vero campo [44 acres in 8 parcels]."[9]
Compton	Transfer of 2 messuages, with 28 acres of arable *in campo aquilonali* and 26 *in campo australi*.[10]
Eatington	Transfer of a croft, with a " culturam... in uno campo ... et culturam ... in alio campo."[11]

[1] C. P. Recov. Ro., 57 Geo. III, Trin. 1817.
[2] K. B. Plea Ro., 3-4 Geo. IV, Hil. 1818.
[3] Cott. MS., Vesp. F XV. 23 Hen. VI.
[4] Rents. & Survs., Portf. 3/57. 1640.
[5] C. P. Recov. Ro., 51 Geo. III, Hil. 1810.
[6] C. Inq. p. Mort., Edw. III, F. 56 (1). 12 Edw. III.
[7] Add. MS. 6165, f. 110. 11 Rich. II.
[8] Cott. MS., Vitel. D XVIII, f. 71 [XIII cen.]
[9] Cott. MS., Vitel. A I, f. 112 [XIII cen.]
[10] Ped. Fin., 242-5-28. 4 John.
[11] Harl. MS. 4028. [31-44 Hen. III.]

Township	Description
Harbury	Transfer of a messuage and a half-virgate, scil., *in campo versus Warewich* 4 acres in 12 parcels *in campo versus [La]dbr[oke]* 3½ acres in 9 parcels.[1]
Hodnel	Transfer of 4 acres *in uno campo* in 8 parcels and 4 acres *in alio campo* in 8 parcels.[2]
Kenilworth	Transfer of a toft and 9½ acres of arable, viz., 5¼ acres in various furlongs in 18 parcels and 4¼ acres *in alio campo* similarly subdivided.[3]
Ladbroke	Transfer of 15 acres of demesne arable, viz., 7½ *in una parte campi* in 3 parcels 7½ *in altera parte campi* in 3 parcels.[4]
Long Lawford	Retention from a virgate of 4 acres, scil., " duas in uno campo et duas in alio." Retention from a virgate of 4 acres, scil., " duas in una parte campi et duas in altera."[5]
Radburn	Transfer of 17 acres of arable, scil., 8½ *in uno campo* et 8½ *in altero*. Several other charters make a similar division.[6]
Tysoe	Transfer of two virgates of demesne, each formerly held by a tenant who had " unam dimidiam virgatam in uno campo et unam dimidiam virgatam in alio campo."[7]
Westcote	Transfer of " sex acras in campo versus Tysho et sex in campo versus Radeweye."[8]

WARWICKSHIRE. THREE-FIELD TOWNSHIPS

Binley	Transfer of 9 selions of arable *in campo occidentali* in 5 places 17 selions of arable *in campo australi* in 3 places and 17 selions of arable *in campo aquilonali* in 4 places.[9]
Long Lawford	Grant of 49 acres of arable, viz., *in orientali campo* 19½ acres in 19 parcels *in campo occidentali* 15 acres in 15 parcels *in campo meridiano* 14½ acres in 10 parcels.[10]
Weston under Wetherley	The enclosure award and plan show six fields approximately equal and named Wind Mill Hill, Heeth, Cross of the Hand, Northalls, Carr, and Pinwell.[11]
Willenhall	A rental stating that the virgate, the half-virgate, and the quarter-virgate of three tenants lie " in tribus campis."[12]

[1] Cott. MS., Vitel. D XVIII, f. 81b. [XIII cen.]
[2] Ibid., f. 74. [XIII cen.]
[3] Stone MS. 937, f. 106b. [XIII cen.]
[4] Cott. MS., Vitel. A I, f. 135. [XV cen. copy.]
[5] Cott. MS., Calig. A XIII, ff. 135b, 149. [XIII cen.]
[6] Cott. MS., Vitel. A I, f. 123b. [XV cen. copy.]
[7] Cott. MS., Vesp. E XXIV, f. 3b. [Edw. I.]
[8] Ped. Fin., 242–6–60. 4 John.
[9] Cott. MS., Vitel. A I, f. 46b. [XV cen. copy.]
[10] Cott. MS., Calig. A XIII, f. 136b. [XIV cen.]
[11] C. P. Recov. Ro., 20 Geo. III, Easter. 1780.
[12] Lansdowne MS. 400, ff. 41, 41b. 12 Hen. IV.

APPENDIX II

WILTSHIRE. TWO-FIELD TOWNSHIPS

Township	Description
Axford	Survey of the township showing the tenants' arable divided between North field and South field.[1]
Badbury Damerham, South Grittleton Kington Nettleton Winterbourne	Surveys of Glastonbury manors (cf. Appendix I). In each case the holdings of the customary tenants are divided between two fields, named respectively East and West, East and West, North and West, East and West, North and South, East and West.[2]
Berwick [St. James]	Extent of the demesne arable, which comprises *in campo versus Austrum* 99 acres and *in campo Boriali* 96 acres.[3]
Marden	Transfer of demesne arable, scil., "in campo orientali viginti acre et in campo occidentali viginti acre."[4]
Overton, West	Survey of the township, showing tenants' arable divided between South field and West field.[5]
Sherston	Transfer of a virgate, scil., 21 acres *in uno campo* and 21 *in alio campo*.[6]
Stanton [St. Quenton]	Survey of the township, showing the tenants' arable divided between North field and South field, though a small East field also appears.[7]
Swallowcliff	Confirmation to the church of 6 acres "in utroque campo."[8]
Warminster	Grant of a half-virgate, the arable of which comprises "septem acras in campo inter Wermenstre et Bissopestre et septem acras in campo inter Upton et Wermenstre."[9]
Winterbourne	Transfer of 80 acres *in una parte ville* and 82 acres *in altera parte ville*.[10]
Yatesbury	Transfer of 70 acres of arable, except a croft and "exceptis xv acris terre in uno campo et in alio campo xi acris terre...."[11]

[1] C. R. Straton, *Pembroke Lands*, i. 163–167. 9 Eliz.
[2] Harl. MS. 3961. 10 Hen. VIII.
[3] Inq. p. Mort., quoted by R. C. Hoare, *Modern Wiltshire*, ii. 25. 42 Hen. III.
[4] Ped. Fin., 250-2-7. 3 John.
[5] Straton, *Pembroke Lands*, i. 143–145. 9 Eliz.
[6] *Reg. Malmesburiense* (Rolls Series), ii. 26. [1205-22.]
[7] Straton, *Pembroke Lands*, i. 132–136. 9 Eliz.
[8] *Reg. Sarisberiense* (Rolls Series), i. 342. [XIII cen. copy.]
[9] Ped. Fin., 250-4-1. 2 Hen. III.
[10] Anc. Deeds, A 244. [XIII cen.]
[11] Ped. Fin., 250-3-16. 7 John.

APPENDIX II

WILTSHIRE. TWO-FIELD TOWNSHIPS — Cont.

Township	Description
Brokenborough Newnton Cowfold with Norton	Extents of the estates of Malmesbury abbey. Of the demesne arable in each case "medietas potest quolibet anno seminari ... et alia medietas nihil valet quia iacet in communi et ad warectam."[1]
Heytesbury, East	Extent. "Sunt ibidem cccclx acre terre arabilis de quibus possunt seminari per annum cclx ... et quum non seminantur nihil valent quia iacent in communi."[2]
Heytesbury, Great	Extent. "Sunt in dominico cccc acre terre arabilis ... et quum non seminantur nihil valent quia iacent in communi de quibus possunt seminari cc acre."[3]
Colerne	Extent of 100 acres of demesne arable, the phraseology being the same as the preceding.[4]
Hurdcott	Extent. "Sunt in dominico clx acre terre arabilis de quibus possunt seminari per annum iiii^{xx} ... et quum non seminantur nihil valent quia iacent in communi."[5]
Maddington	Extent of 100 acres of arable, "de quibus possunt seminari per annum l acre et quum non seminantur nihil valent quia iacent in communi.[6]
Sherston	Extent. "Sunt ibidem in dominico Dc acre terre arabilis de quibus possunt seminari per annum ccc acre ... et quando iacent ad warectam nihil valent quia iacent in communi."[7]

WILTSHIRE. THREE-FIELD TOWNSHIPS

Aldbourne	The enclosure award and plan show six open arable fields named South, Windmill, Rooksbury, West, North, and East.[8]
Manors of the Earl of Pembroke, for the most part near Wilton	Surveys of townships showing the arable of the tenants' holdings divided among three fields, as follows: *Alvediston*, South, Middle, and Home fields; "*Aven*," South, Middle, and North fields; *Broad Chalke*, East, Middle, and West fields; *Burcombe*, East, West, and Wood fields; *Chilhampton*, North, West, and South fields; *Dichampton*, North, Middle, and South fields; *Dinton*, West, Middle, and East fields; *East Overton*, North, East, and South fields; *Fuggleston*, East, West, and North fields; *Newton Toney*, Woodburghe, Bush, and

[1] Add. MS. 6165, ff. 57–59. 19 Rich. II.
[2] C. Inq. p. Mort., Edw. III, F. 56 (1). 12 Edw. III.
[3] Ibid., F, 63 (12). 14 Edw. III.
[4] Ibid.
[5] Ibid., F. 56 (1). 12 Edw. III.
[6] Ibid., F. 62 (6). 14 Edw. III.
[7] Ibid., F. 65 (3). 15 Edw. III.
[8] C. P. Recov. Ro., 53 Geo. III, Trin. 1809.

APPENDIX II 503

Township	Description
Manors of the Earle of Pembroke (*continued*)	Hoome fields; *North Ugford*, North, Middle, and West fields; *Quidhampton*, East, Middle, and West fields; *South Newton*, South, Middle, and North fields; *Stoford*, East, Middle, and West fields (plan in ii. 542); *Washerne*, East, Middle, and West fields; *Wylye*, East, Middle, and West fields.[1]
Steeple Ashton and tithings, Hinton, Southwick, Semington, North Ashton, North Bradley	Brief survey, often showing that the townships lay in three common arable fields.[2]
Bremhill and Foxham, Charlton, Crudwell, "Kemele", Purton, Sutton	Extents of the manors of Malmesbury abbey. Of the demesne arable in each instance "due partes possunt quolibet anno seminari ... et tertia pars nihil valet quia iacet ad warectam et in communi."[3]
Castle Combe	Extent of "due carucate terre arabilis quarum ... tertia pars per annum nihil valet quia iacet ad warectam et in communi."[4]
Durrington	Extent of the demesne arable, "quarum due partes possunt seminari per annum ... et tertia pars quum non seminatur nihil valet quia iacet in communi."[5]
Sharncott	Extent of the demesne arable, "de quibus due partes possunt seminari per annum ... Et tertia pars que non seminatur nihil valet quia iacet in communi."[6]

WORCESTERSHIRE. TWO-FIELD TOWNSHIPS

Broadway	Transfer of a messuage and 2½ acres of arable *in uno campo* and 2½ *in alio campo*.[7]
Hampton	Transfer of a messuage and 30 acres in the fields, "quarum xiiii iacent in uno campo et xvi in alio."[8]
Hill	Grant of 6 acres of arable from half a virgate, scil., *in uno campo* 3 acres in 5 parcels *in alio campo* 3 acres in 5 parcels.[9]
Walcot	Transfer of a half-virgate *in duobus campis*, scil., *in campo iuxta Ellesbergam* 13 acres in 13 parcels *in alio campo* 13 acres in 10 parcels.[10]

[1] Straton, *Pembroke Lands*, vol. i. 9 Eliz.
[2] Land Rev., M. B. 191, ff. 145–158.
[3] Add. MS. 6165, ff. 57–59. 19 Rich. II.
[4] Add. MS. 28206. 46 Edw. III.
[5] C. Inq. p. Mort., Edw. III, F. 40 (10). 8 Edw. III.
[6] Ibid., F. 39 (6). 8 Edw. III.
[7] Exch. Aug. Of., M. B. 61, f. 15b. [XIV cen. copy.]
[8] Cott. MS., Vesp. B XXIV, f. 5. [XIII cen.]
[9] Ped. Fin., 258-3-47. 11 Hen. III.
[10] Exch. Aug. Of., M. B. 61, f. 93b. [XIV cen. copy.]

APPENDIX II

WORCESTERSHIRE. THREE-FIELD TOWNSHIPS

Township	Description
Huddington	At the end of a survey are described three small, non-adjacent arable common fields, viz., Shatherlong field containing 36 acres, Badney or Windmill field containing 20 acres, and Hill field containing 30 acres.[1]
Shurnock	Transfer of a half-virgate of arable, scil., " illam medietatem que ubique iacet in campis de Suthecot, Hulfeld, et Denefeld versus umbram." [2]
Manors of the bishop of Worcester, viz., North Wick, Wick Episcopi, Kempsey, Ripple, Bredon, Fladbury, Blockley, Tredington, Hambury, Alvechurch, Aston Episcopi, Knightwick.	Extents valuing two-thirds of the demesne arable in each instance, but stating that " tertia pars nihil [valet] quia ad warectam et in communi." [3]

YORKSHIRE. TWO-FIELD TOWNSHIPS

Ainderby	Memorandum of the lands of Malton priory. " Item in Aymunderby in campo occidentali iii bovate in campo orientali viii bovate." [4]
Castley	" Thabbot of Fontaunce hath in ye felde of Casteley xii acres of land of the which " 3 acres lie in the South field in 4 parcels and 8 acres lie in the North field in 11 parcels.[5]
Coniston	Transfer of 15¾ acres, viz., *in campo de Conyngston versus le North* 10¼ acres in 5 places *in campo versus le Suth* 5½ acres in one place.[6]
Kilnsey	Extent of the demesne arable, which comprises *in campo orientali* 102 acres and *in campo occidentali* 111¾ acres.[7]
Marton	Transfer of 40 acres of arable, "viginti, scilicet, ex una parte ville et viginti ex alia." Transfer of " iiii acre ex una parte ville versus boream in Fourtenerode et quatuor ex illa parte versus meridiem [in 3 parcels]." [8]

[1] Exch. Aug. Of., Parl. Survs., Worcs. 6. 1650.
[2] Ped. Fin., 258-5-3. 21 Hen. III.
[3] Add. MS. 6165, ff. 81-83. 38 Edw. III.
[4] Cott. MS., Claud. D XI, f. 279. [XIV cen. copy.]
[5] Add. MS. 18276, f. 36b. 1468.
[6] Ibid., f. 39. 1509.
[7] Rents. & Survs., Portf. 17/4. Edw. III.
[8] *Cartl. Prior. de Gyseburne* (Surtees Soc., 1891), ii. 9. [Early XIII cen.]

APPENDIX II

Township	Description
Marton	Transfer of a toft and 20 acres of arable, scil., "decem ex una parte ville et ex alia parte ville decem...."[1]
Owthorn	Extent of the demesne arable, which comprises *in campo ex parte boriali dicte ville* $61\frac{1}{8}$ acres *in campo versus austrum* 95 acres, 10 perches.[2]
Richmond	Extent of the demesne arable, which comprises *in campo occidentali per* $167\frac{1}{4}$ acres *in campo orientali per* $145\frac{3}{8}$ acres.[3]
Skeckling	Extent of the demesne arable, which comprises *in campo orientali* $67\frac{1}{4}$ acres *in campo occidentali* $70\frac{1}{4}$ acres *in duabus culturis de forland quod dicitur Gildcrossewages* $10\frac{3}{4}$ acres.[4]
Thoralby	Transfer of 40 acres of arable, viz., 36 *in septentrionali parte ville* and 4 *in australi*.[5]
Walkington	One holding of two oxgangs comprises 15 acres of arable in the South field and 15 acres of arable in the North field.[6]
Ganstead	Extent of the demesne arable, "de quibus medietas quolibet anno iacet in warecto et in communi."[7]
Hunmanby	Extent of the demesne arable. "Et qualibet bovata continet xvi acre terre unde possunt seminari quolibet anno ... viii acre ... et residuum quod iacet in warecto nihil valet per annum pro eo quod iacet in communi."[8]

YORKSHIRE. THREE-FIELD TOWNSHIPS

Austerfield	Terriers of five leaseholds which, apart from small enclosures, divide their arable almost equally between West field, Riddinge field, and Lowe field (e.g., $6\frac{3}{4}$, 7, $6\frac{1}{4}$ acres).[9]
Bainton	A survey in which the holdings, apart from the messuages and small closes, are described as "arrable in the 3 feilds."[10]
Baldersby	"Mensuratio campi de Balderby. Summa DCXXVI acre." The rubrics in the margin (perhaps of the fifteenth century) are *Campus borialis, Suth Campus*, and *West Campus*; but the South field is not

---

[1] Egerton MS. 2823. [XV cen. copy.]
[2] Rents. & Survs., Portf. 17/4. Edw. III.
[3] Ibid.
[4] Ibid.
[5] Cott. MS., Claud. D XI, f. 184b. [XIV cen. copy.]
[6] Land Rev., M. B. 229, f. 192. 1608.
[7] C. Inq. p. Mort., Edw. III, F. 40 (7). 8 Edw. III.
[8] Ibid., F. 44 (1). 9 Edw. III.
[9] Land Rev., M. B. 229, ff. 180–185. 1608.
[10] Ibid., ff. 159–162. 1608.

Township	Description
Baldersby (continued)	great in extent, and in all three fields certain " flats " are sown with rye.[1]
Barwick [in Elmet]	Extent of the demesne arable, which comprises 40 acres *in campo orientali* 48 *in campo occidentali* 50 *in campo boriali* and 4 *in Scolesker*.[2]
Burstwick	Extent of the demesne arable, which comprises 130½ acres in East field 116¼ in Middle field and 157½ in West field.[3]
Catterton	Terriers of three farmholds each divided among three fields. The second comprises in North field 15½ acres in 6 places in the Tofte field 13½ acres in 9 places in Gosbar' field 10 acres in 5 places.[4]
Darrington	Grant of 2¼ acres from a bovate: ¼ acre in a croft, and " i acram in cultura de Nortfeld et dimidiam acram in cultura de Sudfeld et dimidiam acram in cultura de Westfeld." [5]
Deighton, North	Transfer of 2 acres, scil., 1 acre in North field and 1 *in campo occidentali*, with a *cultura in campo orientali*. At the end of the cartulary is a sixteenth-century terrier of the demesne arable, assigning to West field 11¾ acres, to South field 11¾ acres, to Hoberkes 7¾ acres, to North field 2 acres, and to Oxenclosse 8 acres.[6]
Frickley	Terrier of 3 messuages, 2 crofts, 3 doles of meadow, 6 acres of enclosed arable, and in the Mylne field 10 acres in 9 parcels in the Clough field 7½ acres in 12 parcels in the Kyrke field 7 acres in 9 parcels.[7]
Houghton cum Castleforth	A survey showing the arable acres of the tenants' holdings pretty evenly divided among Kirk field, High (or Meare) field, and Park (or West) field, (e.g., 30, 40, 40; 5, 5, 6; 16, 16, 20, etc.).[8]
Hutton, Sheriff	A survey showing the arable acres of the tenants' holdings equally divided among Dudhill field, Dycegate and Reddinge, and West field (e.g., 12, 12, 12).[9]
Kippax	Extent of the demesne arable, which comprises 23 acres in the South field, 22 in the North field, and 22 in the West field.[10]

[1] Cott. MS., Tib. C XII, f. 193. 1296.
[2] Exch. Treas. Recpt., M. B. 176. 15 Edw. III.
[3] Rents. & Survs., Portf. 17/4. Edw. III.
[4] Cott. MS., Vesp. A IV, f. 188. [XVI cen.]
[5] *Pedes Finium Ebor. reg. Johanne* (Surtees Soc., 1894), p. 136. 1208.
[6] Cott. MS., Vesp. A IV, ff. 63, 190. [1285.]
[7] Rents. & Survs., Portf. 22/41. 5 Hen. VI.
[8] Land Rev., M. B. 229, ff. 27-38. 1608.
[9] Land Rev., M. B. 193, f. 47b. 5 Jas. I.
[10] Exch. Treas. Recpt., M. B. 176. 15 Edw. III.

APPENDIX II

Township	Description
Kirby Grindalyth	Map showing three large fields, viz., East field containing 277 acres, Middle field containing 316 acres, and West field with Garth End field containing 418 acres (265 + 153).[1]
Kirkby-Wiske	Transfer of 4 selions *in campo borientali* in 3 parcels 4 selions *in campo occidentali* in 3 parcels 3 selions *in campo australi* in 3 parcels.[2]
Lund	Several leaseholds consist largely of arable, which is said to lie " in ye three fields."[3]
Monkhill	A tenant's holding consists of a messuage, a close of ½ acre, 2 acres of meadow in the town " Ings," and 6 acres of arable in Kirk field, 7 in Middle field, and 7 in North field.[4]
Norton	A valuation of the demesne arable, which is rubricated as follows: in Chapelcroft, East field, and Middle field 98¼ acres in South field in 8 flats, 98 acres in West field in 17 flats, 92½ acres.[5]
Nun Monketon	" To the towne are belongynge thre fieldes ... the towne fielde ... the middel fielde ... the west fielde."[6]
Pickering	" On each side of the brook lay a suite of common fields, three in number; ... each bovate on one side of the township contained 24 acres, on the other 12 acres."[7]
Rokeby and Smythorp	Exchange of 66 acres of arable, " que iacent in tribus campis de Rokeby et Smythorp," viz., in Blacker field and Langland field 22½ acres in Eskelflat field and Thornholm field 22¼ acres in More field and Tresholm field 22¼ acres.[8]
Skirpenbeck	Of the 14 bovates which belong to Whitby abbey 6 are consolidated, but 8 lie " in flattes in campo dicte ville quarum prima iacet in Northfield super Wolfhow que continet xxii acras terre ... in campo australi a flatt vocata Blaland flatt que continet xvi acras terre et dimidiam ... in campo occidentali ... a flatt vocata Okflat ... que continet ii acras terre."[9]
Sneaton	Transfer of 21 acres of arable, viz., in West field 9¼ acres in 6 places

[1] Add. MS. 36899 A. 1755.
[2] Add. MS. 18276. f. 105b. 1266.
[3] Land Rev., M. B. 229, ff. 163-164. 1608.
[4] Ibid., f. 44. [1608.]
[5] Rents. & Survs., Ro. 753. Edw. I.
[6] Add. MS. 4781. [Hen. VIII.]
[7] William Marshall, *Rural Economy of Yorkshire* (2 vols., London, 1788), i. 50. c. 1788.
[8] Bodl., Rawl. B 449, f. 84. [c. 1269.]
[9] *Cartl. Abbat. de Whitby* (Surtees Soc., 1878), i. 328. 1446.

Township	Description
Sneaton (*continued*)	in Middle field 6⅛ acres in 6 places super Heydun 5¾ acres.¹
Studley	Terriers of two holdings. The first consists of a messuage, a toft, and a croft of 3 acres, with 8 acres in 10 parcels in the South field 2⅛ acres in 4 parcels in Miln field 7⅛ acres in 11 parcels in North field 2¼ acres in 4 parcels in West field. The second tenement consists of a messuage, a toft, and a croft of 2 acres, with 4⅛ acres in South field, 1⅜ in Miln field, and 3⅜ in North field.²
Tibthorpe	A survey in which several tenants have their arable "in the three fields."³
Waghen	Transfer of three bovates, scil., *in orientali campo* 10 selions in 6 parcels *in campo australi* 18 selions in 6 parcels *in campo aquilonali* 9 selions in 6 parcels.⁴
Wetwang	A survey in which the lessee of the demesne and one other tenant have their arable "in the three fields."⁵
Whixley	Transfer of 6 acres "quarum quedam pars iacet in Westfeld . . . et quedam pars in Estfeld . . . et tertia pars in Midelfeld. . . ."⁶
Wighill	A terrier of the arable belonging to the prior of Helaugh Park, which comprised in North field 23¾ acres in West field 30½ acres in East field 23¾ acres.⁷
Wombwell	A terrier of the arable belonging to the prior of Helaugh Park, viz., *in campo boriali* 4½ acres and 2 "pecie" in 9 parcels *in campo occidentali* 9⅛ acres in 13 parcels *in campo australi* 43⅛ acres in 71 parcels.⁸
Boynton	Extent of the demesne arable, of which "quelibet bovata continet xiii acras et de qualibet bovata terre possunt seminari per annum ix acras pro utraque semente equaliter . . . et residuum quod iacet in warecto nihil valet per annum pro eo quod iacet in communi."⁹
Kirkby Malzeard	Extent. "Sunt ibidem xx acre terre que possunt seminari cum semine hiemali . . . sunt xx acre terre que

---

¹ Cott. MS., Claud. D XI, f. 133*b*. [XIV cen. copy.]
² Add. MS. 18276, f. 216. 21 Hen. VII.
³ Land Rev., M. B. 229, ff. 155–158. 1608.
⁴ Cott. MS., Otho C VIII, f. 86. [XV cen. copy.]
⁵ Land Rev., M. B. 229, ff. 145–146. 1608.
⁶ Add. MS. 18276. *c.* 1255.
⁷ Cott. MS., Vesp. A IV, f. 191*b*. [XVI cen.]
⁸ Cott. MS., Vesp. A, f. 184. [Hen. VII.]
⁹ C. Inq. p. Mort., Edw. III, F. 2 (7). 1 Edw. III.

Township	Description
Kirkby Malzeard (*continued*)	possunt seminari cum semine estivali ... sunt xx acre terre in warecto quarum herbagium nichil valet per annum quia iacent in communi."[1]
Scamston	Extent of the demesne arable, " de quibus tertia pars quolibet anno iacet in warecto et pastura in eodem warecto nihil valet per annum quia iacet in communi cum tenentibus ville."[2]
Thorganby	Extent of "una bovata ... de qua quidem bovata terre tertia pars iacet quolibet anno in warecto et pastura in warecto nichil valet per annum quia iacet in communi."[3]
Thorp Arch	Extent of the demesne arable, "de quibus due partes seminabantur ... et tertia pars earundem quolibet anno iacet in warecto et pastura nichil valet quia iacet in communi."[4]
Upleatham	Extent of the demesne arable, " de quibus duo partes possunt quolibet anno seminari cum utroque semine et tertia pars iacet quolibet anno in warecto et pastura inde nichil valet quia iacet in communi."[5]
Sherbourn	Extent, with phraseology as above.[6]
Everingham	Extent, with phraseology as above.[7]
Walton	Extent of the demesne arable, " de quibus tertia pars quolibet anno iacet in warecto et pastura eiusdem nichil valet per annum quia iacet in communi."[8]

[1] C. Inq. p. Mort., Edw. III, F. 5 (5). 1 Edw. III.
[2] Ibid., F. 44 (6). 9 Edw. III.
[3] Ibid., F. 38 (10). 8 Edw. III.
[4] Ibid., F. 59 (15). 13 Edw. III.
[5] Ibid., F. 63 (1). 14 Edw. III.
[6] Ibid., F. 65 (8). 15 Edw. III.
[7] Ibid.
[8] Ibid., F. 39 (10). 8 Edw. III.

APPENDIX III

SUMMARIES OF TUDOR AND JACOBEAN SURVEYS WHICH ILLUSTRATE IRREGULAR FIELDS WITHIN THE AREA OF THE TWO- AND THREE-FIELD SYSTEM

Areas are in acres unless otherwise specified. Messuages are indicated by m., virgates by virg., tenements by tent., cottages by cott.

STONESFIELD, OXFORDSHIRE
Land Rev., M. B. 224, ff. 162–180. 4 Jas. I

		Arable in the Open Common Fields			
Custumarii	Enclosed	Church field	Callowe	Home field	Gennetts Sarte
John Kerke, m., ½ virg.....	2	½	1¾	12	..
Wm. Hedges, m., ½ virg....	1	½	1¾	9½	..
Wm. Hedges, cott., ¼ virg. .	1½	2½	¾	2½	..
Rich. Nashe, m., ½ virg.....	1¾	1	1¾	11⅜	..
Geo. Owen, gent., m., ½ virg.	¾	½	1¾	11½	..
Wm. Larder, m., ½ virg.....	1	..	1¾	13	..
Ric. Keeth, cott.	¼	..	1	1½	..
Wm. Hicks, m., ½ virg.....	1½	1	1¾	23	..
Thos. Sayward, m., ½ virg. .	¾	4½	1¾	12	..
Robt. Jeames, m., 1¼ virg...	1½	1¼	3½	22	..
Liberi Tenentes					
Jac. Lardner, m.	½	2	½	..	½
Rich. Dewe, m....	2¾	4	1¾	9¼	8
Wm. Dutton, m.........	½	..	¾	..	2
Rich. Meede, m.........	½	..	1½	2	..

There are several other copyholders and freeholders. The customary tenants have stinted common of pasture for sheep in the common fields and meadows.

WOOTTON, OXFORDSHIRE
Land Rev., M. B. 224, ff. 181–206. 4 Jas. I

	Enclosed		Arable in the Open Common Fields		Common Meadow
Custumarii	Arab.	Past.	North field or end	West field or end	
Jo. Gregory, m., 3 virg.........	2	3	..	70	1
Jo. Gregory, m., 2 virg.........	1½	..	40	..	½
Egidius Sowthram, m., 2 virg.....	1	1½	50	..	2¼
Robt. Parram, m., 2 virg.	1½	2¼	..	70	2½
Chris. Castell, 2 m., 1 virg.......	2⅛	3	..	40	..
Wm. Horne, Jr., m..............	½	¼	30	..	¼
Jo. Symonds, m...............	¼	..	..	20	½

There are three other copyholds and several freeholds. Two customary tenants are said to have common of pasture, but no locality is mentioned.

Long Coombe, Oxfordshire

Land Rev., M. B. 224, ff. 58–94. 4 Jas. I

Custumarii	Enclosed Arab.	Enclosed Md.	Enclosed Past.	Arable in the Open Common Fields: West field	Arable in the Open Common Fields: Land field	Arable in the Open Common Fields: East end	Common Meadow
Joanna Woodward, m., ½ virg.	⅝	..	1½	..	9	..	1½
Edm. Haukings, m., ½ virg. .	½	3	..	5	..	..	3½
Jo. Newman, m., ½ virg.....	½	2½	1½	¾	..	8	2½
Sara Payne, m., 1 virg......	½	..	4½	10	4	..	4¼
Wm. Bolton, m., ½ virg.....	½	..	8	4	..	9	3
Rich. Swift, m., ½ virg......	½	2	..	1½	6	..	1½
Jo. Hurst, ½ virg...........	..	2	..	2	4	..	5¼
Wm. Seacolle, 4 m., 2 virg. .	10½	20	..	2	..	13	10½
Robt. Newman, m., ½ virg. .	¼	..	2	2	6	..	1½
Wm. Bowden 2 m., ½ virg. .	½	..	2½	..	..	12	2½
Jo. Maye, m., ½ virg.	⅛	..	½	9	..	..	1½

There are several other holdings. The free tenants have open-field arable in Over field. The customary tenants have common of pasture " in campis " (f. 81), stinted for sheep.

Ramsden, Oxfordshire

Land Rev., M. B. 189. 6 Edw. VI

Custumarii	Enclosed	Arable in the Open Common Fields: Olde field	Gode field	Swinepit field	Shurt-lake	Herwell Serte	Lucerte
Jo. Lardener, m...........	23½	24	9½	..	..	..	..
Wm. Cottes, m., 2 virg.	3	23	13	2	3½	3	6
Elizeas Kyrbey, m., 2 virg. .	3	43	12	8	..	3	..
Robt. Camden............	..	23	18½	..	8	½	..
Wm. Lee, m..............	2½	9½	5½	1	½	1	..
Jo. Lardner, cott..........	..	..	..	..	2	..	..

Hampton-in-Arden, Warwickshire

Land Rev., M. B. 228, ff. 1–35. 6 Jas. I

Custumarii	Enclosed	Arable in the Open Common Fields								Common Meadow
		Clay field	Little field	Mill field		Sporley field	Rodnell field	le Heath	Inidge	
Wm. Fisher, m., 1 virg.	¼	6	3	4		2	..	..	..	2
Thos. Taylor, m., ¾ virg.	¼	3	..	3 and Infield		2½	1	½	..	1½
Robt. Broadnock, m., 1 virg.	½	4	4	5	2	2	..	..	..	2
Wm. Biddle, m., 1 virg.	¼	4	4	5 and Infield		2	..	..	..	3
Rich. Feild, m., ½ virg.	..	..	¼	3	½	3½	½	..	¼	¾
Hen. Marshe, gent., m., 1½ virg.	1	4	4	5½	2	4	1	3½	1	3
Hen. Fentam, m., 1 virg.	⅜	3	3	4 and Infield		2½	..	..	..	1⅞
Wm. Taylor, m., 1 virg.	½	4	3	3 "	"	..	..	2	..	1¼
Thos. Burge, m., 1 virg.	¼	4	4	4 "	"	3	..	..	..	1⅜
Thos. Barlowe, m., ½ virg.	½	..	..	4¾ "	"	4	¼	..	2	1½
Thos. Gybbines, m., 1 virg.	¼	..	2	4½ "	"	6	..	..	1½	1½
Thos. Fentam, m., 1½ virg.	½	6	6	7	1	5½	½	..	..	3
Rich. Fwlford, m., ¾ virg.	⅜	..	..	1	2½	5½	..	..	2	1½
Jos. Smith, m., 1 virg.	¼	..	..	2	2	7	1	..	4	2
Ric. Smith, m., ¾ virg.	½	..	..	3	2	4	1	1	3	1½

There are a dozen other copyholds, and a few freeholds and leaseholds. For each virgate there is "communia pasture pro xv averiis, iiii equis, et xl ovibus in communibus campis, pratis, et vastis manerii predicti."

Langdon and Widney, Part of the Manor of Knoll, Warwickshire

Land Rev., M. B. 228, ff. 65–136. 3 Jas. I

Liberi Tenentes	Enclosed Arab.	Enclosed Md.	Enclosed Past.	Arable in the Open Common Fields: Berye field	Seede furlong	What-croft	Hen field	Unspecified, etc.	Common Meadow
Robt. Middlemore, armig., m.	3	2½	6	1½	1½	..	..	34	8
Rad. Vyne, armig., m.	4	..	6	9	14	4¾	4½	9	..
Chas. Waring, armig., m.	2	14	158	8	4	2	..	..	..
Chas. Waring, 3 m.	2½	9½	27	1	..	..	..	..	..
Chas. Waring, m.	½	5	38	6	4	1	..	10½	1
Regin. Heald	..	..	..	..	..	..	2	..	1
Robt. Higgenson, gent., m.	2	7	90½	4	10	3½	..	17½	½
Thos. Holbach, m.	⅜	..	..	7	6	1½	6	4½	1½
Thos. Palmer, m.	¾	6	23	½	2½	..	..	..	..
Jo. Walton	..	..	..	1	2½	..	..	4	..
Wm. Huddesford, m.	3½	..	4	..	3	..	..	2½	1

Only the freeholds above named include common fields. The others, of which there are several, consist, like the Knoll copyholds, largely of enclosed pasture and meadow.

Wootton-under-Wever, Staffordshire

Rents. & Survs., Portf. 14/83. 1 Edw. VI

Tenants by Copy	Enclosed		Arable in the Open Common Fields					Common Meadow
	Croft	Past.	Tynsel field	Dale field	Pates field	South field	Legh.	
Roger Lees, m.	$1\frac{1}{4}$	$3\frac{1}{2}$	4	4	3	2	..	..
Annys Dole, m.	$\frac{1}{8}$	$1\frac{1}{8}$	11	$3\frac{1}{2}$	8	$4\frac{1}{2}$	$2\frac{1}{2}$	3
Rich. Leithe, m.	$1\frac{1}{2}$	4	..	$40\frac{1}{2}$	7	$6\frac{1}{2}$	..	$2\frac{3}{4}$
Jas. Orpe, m.	..	$1\frac{1}{8}$	4	2	1	4	1	$\frac{3}{8}$
Robt. Highynbothome, m.	$\frac{1}{16}$	$4\frac{1}{4}$	3	..	3	3	..	$1\frac{1}{2}$
Jo. Bull, m.	$\frac{1}{2}$	2	6	4	3	3	$\frac{1}{2}$	$\frac{1}{4}$
Thos. Bothome, m.	$\frac{1}{8}$	$1\frac{1}{2}$	6	5	3	3	1	1
Walt. Lees, m.	1	$6\frac{1}{4}$	3	3	2	6	2	1
Thos. Bull, cott.	$\frac{1}{8}$	$\frac{1}{16}$	$\frac{1}{2}$	$\frac{1}{2}$	1	1	..	$\frac{1}{4}$
Thos. Torner, m.	$\frac{1}{2}$	$5\frac{3}{4}$	4	3	4	3	5	$\frac{1}{4}$
Isabell Aynesworth, m.	$\frac{1}{2}$	3	$6\frac{1}{2}$	2	3	2	$1\frac{1}{2}$ (in "le heath")	
Jo. Heathe, cott.	$\frac{1}{8}$	..	6 [1]	3 [1]	3 [1]	6 [1]	..	..
Robt. Trotton, cott.	$\frac{1}{8}$	$\frac{1}{8}$	3	2	1	$1\frac{1}{2}$	$\frac{1}{4}$	$\frac{1}{4}$
Jo. Orpe, m.	..	1	4	..	1	4	1	$\frac{3}{8}$
Thos. Aynesworth, m.	2	$4\frac{1}{2}$	..	..	$\frac{1}{2}$	1	..	3

The list of copyholders is complete. There are five insignificant free tenants, whose lands are not specified.

[1] "Rydges" instead of acres.

ROCESTER, STAFFORDSHIRE
Land Rev, M. B. 183, ff. 128-131. [Hen. VIII]

Tenants at Will	Enclosed	Arable in the Open Common Fields					Common Meadow
		Bowhall field	Newton field	Barowell field	Miscellaneous		
Jo. Bakyn, m.	2 crofts	......	6 lands	20 lands	Whytefield 6 lands		½ acre
Jo. Stathn, m.	1 "	......	2 "	8 "	Wygley 4 acres		1 acre
Hen. Langtonhouse, cott.	2 "	3 lands	8 "	2 "	......		1½ acre, 2 doles
Robt. Grafton, m.	1 "	......	5 "	4 "	......		2 acres
Thos. Annsell, cott.	2 "	3 lands	......	4 "	Wygley 2 lands		¾ acre
Thos. Rige, m.	3 "	......	18 lands	10 acres	......		1 acre, 3 lands
Rog. Jonson, cott.	1 "	6 lands	......	2 lands	......		¾ acre
Jo. Boche, cott.	1 "	......	7 lands	......	......		......

Only the above holdings have acres in the fields. There are several small tenements, each comprising a messuage and some acres of common meadow.

Welford, Gloucestershire

Exch. K. R., M.B. 39, ff. 183–186. 6 Edw. VI

Copyholders	Enclosed	Arable in the Open Common Fields				Common Meadow
		Sholebreade field	Stabroke field	Middle Barrow field	West field	
Geo. Holtam, m., 1½ virg.	1 ferendell [1]	6	6	7	9	1
Jac. Robins, 2 m., 2 virg.	1 "	8	9	8	8	1
Walt. Bromley, m., 1 virg.	1 "	6	5	6	6	1
Thos. Warde, m., 1 virg.	....	5	6	6	6	2
Jo. Hamlins, m., ½ virg.	1 ferendell	3	3	3	4½	1¼
Thos. Hewes, 2 m., 2 virg.	1 "	10	10	10	10	4

There are several similar holdings. George Holtam has common of pasture "pro viii animalibus, xlv bidentibus," and the other tenants fare proportionally; but no locality is mentioned.

Marston Sicca, Gloucestershire

Exch. K. R., M. B. 39, ff. 145–147b.

Custumarii	Enclosed	Arable in the Open Common Fields				Common Meadow
		Natte furlong	West field	Lowe field	Nylls & Hadland	
Ric. Cowper, m., 1½ virg.	¾	7	7	7	7	3
Thos. Warne, m., 2 virg.	¾	8	8	8	8	4
Galf. Beke, 2 m., 2½ virg.	1½	10	10	10	10	5
Robt. Martin, 3 m., 3 virg.	2¼	13	13	13	13	6
Hen. Cowper, 5 m., 5 virg.	3	15	15	15	15	10
Alicia Tommes, m., 4 virg.	1¼	16	16	16	16	4

There are several similar holdings. Richard Cowper has common of pasture "pro lx bidentibus, vi animalibus, iii equis," and the other tenants fare proportionally.

[1] This is the home close.

Clapton, Part of the Manor of Ham, Gloucestershire — Exch. K. R., M. B. 39, ff. 58–60. [Eliz.]

Copyholders	Enclosed Arab.	Enclosed Past.	Arable in the Open Common Fields: Redecroft	Baucroft	Litlecroft	Ridge field	Lake field	Lypiatts field	Miscellaneous	Common Meadow
Johanna Baker, m., 1 virg.	2	15	1½	3½	2	3	6½	7½	Cleves 1¼, Ricks ½, Goldhill field 1, Woodwynnehomes ½, Longcatmershe 1 selion	..
Wm. Nicols, m.	6	8	..	..	2	2	10	9½	Clapham hill 1	2
Wm. Mallet, m.	5	7	2	4	2	..	8	..	Churgaston 1½, Irelands hill ¾, Clappe Ridge ½, Ricks 1½	1½
Thos. Pers, m., 1 virg.	8	..	2½	..	2	..	9	9½	Prestcroft 6, Yelons 3	2
Rich. Hicks, m., 1 virg.	3	13	¾	3	1¼	1¾	13	12	Ricks 2, Prest croft 4, Brickhill 1, Clappe Ridge 2½	1¾
Jo. Wyllis, m., ¼ virg.	½	1½	..	..	..	..	2½	1½	Longacre 1¼, Hunger field 1, Ricks 2, Clappe ruge 2	½
Thos. Wymter, m., 1½ virg.	9	15	3	1	½	1	6½	8	Lobthorne 2, Cleves 1¾, Roweland ½, Hennegaston ½, Homes field 1½, Longacre 1, Clap ridge 7	½
Thos. Hurne, m.	8½	..	3½	5½	3	1	11	..	Prestcroft 1⅝, Homes field ¾, Paddams Down 1¾, Mersheland 1¾, Clap Ridge 3, Ante tenement 3	3
Ric. Samger, m., 1 virg.	2	11	..	..	6½	..	..	5	Perham downe 2, Riks 1	3¼
Wm. Pers, m., 1 virg.	3	12	4	2	1	..	6	6	Homes field 4½, Perham downe 3, Shurmans field 2½, Ricks 3, Cleves 3	..
Katerina Hurne, m.	1¼	5	4	1½	1¼	..	9	9	Hunger field 1½, Shurmans field 3½, Severne field 2½, Churgaston ¾, le Ricks 5, Clapham 4	1½

There are three other copyholders. No statements regarding pasture are made.

FROCESTER, GLOUCESTERSHIRE

Rents. & Survs., Portf. 2/46, ff. 44–66. 1 Edw. VI

Custumarii	Enclosed Arab.	Enclosed Md.	Enclosed Past.	Arable in the Open Common Fields: Nochold field	Up field	West field	Chargest field	South field	Broadcroft	Lang Furlong	Lyde field
Wm. Shipman, 2 m., ½ virg.	..	1	15½	4	...	1	1½	5	3	3	½
Wm. Symons, m., ¼ virg.[1]	..	1½	7¼	1	..	..	..	1	..	1	..
Jo. Pegler, 2 m., 1 virg.	..	6	11	..	..	7	..	7	..	7	..
Jo. Brotone, m., ½ virg.	..	1	14¾	..	3	2	1	4	..	..	..
Thos. Chundelor, m., ½ virg.	..	..	20¾	..	2	2½	..	2½	..	..	..
Wm. Mayowe, m., 1 virg.	8	..	10½	..	8	1	..	4	..	..	..
Jo. Warner, m., 1 virg.	9	3	15	½	2	..	..	7	..	..	..
Thos. Whiteworth, m., ¼ virg.[1]	2	½	1	..	½	..	½	1½	1	1	..
Thos. Whiteworth, m., 1 virg.	..	8½	10	1½	8 (past.)	5	..	9½	1	..	..
Jo. Taylour, m., 1 virg.	..	..	8	..	½	5	..	4	3	1	..
Ric. Ellonde, m., ½ virg.	..	..	17	..	..	3	..	..	..	..	2
Joanna Horewode, m., 1 virg.	..	2	32	..	..	5	..	2½	..	..	2
Joh. Wyley, m., 1 virg.	..	4½	10	..	..	5	..	6	3	..	..
Jo. Wilkins, m., ½ virg.	4	4	4½	..	..	..	¾	6¼	..	1	1¼
Jo. Benett, m., cott.	..	..	5	..	½	..	..	6	2	..	..

[1] The fourth part of a virgate is called a "ferendell."

OXLYNCH, A TITHING OF STANDISH, GLOUCESTERSHIRE — Rents. & Survs., Portf. 2/16, ff. 1–40. 1 Edw. VI

Custumarii	Enclosed Pasture	Arable in the Open Common Fields: Grete Combe	Stony field	Lytelcombe	Dawhill field	Northe field	Miscellaneous	Common Meadow
Eliz. Holder, m., 1 virg.	16	10	½	4	..	..	Admorley field 17½	½
Walt. Borde, m., ½ virg.	13½	..	9	..	5½	7	......	..
Thos. Ricardes, 2 m., ¾ virg.	20½	..	7	..	3	10½	......	¼
Ric. Gardiner, m., ½ virg.	10	8	..	4	..	..	Roulet field 6	½
Ric. Watkins, 2 m., ¾ virg.	12	8	..	8	..	..	Waywardon 1 Admorley field 7	..
Eliz. Marshe, m., ½ virg.	6	4	4	7	..	..	Admorley field ½	½
John Berde, m., ½ virg.	8	½	7	..	6	9	......	¼
John Gabbe, m., ½ virg.	10	..	8	..	..	8	Riddyng field 9	..
Margar. Gibbs, m., ½ virg.	10½	..	6	..	6	10	......	1
Ric. Chewe, m., ½ virg.	5½	½	6	..	5½	9	......	½

HORTON, GLOUCESTERSHIRE — Rents. & Survs., Portf. 2/46, ff. 92–104. 1 Edw. VI

Custumarii	Enclosed Md.	Enclosed Past.	Arable in the Open Common Fields: In field	Yarlinge field	Mershe field	Miscellaneous	Common Meadow
David Luce, m., 1 virg.	5	14	..	..	16	the great field 16.	..
Rich a deane, m., ½ virg.	11	8	17	..	..	alius campus vocatus In field 20	..
Jo. Wichewell, tent., ½ virg.	..	4	9½	9	½	......	1½
Wm. Hicks and Nich. Smyth, tent., 1 virg.	6	12	8	23	14	in le gaso 2	3½
J. Dalyn, m., ½ virg.	..	3 +	..	..	..	End field 10, Castle field 11	1½
Johanna Whityng, m., 1 virg.	2½	5	24	..	24	......	3
Wm. ——, m., 1 virg.	2	2	3	29	..	......	4
Thos. Hobbes, m., ½ virg.	6	..	..	..	..	End field 12	..
Jo. Pope, tent., 1 virg.	3	2	..	..	13, 14	......	..
Jo. Hatheway, m., 1 virg.	4	5½	..	15	15	......	4
—— ——, m., 1 virg.	2	3	..	13	13	......	2½
Edw. ——, m., 1 virg.	3½	7½	..	13	13	......	..

YATE, GLOUCESTERSHIRE

Rents. & Survs., Portf. 2/46, ff. 67–86. 1 Edw. VI

Custumarii	Enclosed Md. and Past.	Arable in the Open Common Fields					Common Meadow
		West field	North field	Up field	Miscellaneous	
Jo. Belsire, m.	77	..	1½	½	The Breche 1	1
Isabella Byrchold, m., 1 lundmar'	13	..	..	6	Whele 3¾, Coulmede 1, Northmedowe ½	4½
Jo. Hoper, m.	5¼	3½	..	..	communis vocata Rogehill 10	2
Thos. Colyns, m.	10	¼	..	8	Whetlands 1½, Coulmede 7, lynehollys 2, Cotestrote 1	6¾
Laur. Voillys, m., 1 ferendell	12	..	4	1	Whylie 3, Coulmede 2	3
Wm. Tremplyn, m.	60	..	3¼	..	Knewewey 2	4
Wm. Nele, m.	2 (with 9 demesne)	13	9	14	......	6
Jo. Coper, m., 1 ferendell	6½	..	..	3	Lynalds 1, Wheley 1	2
Jo. Coper	..	1	2	5	Lynalds 2½, Dyngleigh 3	..
Robt. Browne, cott.	12	4	6	9	Wheley 2	5
Thos. Taylor, m.	14	..	..	5	le hayes ½, Wheley 2½	4
Hen. Smythe, m., 1 ferendell	4	8	2	..	Nocke field 2	3½
Rich. Shorte, m.	3½	7	1½	..	Dunsley 1, Wheley 2, Nuppe field 10	4
Manricius Dymbury, m., 2 ferendells	28	..	..	..	the Leyghe 6, Winesworthye 2	7
Wm. Dymerey, m., 1 ferendell	10	..	..	..	the Leyghe 12	3½

MIDDLETON, A MEMBER OF STOCKTON, HEREFORDSHIRE

Land Rev., M. B. 217, ff. 331–341. 6 Jas. I

Custumarii	Arab.	Enclosed Md.	Past.	Arable in the Open Common Fields
Jo. Yarlech, m.	$\frac{1}{4}$	2	$11\frac{1}{2}$	Grittiehill field 9, Hascrofte 10, Withers field 11
Wm. Smith, m.	$\frac{1}{8}$	4	12	Grittiehill field 10, Withers field 12, le Sedge and littlecroft 13
Thos. Cocke, m.	2	3	14	Withers field 6, le Ford field 10, le Ferfords field 8, Hollow furlong 2, Pirry furlong 3, le Rylands 2
Jo. Carpenter, m.	$15\frac{1}{2}$	$4\frac{3}{4}$	9	Combes field 14, Coggell field 14
Robt. Fulwood, gent., m.	3	5	20	Norman field 11, Crosse field 5, Weyland 3, Litle field 7, Withie crofte 4
Robt. Fulwood, gent., m.	2	$4\frac{1}{4}$	11	Normans field 2, Weyland field 2, Litle field 4, Withie crofte 3
Thos. Yappe, m.	$\frac{1}{4}$	3	$3\frac{1}{2}$	West field 8, Low crofte 3, Church field 4, Norman field 3
Johanna Goughe, m.	$\frac{1}{4}$	6	$6\frac{1}{2}$	Cookle field 18, Combes field 15, Stonall field 6, Lower field 7
John Phillipps, m.	16	9	$8\frac{1}{2}$	The field at the gate 15, Stonewall field, Nun field 7 (not said to be common)
Will Winde, m.	$\frac{1}{4}$	3	$4\frac{3}{4}$	Withers field 7, Grityhill field 4, Hayfurlong $3\frac{1}{2}$, Grimstie $\frac{1}{2}$, Hessich 3, Plumppitts 1, Millepowle 1, Brierland 1, le Heath 1
Phil. Winde, m.	1	3	6	Withers field 6, Asheley field 7, West field 7
Rowland Pitt, m.	$\frac{1}{2}$	$1\frac{1}{2}$	5	Gate field 10, le West field $1\frac{1}{2}$, le querrell field 6, le old field 5, le Flemings field 12
Jocasa Maunde, m.	$2\frac{1}{2}$	$3\frac{3}{4}$	$6\frac{1}{4}$	West field 6, Lowcroft 3, Church field 9, Widdie croft 2, Cros field 2, Litle field and Drane field 4. (The last four are not said to be common.)

HOPE-UNDER-DINMORE, A MEMBER OF IVINGTON, HEREFORDSHIRE — Land Rev., M. B. 217, ff. 49–59. 6 Ias. J

Custumarii	Enclosed Arab.	Enclosed Md.	Enclosed Past.	Arable in the Open Common Fields: Over field	Presthay field	Downe field	Bakenhop field	Brownslonde	Avell field	Southhope field	Miscellaneous	Common Meadow
Joh. Bedowes, m.	1	..	1¾	5	5	1	2	..	..	..	......	½
Joh. Sacker;[1] m.	1	2	21	16	3	4	6	1	..	..	Helde field 12	1½
Jo. Blacke, m.	½	..	6	3	..	4	..	..	3	..	......	..
Wm. Cooke, m.	1	..	3	..	2	2	..	..	..	..	......	..
Wm. Goodman, m.	1	7	18	14	2	11	3	..	12	3	......	1¼
Robt. Davys, m.	3½	..	4½	..	1	½	..	..	..	..	......	..
Rich. Davies, m.	½	2	13½	12	3½	6	..	10	..	4½	Aw field 6	1½
Christ. Higgins, m.	8	3	5	14	..	..	..	13	21	5	Litle field 4	3
Thos. Hoppeswood, m.	7	3½	25	17½	2	13	..	1	..	..	Aw field 6	1½
Jo. Higgins, m.	1	1	2	2	..	2	..	¼	..	..	......	..

BRIERLEY, A MEMBER OF IVINGTON, HEREFORDSHIRE — Land Rev., M. B. 217, ff. 77–84. 6 Jas. I

Custumarii	Enclosed Arab.	Enclosed Md.	Enclosed Past.	Arable in the Open Common Fields: le Much Howe	Littlehowe	Weste field	Great Lowes	Pillcrofte	Ryelands	Grove field	Miscellaneous	Common Meadow
Thos. Badnedge, m.	1	2¼	24	11	1	4	3	3	3	..	Pitt field 9	2
Walter Bedford, m.	1	1	8¼	6	1½	2½	1½	3	4	8	Pitt field 2, Coppgrove 3	2¾
Thos. Whetstone,[2] m.	gard.	..	4¼	4	1	2	½	1½	6	11½	......	2
Marg. Avye, m.	¼	3	7½	6 and Littlehowe		5	..	10 and Rylands		9	......	2½
Thos. Badnedge, m.	¾	3	9	6	2	8	2	12 and Rylands		12	......	7
Thos. Parkes, m.	4	6	18	10	6	1	7	4	5	16	Bury field 2, Ketchlowe field 2	.. ..
Thos. Langford, m.	2	..	6¾	8	..	3	..	..	6	8	......	1
Thos. Tompkyns, m.	1½	2¼	4¾	10	..	5	5	4	6	15	......	4½

[1] John Sacker has "communia pastura in communibus campis predictis pro omnibus avariis suis et in vastis vocatis Buskewood et Wynstleyeshill pro lx ovibus."
[2] Thos. Whetstone has "communia pasture in communibus campis predictis."

STOKE EDITH, HEREFORDSHIRE

Add. MS. 27605, ff. 67–95. 40 Eliz.

	Enclosed	Arable in the Open Common Fields								Open Field	
		Knotwell	Stokes	Pyryan field	Manncroft	Cowlsmore field	Pyrtonshill	Prillhill	Miscellaneous	Md.	Past.
Demesne	238	71	60	39	41	. .	. .	. .	. .	. .	. .
Copyholders											
Wm. Hodges, m. . .	. . .	. .	. .	7, and 1 butt		6, and 4 ridges	12, and 4 ridges, 4 butts	. .	Orgonscroft 2	2	4
Jo. Turnor, m.	. . .	. .	. .	9	. .	11	. .	. .	Dodnarshill 6	2½	4
Jo. Hodges, m.	½	. .	. .	5	. .	7½	11	. .	Dodnarshill ½, Orgons croft 3	2	4½
Marg. Garnors, m. .	1	2	1	. .	. .	. .	1	1	. . . . . .	½	. .
Thos. Jeffreys, 2 m.	1	1	1	1	. .	. .	. .	1	. . . . . .	½	. .
Wm. Fryer, 4 m. . .	3¾	9	9	3	2	. .	1	8	Oldhill 4, Bicknapesland 2	2	4½
Edm. higgs, m. . . .	½	2	2	1	1	. .	. .	2	Oldhill 2	½	2
Thos. danford, m. .	½	2	2	. .	1	. .	1	. .	Oldhill 1	. .	. .
Roger Danford, m. .	½	½	4	½	1	. .	. .	. .	Oldhill 1	½	3
Roger Jeffreys, m. .	4½	. .	. .	. .	. .	. .	10	2	Oldhill 2	. .	9
Alice Nurburye, m.	7	. .	. .	6, and 9 ridges		7	½, and 4 ridges	. .	Ashcroft 4, Unspecified 8, and 35 ridges	3½	1½
Jo. Taylor, 2 m. . . .	½	2	1½	1	½	. .	1	1	Oldhill ½	½	6
Roger Careles, m. . .	1	. .	. .	6	. .	. .	7	. .	Orgons Croft 6, Dodnors Hill 3	2¾	6½

Kingsbury, Somerset

Land Rev., M. B. 202, ff. 199–253. [3–5 Jas. I]

Custumarii	Enclosed Arab.	Enclosed Md.	Enclosed Past.	Arable in the Open Common Field: Byneworth	Kylworth	Norton	Hill field	Miscellaneous	Common Meadow
Wm. Gibbs, m.	1½	2	1	2	5½	..	..	......	1½
Robt. Seagar, m.	¼	..	2¾	6	1	1	2	......	½
Thos. Lye, m.	¼	..	..	1	..	1	2	Tunnland 1	½
Ric. Phillips, m.	1½	..	..	2¼	..	..	1	Tunnland 1¾	½
Jo. Crofts, m.	25½	..	14	..	..	..	..	......	1¾
Geo. Louche, m.	2	..	1	1½	3	1	2	Deanland 1	½
Jo. Drayton, m.	½	1	..	3	2	1¾	4	......	½
Jo. Axe, m.	¾	..	..	3	1½	..	..	Twynefurlong 1	1
Hen. Towchin, m.	¼	..	1	2	1	..	..	Bushoppshill 1, Parne field 1	½
Johanne Rapson, m.	½	..	..	..	2	..	¾	......	2
Jo. Towchin, m.	12	..	7½	..	..	..	..	Dedlond 1	1
Thos. Rooke, m., ½ virg.	1	..	22¼	..	..	..	..	......	..
Jo. Clark, m., 3 ferdells	9½	1	11	..	..	..	..	Gosen field 8	1
Jo. Humfrye, m., 3 ferdells	15½	3	7½	..	4	..	2	Dedland 5	1

Holdings are frequently described as 5, 10, or 20 acres "de antiquo austro." Free and customary tenants have common in all or in some of the following pastures: Horsey, Westmore, Chaldworthmead, Leachleaze, Foreleaze, Sheepleaze, Sharpham.

East Brent, Somerset

Land Rev., M. B. 225, ff. 53–114. 4 Jas. I

Custumarii, Tithing of Burton	Arab.	Enclosed Md.	Past.	Arable in the Open Common Fields: Super le Downe	Sharpham alias West field	Yea field	Miscellaneous	Common Meadow
Jo. Whippie, m., ½ virg.	1½	22	29½	1½	..	..	......	1
Jac. Shew, m.	1	16½	17	3	11	..	......	1¼
Math. Knowles, m.	2½	9	34⅞	3	¼	..	......	..
Timoth. Ill, m.	7	..	2	1	½	2	Bicknell field 3¾	2¼
Isabella Lacye, m.	3	..	17¾	..	7	..	Bicknell field ¾	..
Rich. Tyll, m.	¼	..	16	..	..	10½	Lympesham field 5	..
Edw. Wal, m.	4	..	17	..	1	1½	Lympesham field 3, Horsecroft 1	6

Custumarii, Tithing of Singhampton	Arab.	Enclosed Md.	Past.	Arable in the Open Common Fields: Super le Downe	East field	Hardland	Miscellaneous	Common Meadow
Wm. Mason, m.	½	..	18¾	..	5	..	......	..
Thos. Herse, m.	8	..	..	..	2½	..	West field 4, Myl field 5	5
Nich. Isger, m.	4¾	..	22½	..	8½	¼	Myl field 3, North Ewe field 3	7½
John Dod, m.	½	..	13	1	½	4	......	..
Wm. Purnell, m.	¾	..	11	..	1	2	West field 2	7½
Kath. Browning, m.	1½	..	25½	..	1½	¾	North Ewe field 4	5
Jo. Martin, m.	1½	..	14	1	1	4	......	2¾

Copyholders of Burton have unstinted common of pasture in Markemoore alias Farlemore; those of Singhampton in Thurlmore.

NORTON ST. PHILIP, SOMERSET — Land Rev., M. B. 202, ff. 167–197. 3 Jas. I

Custumarii	Enclosed Arab.	Enclosed Md.	Enclosed Past.	Arable in the Open Common Fields South field	Arable in the Open Common Fields North field
Alicia Aprise, m., 1 virg.	8½, 6[1]	11	2½	3½	5
Jo. Davison, m.	2½[1]	..	1½	..	1
Edw. Thomas, m.	16¼[1]	..	2½	..	..
John Butcher, m.	¼	..	4¼, 1[1]	..	..
Johanna Butcher, m.	1	..	1¼, 2½[2]	3	..
Ric. Tovy, m.	2½	7½	8½	..	..
Edw. Aprise, m.	5½	..	2½	3 and North	
Thos. Methwine, m.	¼	..	4½	..	2
Ric. Tovy, m.	½, 4[1]	2	12½	..	..
Jo. Buddock, m.	2½, 3[1]	2	15	..	..
John Pardies, m.	¾	¼	1¼	2	2

The holdings are usually small, those noted above being among the largest. Many of the parcels of enclosed pasture are "in goddes peece," which was in the North field (f. 192). There is no reference to common of pasture.

WEST PENNARD, SOMERSET — Land Rev., M. B. 202, ff. 105–159. 3 Jas. I

Custumarii	Enclosed Arab.	Enclosed Md.	Enclosed Past.	Arable in the Open Common Fields: Esterne Downe	Westerne Downe	Lyttle field	South field	Breach field	Westmore field	Eastmore field	Common Meadow
Jo. Galler, m.	1½	3½	20	10	7	½	..	..	..	..	5
Radus. Ball, m.	1	5	..	5	5	2	..	..	..	..	11¾
Robt. Shepheard, m.	1½	..	30½	9	15	..	..	..	..	..	12
Jo. Frye, Sen., m.	8½	33	2½	14¼	21¼	2	..	..	..	..	2
Johanna Hole, vidua, m.	1	4¼	5	..	7	..	4	..	..	..	1¾
Eliz. Walter, m.	4½	..	30	..	3¼	..	2	..	..	..	3½
Wm. Sheppard, Sen., m.	1	..	2	5½	5	..	4	..	..	..	7
Wm. Grymstede, m.	½	..	13½	4½	3½	..	10	13	..	..	4¾
Edw. Slade, m.	1¾	..	19½	..	..	..	5	9	..	..	2½
Jo. Frye, m.	½	..	2	3¾	5¼	..	..	..	..	..	2
Thos. Dunkerton, m.	¾	1	7	..	..	..	..	4	5	6	2
Thos. Frye, m.	½	..	17½	..	..	..	..	5	4	5	2
Edw. Carter, m.	½	5	5	..	2	..	..	..	13	4	3½
Jo. Dunkerton, m.	½	3	9½	..	..	5	..	5	5½	2	2

Freeholders and copyholders usually have unstinted common of pasture over all or a part of the following commons: Sedgemoore, Comon moore, Kennard more, Litlemore, Basborowe wood, Crannell more, Edith more (f. 106).

[1] Enclosed, in the South field. [2] Enclosed, in the North field.

Curry Mallett, Somerset

Rents. & Survs., Ro. 566. 1610

Custumarii	Enclosed Arab.	Enclosed Md.	Enclosed Past.	Arable in the Open Common Fields: Campus Occidentalis	Campus Orientalis	Slade	Breache	Eyeberie	Miscellaneous	Common Meadow
Jo. Polman, m.	1½	..	..	3¾	3	2	..	1	......	⅜
Jo. Uttermeare, m.	7½	..	..	8	..	4	5	..	Woodlees ½	2½
Hugo House, m.	7	..	..	6	3	1½	2½	5	......	1½
Thos. Middle, m.	11¼	..	..	5	4	..	..	..	Headmead 1	1½
Eliz. Meade, m.	13½	..	..	..	10	..	..	12	......	1
Math. Masters, m.	19¾ [1]	3½	..	1	4	..	..	..	......	2
Jerome Howse, m.	15	4¼	1	10	1¼	..	..	..	......	½
Jo. Collier, m.	1	4	16	11	10	3	..	3	The eighte part of the Downe 10	1
Jo. Baller, m.	26¼	..	..	6	2	..	..	..	......	..
Rich. Collier, m.	20½.	..	7	1½	..	6	..	..	Campus borialis 1	2
John Meade, m.	20	..	1	¾	2	..	..	..	......	1½
Agnes Hawker, m.	9½	..	2¾	2	5	2	..	2½	Ruddlonde 2	1¼
Philip Tayler, m.	10¼	..	..	..	6	..	..	..	......	¾

The tenants have common "sans stint" in Sedgemore.

[1] Including a close of 5 acres "in campo orientali."

CORSTON, SOMERSET — Land Rev., M. B. 225, ff. 41–50. 6 Jas. I

Copyholders	Enclosed			Arable in the Open Common Fields		Common Meadow
	Arab.	Md.	Past.	North field	South field	
Thos. Coxe, m.	¾	..	7½	28	27	2½
Jo. Holbye, m., cott.	2	..	13	26	23	5
Flower Forde, m.	1	..	9	24	24	4
Thos. Weekes, m.	2	..	9	18	11	2
Edw. Maynard, m.	7½	6	9	23	5½	..
Ric. Wade, m.	1	1	5	7	6	1½
Marie Bilbie, m.	2	6½	9½	22	15	..
Agnes Bushe, m.	1	..	2½	10	6	¾
Edw. Curwell, m.	4	..	9	12	6	..
Edw. Curwell, m.	8	4	8½	11½	6	..
Robt. Baber, gent., 1½ yl.	1	5	30	37½	8	½

There are three other copyholders. The tenants have "common on the Berrie."

BRUTON, SOMERSET — Bodl., Rawl. B, 416 B. 1684

Copyholders	Enclosed		Arable in the Open Common Fields		
	Arab.	Md., Past., & Wood.	North field	South field	Miscell.
H. Albin	..	10	4	4	2
Hannah Albin, m.	..	..	10½	4	..
Wm. Dymond, m.	..	2	2	2	..
Florence Dymond.	..	..	..	4	..
Jas. Allyn, m.	..	4¾	4	..	2½
Elinor Stevens, m.	3	17	1	3	..
Wm. Whittacre, m.	..	..	2	..	..
T. and J. Stevens	..	..	1	1	..

Leaseholders	Enclosed		Arable in the Open Common Fields		
	Arab.	Md., Past., & Wood.	North field	South field	Miscell.
Edw. Moore	100	66	33½	2	..
Jo. Ludiwell	..	6	2	2	..
Jas. Albin	..	32	5	2	..
Wm. Millard	..	5	..	2	6
Alex. Whittacre	5	..	2	3	..

APPENDIX III

CHRISTIAN MALFORD, WILTSHIRE
Harl. MS. 3961, ff. 64–84. 10 Hen. VIII

Custumarii	Enclosed	Campus Borialis	Campus Occidentalis	Arable in the Open Common Fields					Miscellaneous	Common Meadow
				Little field	Benehul field	Middel field	Wodefurlong		
Isabelle Boxe, 2 tofts, 2 virg....	40	$26\frac{1}{4}$ (25)	....	$14\frac{1}{8}$ (15)	5 (7)	....	13 (8)	Brodecroft (10 (2), Estwodehegge 4	8 (18)
Wm. Wastefeld, m., 1 virgs. ..	45	26 (13)	8 (5)	....	....	$5\frac{1}{4}$ (6)	7 (4)	....	$4\frac{3}{4}$ (8)
Rich. Peers, m., 1 virg........	$38\frac{1}{2}$	$15\frac{1}{4}$ (16)	$3\frac{1}{4}$ (3)	....	....	....	2 (2)	....	$2\frac{5}{8}$ (3)
Jo. Hatherell, m., 1 virg.	25	4 (3)	....	$6\frac{1}{2}$ (4)	$5\frac{1}{2}$ (5)	2 (2)	6 (2)	....	$\frac{5}{8}$ (3)
Robt. Snell, m., 1 virg........	10	....	5 (4)	$7\frac{3}{8}$ (6)	$4\frac{3}{4}$ (6)	....	1	....	$1\frac{7}{8}$ (7)
Geo. Snell, m., 1 virg........	$23\frac{1}{2}$	....	$3\frac{1}{2}$ (3)	$8\frac{1}{4}$ (8)	5 (5)	....	$1\frac{1}{2}$ (2)	....	$1\frac{3}{8}$ (4)
Helena Cocke, m., $1\frac{1}{2}$ virg....	$29\frac{7}{8}$	....	....	....	....	2	....	....	....
Jo. Stokehame, m., 1 virg.....	31	$14\frac{3}{4}$ (14)	....	....	....	$5\frac{1}{2}$ (4)	4 (3)	Wodecroft 10 (3)	4 (5)
Wm. Say, m., 1 virg..........	$28\frac{1}{4}$	$14\frac{1}{4}$ (10)	....	4 (4)	4 (3)	....	4 (3)	....	$2\frac{3}{8}$ (5)
Rich. Aldey, m., 1 virg.......	25	7 (3)	$4\frac{3}{4}$ (4)	$2\frac{1}{8}$ (3)	$5\frac{3}{4}$ (6)	$3\frac{3}{8}$ (5)	6 (3)	Brodecroft 5	$1\frac{1}{4}$ (6)
Nich. Ryche, m., 1 virg.......	21	....	$5\frac{1}{4}$ (4)	$6\frac{1}{2}$ (5)	$17\frac{5}{8}$ (11)	$5\frac{1}{4}$ (6)	$4\frac{1}{2}$ (5)	....	$1\frac{1}{2}$ (5)
Robt. Batyn, m., 1 virg.......	34	....	....	....	$9\frac{3}{8}$ (6)	....	3 (1)	Woodhegge 8	$2\frac{1}{4}$
Wm. Rymell, m., 1 virg.......	$8\frac{1}{2}$	$11\frac{1}{4}$ (6)	....	2 (3)	....	....	$1\frac{1}{2}$	Woodhegge 8	$1\frac{1}{8}$
Jo. Gangell, m., $\frac{1}{2}$ virg........	22	....	2 (1)	....	....	$3\frac{1}{4}$ (3)	$1\frac{1}{2}$	Brodecroft 5, Woodcroft 10	$1\frac{3}{8}$
Jo. Bosse, m., $\frac{1}{4}$ virg..........	$20\frac{1}{4}$	$5\frac{3}{4}$ (2)	4 (3)	1 (2)	....	$3\frac{1}{2}$ (3)	$\frac{1}{4}$	Brodecrofts field 5, Woodhegge 3	$\frac{3}{8}$

The figures in parentheses indicate the number of parcels. There are many other copyholds divided irregularly among the fields.

APPENDIX III

NITON, ISLE OF WIGHT — Exch. Aug. Of., M. B. 421, ff. 32–46. 6 Jas. I

Copyholders	Enclosed	Arable in the Open Common Fields		Copyholders	Enclosed	Arable in the Open Common Fields	
		West field	East field			West field	East field
Jo. Harvey, m.	$7\frac{5}{8}$	$9\frac{1}{2}$	..	Wm. Trefford, m.	12	8	..
Judith Sunddle, m.	$6\frac{1}{4}$	$6\frac{3}{4}$	..	Jo. Karvy, m.	$6\frac{1}{4}$	$10\frac{1}{2}$	..
Wm. Peare, gent., m.	$37\frac{3}{4}$	$1\frac{1}{2}$	9	Jane Speed, m.	4	..	..
Rich. Munt, m.	13	$4\frac{1}{4}$	5	Wm. Leoper, m.	..	2	..
Jo. Downer, m.	$5\frac{1}{8}$	1	2	Wm. Spanner, m.	$9\frac{1}{4}$	1	10
Thos. Orchard, m.	$3\frac{3}{4}$	5	$3\frac{1}{2}$	Danl. Haward, m.	$21\frac{1}{4}$	$3\frac{1}{2}$	3

There are eight other copyholders. All tenants have rights of common (stinted for sheep) in five designated commons.

WILLERBY, YORKSHIRE — Land Rev., M. B. 229, ff. 90–96. 6 Jas. I

Leaseholders	Enclosed Past.	Arable in the Open Common Fields						Common Meadow	Pasture in Willerby Carr
		West field	Toffindale	Lowe field	Kirkgate field	Langland field	Elleylund field		
Wm. Rowley, m., 5 oxgs.	$9\frac{1}{4}$	$9\frac{3}{4}$	$4\frac{1}{2}$	$14\frac{1}{2}$	$15\frac{1}{4}$	17	$\frac{1}{2}$	$6\frac{1}{4}$	14 gates
Wm. Wilkinson, m., 4 oxgs.	3	$9\frac{1}{2}$	4	$10\frac{1}{4}$	13	$12\frac{1}{4}$	$1\frac{3}{4}$	5	12 acres
Wm. Wright, m., $1\frac{1}{2}$ oxgs.	$1\frac{1}{2}$	2	1	$8\frac{1}{4}$	1	$4\frac{1}{4}$	$1\frac{3}{4}$	$2\frac{1}{4}$	5 acres
Robt. Lillforthe, m.	$2\frac{3}{4}$	$\frac{3}{4}$	$\frac{3}{4}$	$2\frac{1}{2}$	$1\frac{3}{4}$	$1\frac{1}{2}$	..	..	5 gates
Geo. Wetheroppe, gent., and Hellen									
Skipwith, m., 5 oxgs.	8	$9\frac{1}{4}$	6	16	12	$15\frac{3}{4}$	$4\frac{1}{4}$	6	14 gates
Wm. Gullson, m., 6 oxgs.	$2\frac{1}{4}$	6	8	19	$13\frac{3}{4}$	$18\frac{1}{2}$	$\frac{3}{4}$	$8\frac{3}{4}$	....
Ralphe Risom, 2 m., 5 oxgs.	$2\frac{1}{4}$	$5\frac{1}{4}$	9	15	$12\frac{1}{4}$	14	$1\frac{3}{4}$	$6\frac{1}{4}$	12 gates
Wm. See, m., 2 cott.	9	13	$\frac{3}{4}$	$4\frac{1}{4}$	1	$1\frac{3}{4}$	$28\frac{3}{4}$[1]	8	12 gates
Phil. Risom, m., 4 oxgs.	1	$6\frac{1}{4}$	$3\frac{1}{2}$	$10\frac{1}{4}$	$9\frac{3}{4}$	12	...	5	10 gates

There are no copyholders. Several cottagers have holdings of less than two acres each.

[1] Also in Elley Kirke field $3\frac{1}{2}$ acres, Elley Drubutt field 5 acres, Elley Andrewethorne field $3\frac{1}{2}$ acres each.

BREIGHTON, YORKSHIRE

Land Rev., M. B. 229, ff. 225–232. 1609

Lessees	Enclosed Arab.	Arable in the Open Common Fields: Longland field	Borne and Townsend fields	South field	Hallmore	Car field	Wilderthorne	Miscellaneous	Common Meadow	Pasture "Gates"
Robt. Blancharde, m.	10¼	1½	8	7½	7	1½	..	......	2¾	4
Raphe Rabie, cott....	1	1½	1	..	..	1½	..	......	..	..
Dion. Smithe, m.....	4	6	7	1	14	3½	..	......	3¾	..
Anth. Smythe, m. ...	14	..	9	10	14	2½	..	......	4¾	11
Anth. Smythe, m. ...	23	..	9	8½	7	..	..	......	3¼	..
Hen. Auclus, m......	2½	9	8	..	7	5	4	Northmore 8	2¾	..
R. & J. Bargeman, m.	1	5½	5½	..	7	7 and Wilderthorne close			2¼	2
Rich. Smyth, m.	19	16 and Borne		10 and Hallmore		2½	..	......	4⅜	3½
Franc. Beacham, m. .	½	8	9	..	6½	6	5	in the Lund 2	2¼	4
Robt. Smyth, m.	..	6	10	5	14	5	..	in the Lund 6	3⅞	3
Wm. Binck, m.......	1½	10	9	..	..	8	4	......	2¾	6
Wm. Halley, m......	9	..	7	5	7	..	..	in the Lund 2	6½	4
Alex. Bond, m.......	½	8½	7½	1	7	3	..	......	2¼	7
Anne Beacham, m. ..	5	6	5	1	14	4	4	......	4	6

There are no copyholders. The above list of lessees is complete except for three cottagers.

532 APPENDIX III

OWSTON, ISLE OF AXHOLME, LINCOLNSHIRE
Land Rev., M. B. 256, ff. 56–65, 182–192. 5 Jas. I

		Arable in the Open Common Fields				Common Meadow
Custumarii	Enclosed	Wood field	Syden field	Church field	Selibus field	
Greg. Davies, m.	$1\frac{3}{8}$	2	$\frac{1}{4}$	$1\frac{1}{4}$	1	..
Jo. Phillipson, m.	$\frac{1}{2}$	5	5	5	3	2
Marie Cooke, m., 1 bovate	$1\frac{1}{4}$	$3\frac{1}{4}$	1	5	$2\frac{3}{4}$	3
Jo. Lawghton, 2 m.	$1\frac{5}{8}$	5	$3\frac{1}{2}$	2	$3\frac{1}{2}$	2
Wm. Fishe, m.	$1\frac{3}{4}$	$1\frac{1}{2}$	2	$2\frac{1}{2}$	2	$1\frac{1}{2}$
Jo. Cooke, m.	1	$1\frac{1}{2}$	1	$2\frac{1}{2}$	2	..
Ed. Fayerwether, m., 2 bovates	2	$5\frac{1}{2}$	4	7	7	..
Thos. Johnson, m.	1	2	2	$2\frac{1}{2}$	3	..
Liberi Tenentes						
Thos. Slingsbury, 2 m.	$5\frac{3}{4}$	$2\frac{1}{2}$	2	$1\frac{1}{8}$	$2\frac{1}{2}$	$3\frac{1}{4}$
Geo. Fisshe, m.	1	$1\frac{1}{2}$	$1\frac{1}{4}$	3	$1\frac{1}{2}$	4
Robt. Corringham, m.	$8\frac{1}{4}$	$1\frac{1}{4}$	$1\frac{1}{2}$	4	2	$3\frac{1}{2}$

There are several small copyholds and enclosed freeholds. No statements are made regarding common of pasture.

APPENDIX III

LENTON AND RADFORD, NOTTINGHAMSHIRE
Land Rev., M. B. 211, ff. 106-158. 5 Jas. I

Tenentes per Litteras Patentes	Enclosed Arab.	Enclosed Md. & Past.	Arable in the Open Common Fields Beck field	More field	Red field	Church field	Sand field	Alwell field	Common Meadow
Rich. Grason, m., 1 bovate	$\frac{1}{4}$	..	2	$1\frac{1}{2}$	2	$\frac{1}{2}$	2	1	$3\frac{1}{4}$
Pet. Toole, m., 4 bov.	$1\frac{1}{2}$	1	11	7	11	2	9	8	10
Valentinus Salt, m., 1 bov.	$\frac{1}{2}$	..	$2\frac{1}{4}$	2	$4\frac{1}{4}$	..	1	2	3
Wm. Bosworth, m., 4 bov.	$1\frac{1}{2}$	$1\frac{1}{2}$	10	10	8	..	8	9	$12\frac{1}{4}$
Wm. Hill, m., 2 bov.	1	..	$3\frac{1}{2}$	2	$2\frac{1}{2}$	2	4	..	4
Wm. Hill, m., 1 bov.	$\frac{1}{2}$	..	3	$1\frac{1}{4}$	2	..	2	..	3
Jo. Meryan, m., 3 bov.	$\frac{1}{2}$	5	10	4	8	$\frac{1}{2}$	6	8	7
Wm. Gresley, m., 1 bov.	$\frac{1}{2}$	$\frac{1}{4}$	$2\frac{1}{2}$	$3\frac{1}{2}$	$3\frac{3}{4}$	..	1	4	$3\frac{1}{4}$
Phil. Wilkinson, m. 2 bov.	$\frac{1}{2}$	..	4	2	4	2	4	..	5
Johanna Mottershead, m., 2 bov.	$\frac{1}{2}$	1	$5\frac{1}{4}$	$1\frac{1}{2}$	$5\frac{3}{4}$	$1\frac{1}{4}$	$1\frac{3}{4}$	2	$5\frac{3}{4}$
Jo. Godbehere, m., 4 bov.	$\frac{3}{4}$	..	7	6	8	4	6	..	$10\frac{3}{4}$
Wm. Crompton, m., 4 bov.	2	1	9	4	7	3	8	..	8
Hen. Noten, m.	$\frac{1}{2}$	2	$4\frac{1}{2}$	$2\frac{1}{2}$	4	$2\frac{1}{2}$	5	$6\frac{1}{2}$	$3\frac{1}{2}$
Rich. Daste, m., 3 bov.	$\frac{1}{4}$	..	6	5	6	3	$8\frac{1}{2}$	..	$8\frac{1}{4}$
And. Webster, m.	$\frac{1}{2}$	..	..	..	$\frac{1}{2}$	..	$\frac{1}{2}$	..	..
Jo. Allen, m., 3 bov.	$\frac{1}{4}$	1	$8\frac{3}{4}$	5	5	$1\frac{1}{2}$	8	2	$6\frac{1}{4}$

There are many other similar holdings, together with 88 acres of freehold. The tenants have stinted common of pasture for cattle, horses, and sheep "in campis et communia."

APPENDIX III

EAST BRANDON, DURHAM.
Land Rev., M. B. 192, ff. 40b–43. 5 Jas. I

Tenentes per Litteras Patentes	Enclosed Arab.	Enclosed Md.	Arable in the Open Common Fields					Meadow in the Open Common Fields			Pasture "Gates"
			Water-gate field	Lowe field	Ruden hill	West field	le Crofts	Miscellaneous	Lowe field	West field	Miscellaneous	
Jo. Fletcher, m.	$\frac{3}{4}$	..	..	2	3	..	..	..	20	6	Foot of acre 5	21
Thos. Burlason, m., 1 bovate	1	1	..	..	$\frac{1}{4}$	..	..	..	5	..	East uppe Bancks 4	5
Chris. Hutchinson, m.	..	3	..	..	2	3	2	..	13	..	......	16
Anth. Farer, m.	$\frac{1}{2}$	..	..	..	2	2	2	..	8	..	......	8
Geo. Rippon, m.	..	..	..	..	..	..	..	Langrigge 2	..	..	Langrigge 10, White field 2	$7\frac{1}{2}$
Wm. Briggs, m.	..	..	..	..	..	..	..	Hareham 11	..	..	Hareham 28 md., 16 past.	3
Mich. Brian, m.	4	6	..	..	..	..	..	Farbrome 2	..	..	East field 3	3
Wm. Farrowe, m.	5	$12\frac{1}{2}$	..	..	..	..	..	Broome 3	..	..	Cathills 3, Little Garban $2\frac{1}{4}$	7
Nich. Cathericke, m.	$\frac{1}{4}$	1	2	$5\frac{3}{4}$	$1\frac{1}{2}$	$2\frac{3}{4}$	..	..	7	..	Broomflatts 2	9
Mart. Pinckney, m.	$\frac{1}{4}$	2	$2\frac{1}{2}$	..	$2\frac{1}{2}$	$2\frac{1}{2}$	..	..	$2\frac{1}{2}$	1	......	10
Geo. Mason, m.	$\frac{1}{4}$	..	$\frac{3}{4}$	..	1	1	..	..	..	..	......	3
Robt. Arcle, m.	$\frac{1}{4}$	1	1	..	1	1	..	..	5	..	......	7
Jo. Harrison, m.	..	..	..	..	..	..	..	8[1]	..	..	le great medowe 18, le new medowe $2\frac{1}{2}$	10
Thos. Hall, m.	$\frac{1}{4}$	2	..	..	2	2	..	in le garends 3	10	..	..	10

The tenants have "parcelle pasture," occasionally called closes, estimated in "gates." In addition many of them have "communia sans stinte super communem de Brandon."

[1] "Parcella arabilis iacens in pastura sua."

EGGLESTON, DURHAM.

Land Rev., M. B. 192, ff. 23*b*–27. 5 Jas. I

Tenentes "Termino Expirato"	Enclosed Arab.	Arable in the Open Common Fields East field	Arable in the Open Common Fields Middle field	Arable in the Open Common Fields West field	Meadow in the Open Common Fields East field	Meadow in the Open Common Fields Middle field	Meadow in the Open Common Fields West field
Wm. Whorton, m.	1	1½	2½	½	4	1¼	7½
Jo. Harker, m.	½	..	3	¼	4½	2¾	2¼
Jo. Gibson, m.	¾	..	..	..	1	2	6
Anth. Bland, m.	½	..	4½	..	2½	..	5
Jo. Hodgson, m.	½	¼	3½	¼	4¾	1	4½
Geo. Bailes, m.	1	..	3½	½	¾	½	5¾
Jo. Pinkney, m.	4¼	..	2	..	2	¾	3
Chris. Harrison, m.	2¾	..	..	..	6½	..	2
Jo. Hodgson, Sr., m.	½	..	3¼	..	..	3	9
Jo. Newbie, m.	¼	..	2	2	..	1	8
Mich. Raine, m.	1	..	⅜	..	6	⅝	2½
Thos. Addison, m.	1, ¼ md.	½	3½	..	4	..	6
Wm. Howson, m.	¼	1½	2¼	..	9	¾	4½
Wm. Hodgson, m.	¾	½	..	..	15	1¼	..
Ric. Londsdale, m.	¾	¾	3	1	3	¾	4

There are many similar holdings. The tenants have "communem sans stint."

APPENDIX IV

PARLIAMENTARY ENCLOSURES IN OXFORDSHIRE

TOWNSHIPS IN WHICH MORE THAN THREE-FOURTHS OF THE AREA, EXCLUSIVE OF THE WASTE, WAS ENCLOSED BETWEEN 1758 AND 1864

Date of Enclosure Award	Township	Area in 1902 [1]	Total Area Allotted in the Award	Common or Waste Allotted	Open-field Arable and Meadow Allotted	Old Enclosures Known or Estimated
1768	Adderbury (East and West), Bodicot, and Milton	5275	4310	?	[4310]	[965] [3]
1777	Alkerton	741	689	[32]	[657]	[52] [3]
1816	Arncot	1697	1392	270	1122	305 [6]
1838 } 1859	Ascot-under-Wychwood	1832	1492 } 144	144	1492	196 [6]
1855	Aston and Cote	2982	2690	1464	1226	[292] [5]
1767	Aston, Steeple	1071	989	?	[989]	[82] [4]
1794	Barford, Little	724	681	[15]	[666]	[43] [2]
1796	Barton, Westcote } Barton, Middle	910 } 1326	2018	....	2018	[218] [6]
1794	Bicester, King's End	1455	1329	[300]	[1029]	[126] [2]
1770 } 1851	Black Bourton	2347	2030 } 40	40	2030	[277] [4]
1777	Blackthorn	2026	1772	?	[1772]	[254] [4]
1802	Bloxham	3139	2773	[173]	[2600]	[366] [2]
1778	Bourton, Great and Little	1669	1461	[61]	[1400]	[208] [3]
1802	Brightwell Baldwin	1609	1323	c. 100	[1223]	[286] [3]
1776	Brize Norton	3265	2442	?	[2442]	[823] [3]
1776	Burcott	670	616	?	[616]	[54] [4]
1795 } 1773	Burford } Upton and Signet	753 } 2172	1370 } 803	....	2173	[752] [4]
1804	Cassington	2263	1914	c. 400	1514	[349] [2]
1767 [7]	Chesterton	2340	[1860] [7]	?	[1860]	[480] [2]
1770 } 1849	Chipping Norton, Salford, Over Norton	6365	4805 } 165	335 } 165	4470	[1395] [3]

[1] These areas refer only to the land within the township, not to the water.
[2] Estimated by subtracting the area allotted from the total area of the township.
[3] Estimated by subtracting the area allotted from the total area of the township. An estimate from the tithe allotment is possible but it gives fewer acres of old enclosures.
[4] Estimated from the tithe allotment.
[5] Estimated from the plan.
[6] The area of the old enclosure is stated in a schedule.
[7] The date and area are derived from the petition to parliament, not from the award.
[8] The area in 1808 was 1850 acres.

APPENDIX IV

Date of Enclosure Award	Township	Area in 1902[1]	Total Area Allotted in the Award	Common or Waste Allotted	Open-field Arable and Meadow Allotted	Old Enclosures Known or Estimated
1776	Claydon	1190	1024	....	1024	[166][6]
1854 } 1773	Cottisford } Heath	1062 } 1423	1200 } 804	250	950 } 804	[481][2]
1853	Cowley, Temple } Cowley, St. Johns	908 } 595	908 } 390	200 } 80	708 } 310	[205][5]
1775	Cropredy	1809	1583	20	1563	[226][3]
1808	Deddington	4260	3765	?	[3765]	[495][2]
1794	Duns Tew	1749	1528	c. 200	1328	[221][2]
1773	Epwell	1140	1069	?	[1069]	[71][4]
1849 } 1829	Fencott and } Murcott	1135	982 } 153	40 } 153	942	[20][5]
1808	Fritwell	1742[8]	1757	75	1682	[93][5]
1817 } 1863	Fulbrook	1848	1500[7] } 120	? } 120	1500	[128][2]
1823	Garsington	2233	2040	[135]	[1905]	[193][2]
1846	Grafton	623	512	c. 130	382	[111][5]
1804 } 1824	Headington	1955	1455 } 240	175 } 240	1280	[260][5]
1802	Heyford, Lower and Collcolt	1748	1508	[125]	[1383]	[240][2]
1842	Heyford, Upper	1614	1465	[65]	[1400]	149[6]
1766	Horley } Hornton	1191 } 1422	2289	?	2289	[324][4]
1858	Horsepath	1154	950	240	710	[204][5]
1831	Horton cum Studley	1287	1067	317	750	[220][5]
1778[7]	Idbury, Bowld, & Foscot	1563	1542[7]	?	[1542]	[21][2]
1830	Iffley	388	322	[22]	[300]	66[6]
1808	Islip	1988	1588	c. 300	1288	[400][6]
1799	Kelmscott	1028	960	[200]	760	[68][2]
1767	Kencott	1083	965	232	733	[118][4]
1818	Kidlington and Thrup	2978	2300	[300]	[2000]	[678][2]
1850	Kingham	1872	1798	c. 100	1698	[74][2]
1819	Littlemore	861	770	....	770	[91][5]
1794	Milcombe	1254	1135	[135]	[1000]	[119][2]
1844	Milton, Great	1436	1289	[275]	[1014]	[147][5]

[9] The petition to parliament is the only source of information. It mentions no area, that given above being purely conjectural.

[10] The 58 acres of common were not enclosed by the award.

[11] The 528 acres of common were not enclosed by the award.

[12] This enclosed area is that of the other hamlets of the large parish of Enstone. The common fields of one of them, Radford, were enclosed by agreement in 1773 (cf. John Jordan, *Parochial History of Enstone*, London, 1857, p. 290).

[13] The enclosure relates mainly to the open arable fields of Berrick Prior, Newington having been largely enclosed already.

[14] The area in 1802 was 1065 acres.

APPENDIX IV

Date of Enclosure Award	Township	Area in 1902 [1]	Total Area Allotted in the Award	Common or Waste Allotted	Open-field Arable and Meadow Allotted	Old Enclosures Known or Estimated
1849	Milton-under-Wychwood	2079	1855	600	1255	[224][5]
1729[9]	Mixbury	2445	[2000][9]	?	[2000]	[445][2]
1797[7]	Mollington	1438	1150[7]	?	[1150]	[288][2]
1856	North Stoke	835	775	....	775	60[6]
1791 } 1829	Oddington	1360	923 } 312	312	923	[125][2]
1776	Rollright, Great	2414	1923	[100]	[1823]	[491][4]
1768	Sandford St. Martin	2287	1739	[50]	[1689]	[548][3]
1780	Shennington	1628	1434	[400]	1034	[194][4]
1768[9]	Shipton-on-Cherwell	1100	[900][9]	?	[900]	[200][2]
1790	Sibford Ferris	1007	945	[45]	[900]	[62][2]
1774	Sibford Gower and Birdrup	1758	1665	[65]	1600	[93][2]
1765[7]	Somerton	1960	1800[7]	?	[1800]	[160][2]
1795	South Newington	1436	1367	[15]	[1352]	[69][2]
1856	South Weston	484	434	....	434	[50][5]
1853	Standlake } Brighthampton } Hardwick	2579 } 618 } 438	1761 } 1168 }	[424] } [230] }	1337 } 476 } 462	[706][2]
1774	Stanton Harcourt	3413	2648	?	2648	[765][4]
1804	Stonesfield	817	739	[250]	[489]	[78][2]
1776	Tadmarton	2070	1939	?	[1939]	[131][4]
1822	Taynton	1998	1860	210	1650	[133][5]
1794	Tew, Little	1794	1519	[100]	[1419]	[275][2]
1853	Warborough	1674	1547	175	1372	[127][5]
1762	Wardington, Cotes, and Wilcote	2664 } 318	2411	[50]	[2361]	[571][2]
1801	Wendlebury	1148	1080	[125]	[955]	[68][2]
1777	Westwell	1445	1238	?	1238	[207][5]
1813	Wheatley	988	862	120	742	[126][5]
1796	Wigginton	1187	1124	190	934	[63][2]

APPENDIX IV

Townships in which from One-half to Three-fourths of the Area, Exclusive of the Waste, was Enclosed between 1758 and 1864

Date of Enclosure Award	Township	Area in 1902 [1]	Total Area Allotted in the Award	Common or Waste Allotted	Open-field Arable and Meadow Allotted	Old Enclosures Known or Estimated
1797 } 1851 }	Alvescot	2080	1382 } 40 }	[100?] } 40 }	[1282]	[662] [4]
1814 } 1862 }	Asthall [and Asthalleigh]	2246	1192 } 374 }	374	1192	[680] [2]
1835 } 1858 }	Aston Rowant Kingston Blount Chalford	2920	1516 } 254 }	[365]	[1151] } 254 }	[1150] [2]
1840	Baldon, Toot Baldon, Marsh	1564 } 826 }	1061 } 560 }	46	1061 } 514 }	503 [6] } 266 [6] }
1808	Barford Magna	1132	604	[20]	[584]	528 [2]
1863	Bensington	2908	1522	[60]	[1462]	[1386] [2]
1863	Berrick Salome	603	448	[15]	[433]	[155] [2]
1766 [7]	Bladon	830	[480] [7]	?	[480]	[350] [2]
1845	Brightwell Salome Brightwell Prior	884 } 720 }	736 } 376 }	100 } 57 }	636 } 319 }	[492] [2]
1776	Broadwell Filkins	1776 } 1779 }	2586	[200?]	[2386]	[969] [4]
1814 } 1860 }	Chadlington, East, West and Chilsonback	3446 } 1671 }	3252 } 58 }	[352] } 58 }	2900	[1807] [2]
1845	Chalgrove	2432	1764	[75]	[1689]	[668] [5]
1858	Charlton-on-Otmoor	820	585	5	580	[235] [5]
1854 } 1846 }	Chinnor Hempton Winnall	2711	940 } 735 }	[160] } 58 [10] }	[780] } 735 }	[978] [2]
1788	Churchill	2838	2165	c. 500	1665	[673] [2]
1839	Clanfield	1788	1010	....	1010	[778] [2]
1787	Coggs	2274	1489	250	1239	[785] [2]
1861	Dorchester and Overy	1917	1001	70	931	[916] [5]
1861	Drayton near Dorchester	1287	897	40	857	[390] [5]
1839	Ducklington	1923	1029	....	1029	[894] [2]
1802 } 1803 } 1807 }	Ensham	5410	3342 } 45 } 28 }	1000 } .... } 28 }	2387 } .... } .... }	[1995] [2]
1761 [7]	Fringford	1456	[980] [7]	?	[980]	[476] [2]
1797	Hampton Poyle	800	593	15	578	[207] [5]
1773	Handborough	2248	1385	[50]	1335	[913] [2]
1774	Hook Norton and Southrop	5493	4092	?	4092	[1401] [3]
1815	Kirtlington	3541	2615	[325]	[2290]	[926] [2]
1814	Launton	2816	1901	[300]	[1601]	915 [6]

APPENDIX IV

Date of Enclosure Award	Township	Area in 1902 [1]	Total Area Allotted in the Award	Common or Waste Allotted	Open-field Arable and Meadow Allotted	Old Enclosures Known or Estimated
1815 } 1859	Lewknor } Postcombe	2664	1800 } 22	[250] } 22	[1550]	[842] [5]
1788	Lyneham	1938	1148	[200?]	[948]	[790] [2]
1839	Milton, Little	1342	902	50	852	[440] [5]
1759	North Leigh	2409	1735	380	1355	[674] [2]
1758	Piddington	2351	1420	c. 400	1020	[931] [2]
1852	Shipton-under-Wychwood	2513	1765	568	1197	[748] [5]
1766	Shutford	1360	866	....	866	[484] [2]
1794 } 1774	South Leigh	2359	1267 } 243	306 }	961 } 243	[849] [2]
1803 } 1780	Spelsbury } Dean	4297	1756 } 675	[170] } [30]	[1586] } [645]	[1866] [2]
1778	Stanton St. John	2730	1567	[67]	1500	[1163] [2]
1813	Stoke Talmage	868	489	[60]	[429]	[379] [2]
1772	Swalcliffe	1678	972	....	972	[706] [3]
1815 } 1857	Swinbrook	1710	792 } 60	[528] [11] } 60	} 792	340 [6]
1826	Sydenham	1547	960	....	960	587 [5]
1767	Tew, Great	3002	1778	c. 300	1478	[1224] [2]
1834	Wolvercot	1023	557	....	557	[466?] [2]
1770	Wooton	4211	2366	[300]	2066	[1845] [2]
1805	Wroxton and Balscott	2531	1850	285	1565	[681] [2]

APPENDIX IV

TOWNSHIPS IN WHICH FROM ONE-FOURTH TO ONE-HALF OF THE AREA, EXCLUSIVE OF THE WASTE, WAS ENCLOSED BETWEEN 1758 AND 1882

Date of Enclosure Award	Township	Area in 1902 [1]	Total Area Allotted in the Award	Common or Waste Allotted	Open-field Arable and Meadow Allotted	Old Enclosures Known or Estimated
1831 } 1829	Beckley } Otmoor	3610	690 } 1445	220 } 1445	470	[1475] [2]
1758	Bicester, Market End	2280	1045	[45?]	[1000]	[1235] [2]
1780	Bucknell	1891	848	[20]	[828]	[1043] [4]
1780 [7]	Caversfield } Stratton Audley	1275 } 2300	[1100] [7]	?	[1100]	[2475] [2]
1834 } 1865	Caversham	2363	593 } 252	252	593	[1518] [2]
1882	Crowell	996	358	....	358	[638] [5]
1813	Culham	1963	1100	270	830	[863] [5]
1845	Curbridge	2856	1110	200	910	[1746] [5]
1848	Denton	543	170	....	170	373 [6]
1844 } 1843	Enstone, Church } Enstone, Neat	6243	1138 } 1189	....	2327	[3916] [12]
1763	Merton	1926	c. 540	....	540	[1386] [2]
1760	Neithrop } Banbury	3302 } 79	1038	[75]	[963]	[2343] [2]
1815	Newington } Berrick Prior	2105	c. 636	....	636	[1469] [12]
1862	Ramsden	920	514	c. 325	189	406 [6]
1788	Sarsden	1426	757	185	572	[669] [2]
1837 } 1867	Shiplake	2700	699 } 223	223	699	[1778] [2]
1806	Shirburn	2416	1173	[270]	[903]	[1243] [2]
1853	South Stoke } Woodcote	3355	1728	250	1478	[1627] [2]
1803	Swerford	1923	851	120	731	[1072] [2]
1873	Tackley	2892	1387	....	1387	1505 [2]
1826	Thame, Priest End, North Weston	5187	2330	150	2180	[2857] [2]
1815	Watlington	3685	1534	c. 550	984	[2151] [2]
1806 } 1813	Whitchurch	1965	593 } 115	115	593	[1257] [5]

APPENDIX IV

TOWNSHIPS IN WHICH LESS THAN ONE-FOURTH OF THE AREA, EXCLUSIVE OF THE WASTE, WAS ENCLOSED BETWEEN 1758 AND 1864

Date of Enclosure Award	Township	Area in 1902 [1]	Total Area Allotted in the Award	Common or Waste Allotted	Open-field Arable and Meadow Allotted	Old Enclosures Known or Estimated
1864	Checkendon	3086	575	98	477	[2511] [5]
1802	Drayton near Banbury	926 [14]	199	80	119	866 [6]
1863	Ewelm	2484	458	458	....	[2026] [2]
1788 1812	Goring	4580	878 842	15 842	863	[2860] [2]
1824 1853	Hailey	2814	502 100	 100	502	[2212] [2]
1822	Haseley, Great	3232	495	....	495	[2737] [2]
1856 1863	Ipsden	3418	521 100	 100	521	[2797] [2]
1837 1860	Leafield	900	184 90	90	184	[626] [2]
1823	Noke	790	132	....	132	[658] [5]
1845 1853	Northmoor	2014	430 30	....	430 30	1554 [6]
1805	North Newington Broughton	1106 967	514	c. 100	414	1559 [6]
1852	Pyrton	3305	630	176	454	[2675] [2]
1860	Rotherfield Greys	2605	150	....	150	[2455] [2]
1794	Stoke Lynn and Fewcott	3899	1761	c. 1361	c. 400	[2138] [2]

APPENDIX V

EXTRACTS FROM THE SURVEY OF AN ESTATE LYING IN NEWCHURCH, BILSINGTON, AND ROMNEY MARSH, KENT

Add. MS. 37018, ff. 42–70.[1]

DOLA Godewini iacet in villa de Newecherche et in marisco de Romene inter feodum sacristie ecclesie Christi Cantuariensis et Dolam Mawgeri et capitat ex parte australi ad regiam stratam que ducit a cruce Johanis Cobbe usque Northene. Et predicta Dola continet xl acras terre iacentes coniunctim. Et debet ad terminum sancti Andree Apostoli vs. de redditu assise [rents and services follow in detail]. Summa xviii s. ob. q.

Adam Osbarne tenet de predicta Dola viii acras terre et quartam partem unius acre, viz.,

    dimidiam acram terre iacentem iuxta stratam supradictam et fuit quondam mesuagium Michaelis Galiot

    et i acram terre vocatam longreche

    et i acram et dimidiam terre vocatas hegeton

    et ii acras terre vocatas holland

    et ii acras et tres partes unius acre vocatas Flothame. Et debet ad terminum sancti Andree xii d. q. de redditu assise. . . .

Willielmus atte Mede tenet de predicta Dola iii acras terre et dimidiam, viz.,

    iii partes unius acre iacentes iuxta mesuagium Michaelis Galyot

    et i acram terre et dimidiam iacentes iuxta hegetownys

    et i acram terre et quartam partem unius acre iacentes in holland. Et debet ad terminum sancti Andree v d. q. de redditu assise. . . .

Johanes Northene tenet de predicta Dola i acram terre et dimidiam iuxta hegetownys versus australem. Et debet. . . .

Ricardus de Northene tenet in predicta Dola iiii acras terre, viz.,

    i acram et dimidiam terre vocatas Seadfeld

    et iii partes i acre iacentes iuxta hegetownys

    et i acram et iii partes i acre terre vocatas holland. Et debet. . . .

Johanis Symon de Newcherche tenet de predicta Dola iiii acras terre, viz.,

    iii partes unius acre vocatas longereche

    et i acram et dimidiam vocatas Seadfeld

[1] Cf. above, p. 287.

et i acram et quarta pars [sic] unius acre terre iacentem in holland et dimidiam acram iacentem iuxta. Et debet. . . .

Robertus Londer tenet de predicta Dola quartam partem i acre terre vocatam longereche iuxta feodum Sacristie ecclesie Christi Cantuariensis. Et debet. . . .

Johanes Pundherst tenet de predicta Dola tres partes i acre vocatas longreche. Et debet. . . .

.Heredes hamonis Baron tenent de predicta Dola i acram terre et dimidiam iuxta mesuagium Roberti Salmon versus australem. Et debent. . . .

Gilbertus de Morton tenet de predicta dola ii acras terre iacentes in Holland. Et debet. . . .

Heredes Laurentii Holde tenent de predicta dola ii acras terre et tres partes unius acre, viz.,

> dimidiam acram terre iacentem iuxta Regiam stratam de Northene
>
> et i acram et dimidiam in mesuagio Roberti Salmon
>
> et tres partes i acre in mesuagio Johanis Salmon vocatas Reyetown iuxta stratam. Et debent. . . .

Ricardus Tomelyn tenet de predicta Dola i acram et iii partes i acre iacentes iuxta mesuagium Roberti Salmon versus occidentem. Et predictus Ricardus defendit heredibus laurentii Holde quartam partem unius acre terre iacentem iuxta stratam. Et debet. . . .

Heredes Henrici de Bonyngton tenent de predicta Dola i acram terre et tres partes unius acre vocatas terra[m] Gryffyn et iacet iuxta Sandelyne ex parte boriali. Et debent. . . .

Johanes Freland tenet de predicta Dola viii acras terre iacentes in Holland. Et debet. . . .

Summa acrarum istius Dole xl acre unde portio cuiuslibet acre per annum vd. ob. de redditu, mala, et serviciis. Et plus in toto, viz., ad Natalem Christi i gallum, ii gallinas, et ob. q. de redditu Regis.

Dimidia Dola Mawgeri iacet in Newcherche et in marisco de Romene et inter Dolam Godewini supradicti et Dolam Storni et capitat ex parte australi ad regiam stratam que ducit a cruce Johanis Cobbe usque Northene et predicta dimidia Dola continet xx acras terre. Et debet. . . . Summa viii s. ii d. ob. q.

Heredes Thome Baker tenent de predicta dimidia Dola v acras terre et iii partes unius acre. Et debent. . . .

Adam Osbarn tenet de predicta Dola ix acras et dimidiam acram terre, viz.,

iii acras terre iuxta Twynenton versus North
et iii acras in ii peciis iuxta Twynenton versus South
et iii acras et dimidiam acram terre iacentes iuxta terram Radulphi Claverynge versus South. Et debet. . . .

Idem Adam pro Claverynge de predicta dimidia Dola tenet i acram et iii partes unius acre terre iacentes iuxta predictam terram Ade Osbarne versus North. Et debet. . . .

Heredes Ricardi Pundherst tenent de predicta dimidia Dola iii acras terre vocatas Twyneneton que fuit quondam mesuagium Hamonis Wodman. Et debent. . . .

Summa acrarum istius dimidie Dole xx acre Unde portio cuiuslibet acre per annum v d. ob. de redditu assise, mala, et serviciis. . . .

Alia dimidia Dola Mawgeri iacet in Newcherche in marisco de Romene inter Dolam Godewyni supradicti et Dolam Storni et capitat ad terram Radulphi Claveryng versus North et ad terram Nicholai de Bonyngton versus South et continet predicta dimidia dola xx acras terre iacentes coniunctim. Et debet. . . . Summa v s. iii d. ob.

Johanis Northene tenet in predicta dimidia Dola de terra Ricardi Elys
vi acras terre et l perticas iacentes iuxta terram Adam Osbarne ex parte australi
et iiii acras de terra Ricardi Gryffyn iacentes iuxta predictas vi acras terre ex parte australi
et iii acras terre de terra Ricardi Elys iacentes iuxta predictas iiii acras. Et debet. . . .

Heredes Henrici Bonyngton tenent de predicta dimidia dola
i acram terre et iii partes unius acre terre iacentes iuxta Sandelyne versus North
et iiii acras terre vocatas Sandelyne
et i acram terre et iii partes unius acre terre vocatas Borefeld que iacent iuxta predictam Sandelyne versus australem. Et debent. . . .

Summa acrarum istius dimidie Dole xx acre unde portio cuiuslibet acre per annum iii d. q.

Dola Storni iacet in Bylsington et in Newcherche et in marisco de Romene ex parte boriali et australi ad Regiam stratam que ducit a cruce Johanis Cobbe versus Northene et continet predicta Dola xlii acras terre unde viii acre terre et quarta pars unius acre iacent coniunctim in Bylsyngton et xxxiii acre terre et tres partes unius acre iacent in Newcherche. Et debet predicta dola. . . . Summa xviii s. 1 d. q. Item in predicta dola sunt iiii vetera mesuagia. . . .

Johanis Pundeherst tenet de predicta Dola in Bylsington iii acras iacentes ex parte north iuxta pontem malegare. Et in Newcherche xxii acras terre et dimidiam acram, viz.,

> iii acras terre et quartam partem unius acre iacentes ex parte australi ad Wychysland
>
> et ii acras terre et tres partes unius acre vocatas Wychysland
>
> et i acram et tres partes unius acre vocatas Wylespot et fuit quondam mesuagium Johanis de Northene
>
> et v acras terre et dimidiam acram terre iacentes ex parte North iuxta predictam Wylespot
>
> et viii acras terre et quartam pars [sic] unius acre de terra Edwardi Freland iacentes in Holland. Et debet. . . .

Heredes Richardi Pundherst tenent de predicta Dola iii acras et dimidiam acram terre iacentes in Newcherche ex parte occidentali ad Wychysland et fuit quondam mesuagium Roberti de Northene. Et debent. . . .

Gilbertus de Morton tenet de predicta Dola in Bylsyngton in mesuagio Roberti atte hope i acram et dimidiam vocatas Barattispot iacentes iuxta regiam stratam de Northene versus North. Et debet. .

Heredes Ricardi Thomelyn defendunt de predicta Dola Johani Bordon i acram terram et dimidiam iacentes in Bylsyngton iuxta terram prioris de Bylsyngton vocatam mentere versus West. Et debent. . . .

Ricardus Baker pro Petham tenet de predicta Dola in Newcherche i acram et tres partes i acre terre vocatas Bradfeld iacentes inter terram vocatam Hegemede et terram Johanis Pundherst. Et debet. . . .

Heredes Thome Baker tenent de predicta Dola in Newecherche v acras terre vocatas Hegemede et fuit quondam mesuagium Hamonis atte Hope iacentes iuxta Stormstrete ex parte Est et predicti heredes defendunt heredibus Roberti Holde ii acras vocatas similiter Hegemede iacentes in Newcherche iuxta Stormstrete ex parte Est in uno campo continenti ii acras terre et dimidiam et predicti heredes Thome Baker defendunt Henrico atte Neshe i acram et quartam partem unius acre iacentes in Bylsington iuxta terram prioris de Bylsyngton vocatam mentege vel secundum quosdam lonekyns acre. Et debent. . . .

Prior de Bylsyngton tenet de predicta Dola in Bylsyngton unam acram terre vocatam mentege vel . . . Lonekynes acre. Et debet. . . .

Summa acrarum istius dole xlii acre terre unde portio cuiuslibet acre per annum v d. q.

Dimidia Dola de Westbrege iacet in Newcherche et in marisco de Romene ex parte orientali et occidentali ad Regiam stratam que ducit a Bylsyngton usque Newcherche et continet predicta dimidia Dola xxii acras terre. Et debet. . . . Item in predicta dola est unum vetus mesuagium . . . et debet. . . .

Johanes Pundeherst tenet de predicta dimidia dola iii acras terre in duobus campis vocatis Longefeld iacentes iuxta Stormistrete versus occidentem et iii acras terre vocatas Stretefeld iacentes ex parte orientali iuxta Regiam Stratam. Et debet. . . .

Heredes Ricardi Pundherst tenent de predicta dimidia Dola ix acras iacentes ex parte occidentali ad Regiam stratam, viz.,

    vi acras terre vocatas morefriztege et iii acras vocatas litlefriztege. Et debent. . . .

Heredes Edwardi Godard tenent de predicta dimidia dola iiii acras iacentes ex parte orientali ad Regiam stratam, viz.,

    i acram vocatam Hegespot et fuit quondam mesuagium Johanis atte Bregge

    Et ii acras iuxta predictas ii acras et i acram vocatam Hegespot [sic]. Et debent. . . .

Heredes Ricardi Thomelyn tenent de predicta dimidia Dola iii acras terre iacentes ex parte orientali ad Regiam stratam. Et debent. .

Summa acrarum istius dimidii Dole xxii acre unde portio cuiuslibet acre per annum v d. de redditu assise, mala, et serviciis. . . .

Dola de Kyngessnothe iacet in Bylsyngton et in Newcherch et in marisco de Romene ex parte orientali et occidentali strate que ducit a Bylsyngton usque molendinum Ricardi Staple et continet predicta Dola xlvi acras et dimidiam acram terre unde in Bylsyngton sunt xiiii acre terre et tres partes unius acre terre et in Newcherch sunt xxxi acre et tres partes unius acre terre. Et debet. . . .

Heredes Jacobi de Kyngessnothe tenent totam predictam Dolam unde in Bylsyngton sunt xiiii acre et tres partes unius acre terre, viz.,

    ix acras et dimidiam acram terre vocatas Nessland iacentes ex parte occidentali strate supradicte iuxta mesuagium Radulphi Wolewyke

    et iiii acras et dimidiam acram terre vocatas Pykottismede iacentes ex parte occidentali strate supradicte iuxta terram Johanis Pundherst vocatam Pykottismede.

APPENDIX V

Item predicti heredes tenent tres partes unius acre terre vocatas Alderwynisland iacentes ex parte orientali strate supradicte iuxta Seadfeld et fuit quondam mesuagium Roberti de Kyngessnothe.

Item predicti heredes tenent de predicta dola in Newchurch ex parte orientali strate supradicte xxxi acras terre et tres partes unius acre terre, viz.,

>vii acras terre vocatas Redmede iacentes iuxta Alderwynisland
>
>et v acras terre vocatas Hokydefeld
>
>et ii acras et tres partes unius acre terre vocatas longehamme
>
>et vi acras et dimidiam acram terre vocatas marketfeld
>
>et unam acram et quartam partem unius acre terre iacentes in alia pecia iuxta predictam Marketfeld
>
>et ii acras et dimidiam acram terre vocatas Petfeld
>
>et ii acras et dimidiam acram terre in alia parte iuxta predictam Petfeld
>
>et tres partes unius acre terre vocatas Hegetown
>
>et ii acras et dimidiam acram terre vocatas Bachousefeld
>
>et i acram terre vocatam homstalle iacentem iuxta stratam predictam versus orientem. Et debent. . . .

Summa acrarum istius Dole xlvi acre. . . .

APPENDIX VI

SUMMARIES OF TUDOR AND JACOBEAN SURVEYS WHICH ILLUSTRATE IRREGULAR TOWNSHIP-FIELDS IN THE BASIN OF THE LOWER THAMES

Kew and West Sheen, alias Richmond, Surrey — Land Rev., M. B. 203, ff. 134–146. [*c.* 1600]

Divisions of the Field	Number of Acres and Parcels held by the Tenants														
	Sir Henry Portman	Sir Arthur George	Wm. Portman, Esq.	John Burd, iure uxoris	Stephen Pierce, Gent.	Vuican Jones, Gent.	Payne, Gent., iure uxoris	Robt. Clarke, Esq.	Geo. Charley, Gent.	Mary Crome, vid.	Dame Dorothe Wright	Barth. Smith, iure uxoris	Lott Peerce	The Church land	Thos. Smith
. . . feild															
long furlong	1 (1)	2 (3)	2½ (2)	3 (1)	..	..	..	..	..	..	..	..	..	..	..
parke furlong	7 (2)	3 (3)	12½ (4)	..	..	..	..	..	..	..	..	..	..	..	..
rickhill furlong	2½ (2)	1¾ (2)	11 (5)	..	..	..	..	..	..	..	..	..	..	2 (1)	..
Foxehole	5½ (2)	2 (1)	..	..	1 (1)	..	..	..	..	..	..	..	..	..	..
Keyo heath	15 (1)	..	25 (1)	..	..	..	..	..	..	..	..	..	..	..	..
The lower feild butting upon the baylies banck	..	..	[15] (3)	[7¼] (2)	4 (1)	20 (1)	13 (1)	1½ (1)	1½ (1)	..	..	..	..	..	..
A shott butting upon the highway to Keyo	..	..	..	..	5¾ (1)	..	11 (1)	..	..	4 (1)	..	..	..	..	..
The lower shott in the lower feild butting upon the highway leading to moreslack	..	..	1½ (1)	..	..	8 (1)	..	13 (2)	3½ (1)	2 (1)	5 (1)	2 (1)	5 (1)	..	..
The shott butting upon the parke west	½ (1)	..	18¾ (15)	..	10 (10)	25 (8)	18¾ (10)	5 (7)	2¾ (3)	9¾ (11)	..	2½ (3)	..	..	1½ (2)
Church furlong	2¾ (5)	..	5¼ (8)	2 (2)	9¾ (9)	5½ (6)	11 (8)	5 (4)	6 (6)	2½ (5)	3¾ (5)	1 (2)	1 (1)	2 (3)	3 (4)
Upper Dunstable	4½ (5)	..	4¼ (7)	1¾ (3)	9½ (9)	4 (6)	16½ (13)	3 (3)	2 (4)	2 (4)	2 (3)	..	1 (1)	½ (1)	1 (1)
Lower Dunstable furlong	1½ (3)	..	11¾ (5)	1 (1)	14¾ (11)	7½ (7)	5 (6)	2 (4)	2 (2)	2 (3)	3 (3)	..	1 (1)	..	..
East field shott	½ (1)	..	5¼ (10)	..	6½ (9)	4½ (6)	4¾ (6)	3¼ (7)	1 (2)	4 (7)	1½ (2)	3 (4)	..	½ (1)	2¾ (3)
Long Down	..	..	1¾ (3)	..	2¼ (4)	1 (1)	2 (2)	¾ (2)	½ (1)	1 (2)	..	1 (1)	..	¼ (1)	1 (2)
Short Down	..	..	¼ (1)	..	1 (3)	..	1¼ (2)	¾ (3)	..	½ (1)	..	..	..	½ (1)	..
East Bancroft	..	..	1¾ (4)	..	3½ (4)	1¼ (3)	1½ (2)	2¼ (5)	½ (1)	½ (1)	..	..	..	..	..
West Bancroft	½ (1)	..	1 (2)	..	2¾ (4)	2¼ (4)	1 (1)	1½ (3)	3 (5)	½ (1)	..	¾ (1)	..	¼ (1)	1 (2)
The Maybush shott alias South Downe	1¼ (2)	..	2¼ (4)	..	1½ (2)	4¼ (4)	1½ (2)	..	1 (1)	½ (1)	..	2 (2)	..	¾ (1)	..
[Open field] meadow	20	2	5	1½	2	2	1½	1	..	1	¾	..	..	½	..
Closes of pasture	12 (1)	..	..	..	..	..	..	..	..	..	..	..	..	..	..

CHESHUNT, MANOR OF THEOBALS, CROSBROOKS, AND COLLINS, HERTFORDSHIRE

Land Rev., M. B. 216, ff. 16–31. 19 Jas. I

Leaseholders	Enclosed Arab.	Enclosed Md.	Enclosed Past.	Arable in the Open Common Fields: Alberie	Holbrooke	Brompton	Broad field	Miscellaneous	Common Meadow
Wm. Clerke, gen.	..	..	$13\frac{1}{4}$	8	..	..	..	Grymescroft $8\frac{3}{4}$, Church-field $\frac{3}{4}$	..
Thos. Norton, gen., m.	$\frac{1}{2}$	..	5	..	..	2	..	....	2
M. Jennings, m.	$\frac{1}{4}$	..	$\frac{1}{2}$	..	6[1]	..	$2\frac{1}{2}$	End field 1	4
Roger Holt, m.	$1\frac{1}{2}$	..	..	..	3	2	2	....	$8\frac{5}{8}$
Roger Cooke, m.	$\frac{1}{2}$	..	..	..	12[2]	9	6	Grymescroft $9\frac{3}{4}$	4
Nich. Sliter, m.	$8\frac{3}{4}$	3	46	5	$3\frac{1}{4}$	..	3	Aldwicke $8\frac{1}{4}$, Long fields $13\frac{1}{4}$, Brooke field $1\frac{1}{2}$, Farlie 5	$\frac{3}{4}$
Purfrey, Armiger	..	$2\frac{1}{2}$	..	..	..	..	..	Windhill field 2, Lovedon $10\frac{1}{4}$[3]	29
Edw. Collins, m.	2	3	14	11[4]	..	5	..	Longlands $11\frac{1}{2}$,[4] Allom field 4	$15\frac{1}{2}$
Edw. Collins, m.	..	10	..	..	..	..	..	Long fields, 12,[4] Brooke field $7\frac{1}{2}$[5] Blackdale field $4\frac{1}{4}$, Tennacr feild 8	2
Roger Guildeford	..	..	..	..	3	..	..	....	..
Executors R. Cowlter, m.	1	..	..	..	2	5[1]	2	....	1
John Fitez, m.	$\frac{1}{8}$	..	$\frac{3}{4}$	..	12	$2\frac{1}{2}$	..	....	3
Rich. Petts, m.	$1\frac{1}{2}$	..	..	..	5	4	..	....	3
A. Tarlinge, m.	$\frac{1}{2}$	..	..	..	..	..	$1\frac{1}{2}$	Rowlands field 2	1
Geo. Grymes, m.	$2\frac{1}{2}$	..	$11\frac{1}{4}$	..	11[1]	6	$5\frac{3}{4}$	End field 5, Grymes croft 8	10

[1] In three parcels. [2] In two parcels. [3] In seven parcels. [4] In five parcels. [5] In six parcels.

EDMONTON, MIDDLESEX

Land Rev., M. B. 220, ff. 110–185. 2 Jas. I

Copyholders	Enclosed Arable	Enclosed Pasture	Arable in the Open Common Fields				Miscellaneous	"In magno marisco"
			Langhedge	le Hyde	Oke field	Heg field		
Robt. Estry, m.	3	18	6 (2 parcels)	13 (5 parcels)	6¾	..	..	6
Wm. Smith, m.	1	5	..	5	..	..	..	1½
Jacob. Lockyer, m.	2¾	7¾	5½	2½	1½	..	..	..
Maria Owen, m.	4½	6½	..	1½	..	..	Church feeld 3	1½
Thos. Stebrauk, 2 m.	½	3	..	3	..	..	..	3
Thos. Gray	..	14½	..	6¾	..	5	{ Scots feeld 1 Parti feeld 1½ Pickstones 1½ }	1
Jo. Proctor, 2 m.	¾	..	..	3	..	..	Dedfeeld 3	1
Robt. Hellam, m.	½	5	..	5	..	..	Rushells 3¼	2
Jo. Wright, m.	3	7	..	8	..	..	Hollis feeld 3	2
Heredes Hawes, m.	1¾	7½	..	..	..	3¾	..	..
Rich. Stockden, m.	..	34	..	..	..	..	..	1
Thos. Walkeden	..	..	..	3	..	1	{ Dead feld 1¼ Brome feld 3 Peckshook 1 }	1

FELTHAM, MIDDLESEX — Land Rev., M. B. 220, ff. 78–108. 2 Jas. I

Copyholders	Enclosed	Arable in Open Common Fields: Further field	Middle field	Home field	Meadow
Rich. Welbeloved, m.	4	1½	½	¼	..
Rich. Reade, m.	2½, 2½ (wood)	4⅛	9¼	10½	..
Christopher Tubbes, m.	1½, 1½ (pasture)	4½	4	11¾	..
Thos. Towe, m.	½	1	..	1	..
Baptista Welbeloved	......	2	1½	3½	..
Wm. Robberts, m.	4	32	30	28	..
Rad. Lawrens, m.	8	30	30	30	5½
Jo. Glysson, m.	1	2	1½	2	2
Rich. Pullen	......	¾ ¾	1 1	¾ ½	..

FARNHAM ROYAL, BUCKINGHAMSHIRE — Land Rev., M. B. 200, ff. 57–100. 6 Jas. I

Custumarii	Enclosed: Arable	Enclosed: Pasture	Arable in the Open Common Fields: West field	Hawthorne field	Bidwell field	Dearepath field	Miscellaneous	Common Meadow
Isabella Wright, m.	gard.	3	½	11	6½	..	....	½
Wm. Rolphe, m.	¼	3	1	2½	3	..	....	..
Robt. Walter, m.	½	11	..	3	2	2	le Pease ¾, Upper Studs ½	2½
Jo. Peryman, m.	3	..	..	..	..	..	Litle feild and le Downe 26¼	½
Christofer Redding, m.	30½	10	..	..	..	..	Litle Reding 3½	¼
Jo. Fisher, m.	26	2½	..	9¾	7¼ and Derepath		Unspecified 1	..
Edm. Goodrich, m.	34¼	7	In these four common fields 46				....	..
Jo. Randall, m.	¼	..	½	..	½	..	....	..
Jo. Peryman	22 ("or pasture")	..	..	41¾	..	25½[1]	....	..

Tenants have "communia pasture pro omnibus averiis in communibus campis et le Heath ibidem."

[1] The specification of the arable of this holding in Hawthorne field and Derepath field is as follows: le yards, 11¼; Welcrofts, 3½; Les Downes, 23; Tres les pitles, 3; in le veare ducente a Sipenham versus Farnham, 7; pecia terre nuper inclusa ex boriali parte vie predicte, 20; Little feild, 6½; clausa terre ibidem, 5½.

Sonning, Tithing of Wynnershe, Berkshire

Land Rev., M. B. 202, ff. 66–73. [Eliz.]

Customary Tenants	Arab.	Enclosed Past.	Unspec.	Arable in the Open Common Fields: Char field	Demys field	Whetershe	Benhams	Stonye field	Goswell field	Olde Orchard field	Rudges	Miscellaneous	Common Meadow
Johane Mylam, m., 1 yl.	11	5	3	..	5½	1	1	..	..	..	..	Westreadinges 1	3½
Wm. Headache, m., ½ yl.	7	..	..	..	2	2	1	..	..	..	..	......	2¾
Thos. Crockford, m., 1 yl.	4	..	..	..	..	..	..	6	1½	½	..	Cressfell 7, 2 wood, 2 md.	½
Hamlet Shefforde, m., 2 yl.	..	..	10	¾	..	..	..	11	5	2	2½	......	5
Agnes Astell, m., 1 yl.	19½	½ wood	..	..	..	..	..	..	..	..	..	......	...
Ralphe Balle, m., 1 yl.	9½	2 wood, 3 moor	..	9½	..	..	..	4	3½	3	½	......	4½
Marke Wheteleye, m., ½ yl., etc.	4 / 10	2½ wood	..	..	16½	3	..	..	..	..	..	Weste Rydinge 1, Hayles 3, Hedgemore 3, Hedgemore 3, Stonyham 3, Goldrydinge 3, Malamyn 3	4½
Robt. Shefforde, Jr., 2 yl.	..	6	..	7	8 (the Muse)		..	7½	5	5	9½	Greenhills 3, Vernelle 1, the Brech ½	5
Robt. Phillipps, gent., m., 2 yl.	16	..	..	..	10	2	4	..	..	..	..	Hasill field 2, Rydings 2, Brookspiddle 1	15
Robt. Phillipps, gent., m., 1 yl.	18	1 wood	..	..	..	..	..	..	..	..	..	......	4
Clia Bosworth, m., 1 yl.	11 and wood		..	..	2	1	1	..	..	..	..	Weste Redings ½	
Wm. Maynard, m., ½ yl.	9½	..	..	..	9	5	4	..	..	..	..	Redinge 2½, West Redinge 1, Inward Hales 2	17½
Robt. Collyns, m., 1 yl.	3	..	..	4	..	..	..	6	2	3	1½	Buchers 3	2¾

There is no reference to common of pasture over the fields. In the commons of the manor, of which a list is given (f. 102*b*), and of which the largest, Bullmershehethe, contained 100 acres, the tenants had rights of pasture.

SONNING, A TITHING OF SONNING, BERKSHIRE

Land Rev., M. B. 202, ff. 52–66. [Eliz.]

Custumarii	Enclosed			Arable in the Open Common Fields					Common Meadow
	Arab.	Past.	Unspec.	Char field	Upp field	Downe field	Bullmershe field	Miscellaneous	
Robt. Adams, m., 1 yl.	5½	5	13½	10	..	..	7	Great Puckewell 5 Bramleye 6	9
John Gregory, ½ yl.	11	3	..	..	..	..	..	....	..
Eliz. Langford, m., 1 yl.	..	..	..	5	1½	1	1	Great Puckewell 1 At deane corner ½	..
Andrew Crockford, m., ½ yl.	..	..	⅛	2	2	4	..	Woodley field 1	¾
Thos. Flecher, m., 1 yl.	8	..	10½	5½	..	..	..	Wengell 3, Ridges 1	2
Will. Walles, m., ½ yl.	..	..	..	4 and Up field		4	..	....	1
Ralph Blake, m., ½ yl.	..	..	..	6½	1½	1	½	....	1
Thos. Thorne, gent., m., 1 yl.	..	..	..	2½	..	..	..	Burway fielde 10	5⅛
Thos. Thorne, gent., m., ½ yl.	¼	¼	..	1	4¼	4	½	....	1¾
John Loveioy, m., 1 yl.	½	..	..	8	1	2½	½	....	4½
Thos. Loveioye, m., 1 yl.	2	1	..	6½	1	7	½	Parkworth ½	3½
Ambrose Barker, m., ½ yl.	..	..	..	3¼	1¼	4	¾	....	½
Agnes Hutchins, 2 m., 2½ yl.	..	..	1½	5¾	3½	6	2	....	2

The only statement about common of pasture is that Elizabeth Langford has common for "eight beasts and a Bullock in Burwaye mershe and for one cowe in Sonninge meade (f. 60*b*).

CAVERSHAM, OXFORDSHIRE
Land Rev., M. B. 189, ff. 48–65. 5 Edw. VI

| Custumarii | Enclosed | Arable in the Open Common Fields ||||||||| Common Meadow |
		Hermedene	Barman field	Ley field	Ashecroft field	Le Breche	Grove field	West field	Vernon-hill	Miscellaneous	
John Ireland, m., 4 virg.	44	4	4	2	..	..	..	..	..	4	3
Eliz. Lowgey, m., 2 virg.	51	2½	2	..	1	..	..	..	..	1	1
Rich. Lee, Jr., m., 4 virg.	63	..	5	..	5	..	..	..	..	11¼	1½
Rich. Insalathe, m., 1 virg.	15	2	1	..	4	..	..	..	..	½	¾
Henry Sawyer, m., 1 virg.	25	½	½	2	..	..	..	..	..	1½	¼
Alicia Jenens, m., 2 virg.	2½	..	..	..	..	4	6	6	3½	1	..
Cicilia Atwell, m., 2 virg.	13½	3	3	½	2	½	4	2	..	8	1½
Petrus Atwell, m., 3 virg.	32	6	5	..	..	..	..	1	..	4½	1¾
Johanna Thorn, m., 2 virg.	13	1	..	..	..	3	8	..	5½	14½	1¼
Jo. Flowrey, m., 1 virg.	6½	2	2½	..	..	2	3	1	1	6½	..
Agnes Wapull, m., 2 virg.	50¾	..	12	..	8	3	8	3	3	12	1½
Thos. Alee, m., 1 virg.	1½	..	..	..	..	2	2	3	2	4	..

Ewelm, Oxfordshire

Exch. Aug. Of., M. B. 388, ff. 1–75. 6 Jas. I

Custumarii	Enclosed Arab.	Enclosed Past.	Arable in the Open Common Fields: Grove field	Middle field	Church field[1]	Mundeys field	Miscellaneous
Johes Clarke, m.	½	..	2	1	2	3	....
Rich. Eyre, m.	1	3	..	..	..	..	East feild 20, Berrick feild 20, High feild 20
Thos. Wylles, m.	¼	..	..	½	..	..	East feild 1½, Gravell feild ½
Wm. Poxen, m.	½	2¼	..	7	..	15	Cley 7½, Port feild 18½, Bensington feild 11
Thos. Willis, m.	..	..	4	1½	3	..	....
Thos. Banks, gent. cott.	¼	..	4	1	¾	7½	....
Grif. Powell, m.	..	2¼	..	..	..	..	Hie feild 15 ("pecia terre")
Jac. Poxen, m.	¼	..	7	4	2	1	....
Thos. South, m.	¼	..	..	..	..	..	High feld 10, East feld 14, Koakpet 1½, Little feld ½
Wm. Warner, m.	5⅜	1¾	..	..	14[2]	..	South feld 5½, West feld 8½, the Warden 2
Jac. Nebb, m.	¼	5¼	6½	10	..	1	West feld 7¼, the Warden 1
Edw. Buckland, m.	¾	1⅞	..	..	..	12	Greathome feld 6, Littlehome feld 1, High-feld 1
Walter Palmer, m.	¼	..	5	5	4	..	....
John Atkins, m.	½	1	3	2	4	..	....
Johanna Renell, m.	..	1	8½	8	12	..	....
Prudence Spire, m.	½	..	12½	8½	11	..	....
John Hall, m.	⅛	..	3	½	..	..	the Warden 1, Crofts ⅞

Copyholders have stinted common of pasture "in communibus campis de Ewelme" (ff. 27 *sq.*).

[1] "alias Warren feld." [2] 4 acres by one copy, 10 by another.

APPENDIX VI

WATLINGTON, OXFORDSHIRE
Land Rev., M. B. 202, ff. 1-23. 6 Jas. I

Custumarii	Enclosed Arab. Past.		Arable in the Open Common Fields							Miscellaneous
			Upper field	Clay hill	Edgingdowne field	Middle-hill	East field	Cowberryes field	Brightwell field	Hill field	
Wm. Hambleden, m.	$\frac{1}{2}$	$\frac{1}{2}$	2	8	1	..	..	..	..	..	Pegseare 5
Simon Bartlet, m., 2 cott., 1 virg.	$1\frac{5}{8}$	$1\frac{1}{4}$	20	$11\frac{1}{4}$	$4\frac{1}{2}$	10	$9\frac{1}{4}$	12	10	..	..
John Forde, toft.	gard.	$1\frac{1}{2}$	$5\frac{1}{2}$	..	..	..	..	..	..	..	Lower feld 5
Eliz. Quartermayne, m.	1	$\frac{5}{8}$	$10\frac{1}{2}$ 22[1]	21	13	$4\frac{1}{2}$	..	$13\frac{1}{2}$	$4\frac{1}{2}$	..	..
Rad. Mercer, m.	1	2	12	12 (the Clayes)	..	$14\frac{1}{2}$	$35\frac{1}{4}$	..	..	24	Cuddendon 3
Julian Greendowne, m.	1	..	4	..	..	3	$1\frac{1}{2}$	$1\frac{1}{2}$	..	14	..
Robt. Erestare, m.	$\frac{1}{4}$	..	9	6 (the Claye)	..	$3\frac{1}{2}$	15	3	..	$8\frac{1}{2}$	..
Wm. Johnson, m.	$\frac{1}{2}$	$2\frac{1}{2}$	..	1	..	3	7	..	2	7	Paribane feld $6\frac{1}{2}$
John Adene, m.	$\frac{5}{8}$, 5 next the mill		..	13	..	..	..	13	..	..	Peggsyeere 18
Roger Bartlett, m., $2\frac{1}{2}$ virg., and 4 acres			..	23 (Cley le field)	12	..	..	11	..	$11\frac{1}{2}$	..
John Bowler, m.	1	1	7	16	..	..	..	6	..	..	Cranes feild 2

There are no other customary holdings of importance. Customary tenants have common of pasture in Minigrove and in the common fields.

[1] By another copy.

Bensington, Oxfordshire

Land Rev., M. B. 224, ff. 15–29. 4 Jas. I.

Liberi tenentes	Enclosed Arab.	Enclosed Past.	Arable in the Open Common Fields: le Claye iuxta Ewelm	le Claye iuxta London way	le fowle Sloowe	Super porte-hill	in le Hale	Easte feildes	Roake feildes or hill	Stonye lande	More-lande	Miscellaneous	Common Meadow
Thos. Fortescue, Arm., m.	5½	..	5¾	3¾	1	1½	..	..	..	7	1	apud Roake Eline 1	1½
Robt. Arnolde, m.	1½	..	11½	11½	½	..	½	..	4½	7	3	in Roak Crofts 1 in Bensington Crofts 1	¾
Geo. Penneye, m., 3 cott.	1½	1	2½	3¼	5½	10½	16	..	1	5½	..	in le Crofts 5¾	5½
Geo. Penneye, nuper M. Weathers	..	..	1	4½	2½	2½	1	..	..	5	3½	....	..
Thos. Bennett, m., cott.	2½	..	..	¾	1	1½	2	1¼	½	4½	2	in le Crofts 1¼	1½
Johannes Pawlinge, cott.	½	..	..	4¼	1	..	..	..	..	1½	4½	....	½
Thos. Freeman; m., 2 cott.	½	1¾	..	40	5	..	2	15	2	35	3	Seavenokes 1, Beggars bushe 4, Walton 5	6
Ric. Waterer, m.	½	..	..	..	1	1¾	..	1½	½	4¾	..	Heycroft 1, Pilbrushe 1½ Mill land ende ½	¼
Nich. Smyth, m.	1	..	..	1	1½	3	..	5¼	..	..	1	West feild 1	1½
ad usum Ecclesie de Bensington, m.	2¼	..	1½	3½	1½	12½	7	5	1½	4¼	3½	Churche Feilde 2, West feild ½, Ewelm Home feild ½	2
John Lydall, m., 1 virg.	¼	..	..	..	..	1	..	13	..	..	..	prope Reevewaye 2, prope Blackelandes 1¼, prope Ewell-hed 1, Berricke weste feild 2	
Helinora et Maria Buckland, m.	¼	½	..	..	..	..	..	4½	8½	4½	..	le Crofts 2, Home Feilde 3½ High feild 6	
Thos. Merryweather, m.	1	..	..	..	..	6	..	..	..	1½	..	Crofte lande 4, Churche feilde 2	
Rector et Scolares Collegii Exon. in Oxon., m.	2¾	..	4¾	12½	1¼	6¾	1½	7½	3	13¼	..	le Hill 4, Warborough Home Feilde 2½	6¾

No statement is made regarding common of pasture. Several tenants have an acre or two of pasture in Horselease.

WARBOROUGH, OXFORDSHIRE

Land Rev., M. B. 224, ff. 29–33, 153–159. 4 Jas. I

Custumarii	Enclosed Arab.	Enclosed Past.	Arable in the Open Common Fields: Costoll	le oulde Claye	Chickshill field	Hen field	Linches field	Meade field	Shillingford fields	Common Meadow
Wm. Bartlett, m.	½	⅜	5½	2½	9	9	and feild	..	18	2½
John Cope, m.	½	..	..	1½	1½	4	..	7	..	1
Ric. Wallis, m., 2 cott.	1⅜	1¼	6	7	..	4½	1½	5½	1	2¼
Edw. Frenche, cott.	¼	..	1	..	..	1¼	½	½	..	¾
John Bamon, 3 m.	2½	1	..	15½	15¼	23	4	8¼ [1]	..	5½
Jac. Porter, m., cott.	1½	..	1	7	..	18	..	2½ [1]	..	2
Wm. Arnett, m.	1	..	2	4½	9	5	..	3¼ [1]	7	2
Ric. Butler, m.	¾	½	3¼	3	8½	6	1½	9½	..	3
John Janes. m.	1½	..	½	3½	19	9	1½	10¼	..	4
John Janes. cott.	½	..	6¾	..	..	..	..	..	20½	..
John Webbe	..	..	3	1	..	4	1	1¾	..	1¼
Liberi tenentes										
John Webbe, 3 cott.	..	..	7½	..	2	3½	..	3	..	1
Ric. Harrison, m.	1¼	..	..	1½	..	2	..	6¾	2	4½
John Arnett, m.	..	..	..	7	1	1½	..	1½	..	½
John Smyth, m.	2	..	..	7	3½	½	..	6½	1	½
Ric. Bisleye, m.	½	..	..	3½	1	½	..	11½	..	1

[1] In Meade field and Town hill.

There is no reference in the description of the holdings to common of pasture. In a list of the commons of Warborough (f. 160) it is stated that the "firmarius" of Shillingford ought to have common "pro Rotherbests et ovibus in campis et communiis de Warboroughe cum tenentibus domini Regis ibidem [Warborough was a royal manor], viz., in le Costoll feilde, in Chickeshill feilde, in le olde Claye, super le Townhill, in Henne feilde, in Linches et Shillingford Filds."

INDEX

INDEX

Acre, the standard, 19.
Agriculture, relation of field systems to, 3, 4, 7-12, 403-409.
Alciston, Suss., 33, 34.
Alfriston, Suss., 33, 443.
All Souls College, Oxford, maps of, 34, 77, 274.
Altham, Lancs., 244, 245.
Alvingham, Lincs., 31, 441.
Anglesey, common arable fields in, 183-185.
Anglo-Saxon charters, 51-61, 410; laws, 61, 62.
Anglo Saxons, 71, 298, 304, 409-411, 418.
Ansty, Hants., 33, 443.
Arden, forest of, 86, 87.
Ashbury, Berks., 31.
Ashton Keynes, Wilts., 32, 39, 42, 442.
Assart, 85.
Aston and Cote, Oxons., 118.
Austro, de antiquo, 41, 98.
Avon, valley of the, 31, 88.
Axholme, Isle of, 103, 532.

Bailiffs' accounts, 44.
Barfreston, Kent, 280, 281.
Barking, Essex, 392, 393.
Bawdsey, Suff., 334.
Bedfordshire, 34, 35, 70, 79, 444.
Bensington, Oxons., 387, 558.
Berkshire, 30, 31, 60, 61, 63, 70, 553, 554.
Bicester, Oxons., 79.
Biddletown, Dorset, 80.
Bisley, Surr., 364, 365.
Blatchington, Suss., 33.
Bletchingdon, Oxons., 118.
Borga, 33.
Boundaries, in Anglo-Saxon charters, 51-56.
Bovates, 41, 42.

Bower Henton, Somers., 32, 441.
Brailes, Upper and Nether, Warks., 29, 437.
Brancaster, Norf., 345, 346.
Brandon, East, Durham, 105, 534.
Braunton Great Field, Devon, 262, 263.
Breighton, Yorks., 103, 531.
Brent, East, Somers., 98, 525.
Brixham, Devon, 259, 261.
Bruton, Somers., 100, 528.
Buckinghamshire, 63, 70, 76, 77, 80, 552.
Butts, 19, 163.
Buxton, Norf., 311, 312.

Cambridgeshire, 63, 70, 78.
Campi (fields), 13, 21, 28, 39. *See* Fields.
Carnarton, Corn., 263.
Castle Acre, Norf., 314, 315.
Caversham, Oxons., 386, 555.
Celtic system, 157-205; influence of, in England, 266-271, 404, 405, 412-414, 418.
Chalgrove, Oxons., 18-23, 124.
Charlbury, Oxons., 117.
Charlton Abbots, Gloucs., 30, 438.
Charters, Anglo-Saxon, 51-61, 410.
Cheshire, 64, 404, 412, 414; field system of, 249-258.
Cheshunt, Herts., 376, 550.
Chester, Chesh., 250-252.
Chiltern hills, enclosures in, 119, 120; field system of, 384-387, 401, 417.
Christian Malford, Wilts., 101, 529.
Clapton, Gloucs., 89, 517.
Clifton, Oxons., 116.
Common, or waste, 10, 24, 26, 47, 405, 412-414.
Consolidation of open-field parcels, 175, 176, 256, 257.
Convertible husbandry, 7, 8, 58 n., 100.

Copyholds, 21-23, 25, 27, 28, 41, *et passim.*
Corby, Northants, 44.
Cornwall, 63, 404, 412, 414; field system of, 263–266.
Corsham, Wilts., 74.
Corston, Somers., 100, 528.
Cotswolds, the, 29–31, 70, 88–90, 123, 438.
Cowpen, Northumb., 222, 223.
Crofters, 166–168.
Crofts, held in common, 89.
Crops, succession of, 44, 45; in Oxfordshire, 124, 125, 129; in Scotland, 158–160; in Wales, 200; in Northumberland, 208, 222–225; in Kent, 302; in Norfolk, 318–322, 330, 332, 333; in Suffolk, 331; in Surrey, Hertfordshire, Middlesex, Essex, 396–398.
Crown estates, 23.
Croxton, Lincs., 26.
Culturae, 13, 14. *See* Furlongs.
Culworth, Northants, 80, 477, 482.
Cumberland, 64, 404, 412–414; field system of, 227–242.
Curry Mallett, Somers., 99, 527.
Customary holdings. *See* Copyholds.

Dales, 21, 163.
Damerham, South, Wilts., 24.
Danes, 71, 298, 304, 352–354, 416.
Darliston, Salop, 68.
Day's work, 300, 301.
Deal, Kent, 276.
Demesne, 8, 24, 28 n., 34, 35, 315, 444; tillage of, 43–46.
Denbighshire, 178–183.
Derbyshire, 12, 63, 139.
Devonshire, 63, 404, 412, 414; field system of, 258–266.
Disintegration of holdings, 94–97.
Dola, 286, 287.
Donegal, Ireland, townland in, 191.
Dorset, 30, 32, 63, 70, 80, 439, 442.
Dragga, 15 n.
Drayton, Northants, 78, 477, 482.
Durham, 36, 63, 70, 404, 446, 534, 535; irregular fields in, 105–107, 408.

East Anglia, 48, 416; field system of, 305–354.
Edmonton, Mdx., 381, 382, 551.
Eggleston, Durham, 106, 535.
Egham, Surr., 362–364.
Elloughton, Yorks., 36, 446.
Elmdon, George, note-book of, 316–324.
Enclosure, by agreement, 116–118, 149–152; piecemeal, 145–148, 310–312, 407.
Enclosure awards, 14, 15; for Oxfordshire, 111–114; for Herefordshire, 139–141, 149, 150; for Norfolk, 305–307.
Enclosure maps, 14, 112, *and see table of contents.*
Enclosures, 8–12, 32, 90, 91, 98, 101, 102, 107, 404–408; in Oxfordshire, 110–132; in Herefordshire, 139–141; in Wales, 172, 173; in Northumberland, 206, 207; in Cumberland, 227–229; in Lancashire, 242, 243; in Cheshire, 249; in Kent, 272, 273; in Norfolk, 305–312; in Surrey, 356; in Hertfordshire, 370; in Middlesex, 381; in Essex, 387, 388.
Epping forest, 388.
Eriung, 335–338, 348, 351.
Eskirmaen, Wales, 195.
Essex, 12, 404, 416, 417; field system of, 387–394, 400.
Evenley, Northants, 78, 478, 482.
Every-year lands, 92.
Ewell, Surr., 399, 400.
Ewelm, Oxons., 116, 387, 556.
Extents, 14, 43–47.

Falda, 342–344.
Faldagium, 342.
Farnham Royal, Bucks., 385, 552.
Faughs, 159, 232.
Feet of fines, 13, 62, 68.
Feltham, Mdx., 382, 552.
Ferthing, 298.
Fields, names of, 42, 43, 69; multiplicity of, 89, 93–97, 101, 146–149, 282, 407. *See* Campi, Two-field system, Three-field system, Four-field system, Six fields.

Fingland, Cumb., 232.
Finmere, Oxons., 118.
Fold-courses, 316, 325-329, 341-344, 350.
Folds (falds), 159.
Forest areas, 83-88, 119, 120, 138, 407, 417.
Four-field system, 88, 103, 104, 125, 126, 135-137, 406.
Frampton Cotterell, Gloucs., 91.
Freeholds, 21, 24, 27, 35, 40, 41.
Frocester, Gloucs., 89, 518.
Furlongs, 13, 19, 127, 128, 313.

Gamlingay, Cambs., 44.
Gavelkind, in Wales, 186, 187, 195-199; in Ireland, 191-194; in Kent, 295, 296; in Norfolk, 335-337.
Gavelle (wele), 196-198.
Gedalland, 59, 60, 61.
Gillingham, Dorset, 30, 439.
Gillingham, Kent, 282-286.
Glastonbury manors, 24, 31, 92.
Glebe terriers, 119, 134-136.
Gloucestershire, 30, 61, 70, 438, 516-520; irregular fields in, 88-93.
Gores, 19, 55.
Great Tew, Oxons., 128-130.
Guston, Kent, 275.

Hamlets, 268, 407, 412; in Herefordshire, 95, 153; in Scotland, 167, 168; in Wales, 179; in Ireland, 187-189; in Cumberland, 230, 231; in Devon, 260; in Lancashire, 267.
Hampden-in-Arden, Warks., 86, 512.
Hampshire, 33, 61, 63, 70, 443.
Handborough, Oxons., 27, 28, 40, 430-437.
Hanwell, Oxons., 116.
Hartley, Northumb., 220, 221.
Haverfordwest, Pembrokes., 174.
Hayton, Cumb., 232.
Headland, 19, 55.
Heafodaecer, or headland, 55, 56.
Henley-in-Arden, Warks., 87.
Hennor, Herefs., 37.
Herefordshire, 36, 37, 63-66, 70, 71, 521-523; irregular fields in, 93-97; enclosures in, 139-141; decay of midland system in, 139-153, 407, 411, 447-449.
Hertfordshire, 404, 550; field system of, 369-381, 401, 417.
Highlands, Scottish, field system of, 161.
Hinton St. Mary, Dorset, 32, 442.
Hitchin, Herts., 5, 17, 18, 369.
Hitching, 132, 133, 134.
Holdenby, Northants, 78, 479, 482.
Holkham, Norf., 326-330.
Holme Cultram, Cumb., 230.
Holmer, Herefs., 146.
Hoo St. Mary's, Kent, 277, 278.
Horton, Gloucs., 90, 519.
Houghton Regis, Bedfs., 79, 450, 451.
Humberston, Lincs., 31, 440.
Huntingdonshire, 63, 70.

Ilsington, Devon, 260-262.
Ine's laws, 61, 62.
Infield, 158-161.
Ingleton, Durham, 36, 39, 41, 63, 446.
Inhoc, 92, 93.
Inquisitions post mortem, 44, 46.
Ireland, field system of, 187-194.
Irregularities in midland area, 83-107, 407.
Issacoed, Denbighs., 182.
Iugum, 282-298, 304, 351, 352, 399.
Ivington, Herefs., 94, 522.

Kavels (kenches), 169.
Kent, 404, 415, 543-548; field system of, 272-304.
Kimbolton, Herefs., 37.
Kingham, Oxons., 126.
Kingsbury, Somers., 98, 524.
Kington, Wilts., 24-26, 40, 42, 421-430.
Kislingbury, Northants, 79, 479, 483.
Knoll, Warks., 86, 513.

Lancashire, 64, 404, 412, 414; field system of, 242-249.
Landescore, 261.
Lands, or strips, 19.
Last, The, Salop, 68.

Lazonby, Cumb., 233.
Leaseholds, 28, 36, 41.
Lectum, 196–198.
Leicestershire, 35, 70, 76, 445.
Lenton and Radford, Notts., 104, 533.
Lesbury, Northumb., 207, 209, 212.
Lexham, West, Norf., 314, 317.
Leynes, or fields, 33.
Leys, or lays, 34, 35. *See* Meadows.
Lincolnshire, 26, 31, 63, 70, 71, 75, 103, 440, 441, 532.
Litlington, Cambs., 78, 457, 459.
Long Coombe, Oxons., 84, 511.
Long Houghton, Northumb., 208–211, 226.
Long Lawford, Warks., 78, 500.
Lutterworth, Leics., 35, 445.
Lynches, 56.
Lyng, 328.

Manòrs, in East Anglia, 350–352.
Maps, enclosure, 14, 112, *and see table of contents;* tithe, 14, 15, 18–23.
Marden, Herefs., 95–97, 142, 146–148, 150, 153 n.
Marshall, William, on Gloucestershire, 91, 92; on the Scottish Highlands, 161; on Norfolk, 307.
Marston Sicca, Gloucs., 88, 516.
Martham, Norf., 335–339.
Mawley, Salop, 37, 38, 449.
Meadow strips in arable fields, 35, 106, 408.
Meadows, common, 21, 28, 47.
Mercia, early open fields in, 62.
Merstham, Surr., 366, 367.
Middlesex, 404, 551, 552; field system of, 381–384, 402, 417.
Middleton, Herefs., 93, 521.
Middleton, North, Northumb., 224, 225.
Middleton Stony, Oxons., 117.
Midland system. *See* Two-field system, Three-field system, Six fields.
Monmouthshire, 64, 271.
Morffe forest, 38.
Multiplicity of fields, 89, 93–97, 101, 146–149, 282, 407.

Names of fields, 42, 43, 69.
Nasse, Erwin, on Anglo-Saxon fields, 6, 51–61 *passim.*
Newchurch, Kent, 286, 287, 543–548.
Newington, Kent, 273, 274.
New Shipping, Pembrokes., 175, 176.
Niton, Isle of Wight, 102, 530.
Norfolk, field system of, 305–354.
Normans, 297.
Northamptonshire, 35, 61, 62, 63, 70, 78, 79, 80, 444.
Northumberland, 64, 74, 404, 412–415; field system of, 206–227.
Norton St. Philip, Somers., 99, 526.
Nottinghamshire, 70, 533; irregular fields in, 104.

Oats, cultivation of, in Scotland, 159; in Wales, 200.
Ollands, 319, 320.
Ouse, valley of the, 63, 70.
Outfield, 153–161, 200, 222, 223, 232.
Over Arley, Staffs., 87.
Owston, Lincs., 104, 532.
Oxfordshire, 18–23, 27, 28, 29, 31, 61, 63, 70, 76, 79, 80, 84, 438, 510, 511, 536–542, 555–559; irregular fields in, 84–86; decline of midland system in, 109–137, 407, 408; enclosures in, 110–132.
Oxgangs, 36.
Oxlynch, Gloucs., 89, 519.

Padbury, Bucks., 76.
Pasture, common of, 28, 47, 48.
Pembrokeshire, 172–178; gavelkind in, 186, 187.
Pendicles, 166.
Pennard, West, Somers., 99, 526.
Perticata, 25.
Pickhill and Siswick, Denbighs., 180, 181.
Piddington, Oxons., 76, 488.
Plena terra, 345, 392.
Ploughing, mediaeval, 8, 9.
Poynton, Salop, 67, 68.
Precincts, 313, 314, 322.
Presthope, Salop, 69.
Preston, Northumb., 215, 216.

Quarters, 126–130.

Raines, 228.
Ramsden, Oxons., 85, 511.
Redland district, 128, 131, 133.
Residential townships, 121.
Richmond, Surr., 365, 366, 549.
Riggs, 163, 222, 227-229, 235.
Ringstead, Norf., 345, 346.
Risbury, Herefs., 37, 144, 145, 447.
River valleys, 88-107, 120, 138.
Robèston, Pembrokes., 177, 178.
Rocester, Staffs., 87, 515.
Rolleston, Staffs., 35, 36, 40, 445.
Roman influence on field systems, 5, 12, 415, 418.
Romsley, Salop, 68.
Runrig (rundale), 268, 405, 412-414; in Scotland, 162-167; in Ireland, 191-195; origin of, 190-199; in Wales, 203.

St. Florence, Pembrokes., 176, 177.
St. Margaret at Cliffe, Kent, 275.
Salford, Bedfs., 34, 41, 43, 46, 444.
Scotland, field system of, 157-171, 201, 202.
Seebohm, Frederic, description of Hitchin, Herts., 5, 6, 17; on Anglo-Saxon fields, 51-61 *passim*.
Segregation of strips, 234-237, 245.
Selions, 19, 89; individually named, 254, 255.
Settlement, relation of field systems to, 3, 12, 13, 409-418; types of, *see* Hamlets.
Severn, valley of the, irregular fields in, 38, 88-93, 406, 408.
Shawbury, Salop, 69.
Shipton-under-Wychwood, Oxons., 29, 42, 438.
Shots, 19. *See* Furlongs.
Shropham, Norfolk, 311, 312.
Shropshire, 37, 38, 63, 64, 66-69, 70, 71, 138, 411, 449; irregular fields in, 108.
Six fields, 17, 21, 35, 40.
Somerlie, 319, 324.
Somerset, 30, 32, 63, 70, 138, 139, 439, 441, 524-528; irregular fields in, 97-101.

Sonning, Berks., 385, 386, 553, 554.
Soulby, Cumb., 232.
Staffordshire, 35, 63, 70, 71, 139, 445, 514, 515; irregular fields in, 87.
Stewkley, Bucks., 80, 455, 456.
Sticca, 59.
Stockton, Herefs., 37, 448.
Stoke, South, Oxons., 80, 490.
Stoke, South, Somers., 30, 39, 439.
Stoke Edith, Herefs., 94, 523.
Stoke Prior, Herefs., 37, 447.
Stonebrach, 131, 133.
Stonesfield, Oxons., 84, 510.
Stow, Lincs., 75.
Suffolk. *See* East Anglia.
Sulung, 299, 300.
Surrey, 549; field system of, 355-369, 399, 400, 404, 417.
Surveys, 15, 23-25, 27-36, 83-107, *et passim*.
Sussex, 33, 63, 443.
Sutton, Kent, 276.
Sutton at Hone, Kent, 277.
Syndrig land, 59.

Tallantire, Cumb., 239, 240.
Tees, valley of the, 105.
Tenementum, 300, 334-341, 344, 351.
Terriers, 14, 41, 42, 49, *et passim*; glebe, 119, 134-136.
Thame, Oxons., 124.
Thames, meadow-townships on the, 120; field system of the lower, 355-402.
Thorley, Isle of Wight, 102.
Three-field system, characteristics of, 17-28, 39-48; townships typical of, 27, 28, 32-36; extent of, 34, 39, 62-71; development of, from two-field system, 72-82, 406; deviations from, 83-107; decline of, 124, 125, 135-138, 142-152.
Tithe maps, 14, 15, 18-23.
Trent, valley of the, 70, 87, 103-105, 408.
Two-field system, characteristics of, 17-26, 39-48; townships typical of, 24-26, 29-31; early history of, 50-62; extent of, 62-70; transformation of, into three-field system, 72-82; deviations from, 83-107; decline of, 123-137, 406.

Twyford, Leics., 76, 471, 473.
Typus Collegii of All Souls College, Oxford, 34, 77, 274.

Virgates, 17, 21, 25, 27, 28, 41, 42; in Kent, 298, 299; in Norfolk, 345–347; in Essex, 391–394.

Wales, 64; field system of, 171–187, 200.
Walter of Henley, on thirteenth-century tillage, 71.
Warborough, Oxons., 387, 559.
Warton, Lancs., 245, 246.
Warwick, Cumb., 236, 237.
Warwickshire, 29, 31, 70, 78, 138, 437, 512, 513; irregular fields in, 86, 87.
Watling Street, 70.
Watlington, Oxons., 387, 557.
Wattles, 329, 342, 343, 405.
Weasenham, Norf., 316–326.
Welford, Gloucs., 88, 516.
Welford, Northants, 35, 41, 42, 444.
Wellow, Isle of Wight, 31, 440.
Wessex, early open fields in, 62.

Westmorland, 64, 271.
Weston Birt, Gloucs., 30, 438.
Wight, Isle of, 31, 440, 530; irregular fields in, 102.
Willerby, Yorks., 103, 530.
Wiltshire, 24, 25, 32, 33, 60, 70, 74, 442, 529; irregular fields in, 101.
Wista, 33.
Woodstock forest, 84, 85.
Wootton, Oxons., 84, 510.
Wootton-under-Weaver, Staffs., 87, 514.
Worcestershire, 61, 139.
Wrentham, Suff., 45.
Wrexham, Denbighs., 179, 180.
Wychwood forest, 85.
Wye, Kent, 287–296.
Wye, valley of the, irregular fields in, 93–97.
Wymondham, Norf., 339–341.
Wyre forest, 87.

Yard-lands, 21. *See* Virgates.
Yate, Gloucs., 90, 520.
Yorkshire, 36, 63, 64, 70, 271, 446, 530, 531; irregular fields in, 103.